3rd International Conference on Engineering and Applied Technology (ICEAT 2019)

IOP Conference Series: Materials Science and Engineering
Volume 821

Sorong, Indonesia
30 October - 1 November 2019

Editors:

Dan Mugisidi
Ade FAH Alhashimy
Agus Setyo Munthohar
Rafal Drezewski
Yun Arifatul Fatimah
Agus Aktawan
Andri Pranolo
Novrianti
Leonel Hernandez
Lukas G. Swan
Pedro Hokama
Aster Rahayu
Alfian Ma'arif
Jazaul Ikhsan
Aris Widyo Nugroho
Tri Widayatno
Ahmad Azhari
Adhi Prahara
Rahma Yanda
Alwis Nazir
Roman Volianskyi

ISBN: 978-1-7138-3607-0
ISSN: 1757-8981

Printed from e-media with permission by:

Curran Associates, Inc.
57 Morehouse Lane
Red Hook, NY 12571

proceedings.com

Some format issues inherent in the e-media version may also appear in this print version.

Institute of Physics
Dirac House, Temple Back
Bristol BS1 6BE UK

Phone: 44 1 17 929 7481
Fax: 44 1 17 920 0979

techtracking@iop.org

Additional copies of this publication are available from:

Curran Associates, Inc.
57 Morehouse Lane
Red Hook, NY 12571 USA
Phone: 845-758-0400
Fax: 845-758-2633
Email: curran@proceedings.com
Web: www.proceedings.com

TABLE OF CONTENTS

IOP Conf. Series: Materials Science and Engineering **821** (2020) 011001 doi:10.1088/1757-899X/821/1/011001

Preface

The International Conference on Engineering and Applied Technology (ICEAT) was first launched in 2017 at Mataram hosted by Universitas Muhammdiyah Mataram. The second conference was in 2018 at Banda Aceh hosted by Universitas Muhammdiyah Aceh. The third ICEAT present is hosted by Faculty of Industrial Technology, Universitas Muhammadiyah Sorong, West Papua, Indonesia. The 3rd ICEAT is organized by Science and Technology Association of Muhammadiyah Higher Education (AST-PTM) as well as jointly with Universitas Ahmad Dahlan, Universitas Muhammadiyah Magelang, Universitas Muhammadiyah Malang, Universitas Muhammadiyah Surakarta, Universitas Muhammadiyah Purwokerto, Universitas Muhammadiyah Makassar, Universitas Muhammadiyah Gresik, Universitas Muhammadiyah Jakarta, Universitas Muhammadiyah Yogyakarta, Universitas Muhammadiyah Jember, Universitas Muhammadiyah Surabaya, Universitas Muhammadiyah Sidoarjo, Universitas Muhammadiyah Banjarmasin, and Universitas Muhammadiyah Prof Dr Hamka (UHAMKA). On behalf of the organizing committee, I cordially welcome to all the delegates of the 3rd ICEAT 2019.

Being in the third event, the 3rd ICEAT 2019 is aimed at keeping abreast of the current development and innovation in the advanced of research area on Engineering and Applied Technology as well as providing an engage forum for participants to share knowledge and expertise in related issues. In this conference, submitted papers were limited in five conference tracks. The committee member blindly reviewed as well as provided the technical comments to all the submitted paper before ensuring that submitted paper is qualified. Finally we are accepting 52 papers to be presented in the conference. The accepting papers are covering the scope include: Electronics Engineering, Electrical Engineering, Informatics Engineering, Computer Engineering, Industrial Engineering, Mechanical Engineering, Civil Engineering, Architecture, Chemical Engineering, Applied Science, and Agrotechnology as well.

Last but not least, I do hope through this conference will convey a good opportunity for sharing at the forefront of the field and an informal exchange of ideas, knowledge and friendship in a relaxing environment. Then, our academic and social atmosphere will become more active year to year.

Warm Regards,
General Chair
Aster Rahayu, S.Si., M.Si., Ph.D

IOP Conf. Series: Materials Science and Engineering **821** (2020) 011002 doi:10.1088/1757-899X/821/1/011002

ICEAT 2019 Committee

Organizing Committee

General Chair

- Aster Rahayu (Universitas Ahmad Dahlan, Indonesia)

General Co-Chair

- Yun Arifatul Fatimah (Universitas Muhammadiyah Magelang, Indonesia)
- Andri Pranolo (Universitas Ahmad Dahlan, Indonesia)

Technical Program Committee

- Andri Pranolo (Universitas Ahmad Dahlan, Indonesia)
- Rafał Dreżewski (AGH University of Science and Technology, Poland)
- Shi-Jinn Horng (National Taiwan University of Science and Technology, Taiwan)
- E.P. Nowicki (University of Calgary, Canada)
- Husni Thamrin (Universitas Muhammadiyah Surakarta, Indonesia)
- Dwi Anggraini (Universitas Muhammadiyah Malang, Indonesia)
- Gunawan Ariyanto (Universitas Muhammadiyah Surakarta, Indonesia)
- Zulfatman (Universitas Muhammadiyah Malang, Indonesia)
- Slamet Riyadi (Universitas Muhammadiyah Yogyakarta, Indonesia)
- Yun Arifatul Fatimah (Universitas Muhammadiyah Magelang, Indonesia)
- Aster Rahayu (Universitas Ahmad Dahlan, Indonesia)
- Pedro Hokama (University of Campinas, Brasil)
- Wolfgang Keller (University of Colorado, Colorado, US)
- Hari Prasetyo (Universitas Muhammadiyah Surakarta, Indonesia)
- Eko Setiawan (Universitas Muhammadiyah Surakarta, Indonesia)
- Ilyas Mas'udin (Universitas Muhammadiyah Malang, Indonesia)
- Siti Mahsanah (Universitas Ahmad Dahlan, Indonesia)
- Herry Purnama (Universitas Muhammadiyah Surakarta, Indonesia)
- Dan Mugisidi (UHAMKA Jakarta, Indonesia)
- Lukas G. Swan (Dalhousie University, Canada)
- Sudarisman (Universitas Muhammadiyah Yogyakarta, Indonesia)
- Aris Widyo Nugroho (Universitas Muhammadiyah Yogyakarta, Indonesia)
- Muji Setiyo (Universitas Muhammadiyah Magelang, Indonesia)
- Marwan Effendy (Universitas Muhammadiyah Surakarta, Indonesia)
- Tri Widayatno (Universitas Muhammadiyah Surakarta, Indonesia)
- Tri Yuni Hendrawati (Universitas Muhammadiyah Jakarta, Indonesia)
- Denny Vitasari (Universitas Muhammadiyah Surakarta, Indonesia)
- Nurul Hidayati Fithriyah(Universitas Muhammadiyah Jakarta, Indonesia)
- Haryanto (Universitas Muhammadiyah Purwokerto, Indonesia)
- Agus Setyo Munthohar (Universitas Muhammadiyah Yogyakarta, Indonesia)

IOP Conf. Series: Materials Science and Engineering **821** (2020) 011002 doi:10.1088/1757-899X/821/1/011002

- Rahma Yanda (Gifu University, Japan)
- Alwis Nazir (Universitas Islam Negeri Sultan Syarif Kasim, Indonesia)
- Roman Volianskyi (Dniprovsky State Technical University, Ukraine)
- Sri Sunarjono (Universitas Muhammadiyah Surakarta, Indonesia)
- Nurul Hidayati (Universitas Muhammadiyah Surakarta, Indonesia)
- Samin (Universitas Muhammadiyah Malang, Indonesia)
- Jazaul Ikhsan (Universitas Muhammadiyah Yogyakarta, Indonesia)
- Ade FAH Alhashimy (Universesias Muhammadiyah Sumatera Utara, Indonesia)
- Gunawan (Universitas Muhammadiyah Surabaya, Indonesia)
- Wisnu Setiawan (Universitas Muhammadiyah Surakarta, Indonesia)
- Alfian Ma'arif (Universitas Ahmad Dahlan, Indonesia)
- Novrianti (Gifu University, Japan)

IOP Conf. Series: Materials Science and Engineering **821** (2020) 011003 doi:10.1088/1757-899X/821/1/011003

ICEAT 2019 Editors

- Dan Mugisidi (UHAMKA Jakarta, Indonesia)
- Ade FAH Alhashimy (Universias Muhammadiyah Sumatera Utara, Indonesia)
- Agus Setyo Munthohar (Universitas Muhammadiyah Yogyakarta, Indonesia)
- Rafał Dreżewski (AGH University of Science and Technology, Poland)
- Yun Arifatul Fatimah (Universitas Muhammadiyah Magelang, Indonesia)
- Agus Aktawan (Universitas Ahmad Dahlan, Indonesia)
- Andri Pranolo (Universitas Ahmad Dahlan, Indonesia)
- Novrianti (Gifu University, Japan)
- Leonel Hernandez (ITSA University, Colombia)
- Lukas G. Swan (Dalhousie University, Canada)
- Pedro Hokama (University of Campinas, Brasil)
- Aster Rahayu (Universitas Ahmad Dahlan, Indonesia)
- Alfian Ma'arif (Universitas Ahmad Dahlan, Indonesia)
- Jazaul Ikhsan (Universitas Muhammadiyah Yogyakarta, Indonesia)
- Aris Widyo Nugroho (Universitas Muhammadiyah Yogyakarta, Indonesia)
- Tri Widayatno (Universitas Muhammadiyah Surakarta, Indonesia)
- Ahmad Azhari (Universitas Ahmad Dahlan, Indonesia)
- Adhi Prahara (Universitas Ahmad Dahlan, Indonesia)
- Rahma Yanda (Gifu University, Japan)
- Alwis Nazir (Universitas Islam Negeri Sultan Syarif Kasim, Indonesia)
- Roman Volianskyi (Dniprovsky State Technical University, Ukraine)

IOP Conf. Series: Materials Science and Engineering **821** (2020) 011004 doi:10.1088/1757-899X/821/1/011004

The 2019 3rd International Conference on Engineering and Applied Technology (ICEAT)

30-31 October 2019, Sorong, Indonesia

THE PROCEEDINGS PUBLICATION LICENCE:

Licence terms and conditions

By submitting your paper to the conference organizer, you, as author/representative of all the authors, grant a royalty free license to IOP Publishing Limited (IOP) to use the copyright in the paper for the full term of copyright in all ways otherwise restricted by copyright, including the right to reproduce, distribute and communicate the article to the public under the terms of the Creative Commons Attribution (CC BY) license (http://creativecommons.org/licenses/by/3.0/) and to make any other use which IOP may choose world-wide, by all means, media and formats, whether known or unknown at the date of submission, to the conference organizer.

This license does not transfer the copyright in the paper as submitted which therefore remains with the authors or their employer, as appropriate. IOP encourages authors to use the paper in any way provided that, where possible, he/she displays citation information and the IOP Proceedings License Notice, for electronic use, best efforts are made to include a link to the online abstract in the journal and no author offers the paper to another publisher (prior to withdrawal or rejection) or includes it in another publisher's website.

However, a re-written and extended version of the paper may be published in another journal provided such re-use is within generally accepted ethical scientific limits and provided further citation information and the IOP Proceedings License Notice is displayed if possible, and for electronic use best efforts are made to include a link to the online abstract in the journal.

By granting this license, the author warrants that the paper he/she is submitting is his/her original work, has not been published previously (other than in a research thesis or dissertation which fact has been notified to the conference organizer in writing), all named authors participated sufficiently in the conception and writing of the paper, have received a final version of the paper, agree to its submission and take responsibility for it, and the submission has been approved as necessary by the authorities at the establishment where the research was carried out.

By granting this license, the author also warrants that he/she acts on behalf of, and with the knowledge of, all authors of the paper, that the paper does not infringe any third party rights, it contains nothing libelous, all factual statements are, to the best of the authors' knowledge, true or based on valid research conducted according to accepted norms, and all required permissions have been obtained.

The 2019 3rd International Conference on Engineering and Applied Technology (ICEAT)

30-31 October 2019, Sorong, Indonesia

The IOP Proceedings License Notice

Editor	:	Dan Mugisidi, Ade Fah Alhashimy, Agus Setyo Munthohar, Rafał Dreżewski, Yun Arifatul Fatimah, Aster Rahayu, Agus Aktawan, Andri Pranolo, Leonel Hernandez, Lukas G. Swan, Pedro Hokama, Jazaul Ikhsan, Aris Widyo Nugroho, Tri Widayatno, Ahmad Azhari, Adhi Prahara.
Publisher	:	IOP Publishing Ltd.
Secretariat	:	Fakultas Teknologi Industri Universitas Ahmad Dahlan Jl. Ringroad Selatan, Tamanan, Banguntapan, Bantul Regency, Special Region of Yogyakarta 55191 Email: info2019@ast-ptm.or.id

IOP Conf. Series: Materials Science and Engineering **821** (2020) 011004 doi:10.1088/1757-899X/821/1/011004

The 2019 3rd International Conference on Engineering and Applied Technology (ICEAT)

30-31 October 2019, Sorong, Indonesia

Statement of Peer Review

All papers published in this volume of IOP Conference Series: Materials Science and Engineering has been peer reviewed through processes administered by the Editors. Reviews were conducted by expert referees to the professional and scientific standards expected of a proceedings journal published by IOP Publishing.

Organizers and Sponsors

Organized by

AST PTM (Applied Science and Technology of Muhammadiyah Higher Education), Indonesia

Hosted by

Universitas Muhammadiyah Sorong, Indonesia

Technical Co-Sponsored by

IOPscience, IOP

Supported by

Majelis Pendidikan Tinggi Penelitian dan Pengembangan, PP Muhammadiyah, Indonesia
Universitas Ahmad Dahlan, Indonesia
Universitas Muhammadiyah Yogyakarta, Indonesia
Universitas Muhammadiyah Surakarta, Indonesia
Universitas Muhammadiyah Malang, Indonesia
Universitas Muhammadiyah Purwokerto, Indonesia
Universitas Muhammadiyah Magelang, Indonesia
Universitas Muhammadiyah Jakarta, Indonesia
Universitas Muhammadiyah Banjarmasin, Indonesia
Universitas Muhammadiyah Sidoarjo, Indonesia
Universitas Muhammadiyah Surabaya, Indonesia
Universitas Muhammadiyah Jember, Indonesia
Universitas Muhammadiyah Gresik, Indonesia
Universitas Muhammadiyah Prof. Dr. Hamka, Indonesia
Universitas Muhammadiyah Makassar, Indonesia

Peer review statement

All papers published in this volume of *IOP Conference Series: Materials Science and Engineering* have been peer reviewed through processes administered by the proceedings Editors. Reviews were conducted by expert referees to the professional and scientific standards expected of a proceedings journal published by IOP Publishing.

Growth response of red chilli plants to flowering phase against the application of *Trichoderma* and *Pseudomonas fluorescens* and P fertilizers

Sutarman

Departement of Agrotecnology, Faculty of Agriculture, Universitas Muhammadiyah Sidoarjo, Indonesia

Email: sutarman@umsida.ac.id

Abstract. This study aims to determine the effect of the application of *Trichoderma sp*. Fungi. and *Pseudomonas fluorescens* bacteria as biofertilizer agents and P fertilizer doses as well as interactions between microbial biofertilizer agents and P fertilizer doses on vegetative growth and early appearance of red chilli flower flowers to the initial flowering period. This greenhouse experiment was arranged factorially in a completely randomized design. The first factor was the application of microbial biofertilizer agents consisting of no microbes, Trichoderma sp., And P. fluorescens, while the second factor was the dosage of P fertilizer per 5,000 g of soil planting medium consisting of 0, 0.5, and 1.0 g SP-36. Nine combinations of treatments repeated three times to obtain 27 experimental units. The variables observed were plant height, stem diameter, leaf area, wet weight and stover dry weight at 42 days after planting (DAP) and the beginning of flower emergence. The data were analyzed by 5% ANOVA followed by a 5% HSD Test to determine the difference in influence between treatments. The results showed a very significant interaction effect between the application of microbial agents and the dose of phosphate fertilizer to height, stem diameter, leaf area, wet weight and dry weight of stover and the initial flowering time at 42 DAP. The application of P. fluorescens bacteria as a biofertilizer agent combined with 0.5 g SP-36 fertilizer dose per 5,000 g of planting media showed the highest growth effect of red chili plants at age 42 DAP in terms of plant height, stem diameter, leaf area, stover wet weight, stover dry weight, and earlier flower appearance.

1. Introduction

Red chili is a strategic one that often causes social compilation turmoil that occurs between the demand and supply imbalances. The development of chilli plantations on dry land requires efficient cultivation applications and is based on the utilization of local resources. However, the physiology of chili plants is highly dependent on soil fertility, therefore massive fertilization is carried out on each tanana chilli cultivation [1].

The facts on the ground show that the seedling phase until the plant enters the flowering phase is a period prone to plants in the face of environmental stress both biotic and abiotic. Drought greatly affects the ability to produce phenolic compounds, morphological responses, and gene expression of chilli plants [2]. Manipulation of plant rhizosphere environment to be dominated by effective microbes is expected to help plant health and resistance to environmental stress.

IOP Conf. Series: Materials Science and Engineering **821** (2020) 012001 doi:10.1088/1757-899X/821/1/012001

The use of effective microbial microbes as biofertilizer agents for the cultivation of local chilli plants or utilized as exotic effective microbes in chilli plantations in various areas far from the habitat of the intended microbial origin is a challenge that must be answered by researchers and practitioners of plant cultivation.

Trichoderma and *Psuedomonas fluorenses* are effective microbes which have been isolated from various locations, collected, and developed as biofertilizer agents because of their various abilities to help plant growth and help improve soil fertility physically by improving soil aggregates [3].

Trichoderma is a fungus that is capable of producing secondary metabolites which act as regulators of plant growth for young sprouts and plants [4, 5]. These fungi also produce various enzymes that work extracellularly [6] degrade organic matter that produces nutrients for plants [7]. Bacteria *P. fluorenses* play a role in increasing plant growth and production as well as increasing available P in the soil while reducing the use of phosphate fertilizers [8].

There are many dry land lands in Indonesia that experience degradation in fertility which among them is characterized by low organic matter (<2%) and low P available [9]. *P. fluorescens* bacteria have one role, which is able to produce acid phosphatase enzymes that can facilitate the release of P so that it will increase availability for plants [10].

Giving organic in fertilizer is no guarantee for the restoration of the status of soil organic matter that can sustain plant life. Enriching organic fertilizers with effective microbes is one of the efforts to improve soil fertility both chemically, physically and biologically. The *Trichoderma* formulation in organic fertilizers not only increases the supply of nutrients as a consequence of degradation of organic matter, but also encourages other microbial activities including various types of bacteria that benefit plants while inhibiting pathogenic microorganisms and other harmful ones [11].

On the other hand, the low availability of P requires the provision of inorganic P fertilizer which is ready to be absorbed by plants for optimal crop production purposes. In large-scale plant production activities, P applications require a large additional cost; while the efficiency of production costs while increasing the capacity of indigenous resource support needs to be improved. The application of *Trichoderma* and the bacterium *P. fluorenses* is expected to be able to answer these challenges both applied individually or simultaneously as a form of fertilization.

This study aims to determine the effect of microbial biofertilizer agents especially *Trichoderma* and P. *fluorenses* as well as the application of organic fertilizers in "economical" and "optimal" ways as well as the possible interaction between the two treatment factors on the growth of red chillies until the beginning of flowering phase.

2. Experimental method

Psuedomonas fluorescens (Pf) bacteria used in this experiment were Pf-Kat-01 isolates and *Trichodema sp.* Fungi were Tc-Clkt-01 isolates which are collections of the Microbiology Laboratory of the Universitas Muhammadiyah Sidoarjo (UMSIDA). The experiment was conducted at UMSIDA green house in February-May 2019.

The initial stage is to prepare the seedlings for the red pepper variety Gada. Chilli seeds are placed in the seed tray which contains sterile soil and incubated for 20 days. After that the seeds are ready to be moved into the polybag capacity of 5,000 gr of planting media. On the other hand phosphate fertilizers are prepared, which in this experiment are divided into three types, namely 0, 0.5, and 1.0 g TSP, each of which is given as a mixture in 5,000 g of soil of the planting medium.

The main step is to multiply the microbial biofertilizer agents. To prepare Pf biofertilizer, we do a multiplication of Pf-Kat-01 isolates on Kelman's media with a composition of 10 g peptone, 10 g glucose, 1 g casamino acid, 20 g bar gelatin, and distillate water which is heated and stirred evenly with the total volume 1000 ml before sterilizing in the autoclave. We inoculated bacterial isolates by scraping inoculants using a needle on the surface of the petri dish containing Kelman's media. We harvested the culture after an incubation period of seven days, then crushed it with a blender and diluted it with distilled water up to 103 times, then the suspension was poured into a vessel containing sterile compost and stirred evenly. This bacterial biofertilizer mixture is then incubated for 10 days.

For *Trichoderma* fungi, we propagated using PDA-c media with an incubation period of 10 days and carried out in the same way as preparing a bacterial biofertilizer with the same incubation period.

Before applying as fertilizer, each biofertilizer was examined for spore content through a dilution method to obtain 10^7 cfu / g of conidiospora *Trichoderma* and 10^9 cfu / g of Pf bacterial cells. As a biofertilizer with a sterile compost carrier each prepared to be applied as fertilization at the beginning of planting with a dose of 200 g per polybag capacity of 5,000 g of planting media. The process of examining the biofertilizer propagul agent is carried out for a maximum of 5 days, so that it takes 15 days from the beginning of the incubation of the biofertilizer to be applied as fertilization. At that time the red chilli seeds were ready to be moved and the experiment was ready to be implemented.

The soil used as a planting medium in this experiment came from the village of Randegan, Sidoarjo Regency, East Java, which is known to have low P availability with the following chemical analysis results: pH (H_2O) 5.7, pH (KCl 1N) 5.4, C inorganic 0.64%, N-total 0.09, P (Bray 1) 6.29 $mg.kg^{-1}$, and CEC 34.02 me / 100 mg.

This factorial experiment was arranged in a completely randomized design with the first factor being the application of microbial biofertilizer agents consisting of: without microbes (M0), *Trichoderma sp.* (M1), and *P. fluorescens* (M2); as the second factor, the dosage of P fertilizer consisted of: 0, 0.5, and 1.0 g SP-36 per 5,000 gr of growing media. Repeat three times, to obtain 27 units of experiment. Variables observed were: plant height, stem diameter, leaf area, stover wet weight, and stover dry weight at 42 days after planting (DAP) and the time of flower appearance until the plant was 42 DAP. Data analysis was performed on each observation variable by using variance analysis (ANOVA) at the level of 5 and 1% followed by an Honest 5% Real Difference Test to determine the differences between treatments.

3. Results and discussion

The results of the variance analysis on plant height data showed that there was a very significant interaction effect ($P < 0.01$) between the microbial application of biofertilizer agents and phosphate fertilizer doses at the end of the vegetative phase of red chilli plants (42 DAP). The mean plant height in response to the interaction of the two factors is shown in Table 1. Growth in plant height for all combinations of treatments from 7 DAPs to 42 DAPs is shown in Figure 1.

Table 1.The mean effect of the interaction of application of biofertilizer agents and fertilizer dosage P on plant height of the red chilli plant42 DAP (cm)[*)]

P fertilizer dosage per 5,000 g of planting media soil	Biofertilizer microorganism agents			HSD 5%
	Without m.o.	*Trichoderma*	*P. fluorescens*	
0 g	33,67 a A	33,50 ab A	36,17 b A	
0.5 g	31,33 a B	34,57 a AB	35,93 a A	2,70
1.0 g	32,50 a AB	30,17 b B	33,73 a A	
HSD 5%		2,70		

[*)]The numbers followed by the same lowercase letter in the same column and followed by the same capital letters on the same line show no different based on the 5% HSD test

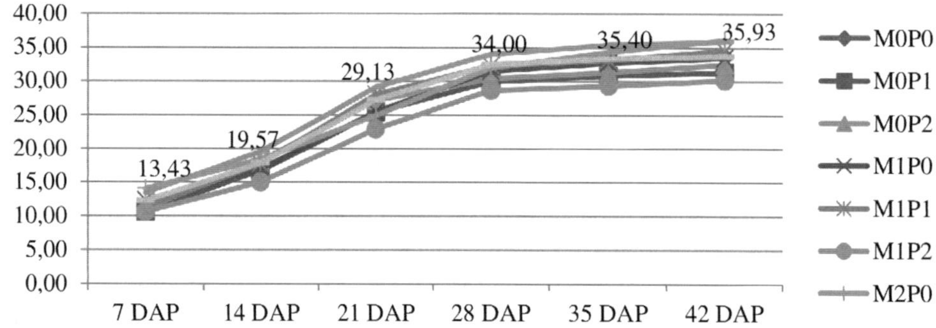

Figure 1. Graph of plant height growth in all treatment combinations 7-42 DAP (cm)

The microbial application of biofertilizer agents and phosphate fertilizer application each had a significant effect ($p < 0.05$) and was very significant ($p < 0.01$) on stem diameter, as well as the interaction of both had a very significant effect on the stem diameter of chili meram plants 0.01). The average stem diameter of the plant in response to the interaction of the two factors is shown in Table 2. Growth of plant stem diameters for all combinations of treatments from 7 DAPs to 42 DAPs is shown in Figure 2.

Table 2. The mean of the interaction effect of the application of biofertilizer agents and P fertilization doses on the stem diameter of the red chilli plant 42 DAP (mm)*)

P fertilizer dosage per 5,000 g of planting media soil	Biofertilizer microorganism agents			HSD 5%
	Without m.o.	*Trichoderma*	*P. fluorescens*	
0 g	1.57 b A	1.58 ab A	1,58 b A	
0.5 g	1.63 ab B	1.82 a A	1,85 a A	0.15
1.0 g	1.70 a A	1.55 b B	1,80 a A	
HSD 5%		0.15		

*)The numbers followed by the same lowercase letter in the same column and followed by the same capital letters on the same line show no different based on the 5% HSD test

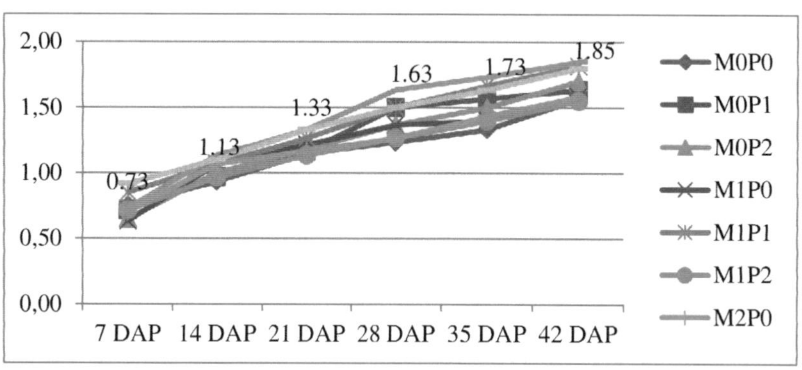

Figure 2. Growth chart of red chili plant stem diameter in all treatment combinations 7-42 DAP (mm)

The application of biofertilizer agents and phosphate fertilizer application had no significant effect ($p > 0.05$) but the interaction had a very significant effect ($p < 0.01$) on the area and red chilli plants. The mean leaf area in response to the interaction of application of biofertilizer agents and phosphate fertilizer application is shown in Table 3.

Table 3. The mean of the interaction effect of the application of biofertilizer agents and P fertilization

P fertilizer dosage per 5,000 g of planting media soil	Biofertilizer microorganism agents			HSD 5%
	Without m.o.	*Trichoderma*	*P. fluorescens*	
0 g	25.93 a	25.20 a	22.93 b	
	A	AB	B	
0.5 g	25.53 a	25.47 a	29.13 a	2.81
	A	A	B	
1.0 g	25.47 a	21.93 b	28.00 a	
	AB	A	A	
HSD 5%		2.81		

doses on the leaf area of red chilli plants 42 DAP (cm^2)*)

*)The numbers followed by the same lowercase letter in the same column and followed by the same capital letters on the same line show no different based on the 5% HSD test

The microbial application of biofertilizer agents, phosphate fertilizer application, and the interaction of both had a very significant effect ($p < 0.01$) on the wet weight of red chili plant stover 42 DAP. The mean wet weight of the interaction effect is shown in Table 4.

Table 4. The mean interaction effect of application of biofertilizer agents and fertilizer dosage P on the wet weight of red chili plant stover 42 DAP (g)*)

P fertilizer dosage per 5,000 g of planting media soil	Biofertilizer microorganism agents			HSD 5%
	Without m.o.	*Trichoderma*	*P. fluorescens*	
0 g	29.67 a	33.33 a	34.57 b	
	A	A	A	
0.5 g	23.67b	33.83 a	46.93 a	4.14
	B	B	A	
1.0 g	28.17 a	23.57 b	35.33 b	
	A	B	A	
HSD 5%		4.14		

*)The numbers followed by the same lowercase letter in the same column and followed by the same capital letters on the same line show no different based on the 5% HSD test

The microbial application of biofertilizer agents has a very significant effect ($p < 0.01$), but phosphate fertilizer application has no significant effect ($p > 0.05$), and both interactions have a very significant effect ($p < 0.01$) on the dry weight of red chili plant stover 42 DAP. The mean dry weight of the interaction effect of the stover is shown in Table 5.

Table 5. The mean interaction effect of application of biofertilizer agents and fertilizer dosage P on the dry weight of red chili plant stover 42 DAP (g)*)

P fertilizer dosage per 5,000 g of planting media soil	Biofertilizer microorganism agents			HSD 5%
	Without m.o.	*Trichoderma*	*P. fluorescens*	
0 g	2.67 a	2.46 a	2.51 b	
	A	A	A	
0.5 g	2.26 a	2.57 a	3.56 a	0.45
	A	A	B	
1.0 g	2.68 a	2.31 a	3.26 a	

	A	A	A	
HSD 5%		0.45		

*)The numbers followed by the same lowercase letter in the same column and followed by the same capital letters on the same line show no different based on the 5% HSD test

The interaction between microbial application of biofertilizer agents and phosphate fertilizer application at the end of the vegetatif phase of red chilli plants (42 DAP) had a very significant effect ($P < 0.01$) on the time of flower appearance. The application of microbes and phosphates as a single factor has no significant effect ($p > 0.05$) on the time of flower appearance. The average time for the appearance of red chilli plants in response to the interaction of the two factors is shown in Table 6.

Table 6. The mean of the interaction effect of the application of biofertilizer agents and P fertilizer dosage on the initial flowering time of red chilli plants up to 42 DAP (days)*)

P fertilizer dosage per 5,000 g of planting media soil	Biofertilizer microorganism agents			HSD 5%
	Without m.o.	*Trichoderma*	*P. fluorescens*	
0 g	35.2 ab	34.2 a	37.8 a	
	AB	A	A	
0.5 g	38.0 a	33.6 a	33.8 b	3.1
	A	B	B	
1.0 g	34.7 a	36.2 a	33.8 b	
	A	A	A	
HSD 5%		3.1		

*)The numbers followed by the same lowercase letter in the same column and followed by the same capital letters on the same line show no different based on the 5% HSD test

Effective microbial application of biofertilizer agents and application of P fertilizers both in combination and in a very real effect on all observation variables except for each treatment factor did not significantly affect leaf area and the beginning of flower appearance.

Although leaf area does not show a difference in response to the application of different biofertilizer agents and application of P fertilizer, the various other growth variables show significant differences. This is given that leaves function as a means of producing energy sources for all metabolism which is more represented in the form of plant height, stem diameter, stover wet weight, stover dry weight, and the initial length of appearance.

The more growth of chilli plants is supported by the role of *Trichoderma* as a nutritional contributor to the degradation of organic matter[12], micronutrients and metabolites which act as growth regulator compounds for plants [13]. All donations from Trichoderma activity are needed for the process of plant growth and induction of flower appearance. The application of cucumber plants also shows the effect of very good growth [14]. The same was shown by *P. fluorescens* given its ability to produce various growth-regulating compounds for plants and produce acid phosphatase enzymes which can help provide P for plants, increasing wet weight and dry weight of plants[15].

Increasing the dose of P will increase plant growth. This very significant response is also supported by the results of soil analysis of the planting media used in this study, namely P available 6.29 $mg.kg^{-1}$ including low category. However, the soil chemical status of the planting medium, especially the relatively high value of soil cation exchange capacity used in this experiment (34.02 me / 100 g) seems to support the normal growth of plants, especially in plant height and stem diameter. Inorganic P is absorbed by plants in the form of orthophosphate ($H_2PO_4^-$ and HPO_4^{2-} play an important role in metabolic processes such as photosynthesis, energy transfer, signal transduction [16] and various processes that can guarantee plant resistance to environmental stress. In all combinations of treatments in this experiment, including treatment without fertilizer and without microbial biofertilizer agents showed the same growth pattern (Figures 1 and 2).

The effect of the interaction of the two factors which is very real with the highest value of each variable is indicated by the combination of the application of P. fluorescens bacteria and the application of "economical" doses or 0.5 g / 5000 g of growing media. This is in line with the evidence of the role of these bacteria as a provider of P for plants [17] which guarantees the adequacy of plants for P elements [18] as shown in the form of plant height (Table 1), stem diameter (Table 2), leaf area (Table 3) , wet stover weight (Table 4), stover dry weight (Table 5), and the initial length of flower appearance (Table 6). Although the average value of all observation variables in the treatment of the combination dose of P is "economical" (0.5 g) and the application of *Trichoderma* is below the average value in the combination of P 0.5 g and application of *P. fluorescens*, but statistically (HSD) the 5% test) on some variables of combining the two combinations was not significantly different, namely 34.57 cm and 35.93 cm, respectively (plant height, Table 1), 1.82 mm and 1.85 mm (stem diameter; Table 2).), as well as 33.6 and 33.8 (the duration of the onset of interest; Table 6). Growth in plant height and stem diameter is largely determined by the activity of growth regulating compounds in the context of cell division and the adequacy of nutrients which in this case can be fulfilled with the help of *Trichoderma* activity [19, 20]. Likewise the process of flower appearance is strongly influenced by the availability of nutrients that can be met by the supply given by *Trichoderma* or by *Pseudomonas* [4, 21]. Both types of microbes have been tested for their ability as phosphate solvent agents [10, 22]. Thus the application of *Trichoderma* and the application of *P. fluorescens* is an alternative to be combined with fertilizer application at a dose of 0.5 g / 5000 grams of planting media in the cultivation of red chillies.

4. Conclusions

The microbial application of biofertilizer agents and P fertilization doses had an effect on increasing the growth of plant height, stem diameter, stover wet weight and dry weight of red chili plant stover at 42 days after planting. The interaction of the two influential factors increases all observation variables and speeds up the initial flowering time. The application of *Pseudomonas fluorescens* bacteria combined with 0.5 g / 5000 g fertilizer media planting media resulted in a response of red chili plants in terms of plant height, stem diameter, leaf area and wet weight and the highest dry stover weight. The response of early flower appearance was indicated by a combination of 0.5 g fertilizer dose with the application of *P. fluorescens* bacteria or with the application of *Trichoderma sp.*

References

[1] Das S, Teja K C, Duary B, Agrawal P K, dan Bhattacharya S S 2016 Impact of nutrient management, soil type and location on the accumulation of capsaicin in Capsicum chinense (Jacq.): One of the hottest chili in the world *Sci. Hortic. (Amsterdam)* vol **213** hal 354–66.

[2] Rodríguez-Calzada T, Qian M, Strid Å, Neugart S, Schreiner M,Torres-Pacheco I, Guevara-González RG 2019 Effect of UV-B radiation on morphology, phenolic compound production, gene expression, and subsequent drought stress responses in chili pepper (Capsicum annuum L.) *Plant Physiol. Biochem.* vol 134, hal. 94–102, 2019.

[3] Six J, Elliott E T, dan Paustian K 2000 Soil macroaggregate turnover and microaggregate formation: a mechanism for C sequestration under no-tillage agriculture *Soil Biol. Biochem.* vol **32** no 14 hal 2099–103.

[4] Gravel V, Antoun H, dan Tweddell R J 2007 Growth stimulation and fruit yield improvement of greenhouse tomato plants by inoculation with Pseudomonas putida or Trichoderma atroviride: Possible role of indole acetic acid (IAA) *Soil Biol. Biochem.* vol **39** no 8 hal 1968–77.

[5] Srivastava R, Khalid A, Singh U S, dan Sharma A K 2010 Evaluation of arbuscular mycorrhizal fungus, fluorescent Pseudomonas and Trichoderma harzianum formulation against Fusarium oxysporum f. sp. lycopersici for the management of tomato wilt *Biol. Control* vol **53** no 1 hal 24–31.

[6] Saravanakumar K, Yu C, Dou K, Wang M, Li Y, dan Chen J 2016 Synergistic effect of

Trichoderma-derived antifungal metabolites and cell wall degrading enzymes on enhanced biocontrol of Fusarium oxysporum f. sp. cucumerinum *Biol. Control* vol **94** hal 37–46.

[7] Chowdappa P, Mohan Kumar S P, Jyothi Lakshmi M, dan Upreti K K 2013 Growth stimulation and induction of systemic resistance in tomato against early and late blight by Bacillus subtilis OTPB1 or Trichoderma harzianum OTPB3 *Biol. Control* vol **65** no 1 hal 109–17.

[8] Ibrahim B 2016 The effort to increase waxy corn production as the main ingredient of corn rice through the application of phosphate solvent extraction and phosphate fertilizer *Ital. Oral Surg.* vol **9** hal 532–37.

[9] Suriadikarta D dan Simanungkalit R D 2006 Pendahuluan," in *Pupuk Organik dan Pupuk Hayati* D Suriadikarta, R D Simanungkalit, R Saraswati, D Setyorini, dan W Hartatik Ed. Bogor: Balai Besar Litbang Sumberdaya Lahan Pertanian, Badan Penelitian dan Pengembangan Petanian hal 1–10.

[10] Sharma S B, Sayyed R Z, Trivedi M H, dan Gobi T A 2013 Phosphate solubilizing microbes: sustainable approach for managing phosphorus deficiency in agricultural soils *Springerplus* vol **2** no 1 hal 587.

[11] Verma M, Brar S K, Tyagi R D, Surampalli R Y, dan Valéro J R 2007 Antagonistic fungi, Trichoderma spp.: Panoply of biological control *Biochem. Eng. J.* vol **37** no 1 hal 1–20.

[12] Vinale F, Sivasithamparam K, Ghisalberti E L, Marra R, Barbetti M J, Li H, Woo S L 2008 A novel role for Trichoderma secondary metabolites in the interactions with plants *Physiol. Mol. Plant Pathol.*vol **72** no 1 hal 80–6.

[13] Siddiquee S, Shafawati S N, dan Naher L 2017 Effective composting of empty fruit bunches using potential Trichoderma strains *Biotechnol. Reports* vol **13** hal 1–7.

[14] Li M, Guang-shu MA, Lian H, Xiao-lin SU, Tian Y, Huang WK, Mei J, and Jiang X L 2019 The effects of Trichoderma on preventing cucumber fusarium wilt and regulating cucumber physiology *J. Integr. Agric.* vol **18** no 3 hal 607–17.

[15] Shanmugaiah V, Balasubramanian N, Gomathinayagam S, Manoharan P T, dan Rajendran A 2009 Effect of single application of Trichoderma viride and Pseudomonas fluorescens on growth promotion in cotton plants *African J. Agric. Res.* vol **4** no 11 hal 1220–25.

[16] Khan M S, Zaidi A, dan Ahmad E 2014 Mechanism of phosphate solubilization and physiological functions of phosphate-solubilizing microorganisms in *Phosphate Solubilizing Microorganisms: Principles and Application of Microphos Technology* M. S. Khan, A. Zaidi, dan J. Musarrat, Ed. Cham: Springer International Publishing hal 31–62.

[17] Ingle K P dan Padole D A 2017 Phosphate solubilizing microbes: An overview 2017 *Int. J. Curr. Microbiol. Appl. Sci.* vol 6 no 1 hal 844–52.

[18] Alori E T, Glick B R, dan Babalola O O 2017 Microbial phosphorus solubilization and its potential for use in sustainable agriculture *Front. Microbiol.* vol **8** hal 971.

[19] Youssef S, Tartoura K, dan Abdelraouf G A 2016 *Evaluation of Trichoderma harzianum and Serratia proteamaculans effect on disease suppression, stimulation of ROS-scavenging enzymes and improving tomato growth infected by Rhizoctonia solani* vol **100**.

[20] Sutarman, Prihatiningrum AE, Sukarno A, and Miftahurrohmat A 2018 Initial growth response of shallot on *Trichoderma* formulated in oyster mushroom cultivation waste *IOP Conf. Series: Materials Science and Engineering* **420** (2018) 012064 doi:10.1088/1757-899X/420/1/012064.

[21] Buysens C, César V, Ferrais F, Dupré de Boulois H, dan Declerck S 2016 Inoculation of Medicago sativa cover crop with Rhizophagus irregularis and Trichoderma harzianum increases the yield of subsequently-grown potato under low nutrient conditions *Appl. Soil Ecol.*vol **105** hal 137–43.

[22] Srinivasan R, Yandigeri M S, Kashyap S, dan Alagawadi A R 2012 Effect of salt on survival and P-solubilization potential of phosphate solubilizing microorganisms from salt affected soils *Saudi J. Biol. Sci.* vol **19** no 4 hal 427–34.

IOP Conf. Series: Materials Science and Engineering **821** (2020) 012002 doi:10.1088/1757-899X/821/1/012002

Utilization of *trichoderma* sp. and *pseudomonas fluorescens* as biofertilizer in shade-resistant soybean

A Miftahurrohmat* **and Sutarman**

Departement of Agrotecnology, Faculty of Agriculture, Universitas Muhammadiyah Sidoarjo, Indonesia

*Email: amiftahurrohmat@umsida.ac.id

Abstract. This study aims to obtain *Trichoderma* fungi and *Pseudomonas fluorescens* bacteria obtained from soybean planting fields and test their ability as biofertilizer agents to assist soybean growth until the end of the vegetative phase. From the soybean crop field, *Trichoderma* fungi and *P. fluorescens* (Pf) were isolated. After purification, observations were made to determine morphological physical characteristics both macroscopically and microscopically. Pf bacterial isolates were tested for their performance characteristics to ensure that bacteria were not pathogenic. For work power testing, as a biofertilizer agent, a factorial experiment was prepared in a completely randomized design (CRD). As the first factor, *Trichoderma* consists of without and with *Trichoderma*, while the second factor is Pf bacteria consisting of without and with Pf. The experiment was repeated four times so that 16 experimental units were obtained. The variables observed were plant height, number of leaves, stem diameter, and time of initial average flower appearance. Data were analyzed by ANOVA followed by HSD testing on a 5% test tarf. The results showed that the fungi *Trichoderma* sp. obtained by having single-celled, smooth-walled conidia measuring 3.2 x 2.0 μm, green with irregular upright hyaline conidiophores. As for P. fluorensence does not cause foul symptoms in test potatoes which are non-pathogenic. Fungi *Trichoderma* sp. TC-Jro-02 isolates and *Pseudomonas fluorescens* bacteria which were applied as biofertilizer agents were each increasing and inhibiting growth in plant height, leaf number, stover dry weight, and dry weight of soybean root of Dena varieties grown in shade up to 30 days after planting. Interactions between them affect the dry weight of stover and the dry weight of plant roots. The bacterial activity of *Pseudomonas fluorescens* inhibits the effect of the performance of *Trichoderma* fungi in supporting the growth of soybean plants in the shade.

1. Introduction

Efforts to increase the production of Indonesian soybeans in order to be able to be independent in fulfilling national needs are carried out not only to increase the genetic capabilities of varieties and cultivation techniques, but also to use dry land that is still widely available. The use of land which is as large as the lower vegetation strata through shading has begun to be tricked by researchers by producing shade resistant varieties. However, the performance of varieties that have been proven to be shade-resistant remains prone to environmental stress due to the lack of soil organic matter and the weak role of the supporting microflora of plant growth.

Trichoderma fungi and *Pseudomonas fluorenses* bacteria which are soil borne microorganisms have the potential to be used to help plant growth on dry land. *Trichoderma*, besides having a role in protecting plants from pathogenic disorders in rhizofer [1], has the ability as a biofertilizer agent because of some of its superior characteristics, namely degrading organic matter to produce nutrients and produce growth regulating compounds for plants [2-4]. *P.fluorescens* bacteria is one type of bacteria that has the ability to produce growth regulating compounds such as auxin [5], producing enzymes that are capable of working in the P-organic mineralization process into P-inorgnaik which is available to plants [6]. Both types of microbes can be used together to help the growth of vegetative plants and maintain plant health so that plants are able to produce well [7-8].

On dry land with limited production resources such as water, nutrition, light intensity that is often very low, good vegetative growth greatly determines the success of growing plants until the production phase. Since sprouts to the final vefetative phase of soybean plants require environmental support both biotic and abiotic. The existence and good performance of microbes in dry land rhizosphere can support plant growth to pass the critical phase at the beginning of growth,

In addition, a test is needed that can ensure that between the two potential microbes the biofertilizer biological agents can synergize with soybean plants, especially in shade resistant varieats. The success of this test can be input into the development of a biofertlizer application technology package on soybean plants on land, especially those under the auspices of the upper strata vegetation canopy such as plantations and plantations.

The aim of this study was to obtain *Trichoderma* fungi and *P. fluorescens* bacteria which were obtained from soybean planting fields and tested their ability as biofertilizer agents to help grow soybean plants until the end of the vegetative phase.

2. Experimental method

2.1. Biofertlizer agent

To get a microbial biofertilizer agent for soybean plants, the location of soil sampling in an area that has a history is often used for planting soybeans with a satisfying appearance of growth and production. The activity of isolating the biofertilizer microbial agent was carried out in the Microbiology laboratory, followed by application testing in the green house of the Universitas Muhammadiyah Sidoarjo from January to May 2019.

Soil samples from the land commonly planted with soybeans were used to obtain *Trichoderma* isolates and *P. fluorescens* (Pf) bacteria. The land is in Ngembat Village, Gondang District, Mojokerto Regency with a lower vegetation strata which is partially covered by pine canopy with an intensity of canopy cover equivalent to 50-80%. The suspension produced from mixing 5 grams of soil that has been diluted in 1,000 ml of distillate water is sprayed into a medium of Potato Dextrose Agar-chloramphenicol (PDA-c) in a petri dish. This PDA-c media is made of 200 g of potatoes extracted through heating in a one liter volume which has been mixed with 20 g of dextrose and 20 g of agar. After 90 hours the incubation period appears various fungal colonies. Carefully the colonies that show the characteristics of the Trichodema colony are touched with the tip of the needle and transferred to the new media. Then after 7 days the incubation period of the *Trichoderma* isolate colonies can be observed. Condiospora is observed under a microscope and the determination of the *Trichoderma* type refers to Gams and Bissett (2002) [9].

The soil suspension from soybean cropping land (in Katerungan Village, Krian District, Sidoarjo Regency, East Java) resulted in a dilution of 10^3 ml of 1 ml sprayed onto the surface of Kelman's media. This medium is a mixture of 10 g peptone, 10 g glucose, 1 g casamino acid, and 20 g agar white bar dissolved in distilled water to a volume of 1000 ml which is stirred until the mixture appears homogeneous. Before being used as a bacterial growth medium, the Kelman media was sterilized in an autoclave for 0.5 hours at 121 °C, 1 atm. After an incubation period of 36 hours of colony that corresponds to the characteristics of Pf, using the tip of the ose needle touched by the center of the colony that appears and is scratched onto the surface of the new media, all activities are carried out in aseptic conditions. The next stage is to ensure that the bacterial colonies that have been isolated do not have pathogenic characters, namely by inoculating them aseptically on sterile potato pieces by soaking them in 50% alcohol for 5 seconds and rinsing them in distlat water 3 times. Non-pathogenic bacteria will not show symptoms of decay in the potato pieces after 72 hours of incubation.

2.2. Biofertilizer application test
For the manufacture of biofertilizers, *Trichoderma* and Pf bacteria as a result of this isolation must be formulated with sterile compost as a carrier agent. Each mature culture is harvested and mixed with sterile compost until evenly distributed and incubated for 14 days. Thus obtained two kinds of biological fertilizers. Before each biofertilizer was used, the contents of conidiospora (for *Trichoderma*) and bacterial cells were examined in each gram of biofertilizer. The examination was carried out using a dilution method to obtain the conidiospora *Trichoderma* content of 10^7 cfu/g, while the content of Pf bacterial cells was 10^9 cfu/g.

Sterile soil that is used as a planting medium is placed on a polybag with a capacity of five kg. In accordance with the treatment designed, the treatment of *Trichoderma* biofertilizer application and application of biofertilizer Pf were each given at a dose of 200 g per polybag. Thus the average content of conidiospora *Trichoderma* and Pf bacterial cells in the soil growing media reached 4 x 10^5 and 4 x 10^7 cfu/g, respectively. Furthermore, the soybean seeds of Dena varieties are planted into each polybag. The position of polybag that has been planted with soybeans is arranged randomly and placed under a paranet shade of 60% or with the intensity of sunlight entering 40%.

This factorial experiment was arranged in a completely randomized design (CRD) with the first factor *Trichoderma* biofertilizer consisting of two levels, namely without *Trichoderma* and with *Trichoderma*, while the second factor was biological fertilizer *P. fluorescens* (Pf) consisting without Pf and with Pf; the experiment was repeated four times. The variables observed in this experiment were plant height, number of leaves, dry weight of plant stover, and dry weight of plant roots at the end of the vegetative growth phase. All data were analyzed using variance analysis (ANOVA) at the level of 5% and 1% to determine the effect of each treatment factor and its interactions. Regarding the observation variable which shows the real effect, then the Honest Real Difference (HSD) test is carried out at the level of 5% to determine the difference between treatments.

3. Results and discussion

3.1. Bioferilizer agent
The results of the isolation of *Trichoderma* fungi from the soil from the land which is usually shaded by pine canopy showed *Trichoderma* sp isolates with the appearance of colonies and microscopic structures as shown in Figure 1.

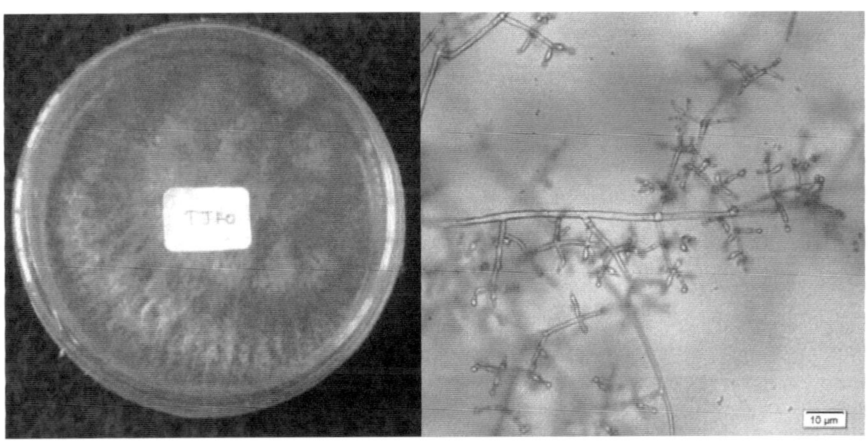

Figure 1. Appearance of colonies (left) and microscopic structures (right) *Trichoderma* isolates Tc-Jro-02

The fungi colony on PDA-c media was originally woven with white hyphae which was immediately followed by the presence of green which was the result of thickening of the ends of hyphae which thickened to hyidiic conidiophores, erect irregular, smooth walls. Single-celled conidia, smooth-walled, and measuring 3.2 x 2.0 μm. When referring to the description shown by Gams and Bissett (2002) [9], the microscopic appearance of this isolate is similar to *T, koningii*. However, the final results of determination should be carried out using molecular markers. For practical purposes in the application experiment, the findings of this finding are referred to as *Trichoderma* sp. Tc-Jro-02 isolate, a code of isolates which later became the collection of the Agricultural Microbiology Laboratory of Muhammadiyah University, Sidoarjo.

The observation of colonies of *P. fluorescens* isolates showed that gram negative bacteria had transparent pink colored colonies. The color formed is the result of pigment production by *P. fluorescens* [10]. Curved or straight stem-shaped cells measuring 0.5-1.0 x 1.5-5.0 μm. The results of testing the inoculation on the surface of fresh potato pieces show no discoloration that indicates the decay process. This shows that *P. fluorescens* isolates are non-pathogenic bacteria and are suitable for use as biofertilizer agents. For practical purposes for application experiments, these bacteria are referred to as *P. fluorescens* Pf-Kat-01 isolates and later become a collection of Microbiology laboratories at the Universitas Muhammadiyah Sidoarjo.

3.2. Biofertilizer application test

The results of variance analysis on plant height data showed no interaction effect between the application of *Trichoderma* and the application of *P. fluorescens* to the height of soybean plants of Dena varieties at the end of the vegetative phase, but each factor had a significant effect ($p < 0.01$). The mean height of soybean plants is shown in Table 1. The *Trichoderma* application increased the growth of plant height by 13.95%, but the application of *P. fluorescens* was 13.09% lower than without *P. fluorescens*.

Table 1. The mean effect of the application of *Trichoderma* and *P. fluorescens* on the height of soybean plants of Dena variety 30 days after planting (DAP) (cm) *)

Trichoderma application	Average plant height (cm)	Increase (decrease) (%)	*P. fluorescens application*	Average plant height (cm)	Increase (decrease) (%)
Without *Trichoderma*	60.50 a	-	Without *P. fluorescens*	69.25 a	-
Trichoderma	68.94 a	13,95	*P. fluorescens*	60.19 b	13.09 (-)
HSD 5%	7.13		HSD 5%	7.13	

*)The numbers followed by the same letter in the same column show no different based on the 5% HSD test

The results of the variance analysis on observational data on the number of leaves of soybean plants in Dena variety showed that the application of *Trichoderma* and the application of *P. fluorescens* had a very significant effect ($p < 0.01$) on the leaf leaves at the end of the vegetative phase, but the interaction was not significant (p> 0, 05). The average number of leaves of soybean plants is shown in Table 2. The *Trichoderma* application increased the growth of plant leaves by 11.68%, but in the application of *P. fluorescens* the number of plant leaves was 13.09% lower than without *P. fluorescens*.

Table 2. The mean effect of the application of *Trichoderma* and *P. fluorescens* on the number of leaves per soybean plant of Dena varieties 30 DAP*)

Trichoderma application	number of leaves	Increase (decrease) (%)	*P. fluorescens application*	number of leaves	Increase (decrease) (%)
Without *Trichoderma*	17.13 b	-	Without *P. fluorescens*	16.38 b	-
Trichoderma	19.13 a	11,68	*P. fluorescens*	19.88 a	21.37 (-)
HSD 5%	1.69		HSD 5%	1.69	

*) The numbers followed by the same letter in the same column show no different based on the 5% HSD test

The application of *Trichoderma* and the application of *P. fluorescens* each had a very significant effect on the dry weight of soybean stover of Dena varieties at the end of the vegetative phase (p <0.01), as well as the interaction was very significant. The average dry weight of soybean stover in response to the combination of treatments is shown in Table 3. Rearata effect of the application of *Trichoderma* and *P. fluorescens* are shown in Table 4. *Trichoderma* application produced 36.80% higher dry stover weight than without *Trichoderma*, on the contrary the application of *P. fluorescens* was 45.058% lower than without *P. fluorescens*.

Table 3. The mean interaction effect of the application of *Trichoderma* and *P. fluorescens* on the dry weight of soybean stover of Dena 30 varieties DAP (g) *)

	Trichoderma application		HSD 5%
P. fluorescens application	Without *Trichoderma*	*Trichoderma*	
Without *P. fluorescens*	1.82 a B	3,05 a A	0,35
P. fluorescens	1.36 b A	1,29 b A	
HSD 5%	0,35		

*) The numbers followed by the same lowercase letter in the same column and followed by the same capital letters on the same line show no different based on the 5% HSD test

Table 4. The mean effect of the application of *Trichoderma* and *P. fluorescens* on the dry weight of soybean stover of Dena varieties 30 DAP (g) *)

Trichoderma application	stover dry weight (g)	Increase (decrease) (%)	*P. fluorescens application*	stover dry weight (g)	Increase (decrease) (%)
Without *Trichoderma*	1.59 b	-	Without *P. fluorescens*	2.43 b	-
Trichoderma	2.17 a	36.80	*P. fluorescens*	1.3 a	45.58 (-)
HSD 5%	1.69		HSD 5%	1.69	

*) The numbers followed by the same letter in the same column show no different based on the 5% HSD test

The application of *Trichoderma* and the application of *P. fluorescens* each had a very significant effect on the dry weight of soybean root Dena varieties at the end of the vegetative phase (p <0.01), while the interaction between the two had a significant effect (p <0.05). The mean interaction effects on the dry weight of soybean root as shown in Table 5. The average effect of the application of *Trichoderma* and *P. fluorescens* is shown in Table 6. *Trichoderma* application produced 107.98% root dry weight higher than without *Trichoderma*, while the application of *P. fluorescens* 59.38% lower than without *P. fluorescens*.

Table 5. The average effect of the interaction of the application of *Trichoderma* and *P. fluorescens* on the dry weight of soybean roots of Dena varieties 30 DAP (g) *)

P. fluorescens application	*Trichoderma* application: Without *Trichoderma*	*Trichoderma* application: *Trichoderma*	HSD 5%
Without *P. fluorescens*	0.48 a B	0.96 a A	0,17
P. fluorescens	0.18 b B	0.41 b A	
HSD 5%	0,17		

*)The numbers followed by the same lowercase letter in the same column and followed by the same capital letters on the same line show no different based on the 5% HSD test

Table 6. The mean effect of the application of *Trichoderma* and *P. fluorescens* on the dry weight of soybean roots of Dena varieties 30 DAP (g) *)

Trichoderma application	dry weight of soybean roots (g)	Increase (decrease) (%)	*P. fluorescens application*	dry weight of soybean roots (g)	Increase (decrease) (%)
Without *Trichoderma*	0.33 b	-	Without *P. fluorescens*	0.72 a	-
Trichoderma	0.68 a	107.98	*P. fluorescens*	0.29 b	59.38 (-)
HSD 5%	0.17		HSD 5%	1.69	

*) The numbers followed by the same letter in the same column show no different based on the 5% HSD test

The application of *Trichoderma* and *P. fluorescens* produced an interaction effect on dry weight of stover and root dry weight. This shows that the two biological agents show their role as biofertilizers. The role of both has not given the effect of interactions on plant height and number of leaves until the end of the vegetative phase of the plant,

Trichoderma produces several enzymes [11] which degradate organic matter to produce nutrients and produce compounds that play a role in promoting plant growth [12]. *Trichoderma* singly increased plant height (Table 1), number of leaves (Table 2), stover dry weight (Table 4) and root dry weight (Table 6), even under shade conditions; the variety used in this experiment proved to be able to grow well at a shade intensity of 60-80% [13, 14].

P. fluorescens is a bacterium that is capable of producing compounds that can facilitate the process of phosphate release in the soil [15]. This bacterium also produces metabolites which act as regulators of plant growth [16]. In this trial, until the vegetative final phase, it appears that bacteria behave as users of the resources produced by *Trichoderma* activity. As shown in Tables 1 to 6, it appears that all growth variables show lower values in the application of these bacteria compared to without the application of *P. fluorescens* bacteria as biofertilizers. Thus in this experiment the activity of *P. fluorescens* bacteria did not provide benefits for plant growth. In the first 8 weeks vermicomposting process, *P. fluorescens* showed a significant decrease in alkaline phosphatase activity [17].

Based on the facts mentioned above, it appears that the rhizosphere of the soybean varieties of Dena varieties that hold this shade does not provide compatibility for the life and activity of the bacteria *P. fluorescens*, on the other hand it is suitable for the activity of *Trichoderma* fungi. The activity of this fungi in the advanced composting process will produce dissolved compounds that benefit other soil microorganisms [18, 19]. The presence of *P. fluorescens* in soybean rhizosphere in this experiment has suppressed the effect of *Trichoderma* fungi.

4. Conclusion

Application of *Trichoderma* sp. Fungi TC-Jro-02 isolates and application of *Pseudomonas fluorescens* bacteria isolate Pf-Kat-01 as biofertilizer agents had a very significant effect on plant height, leaf number, stover dry weight, and root dry weight, while interactions between them affected the dry weight of stover and root dry weight with soybean varieties Dena. The bacterial activity of *Pseudomonas fluorescens* inhibits the effect of the performance of *Trichoderma* fungi in supporting the growth of soybean plants grown in the shade.

Acknowledgements
The author would like to thank the Ministry of Research and Technology of Indonesian National Education for the Higher Education Excellence Basic Research Grants (Penelitian Dasar Unggulan Perguruan Tinggi) in 2019 that has been given.

References

[1] Srivastava R, Khalid A, Singh U S, and Sharma A K 2010 Evaluation of arbuscular mycorrhizal fungus, fluorescent *Pseudomonas* and *Trichoderma harzianum* formulation against *Fusarium oxysporum* f. sp. lycopersici for the management of tomato wilt *Biological Control* **55** pp 24-31.

[2] Yuan S, Li M, Fang Z, Liu Y, Shi W, Pan B, Wu K, Shi J, Shen B and Shen Q 2016 Biological control of tobacco bacterial wilt using*Trichoderma* harzianum amended bioorganic fertilizer and the arbuscular mycorrhizal fungi Glomus mosseae *Biological Control* **92** pp 164–71.

[3] Saravanakumar K, Yu C, Dou K, Wang M, Li Y and Chen J 2016 Synergistic effect of *Trichoderma*-derived antifungal metabolites and cell wall degrading enzymes on enhanced biocontrol of *Fusarium oxysporum* f. sp. Cucumerinum *Biological Control* **94** pp 37–46.

[4] Chowdappa P, Kumar S P M, Lakshmi M J and Upreti K K 2013 Growth stimulation and induction of systemic resistance in tomato against early and late blight by *Bacillus subtilis* OTPB1 or *Trichoderma harzianum* OTPB3 *Biol. Control* **65** pp 109–17.

[5] Pour M M, Saberi-Riseh R, Mohammadinejad R 2019 A.Hosseini. Investigating the formulation of alginate- gelatin encapsulated *Pseudomonas fluorescens* (VUPF5 and T17-4 strains) for controlling Fusarium solani on potato *International Journal of Biological Macromolecules.* DOI: https://doi.org/10.1016/j.ijbiomac.2019.04.071

[6] Sharma S B, Sayyed R Z, Trivediand M H and Gobi T A 2013 Phosphate solubilizing microbes: sustainable approach for managing phosphorus deficiency in agricultural soils *Springerplus* **2** pp 587–600. doi: 10.1186/2193- 1801-2-587

[7] Gravel V, Antoun H and Tweddell R J 2007 Growth stimulation and fruit yield improvement of greenhouse tomato plants by inoculation with *Pseudomonas putida* or *Trichoderma atroviride*: possible role of indole acetic acid (IAA) *Soil Biol. Biochem.* **39** pp 1968–77.

[8] Buysens C, César V, Ferrais F, De Boulois H D and Declerck S 2016 Inoculation of *Medicago sativa* cover crop with *Rhizophagus irregularis* and *Trichoderma harzianum* increases the yield of subsequently-grown potato under low nutrient conditions *Applied Soil Ecology* **105** pp 137–43.

[9] Gams W and Bissett J 2002 Morphology and identification of *Trichoderma*, In *Trichoderma* and *Gliocladium* vol **1** *Basic biology, taxonomy and genetics* edited by C.P. Kubicek and G.E. Harman (Taylor & Francis Ltd., London, 2002) pp 3-34.

[10] Khana M R, Mohidin F A, Khan U and Ahamad F 2016 NativePseudomonasspp. suppressed the root-knot nematode inin vitro andin vivo, and promoted the nodulation and grain yield in the field grown mungbean *Biological Control* **101** pp 159–68.

[11] Mei L, Guang-shu M, Hua L, Xiao-lin S, Ying T, Wen-kun H, Jie M and Xiliang J 2019 The effects of *Trichoderma*on preventing cucumber fusarium wilt and regulating cucumber physiology *Journal of Integrative Agriculture* **18**(3) pp 607–17.

[12] An-le H, Jia L, Xin-hua W, Quan-guo Z, Wei S, and Jie C 2019 Soil application of *Trichoderma* asperellumGDFS1009 granules promotes growth and resistance to Fusarium graminearumin maize *Journal of Integrative Agriculture* **18**(3) pp 599–606.

[13] Miftahurrohmat A and Sutarman 2018 The morphological response of the soybean growth (Glycine max (i)) until vegetative atGE 3 on various intensities of light *IOP Conf. Series: Materials Science and Engineering* **420** (2108) 012069 doi:10.1088/1757-899X/420/1/012069.

[14] Miftahurrohmat A and Sutarman 2019 Local Soybean (*Glycine max* (L)) Stomatas' morphological and anatomic response in 3rd vegetation stage towards light intensity stress *IOP Conf. Series: Journal of Physics: Conf. Series* **1232.** 012043 doi:10.1088/1742-6596/1232/1/012043.

[15] Widnyana I K and Javandira C 2016 Activities Pseudomonas spp. and Bacillus sp. to stimulate germination and seedling growth of tomato plants *Agriculture and Agricultural Science Procedia* **9** pp 419–23.

[16] Attarzadeh M, Balouchi H, Rajaie M, Dehnavi M M and Salehi A 2019 Growth and nutrient content ofEchinacea purpureaas affected by the combination of phosphorus with arbuscular mycorrhizal fungus and *Pseudomonas florescent* bacterium under different irrigation regimes *Journal of Environmental Management* **231** pp182–88.

[17] Lukashe N S, Mupambwa H A, Green and Mnkeni P N S 2019 Inoculation of fly ash amended vermicompost with phosphate solubilizing bacteria (*Pseudomonas fluorescens*) and its influence on vermi-degradation, nutrient release and biological activity *Waste Management* **83** pp 14–22.

[18] Khan Md Y, Haque Md M, Molla A H, Rahman Md M, Alam M Z 2017 Antioxidant compounds and minerals in tomatoes by *Trichoderma*enriched biofertilizer and their relationship with the soil environments *Journal of Integrative Agriculture* **16**(3) pp 691–703.

[19] Sutarman, N P Maharani, A Wachid, M Abror, A Machfud, A Miftahurrohmat 2019 Effect of Ectomycorrhizal Fungi and *Trichoderma harzianum*on the Clove (*Syzygium aromaticum L.*) Seedlings Performances *IOP Conf. Series: Journal of Physics: Conf. Series* **1232** 012022 doi:10.1088/1742-6596/1232/1/012022.

Analysis of academic service cybersecurity in university based on framework COBIT 5 using CMMI

I Riadi[1,]*, I T R Yanto[1], and E Handoyo[3]

[1] Department of Information System, Faculty of Science and Applied Technology, Universitas Ahmad Dahlan, Yogyakarta, Indonesia
[1] Department of Information System, Faculty of Science and Applied Technology, Universitas Ahmad Dahlan, Yogyakarta, Indonesia
[2] Department of Computer Engineering, Universitas Muhammadiyah Lamonggan, Indonesia

*Email: imam.riadi@is.uad.ac.id

Abstract. A secure academic information system is part of the college. The security of academic information systems is very important to maintain information optimally and safely. Along with the development of technology, academic information systems are often misused by some irresponsible parties that can cause threats. To prevent these things from happening, it is necessary to know the extent to which the security of the academic information system of universities is conducted by evaluating. So the research was conducted to determine the Maturity Level on the governance of the security of University Ahmad Dahlan academic information system by using the COBIT 5 framework on the DSS05 domain. The DSS05 domain on COBIT 5 is a good framework to be used in implementing and evaluating related to the security of academic information systems. Whereas to find out the achievement of evaluation of academic information system security level, CMMI method is needed. The combination of the COBIT 5 framework on the DSS05 domain using the CMMI method in academic information system security is able to provide a level of achievement in the form of a Maturity Level value. The results of the COBIT 5 framework analysis of the DSS05 domain use the CMMI method to get a Maturity level of 4,458 so that it determines the achievement of the evaluation of academic information systems at the tertiary level is Managed and Measurable. This level, universities are increasingly open to technological developments. Universities have applied the quantification concept in each process, and are always monitored and controlled for performance in the security of academic information systems.

1. Introduction

Companies or institutions place information technology as a thing that can support the achievement of the company's strategic plan to achieve the goals of the company or institution's vision, mission and goals. Information technology will get effective results if it uses good governance in its use and is able to be evaluated and evaluated [1]. Information systems are systems that contain SPD networks (systems processing data), which are equipped with communication channels used in data organization systems [2]. There are various concepts of information systems, compatibility is one of the keys to the successful implementation and acceptance of information systems [3]. Along with the development of technology, it is often misused by some irresponsible parties that can cause threats [4]. Academic information systems must provide the security, privacy and integrity of data processed, so that the performance of academic information systems is also an important part that must be considered so that academic information systems can be used optimally and safely [5]. The application of information security systems aims to overcome all problems and constraints, both technically and non-technically which can affect the performance of the system such as availability, confidentiality and integrity factors so that the level of information security can be assessed [6] as in Figure 1.

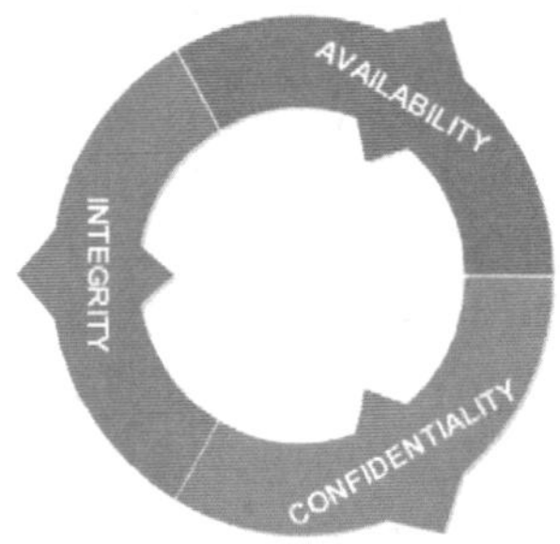

Figure 1 Information security aspects

The existence of a security problem triggers a procedure for controlling access rights to an information system [7]. A good information system security must apply the standard Deming cycle of quality [8]. The security of academic information systems can be audited with various standards such as COBIT, COSO, ITIL, CMM, BS779, ISO 9000. COBIT (Control Objectives for Information and related Technology) is a standard guide to information technology management practices and a set of best practices documentation for IT governance that can help auditors, management, and users to bridge the gap between business risk, control needs, and technical issues [8]. All organizations can adjust COBIT 5 for their various purposes, and are able to evaluate the organization in achieving its intended goals [9]. Domain DSS (Deliver, Service and Support) is related to system delivery and service support needed by the system, which includes service, security and continuity management, service support for users, and data management and operational facilities so that it is more integrated in the domain that provides services well [8].

DSS domains have sub-domain DSS05 wherein this sub-domain is a more intensive procedure for information security. The method that can be used in evaluating the achievement of evaluation is CMMI. Capability Maturity Model Integration (CMMI) is a model approach to assess the scale of capability and maturity of a software organization. The history of CMMI at the beginning was known as the Capability Maturity Model (CMM) which was built and developed by the Software Engineering Institute in Pittsburgh in 1987 [10]. The CMMI method in academic information system security is able to provide a level of achievement in the form of Maturity Level values. So as to be able to give a decision on the extent to which the security process of academic information systems that have been run by universities.

This study aims to conduct an evaluation related to the security management of academic information systems that have been implemented at Ahmad Dahlan University. This study aims to obtain the value of the level of information system security of an institution, so that recommendations and innovations can be made for the security of information systems in these institutions. So that the institution can provide security and comfort for the use of the information system.

2. Methods

2.1. DSS05 framework COBIT 5

The DSS05 sub-domain is part of the DSS domain (Deliver, Service and Support). As in Figure 2.

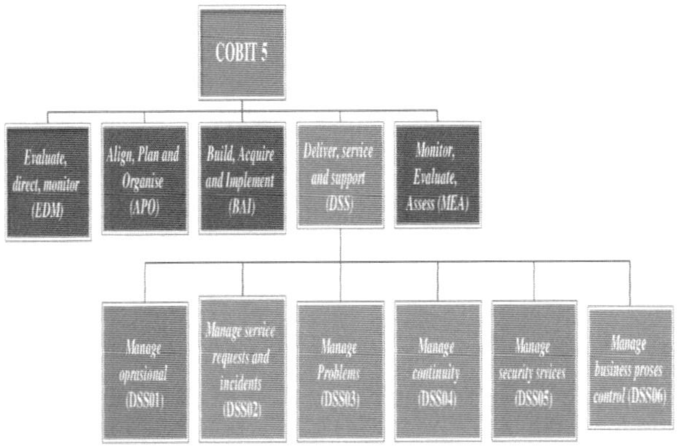

Figure 2. DSS05 scheme

The DSS05 sub-domain is managing security services where these sub-domains are grouped in 7 processes. The seven processes carry out some activities or statements of the 49 statements as follows [11]:

- Protect against malware (DSS05.01) where this process carries out and maintains existing precautions, detective and repairs (especially the latest security patches and virus controls) throughout the company to protect information systems and technology from malware (e.g., Viruses, worms, spyware, spam).
- Manage network and connectivity security (DSS05.02) where this process uses security measures and related management procedures to protect information from all methods of connectivity.
- Manage endpoint security (DSS05.03) where this process provides assurance of end points (e.g., Laptops, desktops, servers, and other cellular and cellular networks or software) guaranteed to be the same or greater than the requirements approved security.
- Manage user identity and logical access (DSS05.04) This process provides certainty for all users to have the right to access information in accordance with business needs. They and coordinate with the business division that manages access rights.
- Manage physical access to IT assets (DSS05.05) this process determines and applies procedures to give, limit and revoke access to physical buildings. Buildings and areas according to business needs, including emergencies. Access to buildings, buildings and areas must be justified, ratified, recorded and monitored.
- Manage sensitive documents and output devices (DSS05.06) where this process establishes physical security. In terms of documents relating to agencies. So that all output documents are standardized in security.
- Monitor the infrastructure for security-related events (DSS 05.07) where this process uses intrusion detection tools, to monitor infrastructure for unauthorized access rights and ensure that every event is integrated with monitoring events and managing events.

2.2 Capability Maturity Model Integration (CMMI)

CMMI has a streamlined assessment process. The assessment was based on questionnaires and was developed specifically to get software that could support process improvement. CMMI is a maturity method that can be used to improve processes within the institution. The purpose of using the CMMI within an institution is to improve the process of developing and improving the software product of the institution [12].

According to [13] CMMI has Capability Level. Capability Level is a model to describe how each core process runs within an institution. Capability Level has 6 levels for each core process, namely. As in Figure 3.

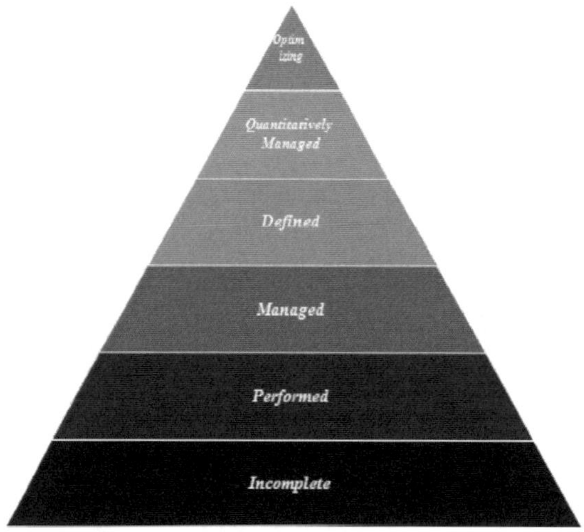

Figure 3. Capability level CMMI

According to [13] The CMMI model places, institutions in 5 Maturity Levels or CMMI levels, namely. As in Figure 4.

Figure 4. Maturity level CMMI

3. Result and discussion

This section will present a structured process method. Analysis of the implementation and measurement of the maturity level of the information system with the framework COBIT 5 sub-domain DSS05 and CMMI.

3.1 Observation of the academic information system process

This process conducts interviews directly with the resource person who has authority in the security of the academic information system at University Ahmad Dahlan, where the results are that BISKOM UAD uses an academic information system to be active in 2008, the beginning of the information system created and developed by vendors (Gama Techno) after that, in 2017 it was able to migrate to a new system where the system was developed by BISKOM UAD itself. Where in this migration is due

IOP Conf. Series: Materials Science and Engineering **821** (2020) 012003 doi:10.1088/1757-899X/821/1/012003

to the development of information systems technology, so it is considered necessary to do the migration to maintain the stability and security of the information system.

The purposes of the academic system of University Ahmad Dahlan are:

- To manage academic activities in University Ahmad Dahlan.
- Providing convenience to the community, namely lecturers, students, staff and BAA in the academic process.

As time goes on the use of information systems also experiences, obstacles, problems and threats to information systems. The problems, obstacles and threats that often occur are as follows:

- There are several systems that have not been well integrated.
- When the online KRS happened the server was down.
- It often happens to forget your username and password.
- The process of data connection or transmission is slow.
- Virus and malware attacks.

The BISKOM UAD standardization and audit process applies ISO 9000 where the standard is used for standards for quality management systems (SMM) which are aggregated with all bureaus in all institutions. This process also discusses the determination of respondents who will provide detailed information related to information on the security of existing academic information systems. The selection of respondent samples using purposive sampling technique, which is the selection of respondents 'samples determined by researchers on the grounds that identification of respondents' samples is done by referring to personal competencies that interact directly with IT governance [14]. Interviews get 2 respondents who are directly concerned with the field of information system security within the institution.

3.2 DSS05 mapping based on the COBIT framework 5

This process is a compilation of DSS05 domain conformity activities with questions to be made in the questionnaire. because of the limitations of our writing, we only list one of the 7 DSS05 sub-domain processes, namely DSS05.01. The DSS05.01 process consists of 6 activities, as in Table 1.

Table 1. Protect against malware activity

Protect against malware (DSS05.01)	
No.	**Activity Questions**
1	Obtain information about malicious software and how to handle it..
2	Install and activate anti-virus on your PC.
3	Is anti virus on the PC always updated.
4	Regularly review and evaluate information about potential malware threats
5	Filter incoming traffic, such as e-mail and downloads, to protect against unsolicited information
6	Conduct periodic training on malware in the use of e-mail and the Internet.

Preparation of Questionnaires with a combination of DSS05 and Capability Level. This process is carried out by questionnaires based on the standard on DSS05 Framework COBIT 5 by combining with the capability level of CMMI standards so that the form of a questionnaire can be obtained that is able to answer the needs of the information system security in the installation. To simplify the reading process, the color differences for each decision are made in Capability Level and Maturity Level as in Table 2.

Table 2. Process color maps

Color	Information	
	Capability Level	*Maturity Level*
Red	*Incomplete*	*Non-Existent Initial*
Purple	*Performed*	*Initial / Ad Hoc*
Yellow	*Managed*	*Repeatable But Invinitve*
Blue	*Defined*	*Define Process*
Orange	*Quantitatively Managed*	*Managed and Measurable*
Green	*Optimizing*	*Optimized*

Where in this questionnaire there are 6 assessments for processes with capability level CMMI as in Table 3.

Table 3. Assessment of IT processes with CMMI capability level

Value	*Capability Level CMMI*	Proses TI
0	*Incomplete*	Are not done
1	*Performed*	Done, not periodically
2	*Managed*	Performed periodically
3	*Defined*	Done with SOP
4	*Quantitatively Managed*	Performed and monitored
5	*Optimizing*	Done, monitored and developed

The assessment of the IT process in Table 3 is combined with the standard COBIT 5 DSS05 framework in Table 1, as in Table 4.

Table 4. Questionnaire form

Protect against malware (DSS05.01)							
Activity		Answer					
		0	1	2	3	4	5
1	Obtain information about malicious software and how to handle it.						
2	Install and activate anti-virus on your PC.						
3	Is anti-virus on PC always updated.						
4	Regularly review and evaluate information about potential malware threats.						
5	Filter incoming traffic, such as e-mail and downloads, to protect against unsolicited information.						
6	Conduct periodic training on malware in the use of e-mail and the Internet.						

3.3 Calculation of security SIA maturity level

This section will explain the results of the analysis of the implementation and measurement of the performance of the maturity level of academic information systems obtained from the results of questionnaires and interviews in accordance with the framework 5 COBIT domain DSS05. To identify the extent to which the company or organization meets good information security standards, can use the framework identification represented at a level of maturity that has a level of grouping capability of the company, as described in Table 5.

Table 5. Value of maturity level criteria

Criteria	Information
0 – 0.50	*Non-Existent Initial*
0.51 – 1.50	*Initial / Ad Hoc*
1.51 – 2.50	*Repeatable But Invinitve*
2.51 – 3.50	*Define Process*
3.51 – 4.50	*Managed and Measurable*
4.51 – 5.00	*Optimized*

The results of the questionnaire that has been given to the respondent and filled in by the respondent get results. Because the page is limited, the data displayed is only on DSS05.01. As in Table 6.

Table 6. The results of the questionnaire

DSS05	Respondent 1	Respondent 2
DSS05.01.1	5	5
DSS05.01.2	5	5
DSS05.01.3	5	5
DSS05.01.4	5	5
DSS05.01.5	5	5
DSS05.01.6	5	5

Furthermore, the correlation between level values and absolute values that are done by calculation in the form of an index uses a mathematical formula. The mathematical equation to determine the index value is as follows [15]:

$$Indeks = \frac{\sum Most\ Question\ Answers}{\sum Questionnaire\ Questions} \quad (1)$$

After getting the index, we can get the current Maturity Level (present). This value is the accumulated value of the process that is running on the institution as in Table 7.

Table 7. Existing maturity value

DSS05	*Value of Maturity Level Existing*
Protect against malware	5.00
Manage network and connectivity security	5.00
Manage endpoint security	4.39
Manage user identity and logical access	4.88
Manage physical access to IT assets	4.64
Manage sensitive documents and output devices	3.10
Monitor the infrastructure for security-related events	4.20

3.4 Gap maturity level calculation

Once the existing Maturity Level values are obtained and Maturity The recommendation level (target) has been determined, then the gap between the current condition and the target to be achieved will be analyzed and identified opportunities from the gap to be optimized. Level gap as in Figure 5.

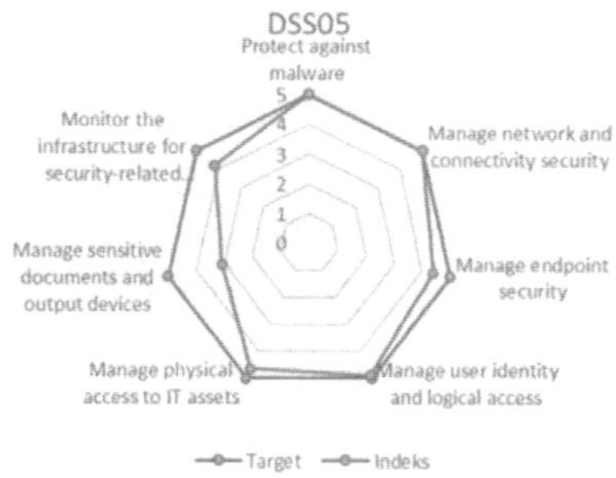

Figure 5. Maturity level gap

3.5 Gap analysis maturity level

Based on Gap analysis obtained from the results of the target level to be achieved and the level achieved on DSS05, as in Figure 5., then here is some Gap Maturity Level Analysis. As in Table 8 as follows.

Table 8. Gap maturity level analysis

DSS05	*Maturity Level*
Protect against malware	*Optimized*
Manage network and connectivity security	*Optimized*
Manage endpoint security	*Managed and Measurable*
Manage user identity and logical access	*Optimized*
Manage physical access to IT assets	*Optimized*
Manage sensitive documents and output devices	*Define*
Monitor the infrastructure for security-related events	*Managed and Measurable*

The overall value of Maturity Level on DSS05 will be calculated on average so that it will get the level of Maturity Level in the organization or institution.

$$Maturity\ Level\ DSS05 = \frac{\sum Maturity\ Level}{many\ processes} \quad (2)$$

$$MLDSS5 = \frac{i(DSS05.01) + i(DSS05.02) + i(DSS05.03) + i(DSS05.04) + i(DSS05.05) + i(DSS05.06) + i(DSS05.07)}{mp}$$

$$MLDSS05 = \frac{5 + 5 + 4{,}388 + 4{,}875 + 4{,}642 + 3{,}1 + 4{,}2}{7}$$

$$Maturity\ Level\ DSS05 = 4{,}458$$

From the calculation results obtained the value of achievement is 4,458 so that it can be set Maturity Level of organization or institution is at the Managed and Measurable level.

3.6 Compilation of IT governance recommendations
After Maturity Level has been determined, the recommendation preparation process will be carried out. Recommendations that can be given to improve the quality of information system security in the agency:

- Protect against malware (DSS05.01) is on the Optimized level where in this level the BISKOM of Ahmad Dahlan University has been able to perform procedures well and is able to develop malware related ones. It is expected that the agency will be able to anticipate the threat of malware more quickly and precisely in detecting malware threats.
- Manage network and connectivity security (DSS05.02) is at the level of Optimized wherein at this level the BISKOM of Ahmad Dahlan University has been able to carry out procedures well and is able to carry out developments related to security of activities. Establish a system that is used to evaluate threats that will arise, documented and monitored. It is expected that the future agencies will be better prepared with the threat of connectivity and be able to quickly provide countermeasures related to connectivity security.
- Manage endpoint security (DSS05.03) in the Managed and Measurable level where in this level the BISKOM of Ahmad Dahlan University has been able to carry out procedures well, only agencies must carry out routine evaluations, at least once a month on information systems that are feared to be potential new threats.
- Manage user identity and logical access (DSS05.04) is on the Optimized level where in this level the BISKOM of Ahmad Dahlan University has been able to carry out procedures properly and is able to develop related access rights of each user. It is expected that companies or institutions are able to provide early warning of the potential security threats to the system and equipment that is owned by all employees.
- Manage physical access to IT assets (DSS05.05) is on the Optimized level where in this level the BISKOM of Ahmad Dahlan University has been able to perform procedures well and is able to carry out development related to physical security. It is hoped that in the future, it will be able to produce and report related to security system trials that are applied and evaluated in a periodic physical shutter.
- Manage sensitive documents and output devices (DSS05.06) in the Define Process level, in this BISKOM, Ahmad Dahlan University has implemented physical security, accounting practices in terms of documents relating to the situation. So that all output documents are standardized in security. It's just hoped that later it will be able to bend documentation and evaluate existing threats.
- Monitor the infrastructure for security-related events (DSS05.07) is in the Managed and Measurable level where in this level the BISKOM of Ahmad Dahlan University has been able to carry out procedures properly using intrusion detection tools, to monitor infrastructure for unauthorized access rights and ensure that every event is integrated with monitoring events and management of events must carry out routine evaluations, at least every semester to the information system which is feared that potential new threats can arise.

4. Conclusion
Sub-domain DSS05 Manage security services is a good procedure to be used in the implementation and mega-audit related to the security of academic information systems and CMMI is a good assessment method in an institution's audit system. Based on the research conducted at the BISKOM, Ahmad Dahlan University received a Maturity Level of 4,458 thus stipulating that the current maturity level is on the Managed and Measurable level. This level, institutions are increasingly made aware of technological developments. Institutions have implemented the quantification concept in each process, and are always monitored and controlled for performance.

References

[1] Riadi I and Handoyo E 2019 Security analysis of GRR Rapid Response Network using COBIT 5 Framework vol **10** no 1 pp 29–39.

[2] Fathoni L F, Muslihudin, Kartika F and Anton Y 2016 Application information system based health services android *J. Ilmu Tek. Elektro Komput. dan Inform* vol 2 no 1 pp 39–48.

[3] Muslimin I, Hadi S P and Nugroho E 2017 An evaluation model using perceived user technology organization fit variable for evaluating the success of information systems vol **4** no 2 pp 86–94.

[4] Riadi Y W I and Yudhana A 2016 Webserver security analysis using the penetration testing method *in Annual Research Seminar 2016* vol **2** no 1 pp 300–4.

[5] Kurniawan E and Riadi I 2003 Security level analysis of academic information systems based on standard ISO 27002:2003 using SSE-CMM *Int J Comput Sci Inf Secur.* vol **16** no 1 pp 139–47.

[6] Rosmiati I. R and Prayudi Y 2016 A maturity level framework for measurement of information security performance *Int J Comput Appl* vol **141** no 8 pp 975–8887.

[7] Hermaduanti N and Riadi I 2016 Automation framework for rogue access point mitigation in ieee 802.1X-based WLAN *J Theor Appl Inf Technol* vol **93** no 2 pp 287–96.

[8] Hicham E, Boulafdour B, Makoudi M and Regragui B 2012 Information security *4th wave," J. Theor. Appl. Inf. Technol* vol **43** no 1 pp 1–7.

[9] Latifi F and Zarrabi H 2017 A COBIT5 framework for IoT risk management *Int J Comput Appl* vol **170** no. 8 pp 40–3.

[10] Konttinen V 2016 Towards Disciplined Software Development no May.

[11] Andry J F 2016 Audit of IT governance based on COBIT 5 assessments: A case study *J. Teknol. dan Sist. Inf* vol 2 no 2 p. 27.

[12] P. D. Syafitri, "Assessment of Quality of Information System Development at Distributor Companies," J. Sist. Inf. Bisnis, vol. 10, no. 01, pp. 15–27, 2016.

[13] CMMI Product Team, CMMI® for Development, Version 1.3. 2010.

[14] P. Rahayu and D. I. Sensuse, "Assessment of e-Government Implementation in the Ministry of Education and Culture's PUSTEKOM based on the PEGI method," J. Sist. Inf. Bisnis, vol. 02, pp. 139–145, 2017.

[15] Prasetyo A and Mariana N 2011 Analysis of information governance (It Governance) in the academic field with Cobit FrameWork Case study at Stikubank University Semarang *J. Teknol. Inf. Din* vol **16** no 2 pp. 139–49.

IOP Conf. Series: Materials Science and Engineering **821** (2020) 012004 doi:10.1088/1757-899X/821/1/012004

Fuzzy feed forward neural network (FFFNN) model for the Jakarta Islamic index (JII) forecasting

N Irsalinda*, **Y S Astuti and Sugiyarto**
Department of Mathematics, Faculty of Science and Applied Technology, Universitas Ahmad Dahlan, Yogyakarta, Indonesia

*Email: nursyiva.irsalinda@math.uad.ac.id

Abstract. Feed Forward Neural Network (FFNN) model is the best model to forecast the time series data. In this research, The Fuzzy Feed Forward Neural Network (FFFNN) with backward propagation method is used to predict the Jakarta Islamic Index (JII) data time series in 2018. Fuzzy is used as input to the FFNN model because it is overcome the weaknesses of the inaccurate results of the FFNN when the data is unclear or incomplete. The purposes of this research are to explain the procedure to generate the FFFNN model. The steps of the FFFNN model prepare the input data to become a fuzzy number using Growth-S Curve fuzzification; the second is divide the data into two training and testing data, the third is determining the best neural network architecture with different neurons and hidden layers to get the best weights that used for the forecasting model. In this research, the best FFFNN model is built by 19 neurons and one hidden layer with 90% and 10% training and testing data, respectively. Therefore with the model obtained, forecasting produces the value of MSE 0.0018 in training and 0.0004 in testing. From the MSE values obtained, it can be concluded that the forecasting using FFFNN model is reasonable to predict.

1. Introduction

Investment is the easiest and intelligent way to accumulate wealth. Financially free people are those who start investing early. They can choose the lifestyle they want by using the investment they have. One of the investment process is by buying shares of companies listed on the stock market. In the investment process, all these investors want to get the most profit and avoid losses. Therefore all investments made need to be planned appropriately. Investors who want to buy company shares need to ensure that the targeted company is a healthy and growing company. That way, the company can generate profits for shareholders through dividends or capital gains. Therefore the investors must do some investment analysis either alone or by using expert analysis services [1].

One of the indices that could be a reference to the condition of the companies listed on the Indonesian stock exchange is the Jakarta Islamic Index (JII). It is established on July 3rd 2000, on the Indonesia Stock Exchange (IDX) (formerly known as Jakarta Stock Exchange) to facilitate the trading of public companies according to Sharia business code [2]. Sharia stocks are securities in the form of shares which are not contrary to Islamic principles in the Capital Market. Therefore for people who want investments without usury and in accordance with Islamic sharia, they can invest in JII. However, to get a significant profit, the investment must be accompanied by an accurate forecasting analysis.

Forecasting is the activity of estimating or predicting what will happen in the future. In this study, forecasting is done to predict the JII index using previous JII indices. The purpose of forecasting is to produce optimum forecasts that do not have errors or as small as possible errors that refer to the Mean Square Error (MSE) forecasts.

One statistical tool commonly chosen in predicting the complex data and having the ability to produce models with high accuracy is Neural Networks. Artificial Neural Network is one of the information processing systems designed to simulate the work of the human brain to solve the problem by learning process through changes in synaptic weight. The neural network model can be divided into two, namely the Feed Forward Neural Network (FFNN) and Recurrent Neural Network (RNN) [3, 4].

FFNN is a neural network model where the learning process goes forward from the input layer to the output layer, while the RNN is a network model that has the characteristics of a feedback connection in the learning process. The algorithm for Artificial Neural Networks operates directly with numbers, so non-numeric data must be converted to numeric data [5].

In Fuzzy Feed Forward Neural Network (FFFNN), the FFNN input network is a fuzzy numbers. The membership value of a fuzzy set is between a closed interval of 0 and 1 [6]. FFNN has been used in some studies as a stock price forecasting tool. Previous research on FFNN, one of which was the FFFNN for forecasting of the Composite Stock Price Index (CSPI) with genetic algorithm variation selection. In this study, the membership function that used is the s-growth fuzzification curve. The activation function in the network architecture is the binary sigmoid function. Then the selection process is roulette wheel selection and rank based on the Genetic Algorithm. Distribution of training data and testing data in this study were 75% and 25%, respectively [7].

Another relevant research is the Application of the Backward propagation Feed Forward Neural Network (FFNN) Method to Predict Share Prices. In this study, the procedure for establishing the FFNN model with the backward propagation algorithm and the results of its application in the weekly period stock data from September 1st 2014 to September 5th 2016. The forecasting results for the period September 13th, 2016 is Rp. 17,405, the difference value is Rp. 45 with Rp. 17,450 as the actual data. The results show that the FFNN method with a backward propagation algorithm is very well used to predict stock prices [8, 9].

From the previous research, we know that the FFNN model is well used to forecast. However, when the data used is incomplete and unclear, it will produce an inaccurate model. To minimizing the inaccuracy of the forecasting model, fuzzy numbers are used in the FFNN input. Fuzzy number generation is done using the growth-S curve fuzzi15fication. This description is the background of this research. The research purposes are to apply the FFFNN and generating the FFFNN model, so that is good to use for forecasting models.

2. Research method

Fuzzy Feed Forward Neural Network (FFFNN) is a neural network with feed-forward type, and input in the form of fuzzy numbers is applied to forecast Jakarta Islamic Index (JII) sharia stock price index with a backward propagation learning algorithm. The forecasting process is explained in the figure below:

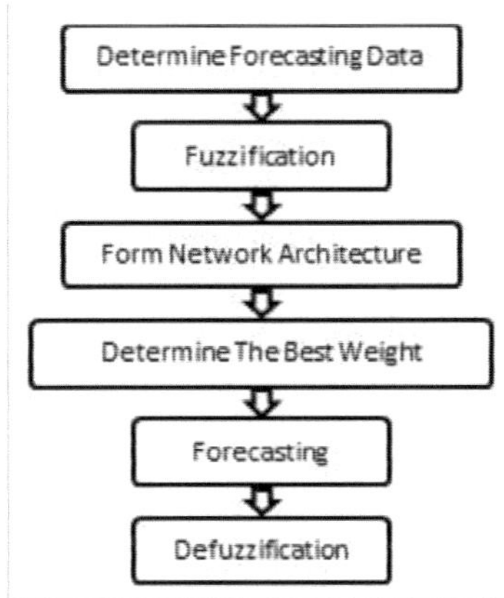

Figure 1. Forecasting process flowchart

The data determined in this research are JII (Jakarta Islamic Index), TASI (Saudi Arabia Stock Price Index), Interest rate, Inflation rate, and SAR (Rupiah's exchange rate to Real Saudi) 172 days in 2018 [6, 7]. The fuzzification is using growth S-curve (Figure 2) and the membership function $\mu_{\tilde{A}(x)}$ obtained by Equation 1 as follows :

$$\mu_{\tilde{A}(x)} = \begin{cases} 0, \; x < a \\ 2\left(\frac{x-a}{b-a}\right)^2, \; a \leq x < \frac{(a+b)}{2} \\ 1 - 2\left(\frac{x-a}{b-a}\right)^2, \; \frac{(a+b)}{2} \leq x < b \\ 1, \; x \geq b \end{cases}, \tag{1}$$

with x are elements of the decision matrix (research data), a and b are the lower and upper value of the data, respectively.

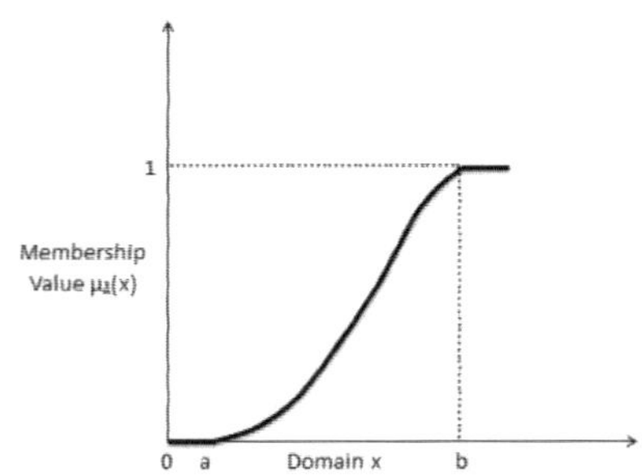

Figure 2. Growth-S Curve

Feed forward neural network (FFNN) is the most frequently used model in the application of time series forecasting [10]. The FFNN model is formulated in Equation 2.

$$y_k = \frac{1}{1+e^{-\left(w_{0k}+\sum_{j=1}^{v}\frac{1}{1+e^{-\left(v_{0j}+\sum_{i-1}^{n}x_i v_{ij}\right)}}\cdot w_{jk}\right)}}, \tag{2}$$

where y_k is k-th output value, v_{0j} = weight of bias hidden layer, v_{ij} = the weight that connecting i-th neuron input to j-th neuron of hidden layer, w_{0k} = weight of bias output layer, w_{jk} = the weight connecting j-th neuron of hidden layer to k-th neuron output, then v and x are the number of hidden layer and the number of input respectively. The architecture of a neural network for forecasting is in Figure 3.

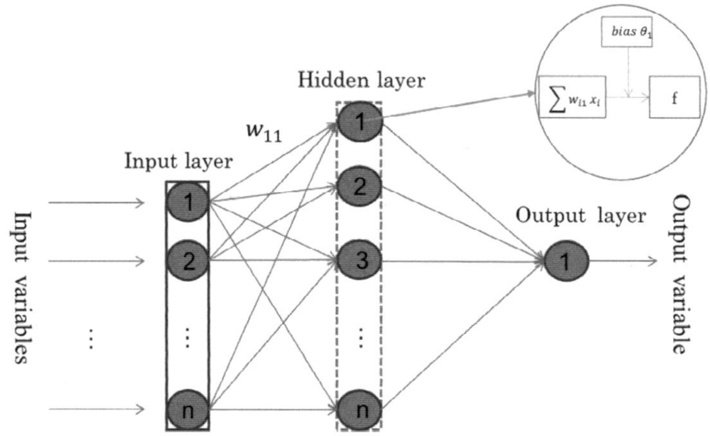

Figure 3. The architecture of neural network used for forecasting
(doi: https://doi.org/10.1371/journal.pone.0155133.g001)

To obtain the best model in Equation 2, then we must get the right weights using back propagation algorithm. The backward propagation algorithm is a network learning algorithm with the activation function is binary sigmoid in each layer. This algorithm includes supervised learning algorithms that are often used on feed forward networks. In general, there are three main stages in the back propagation algorithm which is the input process, error tracking and weight adjustment [11, 12]. The steps of the backward propagation algorithm are as follows:

1. Initialize weights conducted randomly or randomly without using scale factors.
2. The feed forward process is used as an algorithm to calculate the activation values that exist in all neurons, both in the hidden layer and the output layer. Following is the algorithm in the feed forward process:
   1) The input layer (x_i with $i = 1,2, \ldots, m$) received all the input layer neurons are numbered m and forward it to the hidden layer. The input layer is training and testing data.
   2) Hidden layer (z_j with $j = 1,2, \ldots, n$) calculates all input signals are numbered n by its weight and calculates the activation value. It then forward as an input value for the output layer using the following equation:

$$z(in)_{ij} = v_{0j} + \sum x_i v_{ij} \quad (3)$$
$$z_j = f\left(z(in)_{ij}\right) = \frac{1}{1+e^{-z(in)_{ij}}} \quad (4)$$

   The output layer (y_k with $k = 1$) calculates all the hidden layer values with their weights then calculates the activation value as the network output value using the following equation:

$$y(in)_{jk} = w_{0k} + \sum z_j w_{jk}, \quad (5)$$

$$y_k = f(y(in)_{jk}) = \frac{1}{1+e^{-y(in)_{jk}}}. \quad (6)$$

3. Backward Propagation Process
   In this process, a backward calculation is made from the output neurons to the input neurons to produce the appropriate weight value. The output signal generated from the feed-forward process then matched. A difference calculation is made between the target and the output signal

present in the output neuron. Following is the algorithm in the back-propagation process [13-15]:

1) The output layer (y_k with $k = 1$) receives a target pattern that corresponds to the input pattern then calculates k error information using the following equation:

$$\delta_k = (t_k - y_k)\, f'(y(in)_k), \tag{7}$$

t_k is k -th target while y_k is k -th network output. After the error information obtained, then we calculate the difference of weight Δw_{kj} and the correction bias factor Δw_{0j} with the acceleration rate (α) then sent to the hidden layers as follows:

$$\Delta w_{kj} = \alpha\, \delta_k\, z_j, \tag{8}$$

$$\Delta w_{0k} = \alpha\, \delta_k. \tag{9}$$

with δ_k = Correction factor for output to the hidden layer.

2) The hidden layer (z_j with $j = 1, 2, \ldots, n$) calculates the hidden units factor $\delta(in)_j$ based on the error in each hidden unit with Equation (10).

$$\delta(in)_j = \sum \delta_k\, w_{kj}. \tag{10}$$

Then $\delta(in)_j$ multiplied by the derivative of the activation function for calculating the error information δ_j as follows:

$$\delta_j = \delta(in)_j\, f'(z(in)_j). \tag{11}$$

Calculating the correction of the difference of weight in the hidden unit Δv_{ji} and the differences of bias factor correction Δv_{0j} with the Equation (12-13) as follows:

$$\Delta v_{ji} = \alpha\, \delta_j\, x_i \tag{12}$$

$$\Delta v_{0j} = \alpha\, \delta_j \tag{13}$$

3) The weights and biases in the output layer and the hidden layer respectively updated with the following equations:

$$w_{kj}(l+1) = w_{kj}(l) + \Delta w_{kj} \tag{14}$$

$$w_{0j}(l+1) = w_{0j}(l) + \Delta w_{0j} \tag{15}$$

$$v_{ji}(l+1) = v_{ji}(l) + \Delta v_{ji} \tag{16}$$

$$v_{0j}(l+1) = v_{0j}(l) + \Delta v_{0j} \tag{17}$$

where l is the previous weight and $l+1$ is a new weight or the updated weight during the iteration.

2.1. *Determination of the best network architecture*

The data that used in this research are JII (Jakarta Islamic Index) as x_{i1}, TASI (Saudi Arabia Stock Price Index) as x_{i2}, Interest rate as x_{i3}, Inflation rate x_{i4} and SAR (Rupiah's exchange rate to Real Saudi) as x_{i5} and 172 days as i. The data is input in the input layer [16]. The first step is to convert the inputs into fuzzy numbers using Equation (1) so that the data obtained as in Table 1.

IOP Conf. Series: Materials Science and Engineering **821** (2020) 012004 doi:10.1088/1757-899X/821/1/012004

Table 1. Fuzzy numbers of input data

No.	$\mu(x_{i1})$	$\mu(x_{i2})$	$\mu(x_{i3})$	$\mu(x_{i4})$	$\mu(x_{i5})$
1.	0.8981	0.0035	0	0.8177	0.0058
2.	0.9256	0.0025	0	0.8177	0.0097
3.	0.9046	0.0008	0	0.8177	0.0130
4.	0.9077	0.0015	0	0.8177	0.0095
5.	0.8879	0.0400	0	0.8177	0.0007
6.	0.8966	0.0626	0	0.8177	0.0009
7.	0.9359	0.0937	0	0.8177	0.0005
8.	0.9513	0.0996	0	0.8177	0.0028
9.	0.9472	0.0812	0	0.8177	0.0009
10.	0.9467	0.0613	0	0.8177	0.0003
11.	0.9850	0.0690	0	0.8177	0.0005
⋮	⋮	⋮	⋮	⋮	⋮
154.	0.0437	0.3835	0.9592	0.5550	0.9993
155.	0.0344	0.4504	0.9592	0.5550	0.9999
156.	0.0598	0.4281	0.9592	0.5550	0.9996
157.	0.0864	0.5452	1	0.7693	0.9982
158.	0.1039	0.4495	1	0.7693	0.9586
159.	0.1049	0.3830	1	0.7693	0.9317
160.	0.1076	0.3975	1	0.7693	0.8750
161.	0.1281	0.3692	1	0.7693	0.8110
162.	0.1039	0.2637	1	0.7693	0.8660
163.	0.0361	0.3450	1	0.7693	0.9330
164.	0.0621	0.3131	1	0.7693	0.8704
165.	0.0805	0.2064	1	0.7693	0.8732
166.	0.2328	0.0708	1	0.7693	0.7645
167.	0.1909	0.0758	1	0.7693	0.7865
168.	0.1503	0.0819	1	0.7693	0.7682
169.	0.1912	0.0808	1	0.7693	0.7393
170.	0.1760	0.0918	1	0.7693	0.7024
171.	0.1358	0.1257	1	0.7693	0.7264
172.	0.0758	0.1787	1	0.7693	0.6225

The next step is generate the network architecture. In this research, the architecture of the neural network is three-layer with 5 neurons in the input layer, trial-error for 11, 13, 15, 17, and 19 neurons in the hidden layer and a neuron in the output layer. The learning method algorithm of this FFFNN is backward propagation with data training and testing 90%:10%, 80%:20%, 70%:30%, 60%:40%, and 50%:50%. The best architecture is that it has a minimum mean square error (MSE) in learning data and testing data [17-19]. MSE can be calculated using this formula:

$$\mathrm{MSE} = \frac{1}{n}\sum_{t=1}^{n}(Y_t - \hat{Y}_t)^2. \tag{18}$$

where Y_t = real value and $\hat{Y}_t$ = forecasting result value [20]. The results of the trial and error from the variation combination network architecture presented in Table 2.

IOP Conf. Series: Materials Science and Engineering **821** (2020) 012004 doi:10.1088/1757-899X/821/1/012004

Table 2. Trial-error result for variation of neural network architecture

Training and Testing Data Partition(%)	Hidden Layer Neurons	MSE of Training	MSE of Testing
	11	0.0023	0.0009
	13	0.0022	0.0003
90:10	15	0.0021	0.0008
	17	0.0022	0.0004
	19	*0.0018*	*0.0004*
	11	0.0023	0.0008
	13	0.0024	0.0015
80:20	15	0.0022	0.0006
	17	0.0025	0.0017
	19	0.0024	0.0009
	11	0.0024	0.0167
	13	0.0023	0.0023
70:30	15	0.0021	0.0029
	17	0.0021	0.0015
	19	0.0022	0.0028
	11	0.0027	0.0018
	13	0.0023	0.0038
60:40	15	0.0020	0.1655
	17	0.0021	0.0425
	19	0.0018	0.1216
	11	0.0021	0.0117
	13	0.0022	0.1100
50:50	15	0.0021	0.2991
	17	0.0018	0.0128
	19	0.0023	0.0218

From Table 2, the MSE of training and testing data obtained, the best network architecture is an architecture with five neurons in the input layer, 19 neurons in the hidden layer, and a single neuron in the output layer with 90% training and 10% testing data partition. The following is a comparison chart of FFFNN output with the network architecture obtained:

In Figure 4, the result of the training data using backward propagation (green line) and actual training data (magenta) depicted in a coordinate plane. From the graphs, it appears that there is a slight error of 0.0018. Likewise, in Figure 5, the chart of the testing data using the backward propagation method and testing data actual produces an error of 0.0004. The errors are the minimum error of the best architecture in Table 2. Furthermore, this research obtains the best architecture, such as in Figure 6.

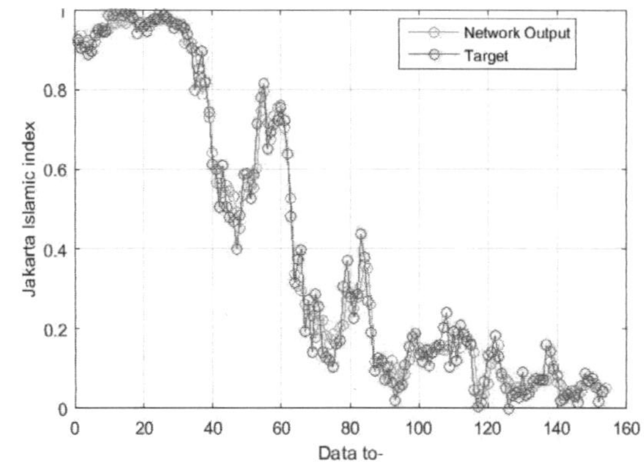

Figure 4. Graphic of backward propagation vs actual training data

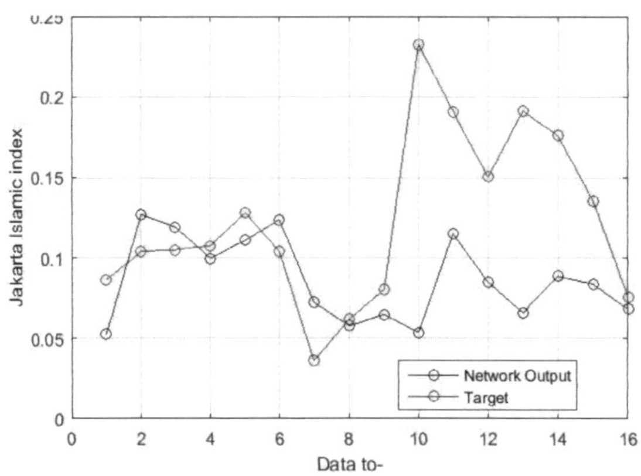

Figure 5. Graphic of backward propagation vs actual testing data

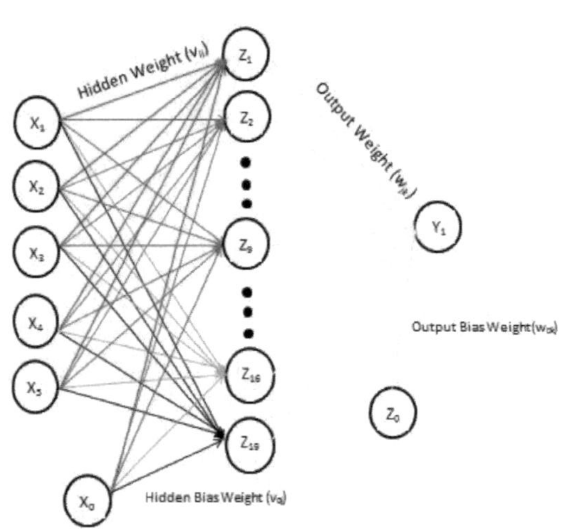

Figure 6. The best architecture in JII forecasting

IOP Conf. Series: Materials Science and Engineering **821** (2020) 012004 doi:10.1088/1757-899X/821/1/012004

The best weights in the best network architecture are calculated using Equation (14) until Equation (17). The results of the weights are as follows:

$$_{ji} = \begin{bmatrix} -2.8497 & -5.5905 & 6.7140 & -3.3334 & 5.2351 \\ 5.6906 & -1.4511 & 6.4145 & -5.2456 & -0.9282 \\ -6.2568 & -1.5037 & 5.2337 & 5.8911 & 2.9811 \\ \vdots & \vdots & \vdots & \vdots & \vdots \\ \vdots & \vdots & \vdots & \vdots & \vdots \\ 1.3915 & 6.7244 & -5.9223 & 2.8275 & 4.4881 \end{bmatrix}$$

$$v_{0j} = \begin{bmatrix} v_{01} & v_{02} & v_{03} & v_{04} & \cdots & \cdots & v_{019} \end{bmatrix}$$
$$= \begin{bmatrix} -7.7416 & 4.6244 & -6.0616 & 1.5558 & \cdots & \cdots & -0.2127 \end{bmatrix}$$

$$w_{jk} = \begin{bmatrix} w_{11} \\ w_{21} \\ w_{31} \\ w_{41} \\ \vdots \\ \vdots \\ w_{191} \end{bmatrix} = \begin{bmatrix} 2.9894 \\ 2.3722 \\ 1.4600 \\ -2.6154 \\ \vdots \\ \vdots \\ 2.2569 \end{bmatrix}$$

$$w_{0k} = \begin{bmatrix} w_{01} \end{bmatrix} = \begin{bmatrix} 0.0706 \end{bmatrix}$$

3. Result and discussion

From the best architecture and weights obtained, the FFFNN model with backward propagation algorithm in JII stock price index forecasting is as follows:

$$y_k = \frac{1}{1+e^{-\left(w_{0k}+\sum_{j=1}^{19}\frac{1}{1+e^{-\left(v_{0j}+\sum_{i=1}^{5}x_i v_{ij}\right)}}\cdot w_{jk}\right)}}, \qquad (19)$$

To forecast day 172 of the JII index, we used data day 171 of JII, TASI, Interest, Inflation and SAR in the form of fuzzy value. JII = 0.1358, TASI = 0.1257, Interest = 1.0000, Inflation = 0.7693 and SAR = 0.7264. After applying these values to FFFNN formed, the result is 0.0683. Then, to get the crisp value of JII, 0.0683 is defuzzification, so that the JII forecasting value is 618.92. Comparison with the real value of the JII index in day 172 is 648.80 have a MSE value of 0.0046.

4. Conclusion

From the discussion, we can conclude that the steps of the FFFNN model prepare the input data to become a fuzzy number using Growth-S Curve fuzzification, the second is divide the data into two training and testing data, the third is determining the best neural network architecture with different neurons and hidden layers to get the best weights that used for forecasting model. In this research, the best FFFNN model is built by 19 neurons and one hidden layer with 90% and 10% training and testing data, respectively. Therefore with the model obtained, forecasting produces the value of MSE 0.0018 in training and 0.0004 in testing. From the MSE values obtained, it can be concluded that the forecasting using FFFNN model is reasonable to predict.

References

[1] Fabozzi F J and Drake P P 2009 *Finance: Capital Markets, Financial Management and Investment Management* (USA: John Wiley).
[2] Tanjung H and Siregar T A 2018 *Journal of Islamic Economics, Finance and Banking* **(1)** pp 147-57.
[3] Dreyfus G 2005 *Neural Networks Methodology and Applications* (Germany: Springer).
[4] Jang J-S R, Sun C T, and Mizutani E 1997 *Neuro-Fuzzy and Soft Computing A Computational Approach to Learning and Machine Intelligence* (Upper Saddle River: Prentice Hall).
[5] Fausset L 1994 *Fundamentals of Neural Networks* (Englewood Cliffs: Prentice Hall).
[6] Herliansyah R and Jamillatuzzahro 2017 *GSTF Journal of Mathematics, Statistics and Operations Research (JMSOR)* **4** (1) pp 8-14.
[7] Deka P and Chandramaoli V 2005 *Journal of Hydrologic Engineering* **3** pp 302-14.
[8] Hansun S 2013 *Ultimatics* **IV** (1) pp 26-30.
[9] Tkacz G 2001 *International Journal of Forecasting* **17** pp 57-69.
[10] Wei W W S 1990 *Time Series Analysis: Univariate and Multivariate Methods* (USA: Addison-Wesley).
[11] Chen X, Racine J, and Swanson N R 2001 *IEEE Transaction on Neural Networks* **12** pp 674-83.
[12] Sudarsono A 2016 *Jurnal Media Infotama* **12** 1.
[13] Purushothaman G and Karayiannis N B 2006 *Journal of Applied Functional Analysis* **1** pp 9-32.
[14] Vasant P M, Nagarajan R and Yacoob S 2004 *Scientiae Mathematicae Japonicae* **10** pp 513-27.
[15] Medasani S, Kim J, and Krishnapuram R 1998 *International Journal of Approximate Reasioning* **9** pp 391-417.
[16] Bakar N A, and Rosbi S 2017 *International Journal of Advanced Engineering Research and Science (IJAERS)* **4** pp 130-37.
[17] Alsharif M H, Younes M K and Kim J 2019 *Symmetry* **11** pp 1-17.
[18] Farhath Z A, Arputhamary B and Arockiam D L 2016 *Int. J. Comput. Sci. Mobile Computer* **5** pp 104-9.
[19] Sotirov S and Atanassov K 2009 *Bulgarian Academy of Science Cybernetics and Information Technologies* **9** pp 62-8.
[20] Ullah H and Bhuiyan M A 2018 *Journal of Science and Technology.*

IOP Conf. Series: Materials Science and Engineering **821** (2020) 012005 doi:10.1088/1757-899X/821/1/012005

Transformation of function, form, zoning, circulation and material of Rumoh Aceh "Study of Aceh traditional architecture in Montasik sub-district, Aceh Besar"

F Azzahra* and Sahriyadi

Department of Architecture, Faculty of Engineering, Universitas Muhammadiyah Aceh, Banda Aceh, Indonesia

*Email: fatimah.azzahra@unmuha.ac.id

Abstract. Along with the time running, traditional architecture starts to be limited by time. While traditional architecture is a self-identity of a region. The location of research is Province of Aceh, namely in Sub-district of Montasik, District of Aceh Besar. The objects of study are traditional house buildings called as Rumoh Aceh. Because, principally, all buildings have similarity; thus, samples were taken randomly from each village. The formulation of questions is to identify transformation of function, form, zoning, circulation and materials of Rumoh Aceh found in the Sub-district of Montasik. The objectives of this study are to understand transformation occurring in Rumoh Aceh in the Sub-district of Montasik over time in order to know causal factors; moreover, it is expected to be reference to the Rumoh Aceh itself. This study used descriptive-qualitative method to adopt phenomena directly (what, who, and where the research location is). This method can interpret and associate data, important situation, attitude, point of view, conflict, and others. Because this study is direct, the techniques of observation and survey (primary data) keep to be choice, such as, interview and documentation. Relevant references to data can be analyzed and developed (secondary data). The results of study indicate that Rumoh Aceh found in Sub-district of Montasik, in terms of functional transformation, is still ideal, but there are changes in function in lower parts of building, that previously functioned as animal breeding area, rice storage and rest area, now it changes into permanent space functioning as room, storage area, bathroom, and cloth drying area. Moreover, we seldom find terrace and gate areas in front of building. While form transformation in Rumoh Aceh tend to be unchanged. Rectangular and triangular form keep to be choice. Transformation of zoning and circulation only occur in lower parts of building because there are additional rooms and functional change of rooms so that automatically circulation system also changes according to needs. Along with time running, transformations of building materials are found in roofs. Before, Rumoh Aceh used local materials, such as, rumbiah leaf plait; now, iron sheets dominate the Rumoh Aceh. Moreover, use of brick materials is widely found in lower parts of building. Current materials, such as, plywood and acrylic, are found in nearly each Rumoh Aceh, especially in door and window blades.

1. Introduction

Traditional architecture is one of national cultures in Indonesia. Various greatest traditional architectures in each region of Indonesia become typical characteristics and identities of local people. As one of cultural forms, traditional architecture derives from one regulation or agreement that generations hold and keep [1]. The conception keeps applicable as far as it is found meeting needs of local people.

Current modernization flow has great effect on existence of traditional architecture in the nation. One of factors of unblocked technology advances is use of technology. Moreover, the number of current building materials seem easy to treat, efficient, effective and handy to work. Therefore, preservation and conservation are very necessary to improve to maintain national architecture masterpiece; in this case, we need to intervene with the government as policy maker.

When speaking of transformation or change in traditional architecture, it is very proper and natural along with time running [2]. The western end of Indonesia, Province of Aceh, District of Aceh Besar has ancestral architecture masterpiece. Variety, width and history are factors of research location selection. In this case, research is limited to determined research location, namely Sub-district of Montasik. Consideration of research location selection is tract record or Aceh-Traditional House Architecture Development the local people call Rumoh Aceh.

Sub-district of Montasik has area of 5,973 Ha where wet area is 1,843 Ha, non-wet area is 700 Ha, and non-farming area is 3,430 Ha. These areas are far from beaches and nearer to mountain area with 39 villages. The northern part is bordered to Sub-districts of Blang Bintang, Masjid Raya, and Ingin Jaya; the southern part is bordered to Sub-districts of Kuta Malaka and Indrapuri; the western part is bordered to Sub-district of Suka Makmur; the eastern part is bordered to Sub-district of Indrapuri. Total residents of Sub-district of Montasik is 20,181 people where 10,268 of them are males and 9,913 of them are females. The dominant livelihood of the people is farming [3].

2. Methodology

This study used descriptive-qualitative method. According Sapitri and Ariska Rati [4], the descriptive-qualitative method is suitable to adopt phenomena directly associated with who, what and where the research location. This method can interpret and associate data, important situation, attitude, point of view, conflict, etc. Because this study characteristic is direct, techniques of observation and survey (primary data) keep being choice, such as, interview and documentation. Relevant references are used in order data can be analyzed and developed (secondary data). If associated with studies by Kothari [5] and Somantri [6], methodology used in this study tends to be relevant and responsible.

If sampel have elements of similarity, in term of physic, location and function as residential house, than a random technique can be used for sampling with descriptive explanation [7]. Thus, samples in this study were taken randomly because their characteristics tend to be similar. Samples were taken from two unit in each village so that total samples are 78 units of Rumoh Aceh.

The research location is in Sub-district of Montasik. Object of observation and sample of study are Rumoh Aceh taken randomly in all villages found in the Sub-district of Montasik. Based on field observation, age of Rumoh Aceh in Sub-district of Montasik is 120 or more years old. Formulation of questions identifies changes in function, form, zoning, circulation and building materials from point of view of architecture as to find clear point that may be reference to increase vision on traditional architecture of Rumoh Aceh that currently starts to be difficult to find.

3. Traditional architectures of Aceh

The Traditional House of Aceh called Rumoh Aceh is a house established by poles so that Rumoh Aceh (staged house) can be seen from the lower, upper, and roof parts (Figure 1). The lower part is pit of house located under floor in open condition (without wall). Height of house floor ranges from 2 m to 3 m. Poles found in Rumoh Aceh is round in ± 30 cm diameter with 16, 20, and 24 poles of Rumoh Aceh consisting of three main spaces, namely:

- Front Space. Front veranda called *Seuramoe Rinyeuen* (veranda with stairs) or called Seuramoe Keue (front veranda). This space functions as guest reception place, siting place to eat when there are events such as *kenduri* and marriage, place where children learn, place for praying, and place to rest.
- Middle Space and *Tungai*. This space functions as a place consisting of two rooms and one corridor. The room located in the west part is *Rumoh Inong*, while in the east part is *Rumoh Anjong*. Rumoh Inong functions as bedroom for the head of the family located in the west part and Rumoh Anjong is a bedroom for daughter located in the east part. If a daughter is married, her place is in *Rumoh Inong*, while the family head moves to *Rumoh Anjong*. The middle room has a higher elevation among the other rooms.
- Back Space or Back Porch usually called as *Seuramoe Likot*. This space functions as kitchen and dining rooms [8]. In addition, Rumoh Aceh has additional rooms such as rice barn area, terrace or open veranda, and gate [9].

Figure 1. Original shape and physic of traditional house in Aceh Besar

4. Discussion and result

4.1. Functional transformation

Rumoh Aceh has are three parts, namely, the lower and middle parts (main building) and the upper part (roof). The lower part functions as an area for animal breeding, rice granary (storage) areas, and areas for relaxing. The main part of Rumoh Aceh is the middle part that functions as a house in general. There are family rooms, bedrooms, kitchens and dining rooms. While the upper part is the roof of the building. Based on the data above, the function of the building and space of Rumoh Aceh in Montasik District did not change much (Figure 2). Maintenance of the main function of the elements of Rumoh Aceh itself such as, the number of building poles as supporting structures is still a standard, namely 16-20 poles, has a ladder as a vertical (linear) circulation system, still has *Seuramoe Rinyeuen* (veranda with stairs) or commonly called *Seuramoe Keue* (front veranda) whose main function is a reception area, party or festivity, religious activities, and family gathering (meeting) place. It still has *Tungai* or living room that functions as a room consisting of two rooms and one corridor. The room, which is located in the west part, is *Rumoh Inong*, while the east part is *Rumoh Anjong. Rumoh Inong* functions as bedroom for the head of the family located in the west part and *Rumoh Anjong* is a daughter bedroom located in the east part. If a daughter is married, she moves to Rumoh Inong, while the family head moves to Rumoh Anjong. It still has Seuramoe Likot or back room that serves as a kitchen or dining area. The functions of space, patterns and circulation systems are very clear in the building. It is relevant when referring to the opinions of Sabila F, Antariksa, and Handajani [10] stating that each room in Rumoh Aceh has different functions,

Figure 2. Current physical shape and spatial pattern of Rumoh Aceh in sub-district of Montasik

different activities, territoriality, self-identity, safety and comfortableness.

However, over time the function from the lower part of the building has changed. Previously, the function of the area under the building was to breed animals, rice barns, and rest areas. Currently this function has been changed as an additional area of room that can function as a bedroom, storage (warehouse), and bathroom or as a cloth drying (clothesline) area. In addition, it is rare to find a terrace and gate area in the front of the building. These results, if related to studies conducted by H. Aiyyub, MN Loebis, and IF Pane [11] on Rumoh Aceh in Blang Baroh Village, Pidie District, Aceh Province is almost the similar, but Rumoh Aceh in Montasik Sub-district, Aceh Besar District tends to have more minimal changes and still retain the main function (Figure 3). The findings of the study by H Aiyyub et al [11] are very large changes in function in every existing space such as spatial typology, spatial configuration, and spatial functions (Figure 4). That is due to cultural shifts and the financial condition of the local people.

4.2. Transformation of form

The transformation of form in architecture is primarily the result of a social cultural process. It includes the most useful changes to the physical environment (Figure 5). One of the changes occurred due to penetration. Transformation in culture can be summarized into a long process that is preceded by the occurrence of inculturation and acculturation, the process of cultural dialogue and synthesis, and followed by various shifts and the development of values to be a new cultural figure [12].

IOP Conf. Series: Materials Science and Engineering **821** (2020) 012005 doi:10.1088/1757-899X/821/1/012005

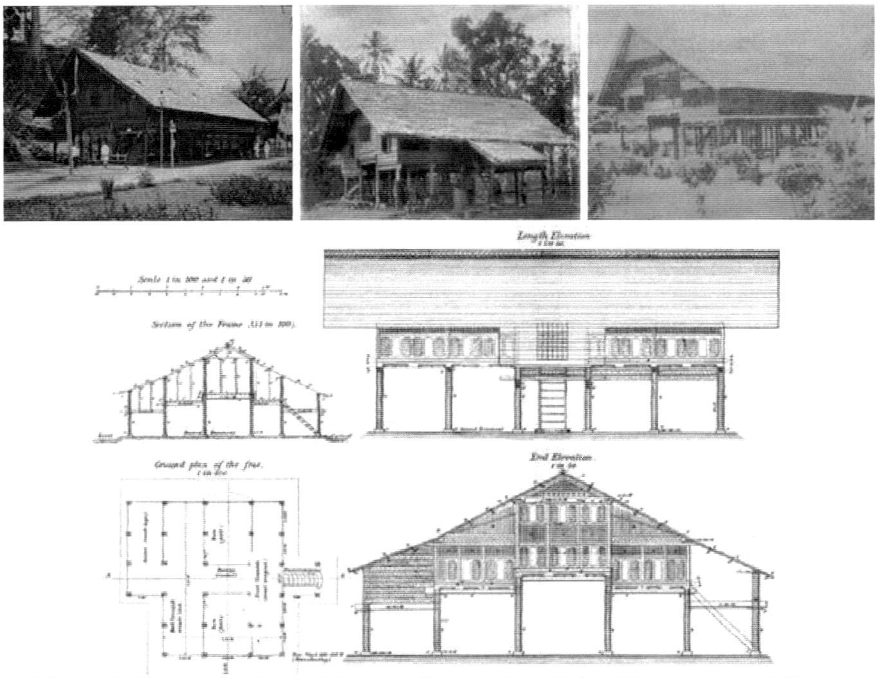

Figure 3. Form, physic, and layout of original traditional house in Aceh Besar

Figure 4. Current form physic of traditional house in Aceh Besar

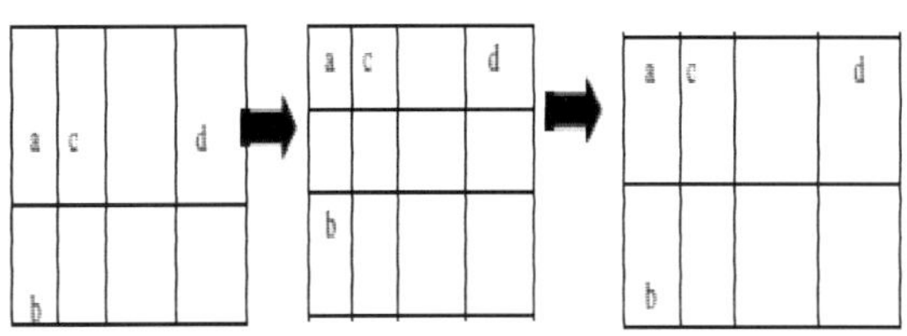

Figure 5. Example of results of transformation

The transformation of a building cannot only be seen in culture or in social terms as well. Changes in a building can also be seen in terms of architectural morphology. The shape morphology cannot be

IOP Conf. Series: Materials Science and Engineering **821** (2020) 012005 doi:10.1088/1757-899X/821/1/012005

separated from transformation. This can be seen in the rectangle. Transformation of the fields is to form a new pattern but still in interrelated types [13].

Layout, spatial patterns, and physical form in Rumoh Aceh from the past until now has not experienced many changes and still maintain the concept of the staged house. The combination of rectangles and triangles (geometry) remains the main shape. The form of floor plan shows side view of buildings and interior complementary elements such as doors and windows (rectangles). While the triangular plane is seen in the front view of the building and the ventilation in the front and back of the building which follows the pattern of the roof shape. In Rumoh Aceh, which is located in Montasik District, it still uses a base of pedestals and main poles with average 20 poles (grid). Rumoh Aceh, which was the object of research, is 120 years old or more.

4.3. Transformation of Zoning and Circulation

Red dash is primary circulation occurring in main part of house, in upper part. Blue dash is secondary circulation occurring in lower part of house (Figure 6). Based on the notes of figures, primary circulation tends to be unchanged over time. While circulation occurring in the lower part of house tends to change because of the functional transformation of lower part of building.

The main functions of the lower part of Rumoh Aceh are resting area, rice barns, and breeding areas. While in Rumoh Aceh in Montasik District, the lower part of the building is widely used for additions such as bedrooms, warehouses and so on. The change in function has caused a change in the circulation system in Rumoh Aceh found in Montasik District (Figure 7).

4.4. Transformation of Building Materials

In terms of changes in building materials, iron sheet (zinc) material on the roof of the building was very striking because Rumoh Aceh actually used roofing material from thatched leaves, which was woven into the roof. The age factor of the building and the current materials are determinants of material use. The lower part of the building, which already uses a lot of concrete (brick) material. The room at the lower part (bottom) of the building can function as an additional room, storage area or warehouse, or bathroom. The walls in some buildings do not use the original material made of wood or hard bark, and some even use plywood and tarpaulin to cover only the wind and heat. In addition, the use of materials such as plywood or acrylic has often been found in door or window.

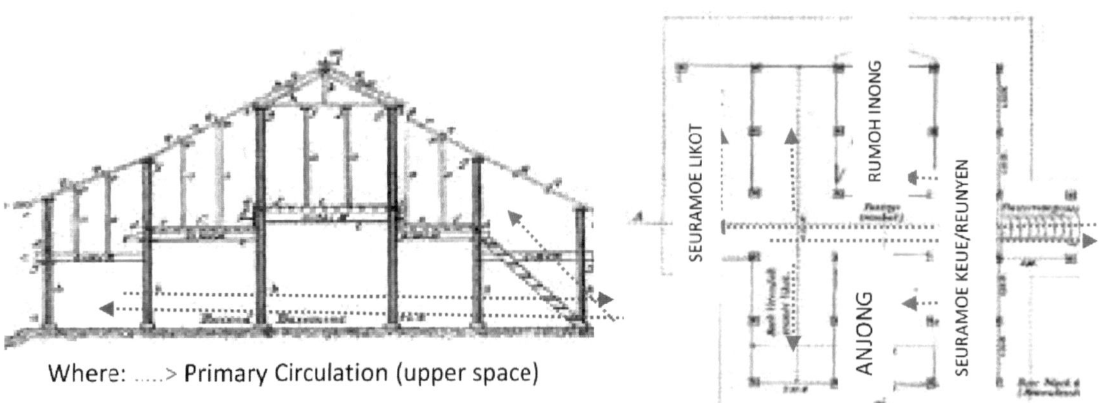

Figure 6. Zoning and circulation of original Rumoh Aceh

IOP Conf. Series: Materials Science and Engineering **821** (2020) 012005 doi:10.1088/1757-899X/821/1/012005

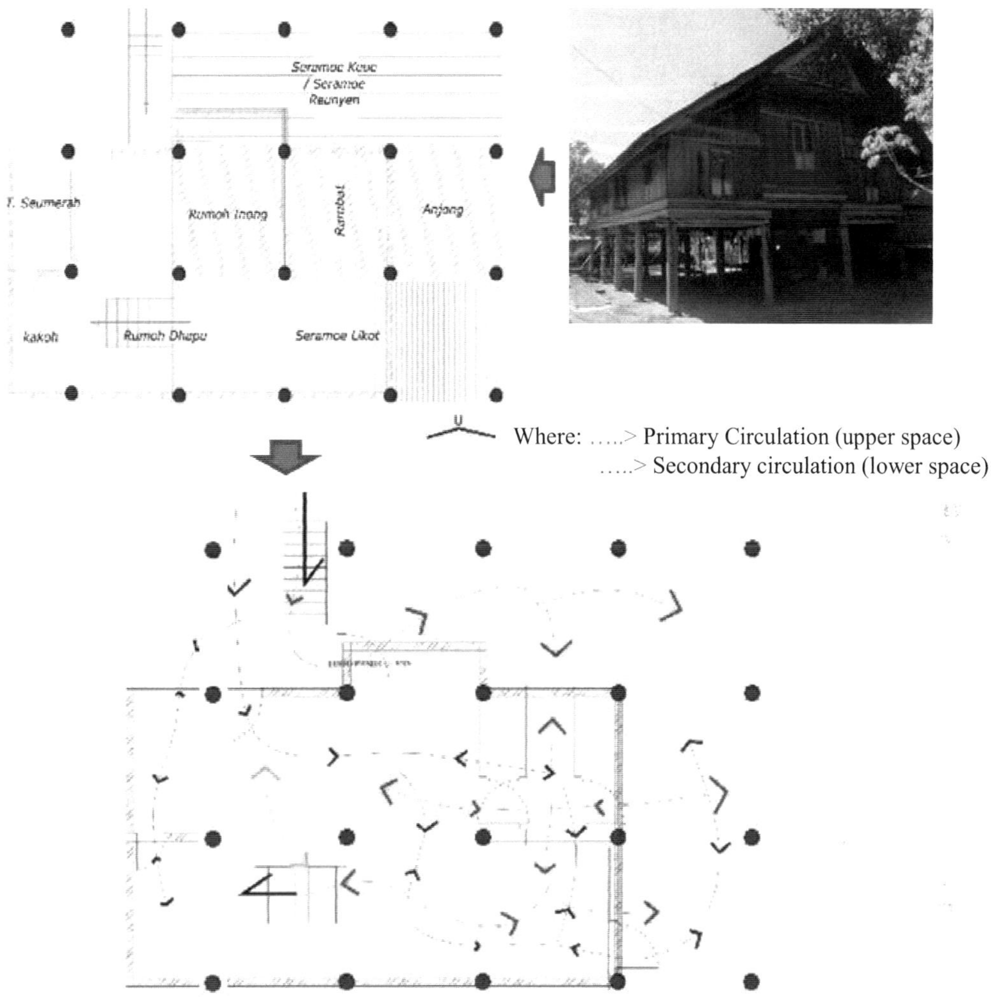

Figure 7. Zoning and Circulation of Rumoh Aceh in Sub-district of Montasik

Oliver Paul in Faisal. G and Wihardyanto. D [14] stated that vernacular buildings consist of rural settlement, types and processes, build form the ground, resources that grow, coping with climate, living space, values, symbols and meanings and Rapoport A theory via Duque Estela [15] suggesting that vernacular buildings grow from traditional people who take advantage of local potential. In addition, Fakriah Nurul [16] and Ulfahmi, Rohidi and Triyanto [17] stated that Rumoh Aceh, in the use of building materials, always uses environment-friendly materials in accordance with environmental conditions such as the use of woven leaves on the roof, the use of logs on the structure poles, the use of bark in some parts of the wall, the use of boards and bamboo as a complement to construction (local wisdom).

If the opinions and theories above are related to the current condition of Rumoh Aceh in Montasik District, it tends to be less relevant because currently the use of material tends to be combined with current material. The main factor of this transformation is the advances of the times and technology (cultural transformation). Technology was created to make it easier for humans in many ways. It has an impact on the use of material in Rumoh Aceh. Over time, the use of current materials are inevitable.

5. Conclusions

Rumoh Aceh in Sub-district of Montasik generally tends to be unchanged. It maintain its function as residential house in general. There are still main elements of Rumoh Aceh itself. For example, the number of poles of building as support of structure are still standardized, namely, from 16 to 20 poles, has stairs as vertical circulation system (linear), still has Seuramoe Rinyeuen (veranda with stairs) or usually called Seuramoe Keue (front veranda) where the main function is guest reception area, party event, and religious agenda, and place to meet family. It still has Tungai or middle room functioning as bedroom. It still has Seuramoe Likot or back room functioning as a kitchen or a dining room. However, along with time running, the function of lower parts of building (staged house) has transformed. Previously, the function of lower area of building was used for animal breeding, rice storage, and rest area. Now, the function has transformed, namely functioning as additional area of rooms that may function as bedroom, storage area (warehouse), bathroom or cloth drying area. Moreover, areas of terrace and gate are seldom found in the front part of building.

Generally, Rumoh Aceh found in Sub-district of Montasik does not experience changes in form. It maintains rectangular geometric form if seen from the top or side and triangular form if seen from front part of building. In addition rectangular form is seen in doors and windows. Ventilation uses roof shape pattern, namely, triangle found in the front and back parts of building. The interior does not experience changes in form, but there are additional electronic units and furniture.

Functional changes dominantly occurring in the lower part of Rumoh Aceh in Sub-district of Montasik result in changes of zoning and circulation sufficiently in this part. However, main parts of Rumoh Aceh found in Sub-district of Montasik keep maintaining part of main rooms so that generally circulation pattern does not experience significant changes.

This part seems to be more prominent. Factor of building age of some 120 years old and need of room are prominent things. It starts from the top of building, namely, roof. In fact, Rumoh Aceh used roof cover of *rumbia* (sago palm), now Rumoh Aceh in Sub-district of Montasik has used materials of iron sheets in general. The lower part of building has owned rooms made of concrete (brick). Moreover, compliment elements of interiors such as doors and windows have used materials of plywood or similar. Because the age of building is old, wood wall made of board or hard wood leather, now some parts of house use wood or common board or plywood, even layered by tarpaulin.

Acknowledgments

Thank you for all my colleagues in the faculty of engineering, University of Muhammadiyah Aceh, to support and to appreciate of this research. Special thanks to my beloved institution, University of Muhammadiyah Aceh which supported financial research. Hopefully this article give great value and benefits for us.

References

[1] Nurhuzna A 2017 Transformasi fungsi dan bentuk arsitektur bugis makasar di pesisir pantai buti merauke *Jurnal Ilmiah Mustek Amin Ha* vol **6** no 2 p 194.

[2] Stephany S 2009 Transformasi tatanan ruang dan bentuk pada interior tongkonan di tanah toraja sulawesi selatan *Journal Dimensi Interior Petra Surabaya Christian University* vol **7** no 1 pp 28-9.

[3] Statistic of Aceh Besar District 2017 Kecamatan Montasik Dalam Angka *Aceh Development Planning Agency* p 3-40.

[4] Sapitri and Ariska R 2016 Acehnese culture as represented on the sign found in Rumoh Aceh *A Thesis Departement in English Faculty of Cultural Studies University of Sumatera Utara* p 16.
[5] Kothari C R 2013 Research Methodology: Methods and Technique *New Age International Publishers* p 24.
[6] Somantri GR 2005 Memahami metode kualitatif *Jurnal Makara Fisipol UI* vol **9** no 2 pp 57-61.
[7] Sahriyadi A, Qurratul and Azzahra F 2019 Post occupancy evaluation after earthquake and tsunami in Meuraxa Sub-district: Case study in Banda Aceh *IOP Conf Ser: Mater Sci Eng* 469 012087 p. 2.
[8] Hadjat A, Melalatoa M J. and Abu R 1981 Arsitektur tradisional Daerah Istimewah Aceh *Proyek Inventarisasi dan Dokumentasi Kebudayaan Daerah* Departemen pendidikan dan Kebudayaan Republik Indonesia pp 21-30.
[9] Sahriyadi 2012 Identifikasi pola ruang, zonasi dan sirkulasi rumah tradisional Aceh di Desa Reudeup Kecamatan Montasik Aceh Besar *Jurnal Arsitektur RUMOH* Universitas Muhammadiyah Aceh. vol **2** no 4 pp 53-5.
[10] Sabilah F, Antariksa, and Handajani 2014 Tipologi tata ruang dalam rumoh Aceh di kawasan Mukim Aceh Lhee Sagoe *Jurnal Arsitektur Universitas Brawijaya* (e-journal) vol **7** no 1 p 16.
[11] Aiyub, Loebis H, MN and Pane I F 2019 Changes of Value and Form on Traditional Architecture Rumoh Aceh in Pidie *IOP Conf. Series: Earth and Environment Science* 126. 2018 012005 pp 1-3.
[12] Krier R 2010 Architectural Composition *Edition Exel Menges* p 46.
[13] Steadman JP, 1983 Architectural Morpholog London Pion Limited p 269-76.
[14] Faisal G and Wihardyanto D 2014 Studi tata ruang rumah tinggal suku talang mamak *Jurnal TESA Arsitektur* vol **12** no p 98.
[15] Duque E 2002 House form and culture revisited *A Subaltern Critique of Rapoport's Reading of Vernacular* pp 1-7. Cited at https://www.researchgate.net/publication/286880177
[16] Fakriah N 2015 Green Materials in Traditional Housing: A Local Wisdom Lesson. Elkawnie *Journal of Islamic Science and Technology* vol **1** no 1 pp 81-3.
[17] Ulfahmi CGM, Rohidi TR and Triyanto 2018 Local wisdom value of rumoh Aceh *Journal of Art Education* Catharsis **7** (3) pp 258-60.

IOP Conf. Series: Materials Science and Engineering **821** (2020) 012006 doi:10.1088/1757-899X/821/1/012006

Spatial behavior pattern of visitors in City Park Case study: Flora and Bungkul Park, Indonesia

V Dharmawan [1,*] and N Rachmaniyah[2]

[1] Department of Architecture, Universitas Muhammadiyah Surabaya- Indonesia
[2] Department of Interior, Institut Teknologi 10 Nopember Surabaya- Indonesia

*Email: vippy.dharmawan@ft.um-surabaya.ac.id

Abstract. Knowledge of park visitor behavior needs to be owned by architects or public facilities designers. This is so that they can predict how visitors will behave in the park they designed. On the other hand, research on the behavior of park visitors is still rare in Indonesia. This research was located in Taman Bungkul and Taman Flora Surabaya-Indonesia and aimed to observe behavior patterns of park visitors. The research uses a qualitative descriptive approach, with behavior mapping techniques whose data are mostly obtained from direct observation in the field. Behavior mapping identified visitor's position, elements and facilities of the park, and the time they spend in the park. The results of the observations revealed that park facilities acted as a prominent stimulus to visitors' behavior pattern, and there was interelation between city park facilities and spatial behavior pattern. In addition, it was also found that the strength of stimulus changed time to time depended on density in the park.

1. Introduction

City Park is an outdoor public open space that is owned and operated by local government. As a public space, it can be accessed by all citizens of the city. There are many benefits of City Park, including promoting human health and wellbeing, and strengthening social cohesion, air quality and carbon sequestration. Most parks in Indonesia are green spaces used for recreation, aesthetic function, or air quality maintenance. Parks may consist of trees and bushes, soil, grassy areas, rocks, but may also consist of buildings and other artifacts, such as fountains, sport arena, monuments, or animal cages.

This article relates to public open space design, particularly to the relationship between city park facilities and spatial behavior pattern. It focuses on the density of visitors and the placement of facilities. The research problem addresses a lack of knowledge about the influence of physical features to people behavior in public space, especially in a city park. The majority of planners and designers recognize that public spaces can influence people's behavior and experiences. In addition, the physical and ambient features of public spaces can facilitate crowding, affect personal space needs, create places that are desirable and attractive, deter people through unpleasant sounds and smells, and cause behavioral changes in response to perceptions of safety [1].

Some research in public space have been conducted. Razavivand Fard [2] studied how people use public open spaces and factors that impact the use of these spaces as well as physical features which are the most influential on the behavior patterns and specific activities related to the space in Kadıköy square, Istanbul. This study reveals that design and quality of public open space influence the use of it, and activities occurring in the place are only meaningful and dynamic as long as humans' everyday life takes place within it.

In context of Indonesia, study of behavior in public spaces have been done quite a lot, for example, a study of visitor behavior in a shopping center that describes the response of visitors to the atmosphere of a shopping center [3]. There are also studies in educational facilities that map early childhood behavior patterns [4], as well as research on the behavior patterns of fast food restaurant visitors [5].

Whereas, research on urban parks rarely discusses behavioral aspects of park visitors. Research conducted by Sasongko [6], for example, is about changing the function of city parks in Semarang. This research shows that changes in the function of the park can result in decreased environmental quality, such as increased air temperature, dust, and dirty. There is also research conducted by Kustianingrum et al [7] which is about the function and role of Taman Ganesha in the city of Bandung. The research by Kustianingrum [7] shows examples of the functions and roles of parks that are well preserved even though they are decades old. Currently, the latest research on the behavior of park visitors is a study conducted by Dharmawan [5] in the play area of Surabaya city parks. This research was conducted to find out the influence of density of visitors on their behavior. The result of this study indicates that when the park is crowded, there will be many unusual behaviors, such as sitting in any place.

Whereas, the research conducted by Rachmaniyah [8] examined the relationship between physical features and human behavior with the stimulus response theory approach. By using behavior mapping techniques, she found that physical features act as stimuli that shape spatial human behavior patterns.

2. Methods

A qualitative approach study was used to reveal the distribution pattern of visitors in Bungkul Park and Flora Park, located in Surabaya. These two parks were selected because both cover a wide area of visitors, and have become one of Surabaya's icon. Most of the data were obtained by direct observation in the field and presented using behavior mapping technique. This was a direct-observation technic that tracks people's behavior in specific space and time. This method was started to be used in the late 1960's to study how physical environment features affect the people's behavior, including the activity level, type of activity, etc. [9]. This technique was developed by Ittelson since the 1970s. It is a very popular mapping technique and often used in behavioral observation. It was also used by Rachmaniyah [8] who examined the behavior of shopping center visitors in a mall in Surabaya. The latest research in Indonesia that uses this technique is research conducted by Fitria [10] who examined the relationship between space settings and their users in the campus of Aisyah University in Yogyakarta.

Field observation was carried out using a type of mapping techniques, namely person centered mapping. This type of mapping focuses on how humans organize themselves in a particular location (Sommer et al, 1980). In this study, visitor distribution patterns were obtained by this technic. The steps taken for mapping were: (1) Doing preliminary observations on both park locations to get an outline of activity patterns and determine observational schedules, (2) Making a basic map of the park, (3) Making observations in the field, and (4) Making data analysis obtained from observations. The observation schedule was grouped into two time categories, i.e. workdays and holiday or weekends. This is because on weekdays the park's situation generally tends to be quiet, while on holidays or weekends the park is generally crowded by visitors.

IOP Conf. Series: Materials Science and Engineering **821** (2020) 012006 doi:10.1088/1757-899X/821/1/012006

3. Results and discussion

The discussion begins with the existing condition of the two parks. Bungkul Park is located in the center of Surabaya. The name Bungkul refers to the existence of Mbah Bungkul grave that located within the park. The park is surrounded by road and accesible from all direction. It has three main facilities: playground area, skateboard area, and open theater area. Playgroud area has several playing equipment for children (swing, slides, see saw, etc.). Skateboard area is an open court that has several tracks for players. Whereas, open theater area is a circular courtyard surrounded by a spectator seats. The second is Flora Park that located in the east of Surabaya near a bus station. This park was formerly called Kebon Bibit because it was used by the city for nurseries of various green plants. This park is unique due to its collection of plants, deer's cages, giant bird nest, and outdoor sport arena that cannot be seen in other parks in Surabaya. In addition, there are also fountain pools, playgrounds, and multi-function courtyard. Unlike Bungkul Park, Flora Park is only open to the public until sunset.

There are various activity in the park, such as standing while enjoying the atmosphere, walking, pushing prams, jogging, sitting, playing ball, playing skateboard, performing, or just watching other visitors' activities. The following table shows the activities that were seen in the park facilities. Visitor activities could be categorized in two groups, i.e. activity with moving position (walking, jogging, playing ball, etc.) and stationary position (sitting, standing, playing see saw, etc.). People in moving position tended to be in pedestrian path, while the other tended to be in area of park facilities. Preliminary observation in both parks showed that people spent more time at the park facilities than on pedestrian path. It revealed that park facilities were the main attraction for visitors. That was why the position of visitors was influenced by existing facilities. They tended to stop or swarm around park facilities. Whereas, on the pedestrian path they tended to move (Table 1).

Furthermore, the observations showed that density of visitors in the park changed from time to time. According to social distance (Hall, 1963), researcher categorized them in three level: high, medium and low density levels. A high density level is specified when the distance between visitors is less than 1.2 m, medium density level when the distance is 1.2 m – 3.7 m, and low density level when the distance is more than 3.7 m.

Table 1. Facilities and kinds of activity

Bungkul Park

Facilities	Kind of Activtiy
Circular Plaza	sitting (+), standing (0), walking around (0), perform (-), push a promm (-), playing ball (-)
Skateboard Arena	sitting (+), standing (0), walking around (-), playing skateboard (0), playing BMX bike (0)
Children Playground	sitting (+), standing (0), walking around (-), playing seesaw, slide, swingset, and others (+)
Culinary/Food Center	sitting (+), standing (-)
Around Fountain	sitting (0), standing (+)
Pedestrian Way	sitting (-), standing (-), walking (+), jogging (0), push a promm (-)

Flora Park

Facilities	Kind of Activtiy
Around Animal Cage	standing (+), feed the animals (0)
Around Fountain	standing (+)
Outbond Area	sitting (+), standing (0), walking around (-), playing outbond equipment (+)
Children Playground	sitting (+), standing (0), walking around (-), playing seesaw, slide, swingset, and others (+)
Multi Recreation Area	sitting (+), standing (0), walking around (-), playing seesaw, slide, swingset, and others (+), perform (0)
Pedestrian Way	sitting (-), standing (-), walking (+), jogging (0), push a promm (-)

(+) *often*
(0) *occasionaly*
(-) *seldom*

High and medium level densities were more often seen in the area of the park facilities. Whereas, in pedestrian areas, the density tended to be low. For example, in the morning when the park was still quiet, low level density was seen in all parts of the park. Then, when the park started to bustle, the visitors would begin to gather around the park's facilities (Table 2). These results indicated that there were several patterns of spatial distribution of park visitors, and park facilities became a stimulus to visitors.

Table 2. Density of visitors

Bungkul Park

Time	Circular Plaza		Skateboard Arena		Children Playground		Culinary/Food Center		Around Fountain		Pedestrian Way	
	w-day	w-end	w-day	w-end	w-day	w-end	w-day	w-end	w-day	w-end	w-day	w-end
05.30-07.00	low	high	low	med	low	high	low	med	low	med	low	med
07.00-09.00	med	high	med	high	med	high	low	high	low	high	low	high
09.00-12.00	low	med	low	med	low	high	low	med	low	high	low	med
12.00-15.30	low	low	low	low	low	med	med	med	low	med	low	low
15.30-21.00	med	med	med	med	med	med	med	med	low	med	low	med

Flora Park

Time	Around Animal Cage (Deer & Bird)		Around Fountain		Outbond Area		Children Playground		Recreation Area		Pedestrian Way	
	w-day	w-end	w-day	w-end	w-day	w-end	w-day	w-end	w-day	w-end	w-day	w-end
05.30-07.00	low	med	low	med	low	med	low	med	low	med	low	low
07.00-09.00	low	high	low	med	low	high	low	high	med	high	low	med
09.00-12.00	low	high	low	med	low	high	low	high	low	high	low	med
12.00-15.30	low	med	low	low	low	med	low	med	low	med	low	low
15.30-17.00	med	med	low	med	low	med	med	med	med	med	low	low

Figure 1 illustrates several visitor distribution patterns in Bungkul Park. The first pattern (Figure 1a) occurred early in the morning of the weekdays. There was low level density in the pedestrian area and all park facilities. The second pattern (Figure 1b) occurred on the afternoon of the weekdays. There was a crowd of visitors with medium level density on the playground and skateboarding arena. Whereas, the pedestrian and plaza areas were in low level density conditions. Next, the third pattern (Figure 1c) occurred on the morning of the weekdays. Crowds with high level density occurred in all facilities, including playgrounds, circular plazas, and skateboard arena. Whereas, in most pedestrian areas, the condition was medium level density. The fourth pattern (Figure 1d) occurred on the morning of the weekends. Crowds with high level density occurred in all parts of the park. There was high density in all facilities and pedestrian areas. Similar distribution pattern can be seen in Flora Park (Figure 2). There was a crowd of visitors with medium level density on the park facilities. Whereas, the pedestrian ways were in low level density conditions.

In line with the previous research conducted by Rachmaniyah [8], this phenomenon showed that the physical element of the environment acted as a stimulus that influenced behavior patterns in space. When the park was quiet, park facilities became a neutral stimulus. Then, when the park started to get crowded, it would be a strong stimulus for park visitors.

IOP Conf. Series: Materials Science and Engineering **821** (2020) 012006 doi:10.1088/1757-899X/821/1/012006

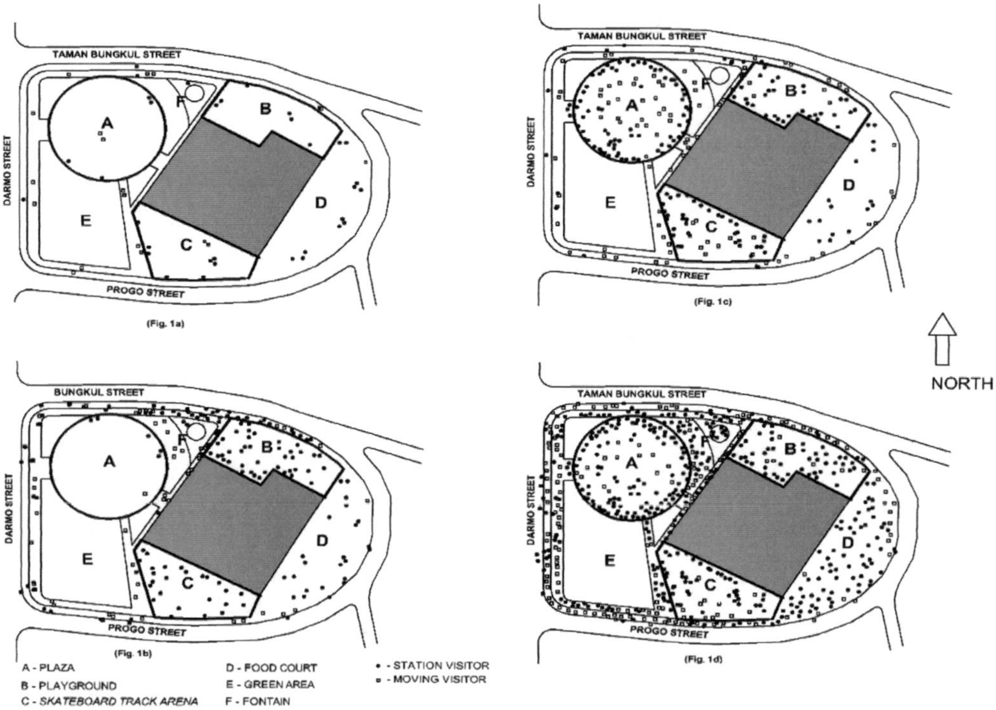

Figure 1. Behavior pattern in Bungkul Park

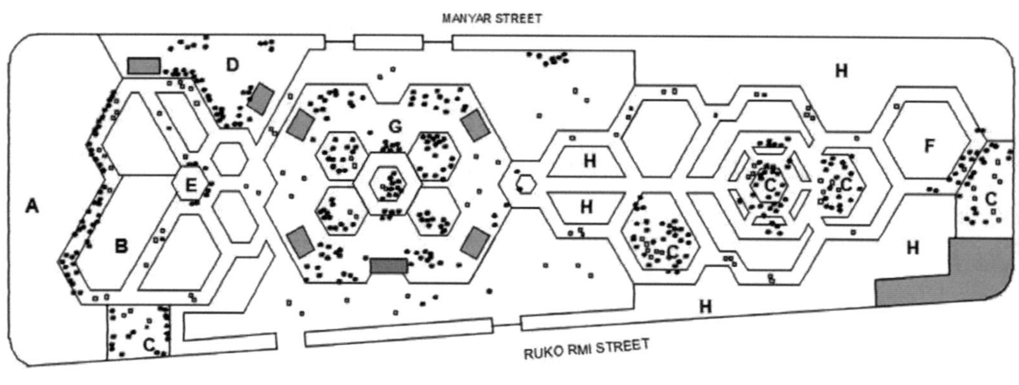

Figure 2. Behavior pattern in Flora Park

4. Conclusion

Visitors spent more time in the facility area, while the density of visitors in the facility area was also higher than pedestrian area. This leads to the conclusion that in public spaces, such as city parks, facilities in the park became a stimulus that shaped visitor behavior. It can also be concluded that the strength of the stimulus in the form of park facilities always changed time to time, depending on the level of visitor density. This finding confirms the result of a study conducted by Rachmaniyah [10] which revealed that existing facilities in public spaces became the dominant stimulus in influencing visitor behavior.

These findings are very useful for designers. It can be a reference in the design process. The fact that facilities in the park became a stimulus that shaped visitor behavior, lead to the conclusion that these facilities are the most taken into account in the city park design process. During the design process the designer can arrange the placement of the facilities as a first step in arranging the park lay out. Furthermore, the other elements of park will follow the pattern of the placement of these facilities.

References

[1] Sangar V A 2007 Human behavior in public spaces *Thesis* (University of New South Wales, Sydney).
[2] Fard H R 2014 Evaluating spatial behavior in the urban public space of Kadikoy square, *2nd International Conference on Architecture and Urban Design* Tirana Albania.
[3] Hijaz T 2007 Desain interior dan perilaku pengunjung di ruang public *Jurnal Dimensi Interior* UK Petra Surabaya.
[4] Makalew L V and Waani J O 2015 Pengamatan arsitektur dan perilaku *Proseding Temu Ilmiah IPLBI* UNSRAT Manado.
[5] Dharmawan 2018 Study of behavior in fastfood restaurant at shopping mall *Jurnal Desain Interior* ITS Surabaya.
[6] Sasongko P D 2002 Kajian Perubahan Fungsi Taman Kota di Kota Semarang *Thesis* UNDIP Semarang.
[7] Kustianingrum D, Angga K S, Rifan A N, Franderdi R 2013 Fungsi dan Aktivitas Taman Ganesha Sebagai Ruang Publik di Kota Bandung *Jurnal Rekakarsa* vol **2** no1 ITENAS Bandung.
[8] Rachmaniyah N 2000 Kajian Perilaku di Ruang Publik Pusat Perbelanjaan *Thesis* ITS Surabaya.
[9] Cosco N G, Moore R C and Islam M Z 2010 Behavior mapping: A method for linking preschool physical activity and outdoor design *Medicine and Science in Sports and Exercise*.
[10] Fitria T A 2018 Keterkaitan perilaku pengguna dengan pola ruang kawasan *Jurnal Arsitektur dan Perencanaan* UNISA Yogyakarta.

The effect of adhesive type and speed pan granulator on the properties of urea slow release fertilizer

R B Pamungkas[1,2,*], B Jos[2], M Djaeni[2] and N S Afridha[1]

[1] Department of Chemical Engineering, Engineering and Science Faculty, University of Muhammadiyah Purwokerto, Indonesia.
[2] Department of Chemical Engineering, Engineering Faculty, Diponegoro University, Semarang, Indonesia.

*Email: rbayup11@gmail.com

Abstract. The application of slow-release fertilizer (SRF) to agriculture provides many benefits. Productivity increases and at the same time, negative impacts decrease. This type of slow-release matrix fertilizer offers an excellent solution for increasing nutrient uptake by plants and restoring agricultural land. The objectives of this research are to study the effect of the type of adhesive and the rotational speed of the granulator on the physical properties of urea slow-release fertilizer (USRF). There are four characteristics used as responses, namely; bulk density, water absorption, granules yield, and durability. The granulation process is carried out using a disc granulator. The study was conducted using one factor at a time method. Research has been carried out by varying the rotation speed of the granulator 5 to 25 with a range of 5 rotations per minute on five adhesives, namely; tapioca flour, glutinous rice flour, rice flour, sago flour, and corn flour. The results showed that the types of adhesives affect the values of bulk density, water absorption, granules yield, and durability. The best type of adhesive is tapioca, however, USRF tapioca adhesive has weakness on water absorbency. Increasing the rotational speed of the granulator causes an increase in the value of bulk density, water absorption, and granules yield while the endurance parameters decrease.

1. Introduction

Indonesia is an agricultural country which has around 15 million hectares [1] of wetland agriculture and the fourth largest population in the world [2]. The population of Indonesia reaches 256 million people [3] and more than 60% of Indonesia's population work in the agricultural sector. An exponential increase in annual population demands reliable food availability. Therefore, agriculture is a very important sector. In 2016, Indonesia has provided 30.1 trillion rupiah in fertilizer subsidies for 3 years in order to encourage food self-sufficiency and increase crop yields [4]. Unfortunately, the production capacity of paddy's is still in low level, around 5,1 tons/ha [1], only. One of the problems is the low efficiency of nitrogen uptake by plants. About 50-70 % of urea fertilizer is lost and has a negative impact on health and the environment [5].

Fertilizer is a material that is added to the growing media that functions as a supplier of various elements needed by plants. The nutrient elements are classified into two, namely macro nutrients (N, P, K,

Ca, Mg, and S), and micro nutrients (Zn, Cu, Mo, Co, B, Mn, and Fe) [4]. Fertilizers are an input that is essential in the process of agricultural production [5]. Fertilizers can be classified into two based on the source of manufacture, namely organic and chemical fertilizers. Organic fertilizer is fertilizer that is partly or wholly derived from plant or animal parts [6]. Some examples of organic fertilizer are compost, livestock manure [7], bird manure, whereas chemical fertilizer is fertilizer made by humans from the processing of mineral materials [6]. Organic fertilizers are environmentally friendly, in addition to having an impact on increasing production synthetic or chemical fertilizers also have a negative impact on health and the environment [8].

Urea is one of the agrochemicals fertilizer which is the main source of nitrogen because it consists a high N nutrient. The level of its utilization by plants only ranges from 30 to 50 %, the rest is lost to the environment through several processes, such as ammonium volatility, alkylation, leaching, and nitrification [9, 10].

Furthermore, it also has detrimental effects in the economic aspect of the issue: material losses, spent energy and human work effort negatively affect the total economic balance of the whole agrochemical production process [5]. Prilled urea is easily solute in the water. Basically, the rate of nutrient release can be reduced by inhibiting the mass transfer process of urea from solid phase into a liquid phase [11]. One method used to overcome it is by modifying the physical form of prilled urea fertilizer into a new type of urea fertilizer that has slow or controlled solubility [12, 13].

There are many factors affects the granulation processes. Granulation needs adhesive to improve the physical properties of the granules, especially the compactness of the granules. The adhesive used as a mixture can be natural or synthetic adhesives. A good adhesive has excellent characteristics such as; good adhesion, not harm plants, be easily found, and be affordable. The second factor which influences granulation is the rotational speed of the pan granulator as it affects the interaction of the type of adhesive and the size of the granule. The theme of granulation research in making slow/controlled-release fertilizer is a concern of many researchers. Many people have researched on granulation with granulators [14-16]. Some factors of concern are the type of adhesive, adhesive concentration, rotation speed, the slope of pan, adhesive flow rate, and rotation time [15]. The response parameters observed were durability, bulk density, yield percentage [15, 16], mechanical stability [15], compressive strengths [17], time of dispersion and elemental concentrations of the granules [17, 18]. However, the literature on granulation matrix consisting of carbon, zeolite with adhesive natural polymer (starch) is very limited. Though the mixed matrix has the ability to form a slow release fertilizer at the same time to remediate agricultural land.

Based on the description above, it is necessary to conduct a study on the effect of adhesive type and the rotational speed of the granulator toward the physical properties of urea slow release granules. The adhesives are made from natural starch which contain a large percentage of amylopectin, easy to find and the price is relatively affordable. The matrix consists of zeolite, carbon with certain ratio and mixed with clay. This study also observes at the effect of the granulator speed. Through this study, it is expected to reveal how the influence of adhesive types and the rotation speed of the granulator affect the quality of urea granule fertilizer.

2. Method

The zeolite powder which has been activated by NaCl 10 grams is mixed with 60-gram carbon which has been modified with cetylpyridinium chloride (CPC). Surfactant modifications of carbon and zeolite activation were carried out by ultrasound model TU-250 Y voltage 220 V / 50 Hz power 250 W. The solid mixture is then put into a solution made from 20-gram Urea in 60 ml demineralised water. The concentrated suspension is stirred evenly and put in the oven for 3 hours. The solid is mashed and mixed with clay powder. The mixture is granulated at certain speeds (5 RPM, 10 RPM, 15 RPM, 20 RPM, and 25 RPM) and spraying a certain type of liquid adhesive. There were five adhesive materials used in this research, namely; tapioca flour, glutinous rice flour, rice flour, sago flour, and maize flour. The granulation process is carried out using a pan granulator. The granulator consists of a rotating pan with adjustable speed motor. The diameter of the pan is 46.5 cm and depth is 16 cm. The results obtained were sieved to get granules product yields. Products which met good quality standards are tested for other physical parameters, namely; bulk density, water absorption, and durability. The effect of factors to response was evaluated statistically by analysis of variance (ANOVA) single factor.

3. Results and discussion

3.1. Effect of adhesive types on the physical properties of SRF granules

There are three (3) forces work together on powder in a pan granulation process, namely frictional force, centrifugal force and gravitational force. The powder rise along with the revolving pan under frictional force, all down by gravitational force and moves to the pan edge because of centrifugal force. The powder rolls in a certain trace under the function of these three forces. The powder collides each other and or collide the pan edge during granulation. Liquid adhesive spray covers the powders and binds the closet neighbor which contact them and compact through the pressing by collision forces and forms nucleus of granules. After the nucleus of granules formed, the next process is growing them into critical size. Critical size is the maximum size of granules. The next collision with other breaks them into small or nucleus of granules. This process continue and gradually attains the certain size.

Figure 1 shows the effect of various adhesives on bulk density, yield granules, durability and water absorption. The type of adhesive as the x-axis is arranged based on the percentage of amylopectin content. Maize starch is on the left side and sticky rice is on the right side, while sticky rice, rice, sago are placed sequentially in the middle. Amylopectin content in cornstarch, sago, rice, tapioca and sticky rice are 70 %, 73 %, 80 %, 91 % and 99 %, respectively.

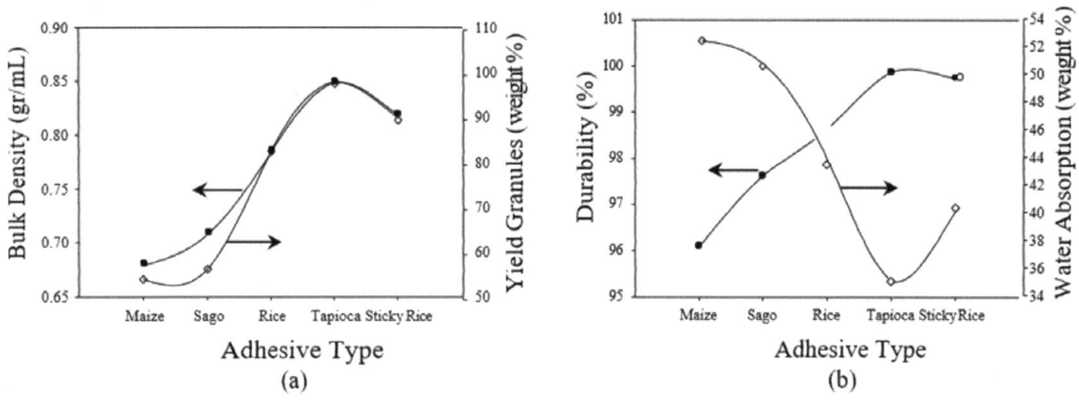

Figure 1. Physical properties of granules SRF at various adhesive type: a) Effects on yield granules and bulk density; (b) Effects on durability and water absorption

The largest **bulk density** value (0.8492 gr/ml) is the granule of tapioca adhesive while the smallest one (0.6659 gr/ml) is granules of maize adhesive as shown at Figure 1(a). Koukkanen (2013) stated that different chemical composition of complex carbohydrates (amylopectin) on flour will affect the bulk density of granules [19].

This study is in line also with Gaudin *et al.* [20]. He stated that starch with a high content of adhesive gives a high value of bulk density. It is indicated that the samples with more starch adhesive content have a large bulk density. In general, increasing amylopectin content in starch increases the value of bulk density. However, the maximum condition does not happen on the starch with maximum amylopectin but tapioca starch adhesives. This happens because adhesives made from tapioca flour have better viscosity values than adhesives from other starches. Single factor ANOVA analysis carried out by putting the percentage of amylopectin on starch as x-axis and bulk density. The results of the single factor ANOVA analysis of the bulk density data obtained a P-value of 0.000015780.

Yield of granules is defined by the percentage of granular 2-5 mm size. The yield of granules is defined by the percentage of grain size of 2-5 mm. As seen in Figure 1(a), the pattern of increasing yield granules is similar to the pattern of increasing bulk density. The highest percentage of granular 2-5 mm size is produced by tapioca flour adhesive that is 98.35 %, and the lowest is from maize adhesive which is only 57.64 %. Tapioca flour has the ability to glue an ingredient because it contains a lot of starch which high content of amylopectin and good viscosity, so that during the granulation process liquid adhesive could be easily sprayed in small spray holes and granules will be easily formed due to the adhesion produced is quite well. The results of the single factor ANOVA analysis of the yield data obtained a P-value of 0.001335. This means that the type of adhesive effects yield of granules.

The effect of adhesive type on durability and water absorption is shown in Figure 1(b). The result of this research shows that the durability pattern similar to the pattern of yield and bulk density. The highest **durability** is produced by tapioca flour adhesive which is 99.87 % and the lowest is maize adhesive which is 96.09 %. On other research shows that tapioca adhesives are better than molasses [21]. The difference is caused by the chemical composition of each different adhesive. Flour chemical composition consists of complex carbohydrates (amylopectin) which are sticky, it can increase bonding between materials in fertilizer granules to increase resistance on pressure and friction that is received by granule fertilizer during carrying and transportation. Data analysis with single factor ANOVA obtained P-value of 0.00000220502.

Figure 1(b) describes the effect of adhesive type on percentage of water absorption. It can be seen that the highest percentage of water absorption is produced by adhesive made from maize, which is 52.48 % and the lowest is Tapioca adhesive type, which is 35.08 %. This is due to the fact that tapioca flour has the ability to glue an ingredient because it contains a lot of starch composed from two kinds of carbohydrates, namely amylose and amylopectin, so that during the granulation process granules will be easily formed due to the good enough adhesion strength produced [21]. This research is in accordance with previous studies which mentioned that there was a tendency that the addition of the adhesive types with good adhesion actually reduced water absorption [22]. Data analysis using single factor ANOVA obtained P-value of 0.00007801.

SEM (Scanning Electron Microscope) test shows that the morphology of granules are different. Figure 2 indicates that the granule with tapioca adhesive looks compact with small porosity, while granule with maize adhesive looks less compact with bigger porosity. Granule with tapioca adhesive looks smoother and flatter surface, slightly fractured, while granule with cornstarch appears to be uneven and consist of large number of fractures with large holes.

IOP Conf. Series: Materials Science and Engineering **821** (2020) 012007 doi:10.1088/1757-899X/821/1/012007

Figure 2. Morphology of granules with the different type of adhesive

3.2. Effect of granulator rotation speed on the physical properties of SRF granules

In this study, the granulation process was carried out by varying speeds of 5.10, 15, 20, 25 RPM on a disc granulator. The process was run out within 20 minutes with the same quantity of tapioca adhesive. The resulting granules are tested for their physical properties. Figure 3 describes the changing of physical properties on various speed rotation of pan granulator. The section (a) shows the effect of speed rotation on bulk density and yield granules, while section (b) illustrates the Effects of granulator speed on durability and water absorption.

Based on the Figure 3(a), it can be seen that the highest bulk density value is resulted at a rotation speed of 25 RPM that is 0.8149 and the lowest is at a speed of 5 RPM which is 0.696. This study is in line with Irshad *et al.* which showed that increasing speed rotation tends to increase bulk density [23]. Besides, the size of the granule also affects the density value. Smaller size of granule and its similarity will increase the density value because the smaller size will fill the space or gap between the particles so that the mass of the granule is greater than the larger granule [8]. Analysis of data using the single factor ANOVA, obtained P-value of 0.000489.

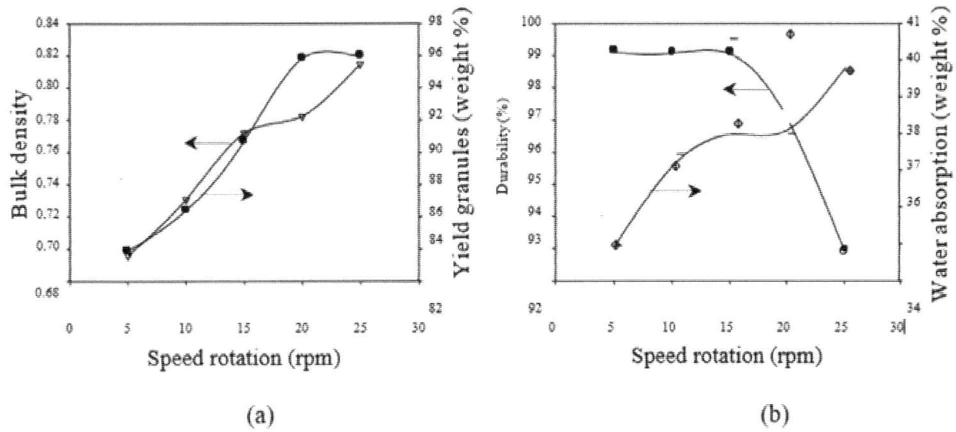

Figure 3. Physical properties of granules SRF at various speed rotation granulator. (a). Effects on bulk density and yield granules; (b). Effects on durability and water absorption.

The percentage of 2-5 mm size granules also affected by speed rotation. Figure 3(a) shows that the percentage of granule in size of 2-5 increased with the increasing of speed rotation. The highest percentage of 2-5 mm size granule is produced by the rotation speed of the 25 rpm granulator that is 95.99 % and the lowest is the rotation speed of the 5 RPM granulator which is 83.89 %. This research is in accordance with other research which states that the average particle diameter increases by the increase up to 25 rpm from granulator pan [23]. In the analysis of data through single factor ANOVA, P-value of 0.00006866 was obtained.

Figure 3(b) illustrates durability and water absorption on various granulator rotational speed. The graph shows that durability decreased with the increase of speed rotation from 5 to 25 rpm. Variations in pan rotation speed effectively change the growth behavior of granulation, which includes not only changes in the granulation growth rate but also the interactions between the basic units that make up the final granulation [16].

The comparison of this study with some standard granules is represented in Table 1. All granule parameters in this study settle within the standard range. It indicates that the range value of factors of this research could achieve good quality of granules.

Table 1. Comparison of test results with quality standards

Parameter	Quality Standard	Results
Bulk Density	The greater the starch content (amylopectin) in the adhesive, the more bulk density will increase [20].	Tapioca (amylopectin 91%) = 0.8492 gr.cm^{-3}. Maize (amylopectin 70%) = 0.6659 gr.cm^{-3}
The percentage of Granule 2-5 mm Size	Min 80% [24]	57% - 98%
Durability	pellet *non wood* ≤ 96% [25]	96% - 99%
Water Absorption	pellets or granules generally range from 20% - 60% [26]	35% - 52%

4. Conclusion

The results of this study indicate that the type of adhesive (tapioca, sticky rice, rice, sago, and corn) affects bulk density, water absorption, granules yield, and durability. Figures 1.a and 1.b show that tapioca adhesives provide granules with three physical properties better than others. Bulk density, yield of granules and durability show the best position, which is the highest value compared to others. While the physical property of water absorbency is in the lowest position. Granules fertilizer should posse a high capacity of water absorbency. Considering experimentally obtained results, it was observed that tapioca adhesive gave more beneficial physical properties compared to others. The results also show that the increasing speed of the granulator rotation causes an increase of three responses, namely; bulk density, water absorption, granules yield; but decreases granule's durability. Based on the data in Table 1, it can be concluded that the granules which were produced in this study meet the standards quality. Moreover, research can be improved by optimizing 2 (two) or more combinations of factors to find the value of the factors that produce the best granule properties.

References

[1]. T Heri, L Nuryati, B Waryanto, and R Widaningsih 2016 *Pusat Data dan Sistem Informasi Pertanian Kementrian Pertanian.*
[2]. H-P Kohler and J R Behrman 2015 Ost 2015 *Concensus.*
[3]. Food and Agriculture Organization of United States 2017 Country Fact Sheet on Food and Agriculture Policy Trends.
[4]. OECD 2015 *Indonesia Policy Brief.*
[5]. K Lubkowski 2014 *Environmental Engineering and Management Journal* **13**(10) pp 2573–81.
[6]. M Q Alamdari and H R Mobasser 2014 *American Journal of Experimental Agriculture* (12) pp 1604-15.
[7]. Y P Timilsena, R Adhikari, T H Muster, and B Adhikari 2015 *Journal of the Science of Food and Agriculture* **95**(6) pp 1131–42.
[8]. N W A Utari, Tamrin, and S Triyono 2015 Jurusan Teknik Pertanian, Fakultas Pertanian: Universitas Lampung.
[9]. A Shaviv 2005 International Workshop of Enhanced-Efficiency Fertilizers. (June) pp 28–30.
[10]. B Ni, M Liu, and S Lü 2009 *Chemical Engineering Journal* **155**(3) pp 892–98.
[11]. R B Pamungkas, A H Mulyadi, dan O D Hajoeningtijas 2017 LPPM Universitas Muhammadiyah Purwokerto.
[12]. P Waluyo Slow Release Fertilizar sebagai Dasar Perumusan SNI Pupuk Urea. http://digilib.bppt.go.id/sampul/IA_98C_11_0216-00001-00001.pdf , accessed at September 18th , 2019
[13]. S Ali and F Danafar 2015 *Life Science Journal* **12**(11) pp 33–45.
[14]. J O Ayorinde, O A Itiola, O A Odeku, and M A Odeniyi 2011 *Brazilian Journal of Pharmaceutical Sciences* vol **47** n 4.
[15]. T Gluba, A Obraniak, and E G Mlynarczyk 2004 *Physicochemical Problems of Mineral Processing* **38** pp 177-86.
[16]. Y Liu, D Scharf, and F J Clemens 2014 *Powder Technology* vol **263** September 2014 pp 159-67.
[17]. J Pesonen, V Kuokkanen, T Kuokkanen, and M Illikainen 2016 *Journal of Environmental Chemical Engineering* **4**(4) pp 4817-21.
[18]. J Pesonen, T Kuokkanen, P Rautio, and U Lassi 2017 *Biomass and Bioenergy* **100** pp 92-7.
[19]. M Kuokkanen 2013 *Dissertation* University of Oulu, Oulu.
[20]. F Gaudin, Y Andres, and PL Cloirec 2008 Packing material formulation for odorous emission biofiltration *Chemosphere* **70** pp 958–66.

[21]. M S Gunawan, O Nawansih, and F Nurainy 2015 Seminar Nasional Sains & Technology VI Lembaga Penelitian dan Pengabdian Universitas Lampung.

[22]. K A Wardhana, R S Soetopo, Saepulloh, P B Asthary, and M N Aini 2015 *Jurnal Selulosa* vol **4** no 2 Desember pp 69–78.

[23]. U Irshad, M N Sharif, R U Khan, and Z H Rizvi 2008 *Journal of Quality and Technology Management* I.C.E.T, Punjab University, Lahore.

[24]. Regulation of the Minister of Agriculture No.70/Permentan/SR.140/10/2011.

[25]. O Zajonca, J Frydrych, and L Jezerska 2014 International Conference on Agricultural and Biosystem Engineering *IERI Procedia* **8** pp 2–10.

[26]. F Rizqa, E Yenie, A Sasmita, and Jom 2018 FTEKNIK vol **5** Edisi 1 Januari s/d Juni.

IOP Conf. Series: Materials Science and Engineering **821** (2020) 012008 doi:10.1088/1757-899X/821/1/012008

Preparation of suppressor column for capillary ion chromatography with conductivity detection

A Rahayu[1,*], J A Fajri[2], L W Lim[3], T Takeuchi[3]

[1] Department of Chemical Engineering, Faculty of Industrial technology, Universitas Ahmad Dahlan
[2] Department of Environmental Engineering, Faculty of Civil Engineering and Planning, Universitas Islam Indonesia
[3] Department of Chemistry and Biomolecular Science, Faculty of Engineering, Gifu University

*Email: aster.rahayu@che.uad.ac.id

Abstract. A capillary suppressor column was prepared by packed column and monolith columns. The packed column prepared using HyperSep SCX as packing material. The sulfonate group was introduced from benzenesulfunic acid that could be used as a suppressor material. The slurry of packing material was packed to capillary fused silica 0.53 mm I.D x 0.66 mm O.D with 30 mm of length. Monolith suppressor column was prepared from methacrylate based modified with Na_2SO_3 for strong cation exchanger and iminodiacetic acid for weak cation exchanger. A suppressor column was connected to a capillary chromatography system for separating 5 inorganic anions (Cl^-, NO_2^-, Br^-, NO_3^- and SO_4^{2-}) using conductivity detection. The suppression system could reduce the background conductivity of the mobile phase sodium bicarbonate and sodium carbonate properly up to 67.72% using packed suppressor column with size 0.53 mm I.D x 0.66 mm O.D with 30 mm length. It column could be used about 1.5 hours before breakthrough.

1. Introduction

Ion chromatography was first introduced in 1975 as an analytical method. In a short time, ion chromatography has also developed rapidly as one of the technologies for separating and determining inorganic ions, especially anions [1, 2]. Inorganic anions were also successfully separated using Bovine Serum Albumin (BSA) and tartaric acid as a stationary phase with a UV detection [3]. Inorganic anions in seawater samples were perfectly analyzed using silica modified with cetyltrimethylammonium as a stationary phase [4]. For the sensitivity of ion detection, conductivity detection methods are often used. In order of its, the suppressor column was needed to reduce the conductivity of the mobile phase and increase the conductivity of the sample signal.

The development of technology for reducing conductivity in the analysis of inorganic ions using ion chromatograms has been carried out [5-8]. The use of suppressor columns is optimized in the determination of inorganic ions by using a contactless conductivity detector (CCD). The suppressor column is the main component related to its function in ion chromatography. In the previous decades, several types of suppressors have been developed to achieve high sensitivity including containerized columns, hollow fiber, and electrolytic mode [9]. However, most of the several types of suppressor columns are designed for conventional system chromatography and less are used in capillary system ion chromatography, whereas in fact there are some advantages when using capillary system ion chromatography such as high sensitivity reducing waste to be generated, minimal use of mobile phase, easy in the operation and optimization of the mobile phase and packaging materials. The packed column is one of the suppressor columns that are widely used in ion chromatographic capillary systems. TSKgel SCX 10 μm of particle size was packed into capillary fused silica (0.32 mm ID × 0.45 mm OD) with 30 mm length as a material for suppressor and has succeeded in reducing 495 μS/ cm conductivity from sodium carbonate/sodium bicarbonate which is used as a mobile phase with slightly high pressure system [5].

In this research, a suppressor column was made using packing material containing sulfonate group that has a larger particle size is used as a suppressor column in order to refrain from high pressure system. The suppressor columns will be used in ion chromatography capillary systems to reduce the conductivity of the mobile phase in the separation of inorganic ions in particular 5 anions.

2. Research methods

2.1. Capillary chromatography system

A capillary chromatography system was comprised of micro feeder (L.TEX Corporation, Tokyo Japan) that uses gas-tight syringes (0.5 ml; Ito, Fuji, Japan) as pumps, Two 6-way switching valves with a volume of 0.2 μl as injectors and suppressor connectors (Upchurch Scientific, Oak Harbor, WA, USA), micro columns prepared from fused-silica capillary tubes (100 mm, 0.32 mm ID x 0.45 mm OD; GL Science, Tokyo, Japan) as separator columns and fused-silica capillary tubes (30 mm; 0.53 mm ID x 0.66 mm OD; GL Science, Tokyo, Japan) as a suppressor column. Tracedec Conductivity Contactless Detector (Istech, Strasshof, Austria) as a detector. Fused silica tube with a size of 50 μm I.D is used as a flow cell in the detector.

2.2. Preparation of separation column

The separation column was prepared using a fused silica capillary tube with a length of 100 mm with a size of 0.32 mm I.D and 0.45 mm O.D. The stationary phase material is put into the capillary column by sprinkling PWxl anion IC using methanol. The Anion PWxl IC stationary separation column is used in anion separation by connecting to the ion chromatography system using UV detector at a wavelength of 210 nm.

2.3. Preparation of suppressor column

Suppressor column was prepared from packed suppressor column and monolith column. A Fused silica with length 30 mm with size 0.53 mm for I.D and 0.66 mm O.D was used as suppressor column that packed using stationary phase material consist of sulfonate functional group. The stationary phase material is put into the capillary column by spreading HyperSep SCX using methanol. As comparison, the monolith suppressor columns were prepared from methacrylate based. Methacrylate based were prepared from glycidyl methacrylate, ethylene glycol dimethacrylate, water, 1-propanol and 1,4-butanediol. Then, the methacrylate based were modified with Na_2SO_3 for strong cation exchanger and Iminodiacetic acid for weak cation exchanger. All the suppressor columns were tested for its ability to reduce the conductivity of the mobile phase in term of sodium bicarbonate and sodium carbonate [7:6] by connecting it to the ion chromatography system in the separation of 5 anions (Cl^-, NO_2^-, Br^-, NO_3^- and SO_4^{2-}) by using a Contactless Conductivity Detector (CCD) conductivity measurement detector.

3. Result and discussions

The ability of the suppressor column has been tested to reduce the conductivity of the mobile phase (sodium bicarbonate and sodium carbonate) which have high conductivity. When monovalent and divalent anion are separated using sodium bicarbonate and sodium carbonate [7:6] as mobile phase without using a suppressor column, then the entire peak of the analyte will be covered by the high conductivity of the mobile phase. This is because the conductivity of the mobile phase is higher than the conductivity of the separated analyte. Figure 1 shows a chromatogram of separation of 5 anions by ion chromatography system without using a suppressor column as well as using a suppressor column (packed column). The existence of suppressor column was important to decrease the conductivity of the mobile phase, so the peaks of the separated analyte can appear. The suppressor column with packing material in the term of HyperSep SCX contains a sulfonate functional group. It was introduced from benzenesulfonic acid. The sulfonate functional group has ability to reduce the conductivity of the mobile phase up to 67.72%.

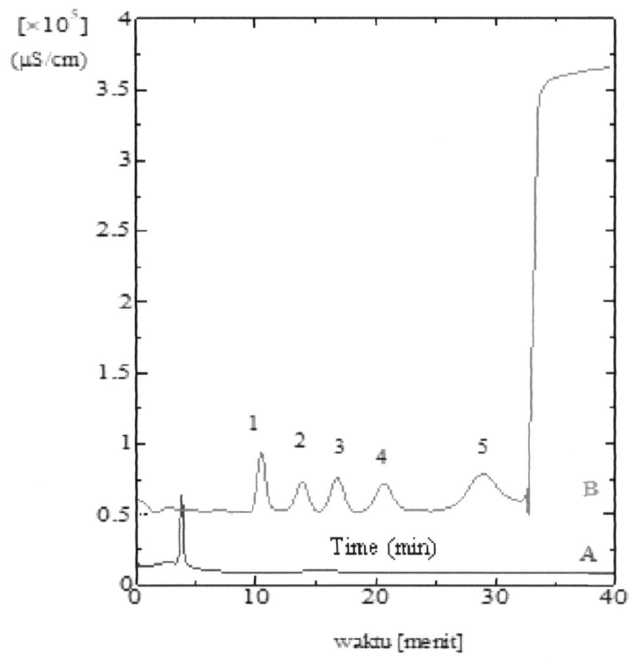

Figure 1. Separation of 5 anions: A. Without using a suppressor column, B. Using a column suppressor (packed column) with HyperSep SCX material in separation (1) Cl^-, (2) NO^{2-}, (3) Br^-, (4) NO^{3-} and (5) SO_4^{2-} by using PWxl IC anion stationary phase with column size capillary 0.32 mm ID x 0.45 mm OD, mobile phase sodium bicarbonate and sodium carbonate [7:6] concentration 1.75 M: 1.5 M, flow rate 3µL/min, contactless conductivity detector (CCD) injection volume 0.02 µL.

Monolith suppressor columns which prepared from methacrylate based formed using glycidyl methacrylate (GMA) and ethylene glycol dimethacrylate (EDMA) to form the epoxy contained in the polymer matrix, then modified with Iminodiacetic acid for attaching carboxyl active site as weak cation exchange and Na_2SO_3 for attaching sulfonate as strong cation exchanger. Figure 2 show the morphology surface of the monolith suppressor column. The polymers were attached tightly into the inner wall of the capillary fused silica tube and the monolith formed packed and filled.

a. Suppressor column strong cation exchanger | b. Suppressor column weak cation exchanger

Figure 2. Scanning electron Microscopy (SEM) photo of monolith suppressor columns with column size capillary 0.32 mm ID x 0.45 mm OD .

The present of the epoxy group acted as the based to attached functional group. Both of monolith column weak cation exchanger (WCX) and strong cation exchanger (SCX) could apply for separating anions as shown in Figure 3. The SCX monolith column and WCX monolith column have ability to reduce the conductivity of the mobile phase 28.18% and 6.81%, respectively.

In the chromatographic system series the weak cation exchange monolith column has the ability to reduce the background conductivity of the mobile phase from 1349 μS/cm - 1257 μS/cm or has a decrease in the conductivity of 6.81% and a system pressure of 3.2 MPa. The ability of a weak cation exchange monolith column is obtained by flowing a mobile phase with a flow rate of 3 μm into a weak cation exchange monolith column with a column length of 3 cm. it was directly connected to the flow cell of the conductivity detector in the ion chromatography system without using a separator column. The use of strong cation exchange monolith columns using Na_2SO_3 which contribute sulfonate groups as suppressor columns has sufficient ability to reduce the background conductivity of sodium bicarbonate and sodium carbonate mobile phase with a lifetime of 1 hour.

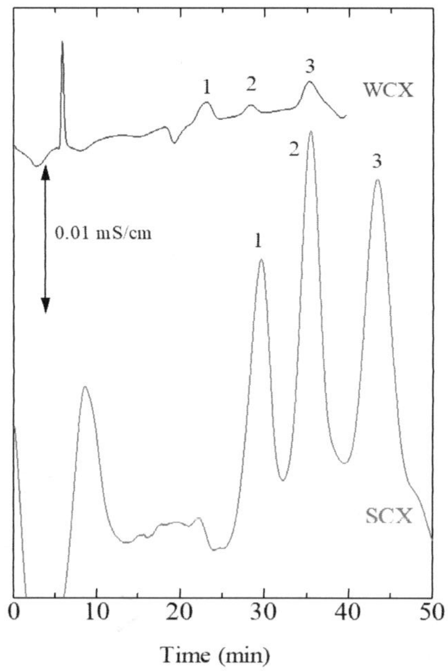

Figure 3. Separation of 5 anions using SCX monolith column suppressor and WCX monolith column suppressor in separation (1) Cl^-, (2) NO^{2-}, (3) Br^-, (4) NO^{3-} and (5) SO_4^{2-} by using PWxl IC anion stationary phase with column size capillary 0.32 mm ID x 0.45 mm OD, mobile phase sodium bicarbonate and sodium carbonate [7:6] concentration 1.75 M: 1.5 M, flow rate 3μL/min, contactless conductivity detector (CCD) injection volume 0.02 μL.

4. Conclusion

The capillary suppressor columns were prepared using HyperSep SCX as packing material for packed column and monolith suppressor column from methacrylate based. The sulfonate group was introduced from benzenesulfunic acid that could be used as a suppressor material. The strong cation exchanger site in monolith suppressor was introduced from Na_2SO_3 and Iminodiacetic acid. The packed suppressor column could reduce background conductivity of the mobile phase sodium bicarbonate and sodium carbonate properly up to 67.72%. The suppressor column with size 0.53 mm I.D x 0.66 mm O.D with 30 mm length could be used about 1.5 hours before breakthrough.

Acknowledgment

The authors would like to express the gratitude to Prof. Toyohide Takeuchi and Prof. Lee Wah Lim Department of Chemistry and Biomolecular Science, Faculty of Engineering, Gifu University, for the support during this work.

IOP Conf. Series: Materials Science and Engineering **821** (2020) 012008 doi:10.1088/1757-899X/821/1/012008

References

[1]. Small H, Stevens T S and Bauman W C 1975 Novel ion exchange chromatographic method using conductimetric detection *Anal. Chem.* **47**(11) pp 1801-9.

[2]. Haddad P R, Jackson P E and Shaw M J 2003 Developments in suppressor technology for inorganic ion analysis by ion chromatography using conductivity detection *J. Chromatography A* **1000** pp 725-42.

[3]. Zein R, Munaf E, Takeuchi T and Miwa T 1996 Microcolumn ion chromatography of inorganic UV-absorbing anions using bovine serum albumin as stationary phases *Analytica Chimica Acta* **335** pp 261-66.

[4]. Jiang X, Lim L W and Takeuchi T 2009 Determination of trace inorganic anions in seawater samples by ion chromatography using silica columns modified with cetyltrimethylammonium ion *Anal Bioanal Chem* **393** pp 387–91.

[5]. Sedyohutomo A, Lim L W and Takeuchi T 2008 Development of packed column suppressor system for capillary ion chromatography and its application to environmental water *J. Chromatography A* 2008 **1203** pp 239-42.

[6]. Zhang X X, Liu Y Q, Yu H and Zhang R Q 2017 Rapid and simultaneous determination of piperidinium and pyrrolidinium ionic liquid cations by ion pair chromatography coupled with direct conductivity detection *Chinese Chemical Letters* **28** pp 126-30.

[7]. Ma Y J, Li M, and Li R S 2013 Fast analysis of thiocyanate by ion-pair chromatography with direct conductivity detection on a monolithic column *Chinese Chemical Letters* **24** pp 1067-69.

[8]. Yang B, Zhang F, Li Y dan Liang X 2011 A cation exchanger resin bead based microscale electrolytic suppressor for capillary ion chromatography *Talanta* **83** pp 1496-500.

[9]. Li Y M, Liao J L, Nakazato K dan Mohammad J 1994 Continuous bads for microchromatography cation exchange chromatography *Analytical Biochemistry* **223** pp 153-6.

Optimization condition of fluoride adsorption using activated clay

Z Mustakim[1], F Y Purwaningtyas[1], E B Leksono[1], Sarto[2], and A Prasetya[2*]

[1]Department of Chemical Engineering, Faculty of Engineering, Universitas Muhammadiyah Gresik, Indonesia
[2]Department of Chemical Engineering, Faculty of Engineering, University of Gadjah Mada, Yogyakarta, Indonesia

*Email: aguspras@ugm.ac.id

Abstract. Drinking water which contain high concentration of fluoride can be harmful to teeth and bones. Clay was cleaned from its impurities then treated for 30 minutes at 700 °C and dissolved in HCl solution for 60 minutes. This research focused on the search for process conditions for absorbing fluoride dissolved in water bodies using local material-based adsorption methods using activated clay. The batch sorption experiment was carried out in 50 ml stirred glass contained 12.17 mg/L fluoride solution at operation temperature 25 °C, 35 °C and 45 °C, pH of 4, with 0.75 gr, 1 gr, and 1.25 gr of adsorbent weight. The solution samples were analysed by using visible spectrophotometer at 570 nm of the wavelength. From this research, the equilibrium reached in 60 minutes with 1.25 gr of adsorbent weight at operation temperature 45 °C with maximum 78% fluoride adsorbed. The isotherm models of fluoride removal mechanism fitted well of the Langmuir and Freundlich isotherm models and followed the pseudo second-order kinetics model.

1. Introduction

Fluoride has potential to contaminate water in areas with metamorphic and granite rock types, which are widely distributed in various regions in Indonesia. Fluoride anion (F-) in the form of organic and inorganic compounds can be found in the surrounding environment [1]. One of the places in Indonesia that has high fluoride concentrations in its body of water is in the crater lake of Ijen in East Java. This high concentration comes from the existing geothermal system in the area. Fluoride concentrations in this area from river water are reported to be 14.2 mg/L, whereas in well water it is reported to be 0.1-4.2 mg/ L [2]. People who live in these areas have potential to be affected fluorosis which affects the health especially the bones, teeth and nervous system.

Fluoride concentrations above threshold levels in groundwater are now a significant environmental health issue because of the negative effects they produce. World Health Organization (WHO) issued a maximum limit of fluoride content in drinking water which is 1.5 mg/L [3, 4]. Dental fluorosis can occur at 1.5-4 mg/L of fluoride concentrations. It is indicated by the presence of white-yellow spots on the teeth and over time turn brown due to excessive fluoride mineralization [5]. Exposure to fluoride with higher concentrations (between 4-10 mg/L) will result in dental and bone fluorosis and at level > 10 mg/L can cause crippling skeletal fluorosis [6, 7]. Fluoride with a high concentration in water and food that mother consumed during pregnancy/breast feeding can also damage fetus [8]. Other studies show that at low concentrations (0.3-1.2 mg/L) fluoride can work to strengthen tooth enamel [9].

Research on fluoride adsorption have been studied. Removal of fluoride by using clay material with 5-10 g/L of adsorbent dose, pH of 2, and contact time 3 hours obtained 40-50% of removal percentage [10]. China clay has been carried out to removal of fluoride with an initial fluoride concentration of 5.1 mg/L and at temperature 29 °C, an optimum condition of fluoride absorption of 93% was obtained at pH of 3, 1.5 g/100 mL dose of adsorbent, contact time for 2 hours and followed the Freundlich isotherm model [11]. Other studies using bentonite-Al^{3+} clay modification obtained optimum conditions for fluoride adsorption at 60 minutes, 2 mg/100 mL adsorbent dose, and 100 mg/L adsorbate concentration at 26 °C, the isotherm model followed the Langmuir and Freundlich isotherms and the reaction kinetics followed the pseudo second order [12]. Chemically activated kaolin with the addition of sulfuric acid is more effective at absorbing fluoride than unactivated kaolin [13] and pyrophyllite clays which were previously heat treated at 400 °C to remove fluoride from liquid waste compared to raw pyrophyllite have 21% higher adsorption capacity [14].

The rate of fluoride removal was calculated by equation:

$$\eta = \frac{(C_0 - C_e)}{Co} x\ 100\% \quad (1)$$

Where Ce and Co (mg/L) are the equilibrium and initial concentration of fluoride and η is the removal rate (%), respectively. The capacity of adsorption (mg/g) at equilibrium as follows:

$$q_e = \frac{(C_0 - C_e)V}{m} \quad (2)$$

where m is the adsorbent mass (g) and V is the volume of the fluoride solution (L). Langmuir and Freundlich isotherms are used to determine the mechanism of fluoride absorption onto adsorbents that can describe the process of adsorption of monolayer and multilayer surfaces [12], respectively. The Langmuir isotherm (Eq. 3) and the Freundlich isotherm (Eq. 4) can be expressed in the form:

$$q_e = \frac{q_m\ K_L\ C_e}{1 + K_L\ C_e} \quad (3)$$

$$q_e = K_f\ C_e^{(1/n)} \quad (4)$$

where q_m is Langmuir constants with respect to adsorption capacity (mg/g), n is intensity of adsorption, K_f and K_L are Freundlich constants (L/mg) and Langmuir constants (L/mg), respectively. The adsorption kinetics model was used to determine the estimated of sorption rate and the characteristics of the adsorption mechanism. In this case, the use of pseudo first order (Eq. 5-6) and pseudo second order kinetics (Eq. 7-8) were tested [15, 16]

$$\frac{dq_t}{dt} = k_1\ (q_e - q_t) \quad (5)$$

$$q_t = q_e(1 - e^{-k_1 t}) \quad (6)$$

$$\frac{dq_t}{dt} = k_2\ (q_e - q_t)^2 \quad (7)$$

$$q_t = \frac{k_2 q_e^2 t}{1 + k_2 q_e t} \quad (8)$$

where q_e and q_t are the amount of solute that is adsorbed at equilibrium (mg/g) and the amount of solute that is adsorbed on the surface of the adsorbent at any time *t* (mg/g), respectively, while k_1 and k_2 are the rate constant of first order adsorption (1/min) and second order adsorption (1/min), respectively.

2. Experimental section

2.1. Materials and instrumentation

The material of this research were clay from Kasongan (Yogyakarta), SPADNS Reagent (Merck, CAS 23647-14-5), Zirconyl Chloride Reagent (Merck, CAS 13520-92-8), HCl (Merck, CAS 7647-01-0), NaF (Merck, CAS 7681-49-4), and aquadest. The instrumentations of this research were visible spectrophotometer, sieve shaker, BET surface area analysis, scanning electron microscopy (SEM), X-ray fluorescence (XRF), X-ray diffraction (XRD), shaker water bath, magnetic stirrer, desiccator, filter paper, oven, and beaker glass.

2.2. Procedure

The clay was cleaned by water from its impurities then baked in oven with 120 °C. After drying, it was mashed and sieved with size 100-200 mesh. It baked in furnace for 30 minutes at 700 °C. It saved in desiccator at room temperature. 100 grams of clay was mixed with HCl 1 M solution I L by using magnetic stirrer for 60 minutes in beaker glass. It was separated from solution using filter paper and washed until pH was neutral with hot water. It was baked for 4 hours in oven 120 °C then saved in desiccator. It was characterized using XRF, SEM, XRD, and BET to examine its chemical composition, surface topography, structure, and specific surface area, respectively. Activated clay with 0.75 gr, 1 gr and 1.25 gr of weight were mixed with 50 ml of fluoride solution contains concentration 12.17 mg/L and pH of 4. It was shaken in shaker water bath at operating temperature of 25 °C, 35 °C, and 45 °C. The solution samples of fluoride were analyzed by using visible spectrophotometer at 570 nm of the wavelength.

3. Results and discussion

3.1. Characterisation of activated clay

The chemical compositions percentage of activated clay of XRF analysis showed the domination of chemical composition of activated clay were Si (55.58%), Al (21.99%), and Fe (19.58%) component. The alumina content in activated clay determines the amount of fluoride that is absorbed [17]. XRD analysis showed the structure of activated clay and the result are shown in Figure 1(a). XRD patterns indicate that the major crystal phases were Dialuminium silicate oxide (Al_2SiO_5), Barium diiron tetra oxide ($BaFe_2O_4$), and Cristobalite (SiO_2). Fluoride from wastewater can be removed by Al-O bond structure, which has larger internal specific surface area through specific adsorption. The exchange of ions on activated clay surfaces can be explained by the ligand exchange mechanism. Hydroxyl ions (OH-) was replaced by Fluorine ions (F-) on the surface of activated clay (Al-OH → Al-F) [14, 18]. Al-F bonds will accumulate on activated clay surfaces until equilibrium is reached. Figure 1(b) shows the scanning electron microscope picture of activated clay. The morphology revealed that the surface structure of activated clay was rough and heterogeneous. The specific surface area of the activated clay was estimated by the BET method and the result was 74.379 m^2/g.

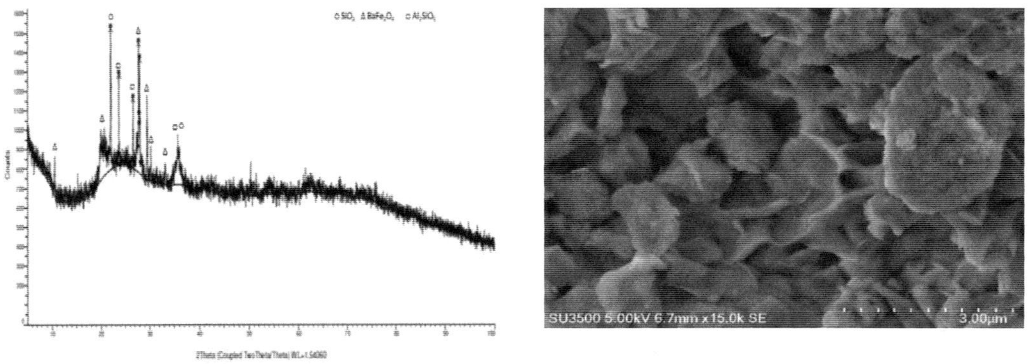

(a) (b)

Figure 1. (a) XRD analysis and (b) SEM analysis of activated clay

3.2. *Effect of adsorbent dose*

The adsorption capacity for fluoride removal by increasing adsorbent dose are provided in Figure 2. The amount of fluoride adsorbed by the adsorbent will increase with the increasing of adsorbent dose in the same operating time at all temperature operation. It can be understood that the adsorption rate increases with increasing adsorbent dose because an increase in the adsorbent dose will increase the contact (area of adsorption) between the adsorbent and the adsorbate [19]. The equilibrium reached in 60 minutes and the highest yield of fluoride removal (78%) occurred at 45 °C at a dose of 1.25 gr (Figure 3).

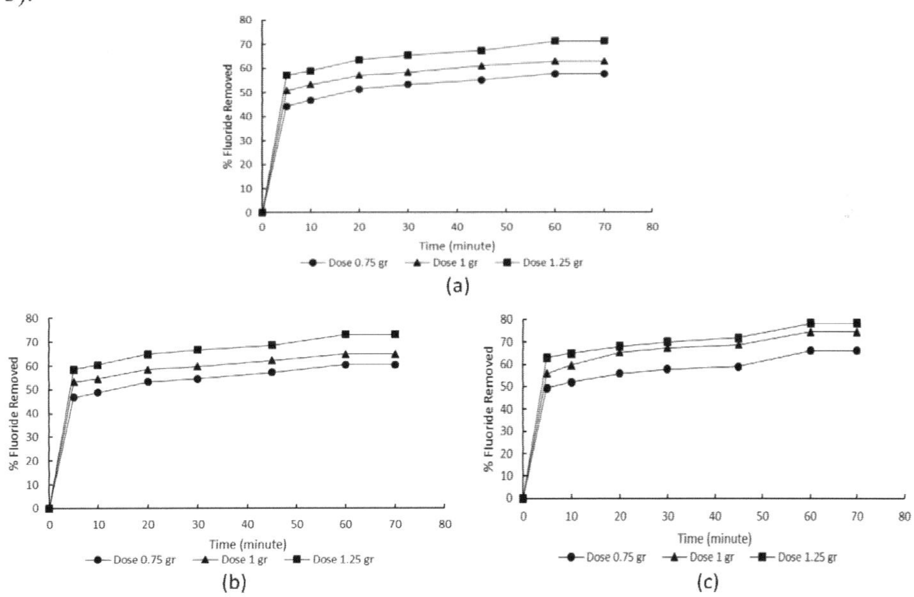

Figure 2. Plots for percent fluoride removal at different dose (a) 25 °C, (b) 35 °C and (c) 45 °C

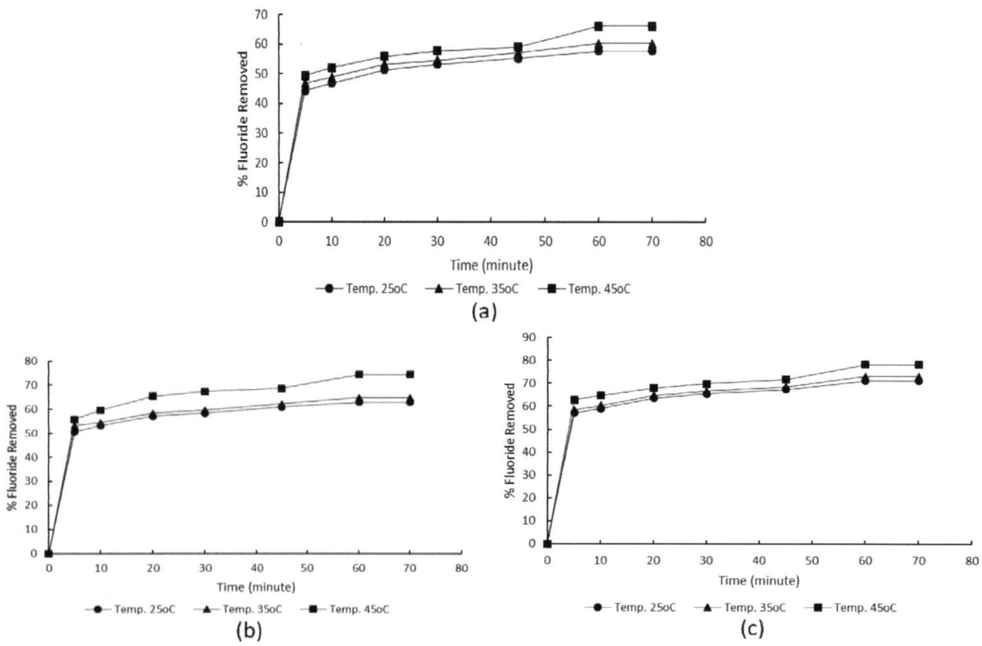

Figure 3. Plots for percent fluoride removal at different temperature (a) 0.75, (b) 1 and (c) 1.25 gr

3.3. Effect of temperature
The adsorption capacity for fluoride removal by increasing temperature are provided in Figure 3. It showed that with an increase in temperature of the solution 25 °C to 45 °C, the adsorption of fluoride increases from 57% to 66% at 0.75 gr of adsorbent dose, 62% to 74% at 1 gr of adsorbent dose, and 71% to 78% at 1.25 gr of adsorbent dose. The increasing of fluoride removal is due to the more active movement of fluoride molecules with increasing operating temperatures which can touch the surface of the adsorbent faster. It is indicated that the adsorption mechanism of fluoride removal is an endothermic reaction [20].

3.4. Isotherm model
The mechanisms of fluoride adsorption onto activated clay using Langmuir and Freundlich adsorption isotherm models were tested. As shown in Figure 4, both of isotherm models could describe the mechanism of fluoride adsorption well. This indicates that the mechanism of fluoride adsorption occurs by monolayer and multilayer adsorption with the value of regression coefficients (R^2) were 0.9774 and 0.9871 of Langmuir and Freundlich adsorption isotherm models, respectively.

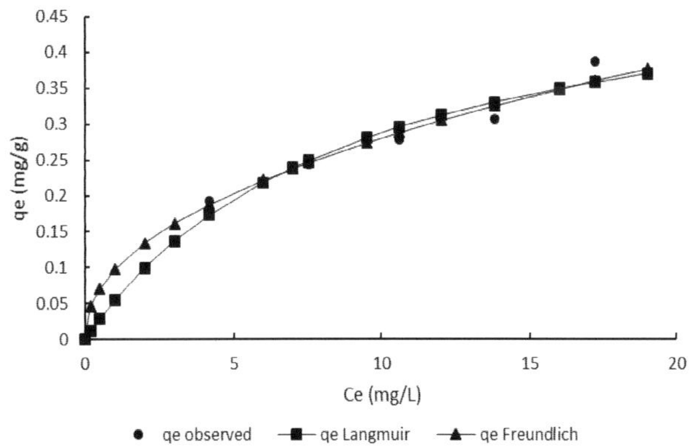

Figure 4. Plots for isotherm models for fluoride removal at 45°C, 1.25 gr dose

3.5. Kinetics model

Figure 5 shows a plot of the kinetic model fit for fluoride adsorption to activated clay with adsorbent dose of 1.25 g/50 ml at 45 °C. This figure shows that the pseudo second-order kinetics model could describe the adsorption mechanism that occurs in fluoride adsorption better than the pseudo first-order kinetics model. The equilibrium reached around 60 minutes with the rate constant of first order sorption and second order were 0.0219 1/min and 0.4697 1/min, respectively.

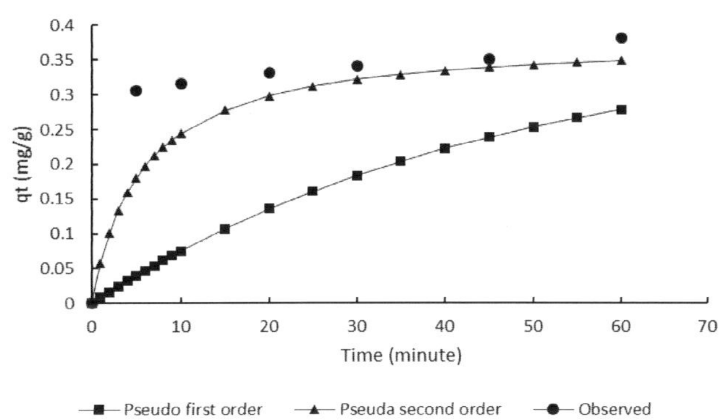

Figure 5. Plots for kinetic models for fluoride removal at 45°C, 1.25 gr dose

4. Conclusion

From this research, the maximum removal efficiency of fluoride reached 78% at a dose of 1.25 gr, at 45 °C with time of equilibrium was 60 minutes. The isotherm models of fluoride removal mechanism fitted well of the Langmuir and Freundlich isotherm models and followed the pseudo second-order kinetics model.

References

[1] Habuda-S M, Ravancic M E and Flanagan A 2014 *J. Material* **7** pp 6317-66.
[2] Heikens A, Sumarti S, Bergen M V, Widianarko B, Fokkert L, Leeuwen K V and Seinen W 2005 *Science of The Total Environment* **346** pp 56-69.
[3] Hosni K and Srasra E 2011 *J. Water Chem. Technol.* **33** (3) pp 164-176.
[4] Saikia P, Bharali R J, and Baruah H K 2017 *J. Analytical Sci. and Technol.* **8** (23).
[5] Dissanayake C B 1991 *Int. J. Environ. Stud.* **38** 137-156.
[6] Valencia-Leal S A, Cortes-Martinez R, and Alfaro-Cuevas-Villanueva R 2012 *Int. J. Eng. Research and Develop.* **5** (4) pp 69-76.

[7] Ramos-Vargas S, Alfaro-Cuevas-Villanueva R, Huirache-Acuna R and Cortes-Martinez R 2018 *Appl. Sci* **8** 1807.
[8] Rao N 2003 *A Review, in Proceedings of the Third International Conference on Environmental and Health*, pp 386-399.
[9] Dysart A 2008 *Investigation of Defluoridation Options for Rural and Remote Communities, Research Report 41 Salisbury SA 5108.*
[10] Panchore K, Sharma S, Sharma A, and Verma S 2016 *Int. J. Advance Sci. Research* **1** (6) pp 28-32.
[11] Lonare R., Inam F, and Deo S, 2009 *Asian J. Chem.* **23** (1) pp 59-62.
[12] Vhahangwele M, Mugera G W, and Tholiso N 2014 *Toxicol. Environ. Chem.* **96** (9), pp. 1294-1309.
[13] Gogoi P K and Baruah R 2008 *Indian J. Chem. Technol.* **15** (5) pp 500-503.
[14] Kim J H, Lee C G, Park J A, Kang J K, Choi N C, and Kim SB 2013 *Desalin. Water Treat.* **51** pp 3408-3416.
[15] Ho Y S 2006 *Water Research* **40** pp 119-125.
[16] Ho Y S and McKay G 1999 *Process Biochem.* **34** pp 451-465.
[17] Chaturvedi A K, Pathak A K, and Singh V N 1988 *Applied Clay Sci.* **3** pp 337-346.
[18] Gupta N, Gupta V, Singh A P, ang Singh R P 2014 *Bonfring Int. J. Industrial and Management Sci.* **4** (2).
[19] Tripathy S S, Bersillon J L, and Gopal K 2006 *Separation Purification Technol.* **50** pp 310-317.
[20] Abe I, Iwasaki S, Tokimoto T, Kawasaki N, Nakamura T, and Tanada S 2004 *J. Colloid and Interface Sci.* **275** pp 35-39.

IOP Conf. Series: Materials Science and Engineering **821** (2020) 012010 doi:10.1088/1757-899X/821/1/012010

Biomass Gasification of Sengon Sawdust to Produce Gas Fuel

Maryudi[1], A Aktawan[1, *], Sunardi[2], K Indarsi[1], E S Handayani[1]

[1] Department of Chemical Engineering, Faculty of Industrial Technology, Universitas Ahmad Dahlan, Yogyakarta, Indonesia

[2] Department Electrical Engineering, Faculty of Industrial Technology, Universitas Ahmad Dahlan, Yogyakarta, Indonesia

*agus.aktawan@che.uad.ac.id

Abstract. Indonesia is a country with extensive agricultural and plantation areas that produce abundant biomass, some of which are rice husks, coconut shells, corn cobs, and wood waste. Wood waste itself has a very large number of species, for example, Sengon sawdust, Teakwood, Mahogany wood, Lamtoro wood and so on. Sengon wood is usually used as material for craftsmen and furniture, but with advanced technology, it can be done using sengon sawdust to produce syngas with the gasification process. The gasification process uses a gasifier tool. In this study, an independent variable was used in the form of sengon sawdust. While the dependent variable that will be obtained is in the form of syngas output, gasification temperature and time needed to produce syngas. The results showed that the more bait sengon sawdust, the more syngas was produced and the longer the time needed. Based on syngas analysis, CO results were obtained 17.235%; CH_4 3.446%; H_2 5.089%. It can be concluded that sengon wood powder waste is one of the biomass that can be used as a gasification raw material to meet energy needs.

1. Introduction

Biomass is a renewable fuel and generally comes from living things (non-fossil) in which is stored energy or in other definitions, biomass is the whole material that comes from living things, including living and dead organic matter, both above ground surface or below the surface of the land. Biomass is a photosynthetic product where the energy absorbed is used to convert carbon dioxide with water into carbon, hydrogen, and oxygen compounds. Biomass is easy to obtain, environmentally friendly and renewable [1]. In general, the potential for biomass energy comes from the waste of seven commodities from the forestry, plantation and agriculture sectors. The biggest potential of biomass waste is from forest wood waste, then followed by the waste of rice, corn, cassava, oil palm, and sugar cane. Overall, the potential for biomass waste energy in Indonesia is estimated at 49,807.43 MW. Of this total, installed capacity is only around 178 MW or 0.36% of the existing potential (Hendrison, 2003). Household waste such as waste cooking oil can be processed into biodiesel which can be used as liquid fuel [2]. the byproducts of processing waste cooking oil into biodiesel produce glycerol which can be processed into triacetin that can be used as a bio additive [3, 4]. Inorganic waste such as plastic can also be processed into liquid fuel through the pyrolysis process [5-7].

Biomass is considered an alternative source of renewable energy for fuels that makes it possible to reduce greenhouse gas emissions. Biomass can be farmed sustainably through a process of fixation cycles and CO_2 release, thereby reducing the problem of global warming. Energy can be obtained by burning biomass directly, it can also be by pyrolysis (without oxygen) or gasification (with limited oxygen) to produce liquid fuel or gas fuel [8-11]. Biomass gasification is one of the most promising technologies because of its ability to quickly convert large quantities and various types of biomass into gas or fuel. The process of biomass gasification is by converting it into a mixture of gases such as carbon monoxide, hydrogen, and hydrocarbons, along with with carbon dioxide and nitrogen [12].

Sengon is a member of the Fabaceae family and is one of the tree species that grows very fast. Its growth for 25 years can reach a height of 45 m with a trunk diameter reaching 100 cm. Given its rapid growth, sengon has been dubbed the miracle tree. At the age of 6 years, the sengon tree can produce 372 m^3 / ha of round wood [13].

The general characteristic of Sengon wood is the color of the terrace wood which is almost white or light brown where the sapwood is generally not different from the terrace wood, the texture is rather rough and evenly distributed, the direction of fibers is straight, wavy wide or combined, the impression of touch on the surface of the wood is rather slippery or slippery, specific gravity 0.33 (0.24-0.29) and strong class IV - V with shrinkage up to furnace 2.5% (radial) and 5.3% (tangential) [14].

According to Sulaiman [15] gasification is an alternative in the framework of energy-saving and diversification programs. Besides gasification will help overcome the problem of handling and utilizing waste from wood products. Gasification is the process of converting solid or liquid fuels to gas fuels without producing waste or solid carbon residues. Gasification is a very important form of conversion because it can be effectively utilized for the decentralization of power generation and the application of heat utilization. Gasification is also an energy conversion technology that can be used for a variety of biomass fuels [16]. Gasification is a process of converting solid fuel to combustible gas (CO, CH_4, and H_2) through the combustion process with a limited air supply (20% - 40% stoichiometric air) [17].

The gasifier is an instrument that can convert various solid or liquid materials such as biomass into gas fuels as shown in Figure 1. The gasifier is a reactor where various complex chemical and physical processes can occur such as drying, heating, pyrolysis, partial oxidation, and reduction. Through gasification carbonate, the solid material is broken down into basic materials such as CO, H_2, CO_2, H_2O, and CH_4. The gases produced can then be used directly for the combustion process or stored in gas cylinders [16].

According to Higman [18], the stages of the gasification process are: Drying, the water content in solid fuels is evaporated by heat absorbed from the oxidation process. This reaction is located at the top of the reactor and is the zone with the lowest temperature in the reactor that is below 150 °C. The drying process is very important so that ignition on the burner can occur faster and more stable. In this reaction, fuel containing water will be removed by evaporating and it takes about 2260 kJ of energy to carry out the process so that it takes up a lot of operating time. Pyrolysis or devolatilization is also called partial gasification. A series of physical and chemical processes occur during the pyrolysis process which starts slowly at $T < 350$ °C and occurs quickly at $T > 700$ °C. The composition of products arranged is a function of temperature, pressure, and gas composition during pyrolysis. The pyrolysis process starts at a temperature of about 230 °C, when components that are thermally unstable, such as lignin in biomass, break and evaporate together with other components. The vaporized liquid product contains tar and PAH (polyaromatic hydrocarbon). Pyrolysis products generally consist of three types, namely light gas (H_2, CO, CO_2, H_2O, and CH_4), tar, and charcoal. Reduction or gasification involves a series of endothermic reactions whose heat is supplied from the heat of the combustion reaction. The products produced in this process are combustible gases, such as H_2, CO, and CH_4. Oxidation or burning of charcoal is the most important reaction that occurs in the gasifier. This process provides all the heat energy needed for endothermic reactions. The oxygen supplied to the gasifier reacts with flammable substances. The results of these reactions are CO_2 and H_2O which are subsequently reduced when in contact with charcoal produced in pyrolysis.

IOP Conf. Series: Materials Science and Engineering **821** (2020) 012010 doi:10.1088/1757-899X/821/1/012010

2. Research Methodology

The research methodology consists of materials used in research, research tools, steps in data retrieval during the gasification process, and gas analysis of the results of gasification.

2.1. Research Materials

Materials used in this research are sengon sawdust obtained from the furniture industry in Kulonprogo.

2.2. Research Instruments

The research tool used in this study is shown in Figure 1 which consists of a gasification reactor equipped with a cyclone and filter, a blower, a high-temperature thermometer, and a scale.

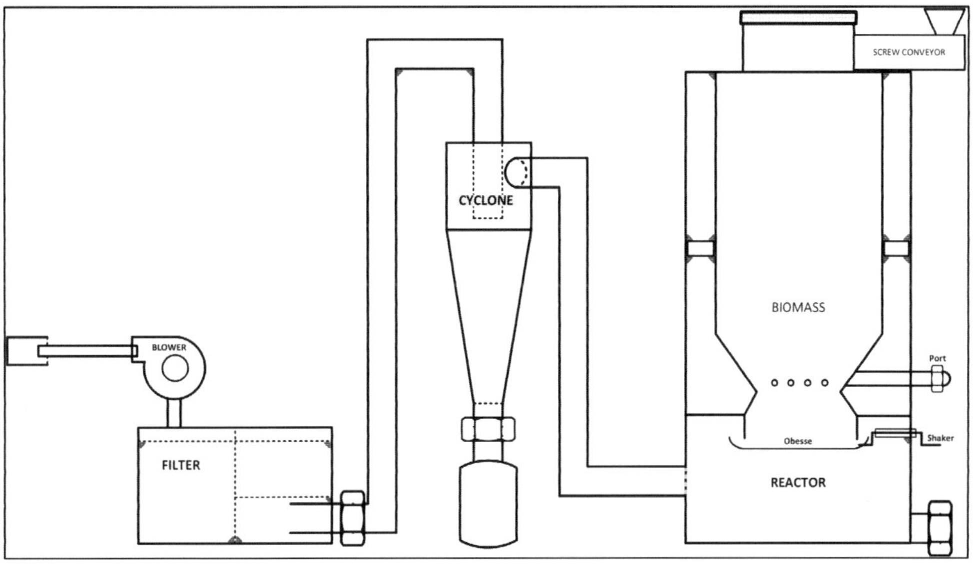

Figure 1. Unit of biomass gasification

This biomass gasification research began by preparing biomass raw materials by drying the raw materials under the hot sun. Weighing biomass using a digital balance sheet. Biomass gasification begins by introducing biomass feedstock into the gasifier. Turn on the blower to drain air into the gasifier. Light the biomass through the combustion pit. Record temperature changes every minute during the gasification process.

The final gasification results in the form of gas coming out of the blower, tar which is accommodated under the cyclone and ash mixed with charcoal retained under the gasification reactor. A mixture of ash and charcoal is taken, then weighed and recorded by weight. The gas (syngas) produced is partially taken with a vacuum and put into a sample tube for gas content testing.

3. Result and Discussion

The results of the research are correlation data on the change in the amount of raw material and the amount of syngas. The other results are in the form of correlation data of changes in the amount of raw material with syngas flame time. Gasification syngas analyzed carbon monoxide, methane, and hydrogen content with gas chromatography-mass spectrophotometry.

3.1. Effect of material weight on the syngas weight produced

The cellulose content in sengon sawdust is around 56.90%. First, the material is put into the gasifier and starts burning. The results of combustion in the form of gas and byproducts in the form of tar. If the gas is able to burn when ignited by fire, then this shows that there is syngas in the gasification gas output. To find out the amount of syngas produced, we need some data taken in the study, namely the weight of the residual ash and the weight of tar. From the two data, we will get the effect of bait weight on the syngas produced as shown in Figure 2.

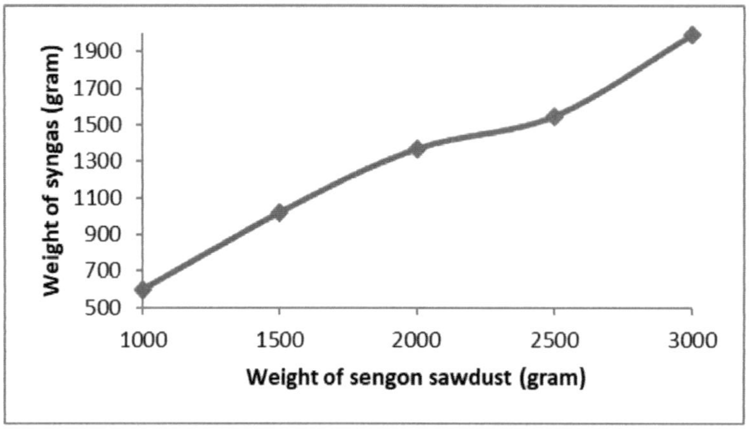

Figure 2. Correlation between the weight of sengon sawdust and weight of syngas produced

From Figure 2 about the effect of sengon wood waste weight on syngas weight with a variable weight of 1000 grams; 1500 grams; 2000 grams; 2500 grams; and 3000 grams produce syngas weight, which is 598,815 grams; 1019,106 grams; 1368,468 grams; 1548,138 grams; and 1991,731 grams. From Figure 6 it can be concluded that the more sengon sawdust waste that is used, the more syngas is produced.

3.2. *Effect of material weight on long time the syngas is on*

In addition to the residual ash weight of the combustion and the weight of tar, other data taken during this study were the time when the combustible gases were able to ignite when ignited by fire. From this time data, it will be known the effect of sengon wood waste feed on the time which shows there is syngas in the combustion gas as shown in Figure 3.

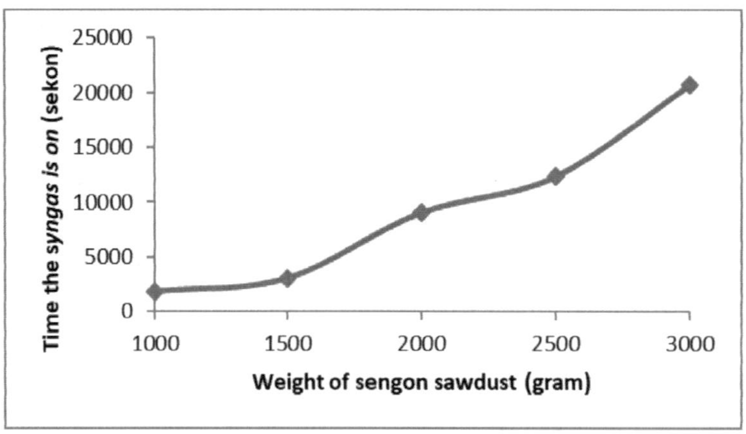

Figure 3. Correlation between the weight of sengon sawdust and time the syngas is on

From Figure 3 about the effect of sengon wood weight on syngas removal time with a variable weight of 1000 grams; 1500 grams; 2000 grams; 2500 grams; and 3000 grams produce syngas with a long time in a row, which is 1800 s; 3000 s; 9000 s; 12300 s; and 20700 s. From these data, graphs can be made of sengon wood powder weight vs. syngas weight. Figure 7 shows that the weight of sengon powder is directly proportional to the time of gas release. The more sengon sawdust used, the longer the time spent on gas.

3.3. Syngas composition

Gasification is a thermochemical conversion process that converts solid fuels to gas in a combustion reactor with an amount of oxidant that is less than needed. Some parameters used to determine the composition of the gas are CO, CH_4, and H_2. To find out the concentration of these compounds in the syngas from sengon wood powder gasification, sampling was carried out in a 10 ml vacuum tube and gas chromatography analysis. Syngas composition shows in Figure 4.

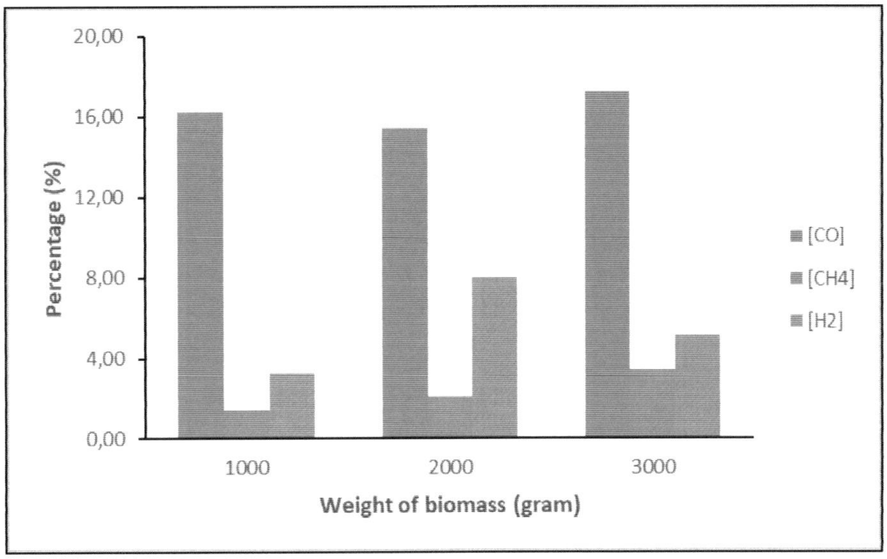

Figure 4. Syngas composition

4. Conclusions

From the research that we have done regarding gasification of biomass from sengon wood waste with the variable in the form of feed weight, it can be concluded that sengon wood waste is one of the biomass that can be used as raw material for gasification in meeting energy needs. This is indicated by the emergence of fire when the combustion gas is able to ignite when triggered by the fire.

The weight of the feed of sengon wood waste which is an independent variable in this study affects other variables in the form of a dependent variable. Some dependent variables in this study are the weight of the syngas produced and the flame time of the syngas. Table 4 shows that the more sengon sawdust waste feed, the more syngas is produced during the gasification process. Figure 7 shows that the more sengon sawdust waste feed, the longer the syngas removal time. In the analysis of syngas 1000 gr samples, it is known that 10 ml of syngas contains 16.214% CO; 1,382% CH_4; 3.244% H_2. For the analysis of 2000 gr syngas samples, it is known that 10 ml of syngas contains 15.402% CO; 2,058% CH_4; 8.002% H_2. While for the analysis of syngas 3000 gr samples it is known that 10 ml of syngas contains 17.235% CO; 3.446% CH_4; 5,089% H_2.

References

[1] M. Lohbeck, L. Poorter, M. Martínez-Ramos, and F. Bongers, "Biomass is the main driver of changes in ecosystem process rates during tropical forest succession," *Ecology*, vol. 96, no. 5, pp. 1242–1252, 2015.

[2] A. Aktawan and Z. Mufrodi, "Small-scale production of biodiesel through transesterification process of waste or used cooking oil Small-scale production of biodiesel through transesterification process of waste or used cooking oil," in *IOP Conference Series: Earth and Environmental Science PAPER*, 2019, pp. 1–7.

[3] A. Aktawan and Z. Mufrodi, "Pembuatan Bioaditif Triasetin dengan Katalis Padat Silica Alumina Abstrak," *J. Bahan Alam Terbarukan*, vol. 5, no. 2, pp. 92–100, 2016.

[4] Z. Mufrodi, E. Astuti, A. Aktawan, and S. Purwono, "The Effect of Recycle Stream on The Selectivity and Yield of the Formation of Triacetin from Glycerol the Effect of Recycle Stream on The Selectivity and Yield of the Formation of Triacetin from Glycerol," in *IOP Conference Series: Earth and Environmental Science*, 2018, pp. 1–6.

[5] Maryudi, S. Salamah, and A. Aktawan, "Product distribution of pyrolysis of polystyrene foam waste using catalyst of natural zeolite and nickel / silica Product distribution of pyrolysis of polystyrene foam waste using catalyst of natural zeolite and nickel / silica," in *IOP Conf. Series: Earth and Environmental Science*, 2018, pp. 1–7.

[6] S. Salamah and A. Aktawan, "Pemurnian Hasil Cair Pirolisis sampah plastik pembungkus dengan Distilasi Batch," vol. 3, no. 1, pp. 31–34, 2016.

[7] I. N. Azizah, N. P. Sari, and Maryudi, "Pengaruh Panjang Kolom Distilasi Bahan Isian Terhadap Hasil Produk Cair Sampah Plastik," *Chem. J. Tek. Kim.*, vol. 2, no. 1, pp. 21–27, 2015.

[8] I. Mufandi, W. Treedet, P. Singbua, and R. Suntivarakon, "Produksi Bio-Oil dari Rumput Gajah dengan Fast Pyrolysis menggunakan Circulating Fluidized Bed Reactor (CFBr) dengan Kapasitas 45 Kg/H," *Chem. J. Tek. Kim.*, vol. 5, no. 2, pp. 1–5, 2018.

[9] Maryudi, "Torrefaction and Densification Characteristics of Empty Fruit Bunch of Oil Palm," *Chem. J. Tek. Kim.*, vol. 1, no. 2, pp. 77–84, 2014.

[10] Maryudi, A. Aktawan, and S. Salamah, "Conversion of Biomass of Bagasse to Syngas through Downdraft Gasification," *J. Bahan Alam Terbarukan*, vol. 7, no. 1, pp. 28–33, 2018.

[11] A. Nurwidayati, P. Ayu, D. Ardiyati, and A. Aktawan, "Gasifikasi Biomassa Serbuk Gergaji Kayu Mahoni (Swietenia Mahagoni) untuk Menghasilkan Bahan Bakar Gas sebagai Sumber Energi Terbarukan," *Chem. J. Tek. Kim.*, vol. 5, no. 2, pp. 31–36, 2018.

[12] S. E. Agustina, "Biomass Potential as Renewable Energy Resources in Agriculture," in *Proceedings of International Seminar on Advanced Agricultural Engineering and Farm Work Operation*, 2004, pp. 47–55.

[13] B. S. Atmosuseno, *Budidaya, Kegunaan, dan Prospek Sengon*. Penebar Swadaya, 1994.

[14] A. Martawijaya, I. Kartasujana, Y. I. Mandang, S. A. Prawira, and K. Kadir, *Atlas Kayu Indonesia jilid II*. Bogor: Badan Litbang Kehutanan Indonesia, 1989.

[15] S. Sulaiman, "Biomassa Gasifikasi. Makalah Pelatihan Biomass Gasifikasi.," Surabaya, 2009.

[16] B. Purwantana, "Pengembangan Gasifier untuk Gasifikasi Limbah Padat Pati Aren (Arenga Pinnata Wurmb)," vol. 27, no. 3, pp. 130–136, 2007.

[17] G. Rinovianto, "Karakteristik Gasfifikasi Pada Updraft Double Gas Outlet Gasifier Menggunakan Bahan Bakar Kayu Karet," Depok, 2012.

[18] C. Higman and M. van der Burgt, "Thermodynamics of Gasification," in *Gasification (Second Edition)*, 2003, pp. 11–31.

IOP Conf. Series: Materials Science and Engineering **821** (2020) 012011 doi:10.1088/1757-899X/821/1/012011

Biodiesel from microalgae *Nannochloropsis oculata* and *Tetraselmis chuii* by sonication technique and K_2CO_3 catalyst

H Purnama[1], Harland[1], and N Hidayati[1*]

[1] Department of Chemical Engineering, Faculty of Engineering, Universitas Muhammadiyah Surakarta, Surakarta, Indonesia

*Email: nur.hidayati@ums.ac.id

Abstract. The way out of the energy crisis is to divert national energy needs with the use of new and renewable energy. Bioenergy such as biodiesel is the expected one of those. Algae which has the potential to be used as raw material for making biodiesel is microalgae because it is easier in its cultivation. The microalgae used in this study were *Nannochloropsis oculata* and *Tetraselmis chuii*. Microalgae that have been harvested in the pasta form are extracted to obtain the oil then processed into biodiesel with an ultrasonic device. The transesterification reaction ran at 65°C with a 1% K_2CO_3 catalyst. The variables applied were the ratio of oil: methanol (1:5; 1:10; 1:15) and reaction time (1 and 3 hours). The biodiesel produced was then tested for acid numbers, %FFA, density, viscosity, and GC-MS analysis. The amount of *N. oculata* and *T. chuii* obtained from harvesting was 591.75 and 460.11 grams, respectively. The yield of oil extracted from *N. oculata* and *T. chuii* was 46.47% and 39.31%. Biodiesel yield obtained from *N. oculata* between 72.19 - 74.33% with the highest yield for N3 sample, meanwhile in *T. chuii* between 69 - 72.95% with the highest gain for T3 sample. In general, biodiesel samples meet national standard SNI for Biodiesel in density, viscosity and acid number. The methyl ester content of the GC-MS test in the best biodiesel samples is N1 sample at 68.4% and T6 sample at 79.49%.

1. Introduction

Increased energy consumption and decreased availability of energy sources that have occurred in the last decade, has the potential to cause fuel scarcity. The solution to overcome the energy crisis is to divert national energy needs by utilizing the availability of new and renewable energy (NRE). Bioenergy such as biodiesel is one of the expected NREs [1, 2]. The obstacle always faced is the availability of sustainable and non-edible raw materials. One of the raw materials that fit the category is algae [3-5].

Algae that have the potential to be used as raw materials are microalgae because they are easier to culture. Microalgae are believed to be able to provide a stock of energy-producing raw materials that are easily applied in life without the need for extensive productive land.

Research for the selection and culture of microalgae has been widely carried out by experts with the aim of producing oil as a substitute for fossil fuels with low operating costs. Much of the oil produced is selected from the type of microalgae and its culture process. Factors that influence it are light intensity, temperature, nutrients, CO_2, pH in the cultivation process and the method of harvesting it [6]. There are several microalgae harvesting techniques, i.e. centrifugation, filtration and flocculation techniques. Centrifugation and filtration techniques have the disadvantage of needing special conditions and expensive costs [7]. The technique of using flocculants is to add what works to influence the pH value in the culture media so that microalgae can settle. Then microalgae deposition in culture is taken by filtering it using a small porous cloth.

The process of extracting and producing biodiesel is chosen using the sonication method for energy efficiency purposes. Ultrasonic waves are able to accelerate the reaction because the effects have a cavitation effect, heat effect, and structural effects that make the penetration of solutes into cells and homogenization can occur more quickly.

In the process of making biodiesel, the catalyst that is generally used is alkaline catalyst. The use of K_2CO_3 base catalyst [8, 9] for biodiesel synthesis is better than KOH. Alkaline catalysts such as NaOH and KOH are very sensitive to water content and free fatty acids. The presence of water can cause saponification of esters to form soap and cause emulsion formation. This situation is detrimental because the catalyst consumption increases and there are difficulties in the process of biodiesel refining.

Microalgae selected as alternative raw materials for biodiesel are *Nannochloropsis oculata* [10] and *Tetraselmis chuii* [11]. A few investigations of various microalgae states that the amount of lipid content in microalgae relies upon the strategy for development including restricting supplement admission, control of temperature and term of light power in the development medium [4, 12]. Growth varies on numerous components and can be improved for temperature, daylight usage, pH control, liquid mechanics, and more [13]. Both microalgae was selected based on data of lipid content, lipid production ability in the culture period and the types of microalgae species that already exist purely for culture. To find out the maximum lipid potential that can be obtained as an alternative raw material for producing biodiesel, it is necessary to conduct research to study it in the producing of biodiesel by sonication and K_2CO_3 catalyst assistance. From this research it is expected that both microalgae with all kinds of limited treatment given will produce optimum biodiesel with good quality.

2. Methodology

The systematic of this research consists of two stages; the first is preparation of raw materials including microalgae cultivation, harvesting and extraction processes; and the second includes the transesterification process, the analysis of biodiesel selected process results, and their comparison with biodiesel quality standards (*Standard Nasional Indonesia*, SNI).

The first microalgae cultivation was carried out in 10L gallons filled with 1L microalgae seedlings and 8L mineral water with 15g of mixed raw salt per 1L of water to form salinity conditions similar to brackish water. Nutrients given at 1 ml/L media culture in the form of a mixture of urea fertilizer 40 mg/L, ZA fertilizer 30 mg/L, and TSP fertilizer 10 mg/L. Provision of nutrients is done every 2 days. Culture media was replaced at harvest time on the day 14th of 18L media in a sterile 19L gallon by taking 2L microalgae on 9L media and the remaining 7L harvested. After the harvest period, the culture media was put into sterile 10L gallons by taking 1L of microalgae on the remaining 18L and 17L media were harvested.

Light intake of culture media uses 13w Philips LED in paired two pcs per gallon. Light exposure given under conditions of 18 hours of light and 6 hours of dark to obtain optimal growth results. For CO_2 intake, a simple series of CO_2 gas producers is made consisting of two 1.5L soda bottles. Each bottle contains a solution of baking soda and citric acid. The ratio of the solution is 1g of citric acid with 2 ml of water and 1g of baking soda with a ml of water. The solution of the citric acid in the pump by squeezing the body of the bottle slowly until the liquid of the citric acid rises through the hose and drips slightly into the bottle containing baking soda will react to produce pressurized CO_2 gas. O_2 intake obtained from aerators that flow through the hose. Tubes in the air supply and CO_2 are fitted with bubble stones to reduce the gas bubbles that come out and enlarge their solutes into the culture media.

Harvesting is carried out by the method of flocculation on every 14 days by adding alum.

Each addition of alum was done manually by stirring for 5 minutes and allowed to stand for 15 minutes until the top culture water deposition appeared clear. The culture media, which has formed the precipitate, are then copied into a bucket that has been covered with a cotton cloth to filter out the microalgae deposits. The microalgae that are held on the cotton cloth is then obtained in the form of a paste. The extraction process takes place in an ultrasonic device [14] with temperatures between 60–65°C and two hours long process and n-hexane solvent with the use of 10 ml for every gram of microalgae. Then the extraction results are transferred into the beaker by using filter paper to separate the pulp from the extracted solution. The extraction product is separated by the solvent using a rotary evaporator at a temperature of 70°C with a speed of 60-90 rpm for three hours.

The transesterification process is conducted in an ultrasonic bath with operating conditions of 65 °C and reaction time of 1 and 3 hours. Firstly, a solution of methanol is made in a beaker with the addition of a 1% by weight mass of catalyst K_2CO_3. Microalgae oil biomass is put into erlenmeyer as much as 15 ml per sample then placed in an ultrasonic device [14]. After that, the mixture of methanol catalyst is put into the erlenmeyer. After the process is complete, the product liquid is transferred into a measuring cup and then into the separating funnel. The liquid is allowed to stand for 24 hours so that the biodiesel solution forms a layer consisting of catalyst, glycerol and biodiesel. Furthermore, biodiesel will be analyzed for acid number, density, viscosity, %yield and biodiesel composition using GC-MS. The results of the analysis obtained were compared with biodiesel standards.

3. Result and discussion

In the preliminary research, the total amount of biomass obtained from the harvesting of poles for 4 months was obtained, and the yield (%) of oil extracted from the microalgae using n-hexane as follows (Table 1).

Table 1. The oil yield (%) of microalgae extraction

Sample	Mass (g)	Mass Product (g)	% yield
N. oculata	591.750	275.003	46.470
T. chuii	460.110	180.893	39.310

Generally, the percentage of extraction yield of *N. oculata* microalgae is higher than that of *T. chuii* microalgae. This is consistent with the productivity and lipid content with lipid content in *Nannochlorpsis sp.* between 12 - 53% of its dry weight and biomass productivity is around 0.17 - 1.43 g/L/day. While the lipid content in *Tetraselmis sp.* between 8.5 - 23% of the dry weight of its time and its biomass productivity is around 0.12 - 0.32 g/L/day. Medipally et.al reported from several sources that microalgae having different content of oil is able to produce oil with yield 30 – 70% [15].

Biodiesel transesterification process with ultrasonic devices and K_2CO_3 catalyst have been tested for physical and chemical properties based on the treatment according to the research design with the following information:

- type of oil = produced from *N. oculata* microalgae and *T. chuii* microalgae
- reaction time = 1 and 3 hours
- oil: methanol = 1: 5, 1:10, and 1:15

Figure 1 shows the yield of biodiesel produced from *N. oculata* and *T. chuii* microalgae oil by varying oil-methanol ratio and reaction time. It is observed that the biodiesel yield reached in the range between 76 and 81%. Generally, oil produced from *N. oculata* microalgae generates more biodiesel than that of from *T. chuii* microalgae. In addition, time of reaction affects negatively to yield. The change in reaction time from 1 to 3 hours decreases biodiesel yield, while the mole ratio of oil to methanol has a slight effect on yield. Maximum biodiesel yield is achieved by oil produced from *N.*

oculata microalgae with an hour reaction and 1:10 oil-methanol ratio. Nair et.al reported that algae biodiesel yield resulted over K_2CO_3/ZnO was about 77% using 7% catalyst loading [16], while another also observed that it was over CuO/zeolite with 3% loading was around 36.8% [17]. It was also reported that conversion of biodiesel from microalga was about 87% over BaO/CaO-ZnO catalyst [18]. A comparison of reaction conditions of those results is detailed in Table 2.

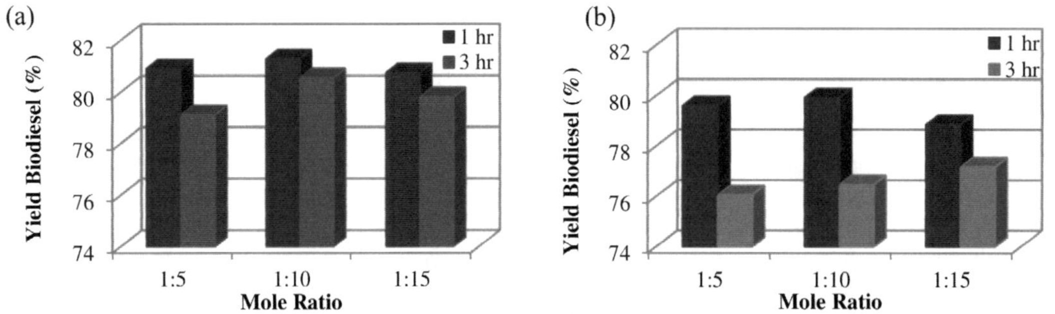

Figure 1. Biodiesel yield (%) varying oil methanol ratio and reaction time and produced from (a) *N. oculata* and (b) *T. chuii* microalgae oil.

Table 2. The comparison of microalgae biodiesel performance over different catalyst

Microalgae	Catalyst	Catalyst Loading (%)	Yield (%)	Reaction time (hours)	Oil-methanol ratio (mole:mole)	Temperature (°C)	Ref..
Spirulina	K_2CO_3/ZnO	7	77	n.a.	1:9	80	[16]
Chlorella vulgaris	CuO/zeolite	3	36.8	4	1:4	60	[17]
S. platensis spirulina	BaO/CaO-ZnO	3	87	2	1:15	65	[18]
N. oculata	K_2CO_3	1	81	1	1:10	65	This research
T. chuii	K_2CO_3	1	79	1	1:10	65	

Figure 2 presents the density measured at 40°C of microalgae biodiesel, and Table 3 is the quality standard of biodiesel based on SNI 7182:2015 in selected parameters according to this research. It is shown that almost all samples fall into the permissible quality range of biodiesel quality standards (0.85 - 0.89 g/ml). Only two samples do not meet the requirement in the biodiesel quality standard.

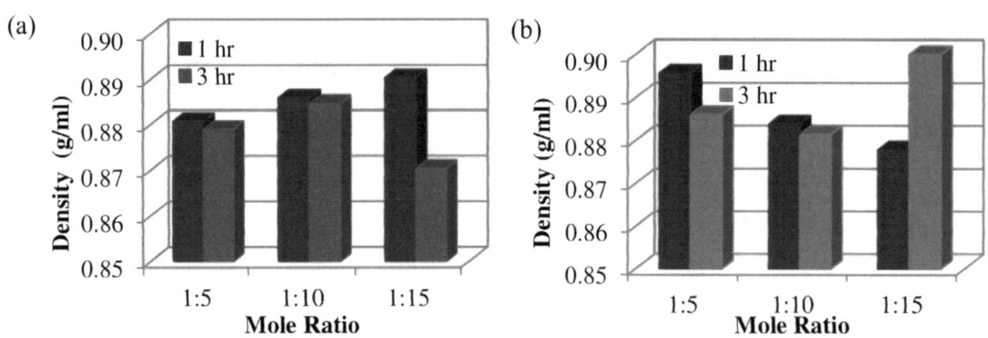

Figure 2. Density of microalgae biodiesel varying oil-methanol ratio and reaction time and produced from (a) *N. oculata* and (b) *T. chuii* microalgae oil.

Table 3. The quality standard of biodiesel according SNI 7182:2015

Parameter	Value
Density (40°C), gr/ml	0.85 – 0.89
Kinematic Viscosity (40°C), mm^2/s	2.3 – 6
Acid Number (max.), mg KOH/g	0.5

he kinematic viscosity data were measured at 40°C and Figure 3 exhibits this result. All samples are in the range of 3.20 – 3.25 mm^2/s and meet the standard of SNI (Table 3). Figure 4 is bar charts showing the acid number all samples. It seems that only 4 of 12 samples of microalgae biodiesel prepared with 3 hours reaction fits the criteria according to SNI. The rest of samples shows higher acid number than the standard. This might be due to the microalgae oil contains high free fatty acid. It can be solved that the oil should be carried out esterification firstly before transesterification reaction.

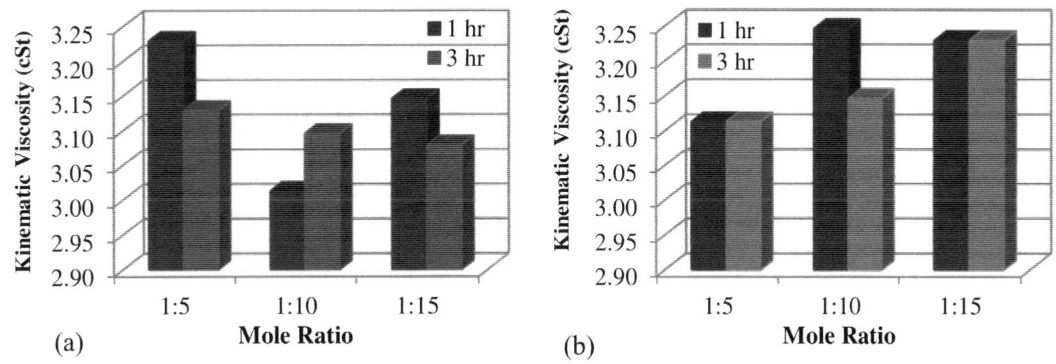

Figure 3. Viscosity of microalgae biodiesel varying oil-methanol ratio and reaction time and produced from (a) N. oculata and (b) T. chuii microalgae oil.

In order to determine the composition of biodiesel, GC-MS measurement was performed using the Shimadzu GCMS-QP2010S instrument. Figure 5 shows the chromatograms of microalgae biodiesel sample. Table 4 lists the biodiesel components and composite ion. It shows that there are 32 peaks detected but only 8 peaks indicates methyl esters. The peak area containing methyl esters is about 79.5%. This result is in accordance with the calculation of acid numbers that shows the acid number is still high. The two major components are methyl palmitate and methyl 9-hexadecenoate.

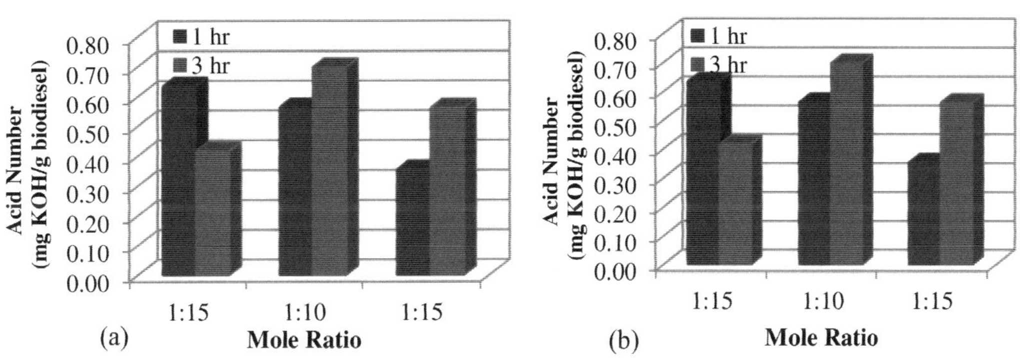

Figure 4. Viscosity of microalgae biodiesel varying oil-methanol ratio and reaction time and produced from (a) *N. oculata* and (b) *T. chuii* microalgae oil

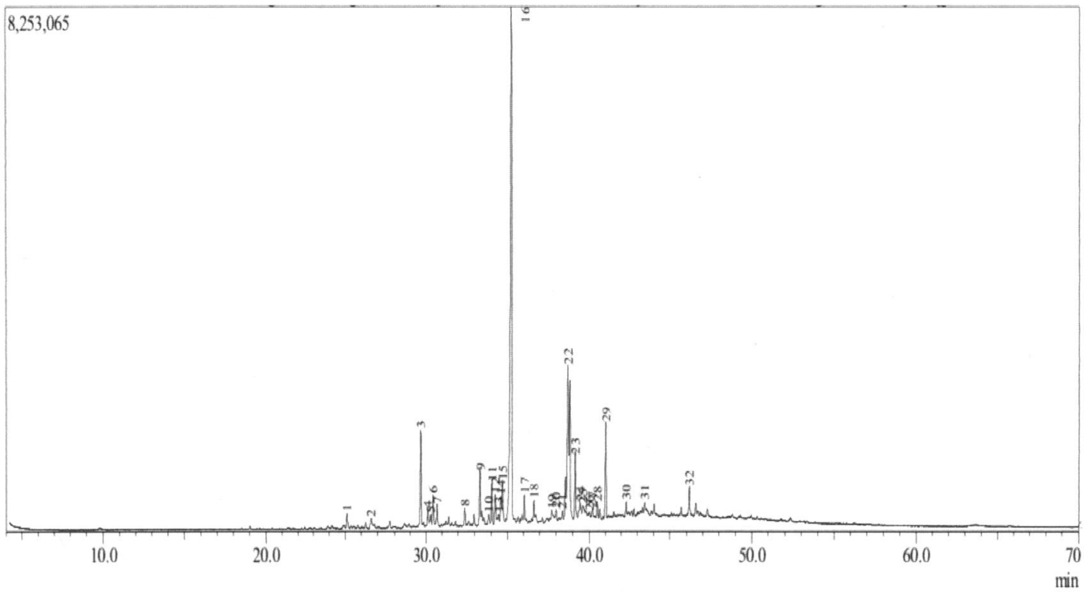

Figure 5. GC-MS chromatogram of sample N1

Table 4. Component and normalized composition of microalgae biodiesel

Component	Composition (wt.%)
methyl myristate	1.37
methyl tetradecanoate	0.68
methyl palmitate	61.73
methyl palmitoleate	4.89
methyl 13-decanoate	0.16
methyl 11-octadecenoate	0.92
methyl heptanoate	0.24
methyl 3-methoxy tetra decanoate	0.94
methyl 7,10,13-hexadecatrienoate	0.70
methyl linoleate	1.98
methyl 9-hexadecenoate	16.79
methyl stearate	1.70
methyl eicosanoate	6.08
methyl behenate	1.81

4. Conclusion

Lipid extraction has been successfully isolated from *N. oculata* and *T. chuii* microalgae and then synthesized into biodiesel using K_2CO_3 heterogeneous catalyst. A high microalgae biodiesel yield was observed during a one-hour reaction. Density and kinematic viscosity are mostly suitable with standard according to the SNI, however, the acid value is higher than the expectation. It can be concluded that *N. oculata* and *T. chuii* microalgae are potential resources for biodiesel production, although developments in the microalgae biodiesel processing field should be taken into account.

References

[1] Dickinson S, Mientus M, Frey D, Arsalon A-H, Ozturk S, Shaikh F, Sengupta D, and Mahmoud M El-H 2017 A review of biodiesel production from microalgae *Clean Technologies and Environmental Policy* vol **19** no 3.

[2] Mata T M, Martins A A, and Caetano N S 2010 Microalgae for biodiesel production and other applications: A review *Renewable and Sustainable Energy Reviews*.

[3] Shuba E S and Kifle D 2018 Microalgae to biofuels: 'Promising' alternative and renewable energy, review *Renewable and Sustainable Energy Reviews* vol **81**.

[4] Paul Abishek M, Patel J, and Prem Rajan A 2014 Algae oil: A sustainable renewable fuel of future *Biotechnol. Res. Int.* vol **2014** pp 1–8.

[5] Hadiyanto, Widayat, and Kumoro A C 2012 Potency of microalgae as biodiesel source in Indonesia *Int. J. Renew. Energy Dev.* vol **1** no 1 pp 23–7.

[6] Islam M A, Heimann K, and Brown R J 2017 Microalgae biodiesel: Current status and future needs for engine performance and emissions *Renewable and Sustainable Energy Reviews* vol **79**.

[7] Gouveia L and Gouveia L 2011 Microalgae as a feedstock for biofuels in *Microalgae as a Feedstock for Biofuels*.

[8] Taslim, Iriany, Bani O, Parinduri S Z D M, and Ningsih P R W 2018 Biodiesel production from rice bran oil by transesterification using heterogeneous catalyst natural zeolite modified with K2CO3 *IOP Conf. Ser. Mater. Sci. Eng.* vol **309** no 1.

[9] Liu X, Xiong X, Liu C, Liu D, Wu A, Hu Q, and Liu C 2010 Preparation of biodiesel by transesterification of rapeseed oil with methanol using solid base catalyst calcined K2CO3/γ-Al2O3 *JAOCS J. Am. Oil Chem. Soc.* vol **87** no 7 pp 817–23.

[10] Zhu Y and Dunford N T 2013 Growth and biomass characteristics of picochlorum oklahomensis and nannochloropsis oculata *JAOCS J. Am. Oil Chem. Soc.* vol **90** no 6 pp 841–49.

[11] Kawaroe M, Prartono T, Sunuddin A, and Saputra D 2016 Marine microalgae tetraselmis suecica as flocculant agent of bio-flocculation method *HAYATI J. Biosci.* vol **23** no 2 pp 62–6.

[12] Van Wagenen J, Miller T W, Hobbs S, Hook P, Crowe B, and Huesemann M 2012 Effects of light and temperature on fatty acid production in Nannochloropsis salina *Energies* vol **5** no 3 pp 731–40.

[13] Perner-Nochta I and Posten C 2007 Simulations of light intensity variation in photobioreactors *J. Biotechnol.*vol **131** no 3 pp 276–85.

[14] Kwangdinata R, Raya I, and Zakir M 2014 Production of biodiesel from lipid of phytoplankton Chaetoceros calcitrans through ultrasonic method *Sci. World J.* vol **2014**.

[15] Medipally S R, Yusoff F, Banerjee S, and Shariff M 2015 Microalgae as sustainable renewable energy feedstock for biofuel production *Biomed Res. Int.*vol **2015** pp 1–13.

[16] Nair J, Murthy Y V V S, Ramesh M, and Edeira G 2019 Synthesis of algae biodiesel using K2 CO3/ Zno heterogeneous base catalyst and its characterisation *RASAYAN J. Chem.* vol **12** no. 4 pp 1757–65.

[17] Dianursanti D, Delaamira M, Bismo S, and Muharam Y 2017 Effect of reaction temperature on biodiesel production from Chlorella vulgaris using CuO/ zeolite as heterogeneous catalyst *IOP Conf. Ser. Environ. Sci.*vol 55 pp 1–7.

[18] Singh R, Kumar A, and Sharma Y C 2017 Biodiesel production from microalgal oil using Barium–Calcium–zinc *Energy and Fuels* vol **33** pp 1175–84.

IOP Conf. Series: Materials Science and Engineering **821** (2020) 012012 doi:10.1088/1757-899X/821/1/012012

Financial feasibility study of batching plant investment on Sigli – Banda Aceh highway construction project

H A. Rani[1,*], R S Bonenehu[2], and M H Mubarak[3]

[1, 2] Department of Civil Engineering, Faculty of Engineering, Universitas Muhammadiyah Aceh, Banda Aceh, Aceh, Indonesia
[3] Department of Business Economy, Faculty of Economy, Universitas Muhammadiyah Sumatera Utara

*Email: hafnidar.ar@unmuha.ac.id

Abstract. The construction of the Sigli - Banda Aceh highway project as long as 74 km stretches across two Regencies namely Pidie and Aceh Besar, generating the large demand for ready mix concrete. It is required the investors in providing ready mix concrete service and wishing adequate mutual exchange from each capital invested. The problem in this study is to investigate how the feasibility and business opportunity of a batching plant investment from its financial aspects. This study aims to determine the financial feasibility and business opportunities on batching plant investment of ready mix concrete at Sigli - Banda Aceh highway project. In this research data was collected and processed to obtain cash flow and Depreciation. The method used to determine the financial feasibility is by obtaining its investment valuation analysis from data processing. Those are Net Present Value (NPV), Annual Equivalent (AE), Internal Rate of Return (IRR), Benefit-cost ratio (BCR) and Payback Period (PBP). From the calculation, it is obtained that the profit obtained from NPV is IDR 12,145,111,515.27, the profit obtained from AE is IDR 3,818,104,138.07, IRR is above 40%, BCR is ≥ 1 as 1,10, PBP is IDR 15,104,717,354.22 in the second year or equal to the investment period. It can be concluded that batching plant investment on Sigli – Banda Aceh highway construction project is feasible to be constructed and become a business opportunity for the investors.

1. Introduction

A business feasibility study has many advantages that according to [1], it is useful for the investor, creditor, company management, government, community, economic development goal. There are several phases of study that must be carried out consisting of finding project ideas, study phase, project evaluation phase, sorting feasible proposal, planning phase of business project implementation, and business project implementation phase. There are several aspects according to [1-4] requiring to be studied in this business feasibility project including market aspect, technical aspect, and financial aspect. The financial aspect is the aspect used to assess the overall company financial [2]. According to [5] and [6], the assessment of the financial aspect consist of investment budget and production cost.

In accordance with its commercial characteristics, investors want adequate mutual exchange from any capital that has been invested. According to [7] and [8], investment in a company is the use of sources

IOP Conf. Series: Materials Science and Engineering **821** (2020) 012012 doi:10.1088/1757-899X/821/1/012012

expected to achieve the return profitable in the future. It means that investing the amount of capital to run the business or project in the future expecting the capital return profitable in the future. According to [9], investment is the capital funding for one or more assets owned in the long period to receive the profit in the future.

The construction of Trans-Sumatra Highway is carried out to provide the benefit especially economic growth in Sumatera, including increasing employments, increasing tax income contribution of the state, facilitating the development of surrounding areas, and improving property value and potential housing construction [10, 11]. The construction of Sigli – Banda Aceh Highway is one of 8 (eight) road sections of Trans-Sumatra Highway which started at the end of 2018. This construction provides many opportunities for entrepreneurs and investors in the construction sector both local and national levels. The requirements of ready mix concrete in this project which reaches 941,272 m^3 make investors interested in investing in ready-mix concrete production.

As stated earlier, investment decision making in a project or business sector such as the construction of the batching plant for ready mix concrete must be based on a financial feasibility study. Thus this research aims to find out financial feasibility and business opportunity in batching plant investment for ready mix concrete. The review is limiting only to the batching plant planning.

2. Data and methodology

2.1. Location
The construction of the batching plant is located on Indrapuri Sub District, at Km 49 of highway project location which is near the quarry. Data related to the factory and ready mix concrete products is obtained by observing the factory and direct interviews with the owner of the ready mix concrete batching plant in Aceh Besar District consisting of PT. Lhoknga Beton. Data related to the highway project is obtained directly from the project owner namely the project location map, the data needs of the ready mix concrete volume, work method information, project implementation time, ready mix concrete specifications.

2.2. Cash flow
According to [12], cash flow is cash in and cash out calculated for the current period. The cash flow period is determined at various time intervals starting from daily, weekly, monthly, quarterly, and annually depending on the data aggregation level required. The cost is cash out and the benefit is cash in.

2.3. Depreciation
According to [13], depreciation is an asset value decrease due to the time period. The asset is divided into the current asset and fixed asset, but the asset of depreciation is only fixed asset including building, machine/equipment, fleet, and others. The depreciation in this study is calculated by using the Straight of line Depreciation Method as the following equation:

$$SLD = \frac{1}{N}(I - S) \tag{1}$$

Description:

SLD = The total of depreciation per annual.
I = The investment (initial asset value)
S = The residual asset value of final production period
N = The asset period of depreciation

2.4. Data processing

Data processing is performed on the financial/financial aspects as follows:

a. Determine the amount of sales value or revenue target, obtained from the multiplication between the volume of ready mix concrete with the unit price of sales for each type of concrete.
b. Determine the amount of the cost of production, obtained from the multiplication of the volume of ready mix concrete with the cost of production for each type of concrete.
c. Determine the amount of operational and maintenance costs during the investment period.
d. Determine the investment value plan and the fund.
e. Determine a bank loan return plan for investment costs.
f. Depreciation calculation using Equation (1).
g. Develop the plan of cash in and cash out.

2.5. Investment assessment

Investment is a long period capital fund realized that the investment will be followed by other periodic expenditures required to be prepared. The expenditures consist of operational cost, maintenance cost, and other costs that cannot be avoided [14].

The investment opportunity management is developed by using the method in analyzing the project. The method is investment criteria. An investment criterion is used in determining a project proposal after evaluation. All criteria use a comparison or relation between all benefits and all costs both in quality and quantity [15, 16]. Investment assessment analysis is carried out on five assessment criteria including Net Present Value (NPV), Annual Equivalent (AE), Internal Rate of Return (IRR), Benefit-cost ratio (BCR) and Payback Period (PBP).

2.6. Net Present Value (NPV) method

NPV is a method used in calculating the net value at present. The present assumption explains the initial time of calculation in accordance with the evaluation time carried out in the annual period of zero (0) of the investment flow cash calculation [13, 17, 18]. The cash flow of benefit calculation is Present Worth of Benefit (PWB), while the cash-out calculation is Present Worth of Cost (PWC), so NPV is the calculation of PWB minus PWC (PWB – PWC).

If: NPV > 0 meaning that investment will be profitable (feasible).
NPV < 0 meaning that investment will not be profitable (unfeasible)

2.7. Annual Equivalent (AE) method

The concept of the AE Method is opposite to NPV Method. The cash flow of the NPV method is in the present position. Meanwhile, the cash flow of the AE Method is distributed equally in each period of total investment both cash in and cash out [13].

The equal distribution output of cash in produces mean benefits per annual called Equivalent Uniform Annual of Benefit (EUAB). While equal distribution output of cash out is called Equivalent Uniform Annual of Cost (EUAC). EUAB is reduced by EUAC called Annual Equivalent (AE).

If: AE > 0 meaning that investment will be profitable (feasible).
AE < 0 meaning that investment will not be profitable (unfeasible)

2.8. Benefit-cost ratio (BCR) method

BCR is one of the methods often used in the initial evaluation phases of investment planning or as additional analysis invalidating evaluation inputs carried out using other methods. BCR method emphasizes the comparison value between benefit aspect received and cost and lost aspects assumed of the investment [13, 19].

If: BCR $\geq$ 1 meaning that investment will be profitable (feasible).
BCR < 1 meaning that investment will not be profitable (unfeasible)

2.9. Internal Rate of Return (IRR) method
Different from the previous method in which the equivalent value of cash flow is found out by using interest rates as the main determining factor, IRR Method will find out the interest rate when NPV is zero [13]. The simple logic explains how the cash flow ability can return the capital and how many obligations it must have complied with. This ability is IRR, while the obligations are Minimum Attractive Rate of Return [20, 21].

2.10. Payback Period (PBP) method
Payback Period Method aims to find out how long the period of the investment will be returned to break-even point condition. PBP Method is used by inserting interest factors in the calculation or named as a discounted payback period. To find out whether it is feasible or not, the return period (k) must lower than the investment period (n) [22].

3. Results and discussion

3.1. Analysis of market price
The market price targeted can be observed from the projection of benefits planning (selling price). The annual benefit is received from the multiplication between the design volume and the selling price of each concrete quality qualification. So it is received that the benefit design of 2019 is IDR 94,574,079,285.71 and 2020 is IDR 47,747,359,642.85. Total benefit is IDR 140,480,158,928.57.

3.2. Production cost
The planning of production cost is received from the multiplication between the volume and the basic unit price of each concrete quality qualification. So the multiplication of each concrete quality is received that is the production cost of 2019 is IDR 80,157,807,392.85 and 2020 is IDR 40,355,095,696.42. The total production cost is IDR 119,408,135,089.28.

3.3. Operational and maintenance costs
Operational and maintenance costs become the routine costs expended monthly during the production period. In this study, operational and maintenance costs of batching plant of ready mix concrete calculated are head of batching plant, production manager, financial administration, operator of batching plant, operator of the loader, operator of the laboratory, mechanic, workers, security, water, and electricity. The total monthly expenditure is IDR 57,500,000.00. So the operational cost required in 2019 is IDR 575,000,000.00 and in 2020 is IDR 690,000,000.00, the initial license cost is IDR 100,000,000.00. The maintenance cost required in 2019 and 2020 are IDR 160,000,000.00 of each.

3.4. Investment cost
The planning of investment cost is IDR 15,185,150,000.00, and 30% of the investment cost is funded by its own capital. While 70% is funded from bank loan capital using effective interest loan type as 7% per annual or 0.58% per month and the return period is 22 months.

The equipment types invested are 1 unit batching plant, 8 units mixer truck of $8m^3$, 959 hours of concrete pump, 1 unit wheel loader, 1 unit weighing scale, and 1 unit strength test. While the building types are land lease for 3 years, 1 unit temporary office, 1 unit laboratory, 1 unit employee mess, and 1

IOP Conf. Series: Materials Science and Engineering **821** (2020) 012012 doi:10.1088/1757-899X/821/1/012012

unit temporary fence. The total capital cost is IDR 4,555,545,000.00, and total bank capital is IDR 10,629,605,000.00 from total investment requirement as IDR 15,185,150,000.00.

3.5. Investment loan return

The investment capital is funded from bank loan using an effective interest loan system which is 70% of the total capital required; the bank loan return consists of the main loan return and loan interest return. The loan interest is 7% per annual or 0.58% per month and the interest return is 22 months. The monthly main repayment is received from the output of the main loan divided by the return period. The amount of interest repayment is calculated by interest percentage per annual or per month multiplied by the remaining main loan in the current period.

The total interest of 2019 is IDR 493,229,777.46, and in 2020 is IDR 219,839,557.95. The total interest is IDR 713,069,335.41. The loan for investment cost is IDR 10,629,605,000.00, and if it uses the bank loan, the investment cost becomes IDR 11,342,674,335.42.

3.6. Depreciation

The depreciation as the set value of batching plant equipment and other supporting tools invested by new purchases must be calculated by the production period completion or investment period. The investment value of equipment purchases using the Straight of line Depreciation (SLD) Method is achieved by the equipment depreciation in the second year of investment period which is IDR 490,133,333.33.

3.7. Cash flow

The components of cash in and cash out are managed in one cash flow. It is summarized that investment cost and license cost planned in 2019 which is IDR 15,285,150,000.00 become cash out that must be expended at the beginning of the year when starting the development of batching plant. At the end of 2019, it is planned the cash in of business income which is IDR 94,574,079,285.71, and cash out of production cost, operational and maintenance costs, and loan interest return which is IDR 81,386,037,170.31. Meanwhile, at the end of 2020, it is planned cash in of business income which is IDR 47,747,359,642.86, and the equipment remaining value recorded as cash in which is IDR 10,971,733,333.33. The production cost, operational and maintenance costs, and loan interest return in 2020 is IDR 41,424,935,254.38 recorded as cash out.

3.8. The assessment of investment criteria

The benefit in 2019 and 2020 and the remaining value of the investment in the second year are recorded as the Present Worth of Benefit (PWB). While investment cost, license cost, production cost, operational and maintenance costs, and loan interest return are recorded as Present Worth of Cost (PWC). The calculation of NPV on two years production of investment period by using the coefficient value of 7% interest table is received PWB value as IDR 139,674,190,305.83 and PWC value as IDR 127,589,167,706.04. NPV value is IDR 12,085,022,599.79 > 0. It means that the investment is feasible.

The benefit in 2019 and 2020 and the remaining value of the investment in the second year which is recorded as Equivalent Uniform Annual of Benefit (EUAB) is achieved IDR 77,255,113,488.98. While investment cost, license cost, production cost, operational and maintenance costs, and loan interest return which are recorded as Equivalent Uniform Annual of Benefit (EUAC) is achieved IDR 73,773,521,509.40. The calculation of AE on two years' production of investment period by using the coefficient value of 7% interest table is received AE value as IDR 3,481,591,979.57 > 0. It means that the investment is feasible.

Benefit-cost ratio is the concept emphasizing the comparison value between benefit aspect received and cost and lost aspects assumed of the investment. BCR Analysis is carried out in the present so PWB and PWC are used to analyze BCR. BCR value is the output of the PWB value divided by PWC value. BCR is

IOP Conf. Series: Materials Science and Engineering **821** (2020) 012012 doi:10.1088/1757-899X/821/1/012012

1.10. If BCR analysis is carried out manually, the values of EUAB and EUAC are used to find out BCR. BCR value is the output of EUAB value divided by EUAC value. BCR is 1.05 $\geq$ 1. It means that the investment is feasible.

Internal Rate of Return is the concept to find out equivalent value by using interest rates as the main determinant of the interest when NPV value is zero. Because the NPV value of 7% interest is still very much as IDR 12,085,022,599.79, it needs to find out NPV value which is close to or equal to zero. By using a maximum interest of 40%, it is received the NPV value as IDR 2,709,954,589.78. Thus NPV is higher than zero and it is feasible.

For PBP Method, cash in and cash out on two years' production of investment period by using the coefficient value of 7% interest table has not resulted in return value in the first year of the investment period. The return period is in the second year of the investment period as IDR 15,076,885,666.18.

4. Conclusion

The investment planning in the batching plant business of ready mix concrete must notice market price as the target of concrete produce selling during the investment period. The market price calculated in batching plant planning of plant ready mix concrete on Sigli – Banda Aceh Highway Project reviewed is IDR 140,480,158,928.57 during two year investment period planned.

The production cost of the market planned is IDR 119,408,135,089.28. The production cost intended is the material cost (fine aggregate, coarse aggregate, and cement), the operational cost of batching plant equipment including loader and transportation cost of ready mix to the location (consist of fuel and driver salary). While the operational cost during the investment period is IDR 1,585,000,000.00 and the maintenance cost is IDR 320,000,000.00.

The initial investment value required is IDR 15,185,150,000.00 received from its own capital as 30% and bank capital loan as 70%. The capital loan return is planned for two years by bank interest of 7% per annual. The bank interest return during the two years period is IDR 713,069,335.41. The depreciation of the batching plant and its supporting equipment during the investment period is IDR 490,133,333.33.

The investment feasibility study by using cash flow during the investment period planned is in 2019 and 2020. The output of the study concludes that investment criteria value is NPV as IDR 12,145,111,515.27 higher than zero meaning the investment is feasible; AE as IDR 3,481,591,979.57 which is higher than zero meaning the investment is feasible; BCR as 1.10 which is higher than 1 meaning the investment is feasible; IRR with NPV value which is zero is higher than 40% of interest meaning that the investment is very feasible; and PBP return period lower than investment period meaning that the investment is feasible.

The benefits of two years investment period are mostly in the remaining equipment value of the investment. Therefore, the equipment economic period is still long and it is possible to continue other businesses in the following years. Finally, it can be concluded that Sigli – Banda Aceh Highway Construction Project becomes the opportunity for the investors and entrepreneurs to invest in batching plant procurement to serve the requirement of ready mix concrete during the project construction.

Acknowledgments
This research was supported by Universitas Muhammadiyah Aceh. We thank our colleagues who provided insight and expertise that assisted this research.

References

[1]. Husein U 2007 *Studi Kelayakan Bisnis* (Edisi ke-3, Gramedia Pustaka Utama, Jakarta).
[2]. Kasmir dan Jakfar 2012 *Studi Kelayakan Bisnis* (Edisi Revisi, Kencana, Jakarta).
[3]. Subagyo A 2008 *Studi Kelayakan Teori dan Aplikasi* (Elex Media Komputindo, Jakarta).
[4]. Widardo I, Afifuddin M and Rani H A 2015 Feasibility Study Investasi Pembangunan Perumahan Honco Lingke Residence di Kota Banda Aceh *Jurnal Teknik Sipil Universitas Syiah Kuala* vol **4**(3) pp 251-60.
[5]. Sören W 2004 Engineering and economic laws of production *International Journal of Production Economics* vol **92**(3) pp 203-06.
[6]. Wenbo F 2018 Financial risk assessment and modeling of PPP based Indian highway infrastructure projects *Research in Transportation Economics* vol **70** pp 148-160.
[7]. Suratman 2001 *Studi Kelayakan Proyek, Teknik dan Prosedur Penyusunan Laporan*, (J&J Learning, Yogyakarta).
[8]. Jiang Y and Bhattacharyya D 2016 Chapter Six - Sustainable Engineering Economic and Profitability Analysis in *Sustainability in the Design, Synthesis, and Analysis of Chemical Engineering Processes* pp 141-67.
[9]. Sunariyah 2003 *Pengantar Pengetahuan Pasar Modal* (UPP AMP YKPN, Yogyakarta).
[10]. Peraturan Presiden (Perpres) Nomor 100 Tahun 2014.
[11]. Peraturan Presiden Nomor 117 Tahun 2015.
[12]. Kotler P 2003 *Marketing Management*, (11th Edition, New Jersey).
[13]. Giatman M 2006 *Ekonomi Teknik*, (Rajagrafindo Persada, Jakarta).
[14]. Robert J, Arnon K, Dominik H and Roman D 2019 Collecting data in the assessment of investments within production *Procedia CIRP* vol **79** pp 466-71.
[15]. Stefan Sievers S, Tim S, Marcel F, Franzen G, Schembecker and Christian B 2017 Fixed capital investment estimation for modular production plants *Chemical Engineering Science* vol **158** pp. 395-410.
[16]. Yeelyong N and Daejun C 2019 Methodology of exergy-based economic analysis incorporating safety investment cost for comparative evaluation in process plant design *Energy*.
[17]. Stefan C 2018 Moments and distribution of the net present value of a serial project *European Journal of Operational Research* vol **267**(3) pp 835-48.
[18]. Ondřej Z 2014 Net Present Value approach: Method for economic assessment of innovation projects *Procedia - Social and Behavioral Sciences* **156** pp 506-12.
[19]. Salvatorecafiso and Carmelo D'A 2016 Assessing the stochastic variability of the benefit-cost ratio in roadway safety management *Accident Analysis & Prevention* vol **93** pp 189-97.
[20]. Marco P and Emanuele B 2012 A note on the sensitivity analysis of the internal rate of return *Discussion International Journal of Production Economics* vol **135**(1) pp 526-29.
[21]. David Promislow S and David S 1996 Postulates for the internal rate of return of an investment project *Journal of Mathematical Economics* vol **26**(3) pp 325-61.
[22]. Deolalkar S P 2016 Chapter 8 - Capital Costs, Savings, and Payback Period in *Designing Green Cement Plants* pp 197-98.

IOP Conf. Series: Materials Science and Engineering **821** (2020) 012013 doi:10.1088/1757-899X/821/1/012013

The insulation properties of foam concrete with the use of foam-agent and fly-ash

E Rommel[1,*], L Prasetyo1[1], Y Rusdianto[1], R Karimah[1], Riyanto A.S[2], N Cahyo S.A[2]

[1] Department of Civil Engineering, Faculty of Engineering, Universitas Muhammadiyah Malang, Indonesia
[2] Post-graduate of Civil Engineering, Universitas Muhammadiyah Malang, Indonesia

*Email: erwin_r@umm.ac.id

Abstract. Foam concrete is an alternative wall material that can be used in tropical buildings because of its insulation properties, good thermal conductivity, and sound absorption coefficient when compared with the normal concrete. The percentage of foam agent used in its production affects density and compressive strength. Furthermore, the use of fly-ash also has an effect on its mechanical and insulation properties. In this study, 40 units of foam concrete specimen (200x200x20) mm were tested for thermal conductivity, and cylinder (dia-100x20) mm for sound absorption. The mixture used for the concrete includes cement/sand ratio of 1:2.75 with w/c 0.425 and addition of 20% fly-ash as a cementitious material and foaming agent (ratio fa/w 1:20) at 1%, 2%, 3% and 4% by weight of cement paste, respectively. The result showed that foam concrete with 20% fly-ash and 2% foaming agent or more has a density below 2000 kg/m3 with a maximum compressive strength of 13.24 MPa. The highest sound absorption coefficient was obtained to be $\alpha = 0.89$ at 20% fly-ash and 4% foaming agent. The thermal conductivity (k) of foam concrete was found to have increased using a 20% fly-ash and 1% foaming agent compared to the mixture without fly-ash at a value of 0,898. It was discovered that the use of fly-ash on foam concrete has the ability to improve thermal conductivity and sound absorption.

1. Introduction

Passive energy-saving houses and buildings made of thermally insulating materials have become popular in recent construction practices in order to address the demanding energy needs and reduce the consumption of hydrocarbon energy resources. Material technology is growing rapidly in the field of construction, including the use of wall materials in constructing buildings in the tropics. The use of lightweight, strong and insulating materials is now common because they provide comfort to residents and of those widely offered in tropical buildings is the foam concrete. It is a mixture of cement, water, and colophony foaming composition and a type of porous concrete with a highly workable low-density material with the ability to incorporate up to 50% entrained air. The concrete is generally self-leveling, self-compacting and can be pumped. It is made through blending, molding, maintaining, and hardening processes. The foamed concrete has the properties of porousness, lightweight, heat preservation, heat insulation, and sound absorption. Furthermore, fly ash foamed concrete can be produced from fly ash, lime, gypsum, and foaming composition. The material can be used to reserve heat in exterior protected construction. Foam concrete is formed from the surfactant reaction which produces air bubbles which are actually beneficial for heat transfer and sound absorption on the wall. The concrete can usually be applied to a wall panel, light concrete brick, ready mix, and special shape. However, concrete wall panels are a type of membrane panel with maximum sound-absorbing ability and curve peaks at a frequency of 250 Hz.

Several treatments have been conducted on foam concrete in order to improve insulation properties such as making pore cells closed on the concrete, giving high temperatures, and lightweight aggregate replacement but the foaming agent and fly-ash have never been applied as cementitious materials in its production.

Therefore, the aim of this study was to determine the effect of fly-ash on the insulating properties of foam concrete. The thermal conductivity was measured by heat transferability of the concrete based on the constituting materials, porosity, the temperature of the surroundings, and the direction of heat current (see Figure 1)

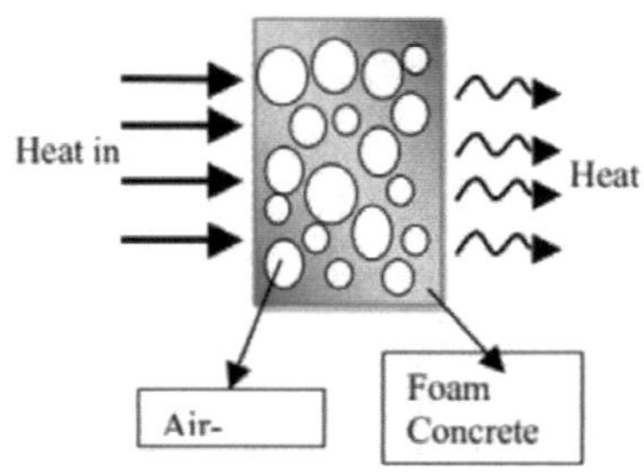

Figure 1. Diagram of air voids in foam concrete

2. Literature review

Foamed concrete, which is also known as a closed-cell structure, is one of the materials in the lightweight category. It has a low density between 300 kg/m^3 to 1600 kg/m^3 and thermal conductivity properties between 0.10 W/m.K to 0.66 W/m.K. It was discovered that its structure plays important roles in providing great thermal insulation performance [1].

A study had been conducted to quantify the thermal properties of lightweight foamed concrete (LFC) at high temperatures with a density of 600 to 1800 kg/m^3 [2]. This was aimed at obtaining data to predict fire resistance of LFC based systems. The concrete is considered to be a two-phase material with solid cement and air pores, therefore, there is an assumption that its thermal conductivity is a function of its porosity and pore size. The temperature on LFC was found to be increasing its heat conductivity. A concrete with a density of 650 kg/m^3 has lower thermal conductivity (average k = 0.3 W/m.°C) compared to the one with 1000 kg/m^3 (average k = 0.2 W/m.°C), however, both will experience a decrease in their thermal conductivity until the temperature approaches 200°C to 1000°C where it continues to increase. It is, however, important to state that these predictions are moderately affected by the air pore diameter, hence, the size of the pore should be determined with good accuracy. For LFC densities of 650 kg/m^3 and 1000 kg/m^3, the average pore diameter may be taken as 0.72 mm and 0.55 mm respectively.

The replacement of normal aggregate with lightweight ones effectively reduces the thermal conductivity depending on the raw materials used. The hollow glass bubble also causes a reduction in thermal conductivity but its impact is less pronounced compared to coarse aggregates. Studies conducted by Yun, Tae Sup et al [3] explained that the most effective insulating additive to achieve a thermal conductivity of ~1.25 W/m.K is through the addition of 20–30% of glass bubbles as well as the

IOP Conf. Series: Materials Science and Engineering **821** (2020) 012013 doi:10.1088/1757-899X/821/1/012013

replacement of cement by fly-ash and incorporation of a water-reducing agent on the thermal insulation and mechanical properties of porous concrete with a density of 600 kg/m3 [4]. It was found that porous concrete achieved optimum performance with a water-binder ratio of 0.32, a fly ash replacement percentage of 30% and the addition of 0.65% of the water-reducing agent. Under these optimal conditions, 28-day compressive strength up to 4.37 MPa, 12% rate of water absorption, and a thermal conductivity coefficient of 0.116 W/m.K were observed in the specimen obtained.

A state of the art research has been conducted on the insulation properties of foam concrete and it has been discovered that covering cell pores in concrete, providing high heat to reduce the coefficient of thermal conductivity, replacing aggregates with hollow glass bubbles, and cement with fly-ash and water reducing-agent have the ability of producing the lowest conductivity coefficient of 0.116 W/m.K

3. Materials and method

In this study, 40 units of (200x200x20) mm and (dia-100x20) mm concrete foam with a mixture of cement/sand at 1:2.75, w/c 0,425 and 20% fly-ash by weight of cement were used as specimen (Table 1). Foaming agent was also added with a ratio of 1:20 (fa/w) both without fly-ash and with fly-ash at 1%, 2%, 3% and 4% of the weight of the cement. Tests were conducted after 28 days of production with reference to ASTM C1044-98 (The Hot-Guarded-Plate Test) and ASTM C384-98 (Impedance Tube Method), for sound absorption test set-up (see Figure 2). The sound absorption test on the specimen was measured with a sound level meter device with respect to the frequency of the sources 250, 500, 1000 and 2000 Hz. In general, the steps of this research are shown in Figure 3.

Table 1. Mix-proportion of specimen

Materials	**Mix-controls**				**Mix-proportion**			
	I	**II**	**III**	**IV**	**I**	**II**	**III**	**IV**
cement	1	1	1	1	0.8	0.8	0.8	0.8
fly-ash/cement ratio	-	-	-	-	0.2	0.2	0.2	0.2
sand	2.75	2.75	2.75	2.75	2.75	2.75	2.75	2.75
water/cement ratio	0.425	0.425	0.425	0.425	0.425	0.425	0.425	0.425
Percentage of foaming-agent by cement weight (%)	1	2	3	4	1	2	3	4

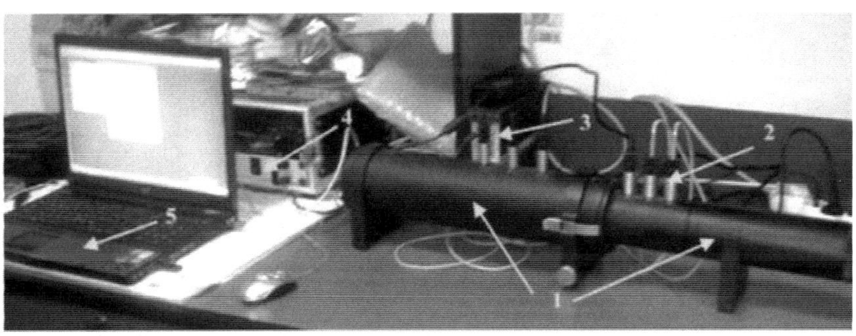

Figure Caption.
1. Impedance Tube
2. Specimen
3. Sound Level meter
4. Power amplifier
5. Data processing

Figure 2. The experimental stand used for the sound absorption coefficient measurement

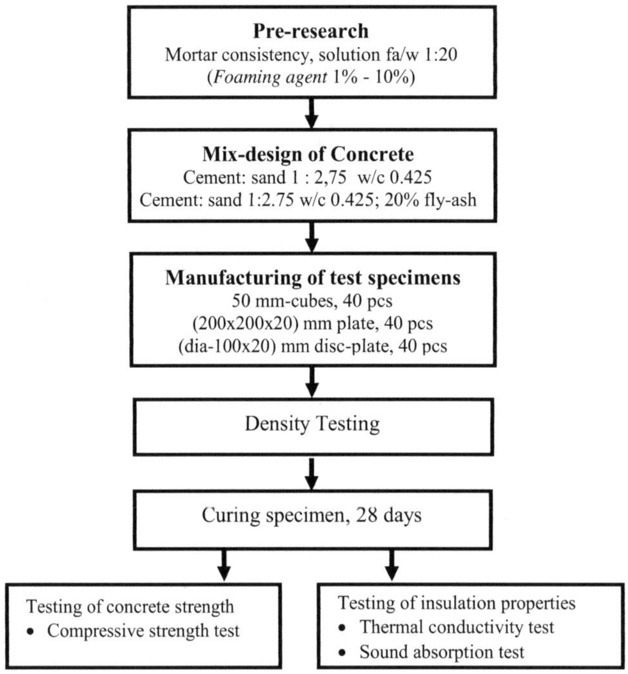

Figure 3. Stages involved in the experimental research

4. Results and discussions

4.1. Strength and density

The density of the foam concrete is influenced by the quantity of foaming agent and fly-ash used such that the more foaming agent produced, the smaller the density (Figure 4). While the use of a 20% fly-ash tends to produce larger density, concrete produced from a 4% foaming agent without using fly-ash has the lowest density of 1560 kg/m^3. It was found that using this material from the range of 2% of the weight of cement has the ability to produce concrete in the lightweight due to the value of its density < 2000 kg/m^3. Furthermore, the largest compressive strength was obtained with the use of a 2% foaming agent, 13 MPa without fly-ash and 13.24 MPa with a 20% fly-ash. The complete test results can be seen in Table 2.

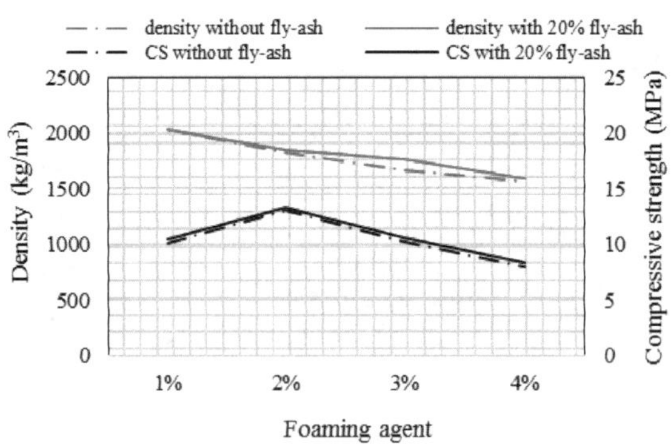

Figure 4. Strength and density of foam concrete

Table 2. Properties of foam concrete

Foam agent	Density (kg/m^3)		Compressive strength, 28 days (MPa)		Thermal conductivity (W/m.K)	
	No fly-ash	20% Fly-ash	No fly-ash	20% Fly-ash	No fly-ash	20% Fly-ash
1%	2031	2031	10.08	10.48	0.838	0.898
2%	1825	1853	13.00	13.24	0.898	0.896
3%	1666	1766	10.22	10.62	0.894	0.893
4%	1560	1585	7.96	8.36	0.885	0.886

Nambiar and Ramamurthy [5] reported that foam concrete density ranged from 400 - 1600 kg/m^3. Furthermore, Cement Ready Mix, one of the ready-mix companies in Mexico, classified foam concrete produced based on its density value, with the Highly Air-Entrained Concrete (HAC) type ranging between 1600 - 2000 kg/m^3. According to SNI 03-2847-2002, the density of lightweight concrete should not be more than 1900 kg/m^3.

The replacement of cement by an appropriate amount of fly ash has the ability to favorably reduce the overall cost of the end product. At an early curing age (7 days), the strength of the porous concrete decreased significantly with an increase in the fly ash content and at a later curing age (28 days), it can be enhanced when more fly ash is added. It has also been found that the compressive strength at 28-day curing would be up to 4.14 MPa when 30% of cement has been replaced by the fly ash [4].

A different case was observed in the research conducted by Kozlowski and Kadela [6] where the foam concrete without fly-ash seemed to show higher strength than the mixtures containing fly-ash. This can be attributed to the slowing down of the hardening process because of the presence of fly-ash. The compressive strength and modulus of elasticity were also slightly decreased by the addition of 5% fly ash.

4.2. *Thermal conductivity*

Materials with high thermal conductivity are good conductors, while those with low thermal conductivity are good heat-insulating materials. The addition of fly-ash to the foam concrete is expected to improve the insulating properties of the foam material. However, pores are formed due to the use of foam agents in the production process to increase the thermal conductivity value of the foam concrete while the use of fly-ash as cementitious only serves to increase adhesion between aggregate and bubbles formed in the concrete matrix.

The air bubbles formed by mixing a foaming agent in the foam-making process affect the heat delivery mechanism through the formation of pores. However, the dominant fly ash grains will fill the pores because their functions are inversely proportional to those of the foaming agents. This is important because the factors affecting the value of thermal conductivity include the type of aggregate and porosity (pore type, volume, spacing, direction, and moisture). The pore volume produced by the foaming agent is not too different from the one produced through the use of fly-ash to increase the adhesiveness between the surface of the aggregate.

In Figure 5, the use of fly-ash in concrete foam does not have any significant effect on the value of thermal conductivity. It was also discovered that increasing the percentage of the foaming agent above 2% does not also have any significant effect. Whereas in conventional concrete or mortar, the usage of 20% fly-ash as cementitious increases the value of thermal conductivity. Therefore, production of foam concrete with the use of foaming agent and 20% cementitious fly ash have better thermal conductivity properties compared with ordinary concrete (k = 1.448 W/m.K) and this is evident from the 0.898, 0.896, 0.893, and 0.886 W/m.K obtained on each use of 1%, 2%, 3%, and 4% foaming agent.

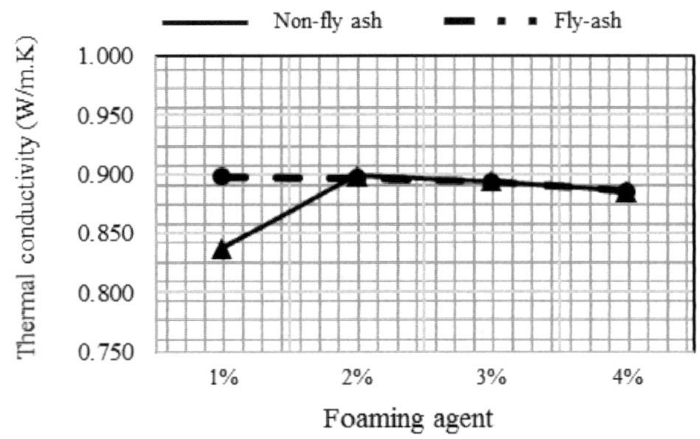

Figure 5. Thermal conductivity of foaming-agent

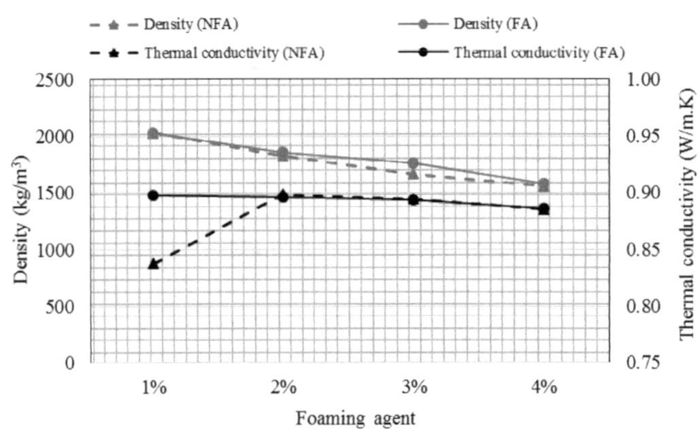

Figure 6. Density and thermal conductivity of foam concrete

Figure 6 shows the relationship between thermal conductivity and density based on the quantity of foaming agent and 20% cementitious fly-ash. It was found that the use of fly-ash does not significantly affect the density of the foam concrete material but very influential in lowering the thermal conductivity when a 1% foaming agent is used in the manufacturing process.

4.3. Sound absorption

The sound absorption coefficient of foam concrete specimens with average was measured at a peak frequency of 2000 Hz with an increase of 100 Hz intervals as shown in Table 3.

Table 3. Sound absorption of foam concrete

Specimen	Foaming agent	Sound absorption coefficient (α)				
		Frequency measurement (Hz)				Average
		250	500	1000	2000	
Foam concrete without fly-ash	1%	0.72	0.57	0.85	0.55	0.67
	2%	0.78	0.67	0.9	0.65	0.75

	3%	0.81	0.71	0.93	0.69	0.79
	4%	0.82	0.61	0.92	0.56	0.73
Foam concrete with 20% fly-ash	1%	0.71	0.63	0.73	0.67	0.69
	2%	0.84	0.82	0.84	0.79	0.82
	3%	0.85	0.82	0.9	0.85	0.86
	4%	0.88	0.83	0.94	0.89	0.89

Figure 7 shows the relationship between the sound absorption coefficient with the addition of the foaming agent and the cementitious fly-ash. The graph shows that the continuous use of a foaming agent increases the sound absorption coefficient. This can be attributed to the addition of more air bubbles to the concrete mix through the inclusion of the material in the production process because a large number of concrete pores formed results in better sound insulation for the concrete. This is similar to a study conducted on three treatments including coarse, smooth, and painted surfaces by Zulfian et al. [7]. The aim was to create a porous surface in order to increase sound absorption value and the coefficient obtained for the rough surface at the lowest frequency of 125 was 0.33, it can significantly increase in its sound absorption coefficient compared to others.

The use of fly-ash as a cementitious material produces better sound absorption compared to those without fly-ash. However, the rate is determined by the quantity of foaming agent used. The highest value of 0.89 was obtained with a 20% fly-ash and a 4% foaming agent.

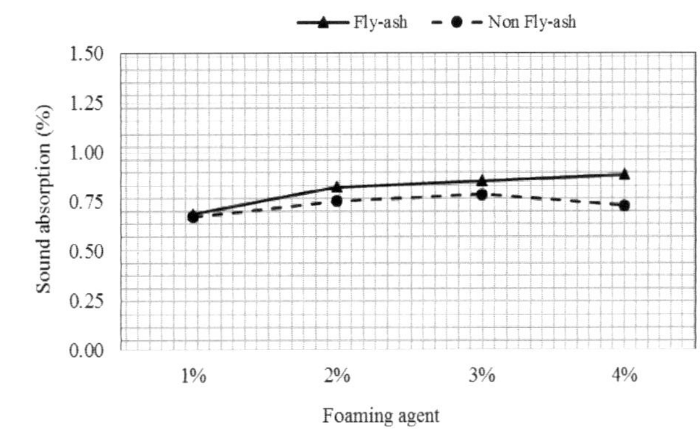

Figure 7. Sound absorption with a foaming agent and fly-ash

5. Conclusion

The use of 20% fly-ash as cementitious material and 2% foaming agent has the ability to produce foam concrete with densities lesser than 2000 kg/m^3 and optimum compressive strength of 13.24 MPa. It was also discovered that 20% fly-ash did not have any significant effect on thermal conductivity (k) of foam concrete, however, the value of k decreased with the addition of foaming agent with the lowest being 0.838 at 1% without fly-ash. Furthermore, sound absorption (α) increased with the use of a 20% fly-ash and peaked to $\alpha_{maximum}$ 0.89 through the use of a 4% foaming agent.

Acknowledgment

The authors would like to express their sincere gratitude to The Directorate Research and Community of Services, The Directorate General of Strengthening Research and Development, and The Ministry of Research, Technology, and Higher Education, Indonesia for financing this research.

References

[1]. Mohd Z, Nooraini, Ismail Abdul R and Ahmad Mujahid Ahmad Z 2009 Foamed concrete: potential application in thermal insulation *Malaysian Technical Universities Conference on Engineering and Technology* Pahang Malaysia.

[2]. Othuman, Azree Md and Wang Y C 2011 Elevated-temperature thermal properties of lightweight foamed concrete *Construction and Building Materials* vol **25** pp 705-16.

[3]. Yun, Tae S, Yeon Jong J, Tong-Seok H and Kwang-Soo Y 2013 Evaluation of thermal conductivity for thermally insulated concretes *Energy and Buildings* vol **61** pp 125-32.

[4]. Sun, Yafei, Peiwei G, Fei G, Haoran L, Lifang Z, and Hongwei L 2017 Thermal conductivity and mechanical properties of porous concrete materials *Materials Letters* vol **209** pp 349-52.

[5]. Nambiar, Kunhanandan E K and Ramamurthy K 2008 Fresh state characteristics of foam concrete *Journal of Materials in Civil Engineering* vol **20** Issue 2 pp 111-17.

[6]. Kozlowski, Marcin and Marta K 2018 Mechanical characterization of lightweight foamed concrete *Advances in Materials Science and Engineering* pp 1-8.

[7]. Zulfian, Abdullah and Fahridhal 2011 Penentuan nilai penyerapan suara pada panel dinding beton busa sebagai panel alternatif yang ramah lingkungan *Jurnal Rekayasa Kimia dan Lingkungan* vol **8** No 1 hal 41-46.

IOP Conf. Series: Materials Science and Engineering **821** (2020) 012014 doi:10.1088/1757-899X/821/1/012014

Steel column planning using the *American Institute of Steel Construction* (AISC) 2010 method

Wahyuni[1,*], K C S Zein, and Muchlis

Department of Civil Engineering, Faculty of Engineering, Universitas Muhammadiyah Aceh, Banda Aceh, Indonesia.

*Email: wahyuni@unmuha.ac.id

Abstract. The latest SNI rules for steel structure building has been published namely SNI 1729-2015 which refers to the AISC 2010. In AISC 2010, there is a new design method called Direct Analysis Method (DAM) as an alternative to the Effective Length Method (ELM). The aim of the study is to plan steel columns using the AISC 2010 method and calculate and also analyse the steel column capacity. In this study, ETABS computer analysis program that have been whole integrated includes the modelling, analysis and design were used, so as to produce safe and economical modelling, analysis and design. The results of trial and error showed that dimensions for the 1st and 2nd floor columns WF profile, 3rd and 4th floor columns used WF 250x250x9x14 profile, and the 5th floor column used WF 175x175x7.5x11 profile. The load that applied was only dead load and live load. ETABS analysis results obtained output in the column design in the form of axial force (P) 1688,971 kN, moment (Mux) 60,257 kN.m, moment (Muy) 23,009 kN.m. The calculation results of steel column design analysis using DAM method obtained the value of the ratio of the strengths of columns 1st and 2nd floors for 0.42, columns 3rd and 4th floors for 0.33 and the 5th floor for 0.15. The steel column material requirements for each 1st and 2nd floor are 352.66 kg, 2nd and 3rd floors are 271.35 kg, and the 5th floor is 150.75 kg. The design of steel columns by means of DAM results in a higher profile capacity, so that the profile designs used can be more economical.

1. Introduction

Column is a compressive structure element that plays an important role of a building, so the collapse in a column is a critical location that can cause the concerned floor collapse and also the total collapse of the entire structure. The latest SNI rules for steel structure building has been published namely SNI 1729-2015 which refers to the AISC 2010 [3]. There is a new design method called Direct Analysis Method (DAM) as an alternative to the Effective Length Method (ELM) [1]. The limitations in using first order with amplification factor or Effective Length Method make this method has begun to be abandoned by number of countries such as the British code, Australian code, Canadian code and Hong Kong code. AISC 2010 begun to adopt the mentioned codes, by using the Direct Analysis Method for second order analysis.

The aim of steel column design by using the AISC 2010 (American Institute of Steel Construction) method with Direct Analysis Method (DAM) is to see how much the value of column strength ratio in the DAM method, and also to analyse the capacity of the steel column by means of DAM. The planning of steel structure and designing steel column using the AISC 2010 method by means of DAM, for 5-storey building with a floor height of 4 m and floor height of two to fifth floor 3.75 m. The

IOP Conf. Series: Materials Science and Engineering **821** (2020) 012014 doi:10.1088/1757-899X/821/1/012014

building is planned to use Wide Flange (WF) steel with BJ-41 quality, which will be analysed using 2016 ETABS software.

The calculation results of analysis of column design using the AISC 2010 method using DAM obtained the value of the capacity ratio in the 1st and 2nd floor column is 0.42, 3rd and 4th floor column is 0.33, and the 5th floor column is 0.15. The steel column requirements for each 1st and 2nd floor are 352. 66 kg; the 2nd and 3rd floors are 271. 35 kg, and the 5th floor is 150. 75 kg. In a different study conducted by Salmon et al. 2009, the different between the Direct Analysis Method (DAM) and Effective Length Method (ELM) was obtained by a difference of 7%. This shows that DAM column design results in a higher profile capacity than ELM column design, so that the profile design used can be more economical.

2. Literature review

2.1. Steel

Generally, the steel construction contains more than 98% iron and usually less than 1% carbon. Although the actual chemical composition varies for desired properties, such as strength and resistance to corrosion, steel can also contain other alloying elements, such as silicon, magnesium, sulphur, phosphorous, copper, chrome, and nickel in various amounts [14]. The mechanical properties of structural steel planning are set as follows:

Modulus of elasticity (E) = 200.000 MPa
Shear modulus (G) = 80.000 MPa
Poisson's ratio (μ) = 0.3
Coefficient of expansion (á) = 12 x 10^{-6}/ ° C

2.2. AISC regulations for steel column design

The basic rule for compressive rods are explained in Chapter E3-1 until E3.4-Design of Members for Compression [1], Nominal compressive strength, *Pn,* is the smallest compressive strength for bending, torque bending and bend torque-bending conditions which depends on the shape of the column cross section as follow:

$$P_n = F_{cr} . A_g \tag{1}$$

where P_n is smallest compressive strength for bending, F_{cr} is critical buckling stress, and A_g is column area. The F_{cr} can be searched based on the column compressive strength curve which is a function of slenderness. The critical buckling stress curve of the column formula, specifically bending only, is:

For KL/r$\leq 4.71\sqrt{E}/Fy)$ or condition of column in-elastic buckling

$$Fa = \left(0.658\ \frac{fy}{fe}\right) . Fy \tag{2}$$

$/r \leq 4.71\sqrt{(E/Fy)}$ or condition of the columns with in-elastic buckling

$$F_{cr} = 0.877\ F_e \tag{3}$$

where *Fe* = strength of elastic critical buckling

$$F_e \frac{\pi^2 E}{(\frac{KL}{r})^2} \tag{4}$$

The three formulas above for the AISC (2005) or (2010) version are the same. So if it is only based on the critical buckling stress of the column, which also directly is the compressive strength of the column, singly or individually or isolated, between the old methods (*Effective Length Method*) and the new one (*Direct Analysis Method*), then there is no difference between it. Identical are the same.

2.3. *Direct Analysis Method (DAM)*

Steel structure planning that is generally slim, requires stability analysis. The result is influenced by the presence of imperfection (nonlinear geometry). Direct Analysis Method (DAM) as a new way of planning on steel structure that has incorporated modern principles in stability analysis [1].

2.4. *Stability design*

DAM is needed to overcome the limitations of elastic structure analysis, which cannot access the stability. By using DAM, the influence of structural loading can be sought by calculating the influence of the imperfection (nonlinear geometry) and inelastic (nonlinear material) that occurs.

2.5. *Notional load and inelastic attenuation*

Notional loads can also be used to anticipate the weakening of flexural stiffness, τb due to in-elastic conditions because of residual stresses. This strategy is suitable to simplify the calculation of DAM on the stem with a large compressive force αPr> 0.5Py, where the value τb< 1.0. Based on the AISC 2010 Chapter C2.3.3, then τb = 1.0 and additional notional load of:

$$Ni = 0.001Yi \tag{5}$$

Information:
Ni = notional load used a the level *i*, kips (N)
Yi = gravity load used at level *i* of the load combination, as appropriate, kips (N)

2.6. *Nominal cross section strength*

If a structural stability using the DAM method is used, then for calculating the nominal structural strength is sufficient to use the usual procedure as used in the ELM method, called Chapter E ~ I for nominal cross-section, or Chapter J ~ K for connection on AISC code (2005 or 2010), except the value of the K on the slenderness of the stem (KL/r) is taken constant at K = 1.

3. Research methods

3.1. *Structure analysis*

The structure analysis of this final project proposal uses the help of a computer program, the 2016 ETABS program.

3.2. *The planning stages*

The initial dimensions of the secondary structure planning will be used as input to plan the primary structure (main beams and columns) of ETABS 2016. The preparation stage, preparation was made to find the data and information that supported the structure design. The structural analysis of the structural model with the help of ETABS 2016 to determine the magnitude of the value of joint displacement, moment, shear force, and the compressive or tensile forces on the portal structure against the working load (external loads and gravity loads).

3.3. *Research object and location*

The research object of this study was the design of steel column structure, multi-story buildings (five floors), which were planned to be built in Banda Aceh.

3.4. *Building plan data*

Function : Office building
The main structure : Steel
Number of floor : Five floors
Steel material data
Column profile : Profile WF (BJ 41)

IOP Conf. Series: Materials Science and Engineering **821** (2020) 012014 doi:10.1088/1757-899X/821/1/012014

fy= 250 Mpa; fu= 410 Mpa
Beam profile : Profile WF (BJ 41) fy= 250 Mpa; fu= 410 Mpa
Modulus steel elasticity (E) : 200000 MPa
Reinforced concrete material data

a. Concrete quality (fc') : 25 Mpa
b. Quality of reinforcing steel : - BJTD 400 Mpa
   - BJTP 240 Mpa

3.5. *Secondary structure planning*

In this study, the secondary structure planning is only floor plates, roof plates and sliding connectors. For beam planning, using WF 150x150x7x10 profile without using joists.

3.6. *Floor and roof plates*

The thickness of the concrete slab according to equation 2.1 for the minimum slab thickness, the thickness of the roof plate is planned to be 100 mm, while the thickness of the floor plate is planned to be 120 mm.

3.7. *Beam planning*

The beam is planned to use the WF 150x150x7x10 profiles, with the data below:

H = 150 mm	Weight = 31,50 kg/m
B = 1500 mm	A = 4014 mm2
t1 = 7 mm	Ix = 1640 cm4
t2 = 10 mm	Iy = 563 cm4
r = 11 mm	

A beam bearing pure Mu's factored bending load must be planned in such a way that the relationship is always fulfilled:

r = 11 ı ım
A bear ı bearii
fulfille l:
$Mu \leq \phi Mn$

liding connector

$Qn \leq Ascxfu$
ber of studs nee

3.8. *Slide connecting planning*

For sliding connector used is the nail type with ds= 19 mm, Terms:
The number of studs needed is 20 studs for the entire span. So, the shear connector is installed with a distance of 20 m by 20 pieces for each span.

3.9. *Dead load*

The dead load calculated on a steel structure is the structure's own weight and the planning dead load. Dead load structure calculation is done automatically with ETABS program and plan dead load is done manually. In the cross section of the column assumed that at the end it is considered pinch-pins.

3.10. *Live load*

The live load used is in accordance with SNI 2013 loading regulations as follows:

Office floor = 2,40 KN/m^2
Corridor = 3,83 KN/m^2
Meeting room = 4,79 KN/m^2
Computer = 4,79 KN/m^2
Reading room = 2,87 KN/m^2

IOP Conf. Series: Materials Science and Engineering **821** (2020) 012014 doi:10.1088/1757-899X/821/1/012014

3.11. *Lead combinations*

Strong needs U to hold the dead load D in this plan, namely:

$$U = 1.4D + 1.4SW$$

Strong needs U to hold dead load D and live load L, namely:

$$U = 1.2D + 1.6L + 1{,}2SW$$

3.12. *Structural modelling*

Modelling structure is done to make it easier to analyse the overall structure to be analysed. Structural modelling includes structural drawings, planned material manufacturing, insertion of dead and live loads and a combination of loading.

3.13. *Column design*

Cross section planning of the column is calculated based on the rod bending moment, shear force and maximum factored load produced by a loading combination

4. Results and discussions

4.1. *Steel analysis and column design*

Using the force outputs from the ETABS program analysis a column will be planned for each floor. In this case, what is being examined is the design of steel columns using the AISC 2010 method by means of *Direct Analysis Method* (DAM) which will simplify the steel planning structure buildings. The maximum internal forces that occur for each floor can be seen in the Table 1.

Table 1 Recapitulation of the design force_column

Story	P	M2	M3	V2	V3
	kN	kN-m	kN-m	kN	kN
1	1688.971	17.032	7.010	5.500	12.971
2	1318.823	60.257	23.009	13.286	33.427
3	928.535	41.079	17.247	9.081	22.802
4	531.795	55.834	22.446	13.398	32.260
5	133.407	19.408	10.173	6.037	10.068

The analysis results of the largest axial force occurred in column C17 on the 1st floor, showed that the largest moment and shear force occurred in column C17 on the 2nd floor, differences in the internal force variations that occur due to the differences in the steel profile used and the load work are not the same in every floor. While the axial force is caused by the 1st floor bears all the burden of the entire floor, so that the largest axial force occurs on the 1st floor.

Based on the calculation results of steel column analysis using DAM, the ratio capacity on the 1st and 2nd floors are 0.4, the 3rd and 4th floors are 0.33, and the 5th floor is 0.15. The smaller the ratio value on each floor due to the smaller steel profile used for the above floor, this is done to produce a more economical plan in accordance with the burden of working on the planned structure.

4.2. *Steel column material requirements*

From the trial and error results obtained an economical profile size for 1st and 2nd floors profile is WF 300x300x10x15, 3rd and 4th used WF profile 250x250x9x14, and the 5th floor used WF profile 175x175x7,5x11. To determine the steel column requirements for each floor, the steel column material requirements calculation is as follows:

IOP Conf. Series: Materials Science and Engineering **821** (2020) 012014 doi:10.1088/1757-899X/821/1/012014

For 1st and 2nd floors

WF 300x300x10x15,

As = 119.80 cm^2

Steel weight = 119.80 x 10^{-4} x 3.75 x 7850 kg/m^3 =352.66 kg

For 3rd and 4th floors.

WF 250x250x9x14

As = 92.18 cm^2

Steel weight = 92.18 x 10^{-4} x 3.75 x 7850 kg/m^3 = 271.35 kg

For 5th floor

WF 175x175x7.5x11,

As = 51.21 cm^2

Steel weight = 51.21 x 10^{-4} x 3.75 x 7850 kg/m^3 =150.75 kg

With the data's previously calculated obtained steel material requirements, material requirements for each column are shown in Table 2 and Table 3:

Table 2. Recapitulation of the column material requirements

No	**Story** **Column Profile** Column material requirements	**As** Cm^2	**Steel** kg
1	WF 300x300x10x15	119.80	352.66
2	WF 300x300x10x15	119.80	352.66
3	WF 250x250x9x14	92.18	271.35
4	WF 250x250x9x14	92.18	271.35
5	WF 175x175x7,5x11	51.21	150.75
Total Volume			1398.77

Table 3. Maximum story displacement

Story	**Elevation**	**Location**	**X-Dir**	**Y-Dir**
	m		mm	mm
Story5	19	Top	0.062	0.072
Story4	15.25	Top	0.103	0.127
Story3	11.5	Top	0.075	0.1
Story2	7.75	Top	0.049	0.073
Story1	4	Top	0.048	0.039
Base	0	Top	0	0

Based on the table, graphs are made so that it can be seen more clearly displacement that occurs in combinations 1 for each floor

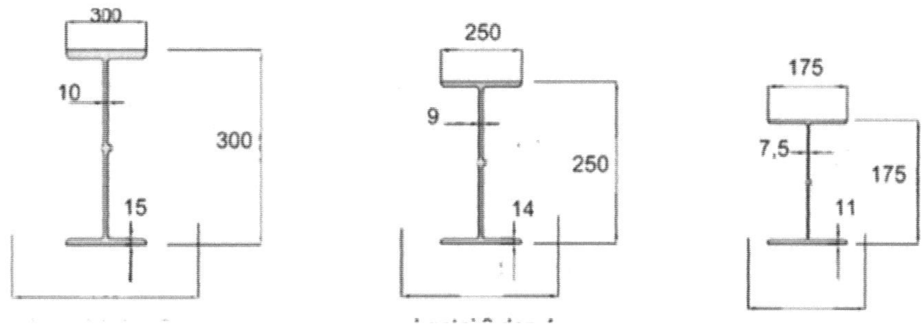

First and second floor Third and fourth floor Fifth floor

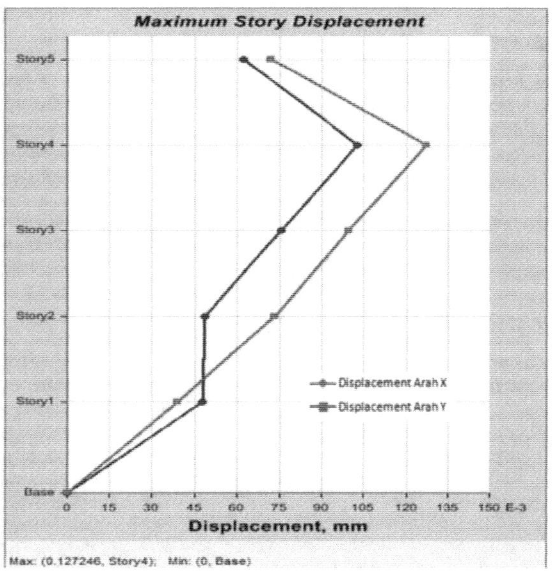

Figure 1. Maximum story displacement in combination 1

Based on the Figure 1, it can be seen that the X direction displacement tends to be smaller than the Y direction. The biggest displacement occurs in the Y direction of 0,127246 mm on the 4th floor as shown in Table 4.

Table 4. Maximum story displacement

Story	Elevation	Location	X-Dir	Y-Dir
	m		mm	mm
Story5	19	Top	0.082	1.372
Story4	15.25	Top	0.1	1.179
Story3	11.5	Top	0.075	0.553
Story2	7.75	Top	0.042	0.283
Story1	4	Top	0.04	0.16
Base	0	Top	0	0

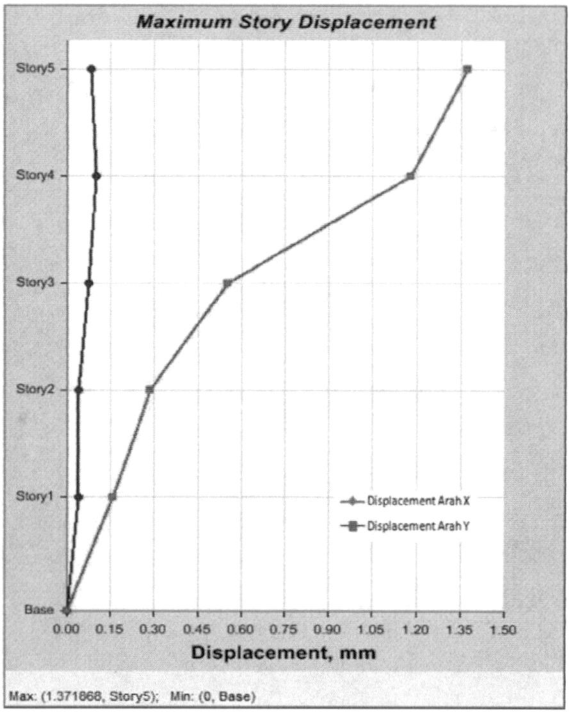

Figure 2. Maximum story displacement in combination 2

From the Figure 2 above can be seen that the largest displacement on the 5th floor is 1.7 mm Y direction. Based on the analysis results of steel column design using the AISC 2010 method using DAM, the capacity ratio in the 1st and 2nd floors are 0.4, the 3rd and 4th floors are 0.33 and the 5th floor column is 0.15. The steel column material requirements for each floor are as follows:

Floors 1-2 = 352.66 kg steel
Floors 3-4 = 271.35 kg steel
Floor 5 = 150.75 kg steel

5. Conclusion

In a different study conducted by Salmon et al. 2009, the difference between the *Direct Analysis Method* (DAM) and *Effective Length Method* (ELM) was obtained by a difference of 7%, so that the profile design of the DAM method was more efficient compared to ELM method.

In different studies for different buildings, when viewed from the forces that occur, the weight of the building and displacement, composite columns tend to resemble steel columns. However, composite columns have advantages compares to steel columns in terms of price. Based on the analysis results of steel column design using the AISC 2010 method using DAM, the capacity ratio in the 1st and 2nd floors are 0.4, the 3rd and 4th floors are 0.33 and the 5th floor column is 0.15

The steel column can be chosen for high-rise building plan, because steel has high strength compared to other materials. The existence ductility in steel makes steel construction cannot collapse suddenly when there is excessive load, this is very beneficial if the buildings experiences sudden large swings like earthquake loads.

Acknowledgements

This research was supported by University Muhammadiyah Aceh. We thank our colleagues who provided insight and expertise that assisted this research.

References

[1]. Anonim 1, 2010. "An American National Standard ANSI/AISC 360-10 : Load Specification for Structural Steel Buildings", American Institute of Steel Construction, Inc., Chicago, Illinois

[2]. Anonim 2, 2002. *Tata Cara Perencanaan PerhitunganStruktur Baja Untuk Bangunan Gedung*(SNI 03-17292002), BadanStandarisasiNasional, Jakarta

[3]. Anonim 3, 2015. *SpesifikasiUntukBangunanGedung Baja Struktural*(SNI 1729-2015), BadanStandarisasiNasional, Jakarta

[4]. Anonim 4, 2013. *PersyaratanBetonStrukturalUntukBangunanGedung*(SNI 2847-2013), BadanStandarisasiNasional, Jakarta

[5]. Anonim 5, 2013. *Beban Minimum UntukPerancanganBangunanGedungdanStruktur Lain* (SNI 1727-2013), BadanStandarisasiNasional, Jakarta

[6]. Bjorhovde, 1986. "*Test of Full Size Gusset Plate Connections,*" Journal of Structural Engineering, ASCE, Vol. 111, No. 3, March, pp. 667-684.

[7]. Dewobroto, W.(2013b*). "KomputerRekayasaStrukturdengan SAP2000"*, Lumina Press, Jakarta

[8]. Dipohusodo, Istimawan, 1994. *StrukturBetonBertulang*. Jakarta: GramediaPustakaUtama.

[9]. Galambos . (1998). *"Guide to Stability Design Criteria for Metal Structure 5th Ed."*, John Wiley & Sons

[10]. Nawy, 1990. *BetonBertulang (SuatuPendekatanDasar)*, Penerbit PT. RafikaAditama, Bandung.

[11]. Tall, Lambert, 1974, *Structural Steel Design,* The Ronald Press Company, New York.

IOP Conf. Series: Materials Science and Engineering **821** (2020) 012015 doi:10.1088/1757-899X/821/1/012015

Effect of coconut-shell ash as filler and plastic bottle as substitution of porous asphalt mixture

T Syammaun*, Meillyta and R Yati

Department of Civil Engineering, Faculty of Engineering, University Muhammadiyah Aceh, Aceh, Indonesia

*Email: tamalkhani@unmuha.ac.id

Abstract. Generally, porous asphalt mixtures are having low stability and highly dependent on the quality of asphalt as an aggregate binder, so the innovation of using additive needs to be examined to produce high quality asphalt. In this study, the additive used in asphalt mixtures is plastic bottle waste (Polyethylene Terephthalate/PET) and the use of coconut-shell ash (CS) as filler. This study aims to find out the influence of using PET waste and CS in asphalt porous mixtures which meets its parameter of Cantabro Loss (CL), Asphalt Flow Down (AFD), Marshall and permeability test on 60/70 asphalt penetration. Its benefit is to discover the influence of adding PET and ATK. In this research, the variation of asphalt content was made to obtain Optimum Asphalt Content (OAC) of 4.5%, 5.0%, and 6.0%. Australian Asphalt Pavement Association (AAPA) method was used to determine OAC including CL, AFD, and Void in Mix (WIM) test. After OAC was determined, OAC specimen with the variation of 0%, 2%, 4% and 6% PET substitute on asphalt weight, and the use of 50% combination of CS and portland cement from the total weight of filler were made. At OAC value of 5.74%, it is obtained that the value of CL, AFD, Marshall and permeability which meet the AAPA specifications is in the addition of 2% and 4% PET in PA mixture.

1. Introduction

Roads infrastructure is to connect one area to another and also to improve the process of economic development in the community. Road damage generally occurs due to existency of puddle on road surface from high rainfall, which reduces rainwater infiltration and accelerates the damage. Porus asphalt is one of pavement type that is developed on the surface layer.

Porous Asphalt (PA) is strongly related to the behavior and properties of asphalt mixtures that use aggregate gradations with a number of coarse fractions above 85% of the total weight of the mixture, which is resulting on porous asphalt structure. The existing cavities provide the ability to drain water both vertically and horizontally so that it will keep the surface layer dry, mean while large cavities can cause a decrease of characteristic value on PA mixture [1, 2]. Generally, PA mixture has low stability and is highly dependent on the quality of asphalt as an aggregate binder, so modified asphalt is needed to produce high quality asphalt [3].

Modified asphalt is an asphalt made by mixing hard asphalt with additive material, whic is intended to improve the physical properties of asphalt. One alternative to deal with the rare of modified asphalt is to use other materials as an additive materials including the utilization of plastic bottle waste (Polyethylene Terephthalate / PET) for asphalt substitution material [4, 5] and Coconut Shell Ash (CA) as filler substitution material in PA mixture so that asphalt mixture characteristics is found out later. This research aims to find out the influence of using PET waste and CS ash on PA mixture which meet CL, AFD, VIM and permebiality coefficient on 60/70 asphalt penetration which is based on AAPA spesification 2014 [6].

2. Materials and methodology

2.1. Material selection

2.1.1. Aggregate
Material used in this investigation is coarse and fine agregrate which come from stone crusher located in Seulimum, Aceh Besar Sub-District. Porous ashpalt gradation (open graded) is used as shown in Table 1. The results of physical properties of tested aggregates are given in Table 2.

Table 1. Gradation of porous asphalt

Sieve Size (mm)	Specification passing range (%)	Used Passing (%)
19.0	100	100
13.2	85 – 100	92.5
9.5	45 – 70	57.5
4.75	10 – 25	17.5
2.36	7 – 15	11
1.18	6 – 12	9
0.6	5 – 10	7.5
0.3	4 – 8	6
0.15	3 – 7	5
0.075	2 – 5	3.5

Table 2. Physical properties of aggregate

No.	Physical Properties	Unit	Value	Requirement
1	Specific gravity	gr/cm3	2.78	Min. 2.5
2	Absorption	%	1.12	Mak. 2
3	Unit weight	Kg/dm3	1.67	Min. 1
4	Hardness	%	8.94	Max. 30
5	Abrasion	%	15.02	Max. 40
6	Flakiness index	%	17.42	Max. 25
7	Elongated index	%	15.43	Max. 25

2.1.2. Asphalt Binder
The conventional 60/70 penetration grade bitumen was used for producing the tested PA mixtures. The main characteristics of the used bitumen are summarized in Table 3.

Table 3. Characteristics of 60/70 conventional bituminous binder

No.	Test	Unit	Value	Requirement
1	Penetration	mm	64	60-70
2	Ductility	cm	120	Min. 100
3	Softening Point	°C	49	Min. 48

2.1.3. Plastic Bottle (PET)
PET waste is obtained from PET bottle. PET is one of polyester material often used as a packaging in beverage industry. PET is easy handled polimer and strong, durable, low gas permeability, thermally and chemically stable [7]. Table 4 shown the physical properties of PET.

Table 4. Properties of plastic bottle [4]

No.	Test	Unit	Value
1	Density	g/cm^3	1.35
2	Tensile Strength	psi	11,500
3	Water absorption	%	0.1
4	Approx. melting temperature	°C	250

2.1.4. Coconut-shell Ash (CA)
Active carbon is a material that has a large amount of cavities which has a function to absorb everything through it. One of them is an activated carbon material derived from animals, plants, or minerals containing carbon, among others: CA, hardwoods, palm oil, some are from coal, bones, sugar cane ash, sawdust and others [8]. Coconut-shell ash is an ash in the form of soil or mineral materials such as silica, which is the residual combustion of charcoal that has become ash, with black to gray colour, usually tasteless and odorless.

3. Experimental plan
This experimental research was made by testing the characteristic of PA with certain proportion variation of PET on 60/70 asphalt penetration using 50% CA as filler. The step in this research is preparing material and tools, examining physical properties of agregates and asphalt, and sieve analysis. As it meets the specified requirement, then the experiment can be continued to specimens manufacturing.

3.1. Sample Preparation
Australian Asphalt Pavement Association (AAPA) 2004, determined that the used asphalt content is at 4.5% - 6% and with 2 x 50 collisions for each specimens test (50 collisions on each surface of specimens). After OAC is obtained, PET is added with the variation of 0%, 2%, 4%, and 6%. PET bottle is substituted in PA mix. Bottle was cutted into small pieces. CA preparation was performed by taking the weighting charcoal sample in order to obtain constant mass around 1 gram and then the sample was heated for 3 hours at 900°C. After finishing, the sample is then cooled and weighed again in order to obtain ash content.

4. Results and discussion
In this study, the characteristic of PA is found by performing test including Cantabro Loss, Asphalt Flow Down, Marshal and permeability. The result of CL, AFD and VIM with the variation of asphalt content to detemine OAC can be seen in Table 5. It is found in this research that 5.74% is an optimum asphalt content.

Table 5. Summary of test result without PET (Determination of OAC)

Test	Unit	Asphalt Content (%)				Requirement
		4.5	5.0	5.5	6.0	
Asphalt Flow Down	%	0.05	0.13	0.19	0.21	Mak. 0.3
Void in Mix	%	23.03	21.57	19.81	19.06	18-25
Cantabro Loss	%	30.40	26.85	19.86	17.89	Mak. 20
Permeabilitas	cm/s	0.18	0.19	0.21	0.24	0.1-0.5

4.1. Effect of PET substitution to asphalt flow down
Figure 1 ilustrates the higher PET percentage cause the decrease on AFD value, excluding variation of 6% PET which increases. This is due to the increasing levels of PET in asphalt which causes smaller level of separation of asphalt in the mixture.

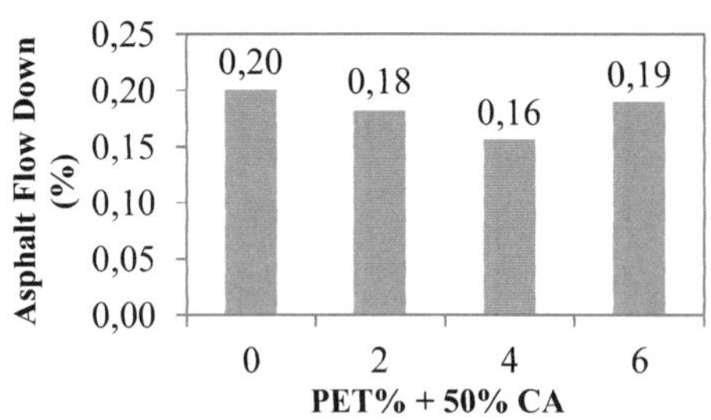

Figure 1. Asphalt flow down value with the different PET contents

4.2. *Effect of PET substitution to void in mix*

From the test result, it can be concluded that VIM value increases for all variation both for with or without variations of PET + 50% ATK. The biggest value is on the variation of 6% PET which is 22.95%. Cracking can be risen by many air void formed as the result of asphalt binders that are not perfectly coated, while asphalt bleeding and rutting may be induced by low air voids. [9]. Furthermore, when PET is used in a mixture, it seems to reduce its compact ability, where this condition allows a higher air void value (Figure 2).

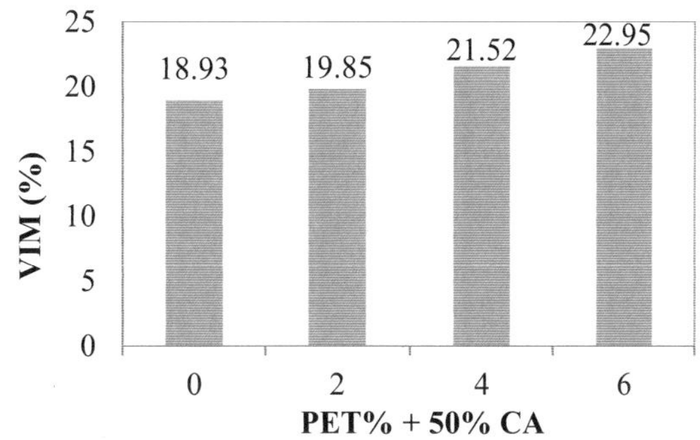

Figure 2. Asphalt void in the mix value with the different PET contents

4.3. *Effect of PET substitution to Cantabro loss*

Based on the cantabro loss test results, Figure 3 shows that the highest the percentage of PET + 50% CA substitution. The highest value of 26.18% is found at variation of 6% PET and the lowest value is 15.21 % at variation of 0% PET. The cantabro loss value specification as implied by AAPA (2004) is 20 maximum, the higher the PET value, the weaker the resistance to destruction. However only the mixture with 2% PET content that satisfy the specification of lower than 20.

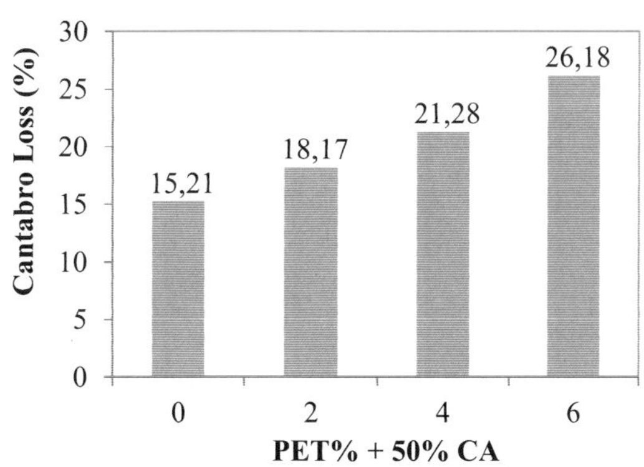

Figure 3. Cantabro loss in the mix value with the different PET contents

4.4. *Effect of PET substitution to permeability*

The results of the permeability test in Figure 4 show the water flow rate is strongly influenced by the number of continuous cavities present in the mixture. This permeability value is caused by the formation of continuous cavities in the mixture which causes the flow of water from the surface vertically to the bottom is faster. It is examined that for all variations of PET content, the permeability testing value meets the specifications, which are 0.1-0.5 cm/ s.

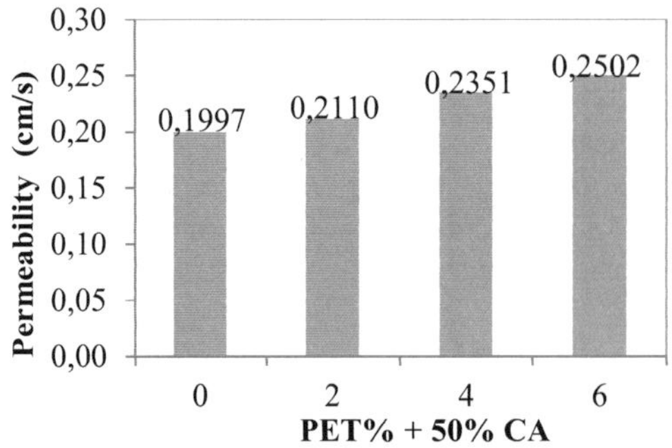

Figure 4. Permeability in the mix value with the different PET contents

4.5. Marshal stability and flow

Figure 5 illustrates the Marshall Stability and Flow value versus PET content for optimum asphalt content. From the results of stability test, on porous asphalt without or with PET + 50% CA, it is shown that there was an increamental of the stability value, but there was a decrease in the variation of the addition of 6% PET. An increase value of stability occurs in variations of 0%, 2% and 4% PET. It increases from 5.17 kN at the variation of 2% PET to 5.46 kN at the variation of 4% PET. However it decreases at the variation of 6% PET to 4.58 kN. The best stability value is at variation of 4% PET with control using 0% variation (without PET mixture). Polymer can increase stability in hot mix asphalt and develop connection between material in the mixture [9, 10]. Furthermore, in the flow test for variations of PET + 50% CA substitution, the value of each PET addition decreases except for variation of 6% PET. However, a high percentage of PET causes the flow to increase while the stability decreases.

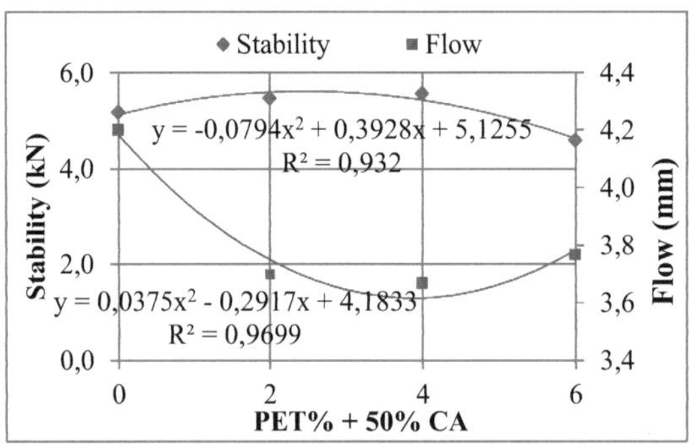

Figure 5. Relationship between stability and flow at different PET contents

4.6. Marshal Quotient

Marshal Quotient (MQ) values is calculated in order to evaluate specimen deformation as it is an indicator to resist deformation of the bituminous mixture [11, 12]. Figure 6 shows the results of the MQ (Marshall Quotient) test of PA on either with or without PET shows the increasing MQ values. The increase in MQ value occurs in variations of 2% and 4% of PET. At the level of 2% PET, it increases from 135.42 kg to 141.63 kg, at the level of 4% it increases to 154.84 kg, while at the variation of 6% PET experienced a decrease of 124.34 kg. The best MQ values are in the 4% variation with controls using 0% content without PET mixture. The higher the MQ value the asphalt will become stiff, easy to crack but has high stability. Vice versa, asphalt will be more susceptible to change in shape due to daily trafic load.

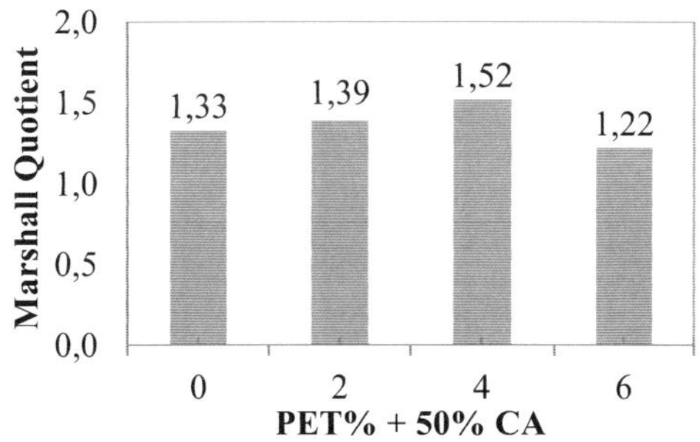

Figure 6. Marshall Quotient in the mix value with the different PET contents

5. Conclusion

There are some summary can be drawned for this experimental results. Therefore, it can be concluded that: First, adding PET to the mixture decreases the AFD, however it experiences an increasing at the variation of 6% PET. The highest VIM values is at 4% PET addition. Second, the increasing variation of PET addition causes greater value of CL, in other words the resistance of the mixture to the release of granules becomes smaller. Third, for a particular PET content, the permeability of the porous mix increases as the amount of PET increases. Fourth, as content of PET + 50% CA increases into the porous asphalt mixture, Marshall Stability also increases, but then decreases after 4% of PET addition. However, the opposite condition occurs in Marshall Flow where the use of PET + 50% CA causes a decrease in value at the beginning followed by an increase when the use of PET increases into the mixture. For last, the incremental of PET content + 50% CA cause the increase of stiffness level in the mixture, which lead an increase in resistance to permanent deformation at the MQ value.

Acknowledgements
This study was supported by the Universitas Muhammadiyah Aceh Indonesia. The authors wish to thank our colleagues who provided insight and expertise that assisted this research.

References

[1]. Cong L, Wang T, Tan L, Yuan J and Shi J 2018 *Construction and Building Materials* vol **169** pp 436–42.
[2]. Haryati Y, Norhidayah A H, Nordiana M, Juraidah A, Nor Hayati A H, Ramadhansyah P J, Khairil M A and Haryati A 2019 *IOP Conference Series: Earth and Environmental Science* 244 012043.
[3]. Syammaun T and Rani H A 2018 *IOP Conference Series: Materials Science and Engineering* IOP Publishing 403 012023.
[4]. Baghaee T M, Karim M R and Syammaun T 2012 *Construction and Building Materials* vol **34** pp 236-42.
[5]. Ahmadinia E E, Zargar M, Karim M R, Abdelaziz M and Shafigh P 2011 *J Mater Design* vol **32** 4844–9.
[6]. Association A.A.P *National asphalt specification* The Association.
[7]. Awaja F and Pavel D 2005 *European Polymer Journal* vol **41**(7) pp 1453-77.
[8]. Shelke A S, Kalyani R, Ninghot, Pooja K P and Shraddha G P 2014 *International Journal of Civil Engineering Research* vol **5**(3) pp 211-14.
[9]. Chen H, Xu Q, Chen Sh and Zhang Z 2009 *Mater Design* vol **30** pp 2595–603.
[10]. Sabina AT K, Sangita, Sharma D K and Sharma B M 2009 *J Sci Indus Res* vol **68** pp 975–9.
[11]. Ahmedzade P, Yilmaz M 2008 *Construct Build Mater* vol **22** pp 481–6.
[12]. Tayfur S, Ozen H and Akso A 2007 *Construct Build Mater* vol **27** pp 328–37.

IOP Conf. Series: Materials Science and Engineering **821** (2020) 012016 doi:10.1088/1757-899X/821/1/012016

Design mixed formula of hot mix asphalt mixture very thin overlay using crumb rubber

S Sunarjono[1,2,*], P A Wijaya[1], N Hidayati[1,2], and A Magfirona[2]

[1] Civil Engineering, Postgraduate Study Program, Universitas Muhammadiyah Surakarta, Surakarta, Indonesia

[2] Transportation Research Center, Faculty of Engineering, Universitas Muhammadiyah Surakarta, Indonesia

*Email: Sri.Sunarjono@ums.ac.id

Abstract. One of the technologies expected to improve the quality of flexible pavement is to use a modified asphalt mixture of Crumb Rubber (CR). The material can be used as a very thin added layer which is then called Very Thin Overlay Modified Crumb Rubber (VTO Mod-CR) and the functions as a leveling layer. This asphalt mixture is expected to have high durability and increase the age service life of flexible pavement. The aim of this study was to analyze the design mix formula (DMF) and the durability of the asphalt mixture. This research carried out variations of asphalt content i.e. 6.0%, 6.5%, 7.0%, 7.5% and 8.0% with each variation of 3 specimens for Marshall Test. Based on the DMF analysis results obtained optimum asphalt content of 7.00%, while the properties of the mixture in the form of bulk density of 2.240, VIM of 3.77%, VMA of 18.15%, VFA of 79.23%, Stability of 1049 kg, Flow of 3.47 mm, Marshall Quotient of 251 kg / mm, Bitumen Film Thickness 16.08 μm. The results of the analysis of the mixture properties have met specifications General of Highways, so it can be used as a reference for AMP. The BFT value obtained meets the minimum specification specifications of 16 μm, this shows that the resistance to water and oxidation is higher, so the pavement becomes durable.

1. Introduction

The first paragraph after a heading is not indented (Bodytext style). The flexible pavement construction so far has relied more heavily on new material and neglected the environmental issue [1-3]. Road pavement construction and reconstruction indicate a considerable consumption of valuable and non-renewable natural resources and, in particular, the component materials of asphalt mixtures [4-6]. On the other hand, the increasing need for transportation and technological advances that are always developing, the quality of transportation infrastructure is demanded to be more reliable to serve traffic, but must also consider environmental and economical aspects [7, 8]. In the last few decades, several developing countries have developed and implemented the innovative design of environmentally friendly asphalt mixtures, one of which is the asphalt modification crumb rubber technology [9-11]. Indonesia began to adopt the technology in 2012, while as a mixture the new asphalt began to be implemented in 2017 [12].

Asphalt Modification of Crumb Rubber (Asphalt Mod-CR) began to be the first choice as a binding material in the flexible pavement because it is believed to show excellent quality, has high durability and a comfortable surface to cross [13-15]. Previous Research showed that having asphalt rubber mix resistance has better resistance to deformation [16]. Asphalt mixture of modified rubber powder or Crumb Rubber Modified (CRM) with a composition of more than 20% has a higher viscosity (viscosity), better flexibility (Marshall Quotient), increase anti-fatigue and anti-aging [17-19]. These aspects are very suitable to meet the needs of road infrastructure with flexible pavement construction such as in Indonesia which is a high-temperature tropical country [20-22].

Utilization of Asphalt Mod-CR can be used as a very thin overlay that also functions as a leveling wear layer as a leveling wear layer. VTO Mod-CR as a surface layer functioned as a level wear out grading which aims to improve the surface level of the road, depending on the elevation of the rigid pavement surface below [23-25]. Hot mix with asphalt rubber has a very good structural value so that it can be applied with a thickness of 50% compared to conventional asphalt [26-28]. The implementation of VTO Mod-CR asphalt mixture starting from the production process, mixing, spreading to compaction requires special treatment compared to conventional asphalt in general [29-31]. In the production process, this asphalt mixture requires a higher mixing temperature than the ordinary asphalt mixture, so it requires fuels that are capable of producing high heat such as kerosene or gas [22, 32, 33]. The mixing process in the mixing unit or Asphalt Mixing Plant (AMP), is also a little longer to ensure that the asphalt mixture is really well mixed and evenly (homogeneous). The consequence of this is that production costs will also increase. Because the thickness of the very thin layer and homogeneity of the mixture must be maintained so that segregation does not occur at the time of the laying, the temperature at the time of the laying and compaction of the mixture must continue to be controlled. Thick beds that range from only 2.5 cm cause the layer to be very sensitive to the ambient temperature which results in a very rapid fall in the temperature of the mixture. Spraying water on the compactor wheel which is usually done in the compaction process must be avoided. The use of compactors with rubber tires is also not permitted because it will cause a tiring trail. Inadequate temperature [34] during spreading and compaction will result in the mixture clumping and segregation so that when compacted it will not get maximum density. Even if compaction continues when the temperature of the mixture has exceeded the minimum limit of compaction temperature, cracking or rupture of the layer will occur. Another significant difference with conventional asphalt is in the manufacture of Mod-CR asphalt Design Mix Formula (DMF) especially in the aspect of the number of collisions of test specimens. The number of collisions of Asphalt Concrete Base, AC Binder, and AC Wearing according to SNI is 2 times 75 collisions back and forth. Meanwhile, modified Crumb Rubber heat asphalt mixture used in VTO (Very Thin Overlay) is 2 times 50 alternating collisions. This is due to the consideration and calculation that the largest aggregate size used is 12.5 mm and will be spread with a very thin layer thickness which can cause the temperature of the mixture to tend to fall faster. The number of collisions 2 times 50 for the VTO is already considered quite perfect in the laboratory as a reference and comparison of the implementation in the field later.

According to DGoH (2015) [35], the value of the degree of density in the field must comply with the specification requirements of at least 98% of laboratory density. The minimum thickness of the layers is 2.5 cm, causing the layer to be very sensitive to the ambient temperature. Thus, based on the background that has been described, it is necessary to study the process of making the design mix of VTO Mod-CR asphalt mix formulas. This study aims to determine the design mix formula (DMF) of VTO Mod-CR asphalt mixture which is used as a reference or reference prior to the implementation of production in AMP and analyze the durability of the asphalt mixture.

IOP Conf. Series: Materials Science and Engineering **821** (2020) 012016 doi:10.1088/1757-899X/821/1/012016

2. Methods

The main material used in this research is coarse aggregate, fine aggregate and crumb rubber asphalt. Coarse aggregates and fine aggregates (stone ash), originating from the village of Ngargosari, Ampel, Boyolali which are produced at PT. Anugrah Batu Alam (ABA). Modified crumb rubber asphalt is made from a mixture of asphalt oil and crumb rubber (rubber powder from used tires), originating from PT. Bintang Djaja, Cilacap.

The DMF creation begins by testing the material properties which serve to determine the character and quality of the material to be used in the mixture. Furthermore, mixed test specimens are made with variations in asphalt content to determine the optimum asphalt content (OAC). Investigation of the properties of the mixture was carried out by the Marshall Test method. Whereas to obtain the value of BFT (Bitumen Film Thickness) obtained based on aggregate gradation data. The outlines of the stages of making DMF VTO Mod-CR is presented in Figure 1. The number of test specimens for Marshall Test is 3 test pieces with 5 variations of asphalt content namely 6.0%, 6.5%, 7.0%, 7.5%, and 8.0%. Optimum asphalt content is used to determine the initial level of asphalt in laboratory planning. Research conducted in the laboratory is used to obtain asphalt contents used in planning flexible pavement in the field. The asphalt content of each plan differs due to variations in grain size in each

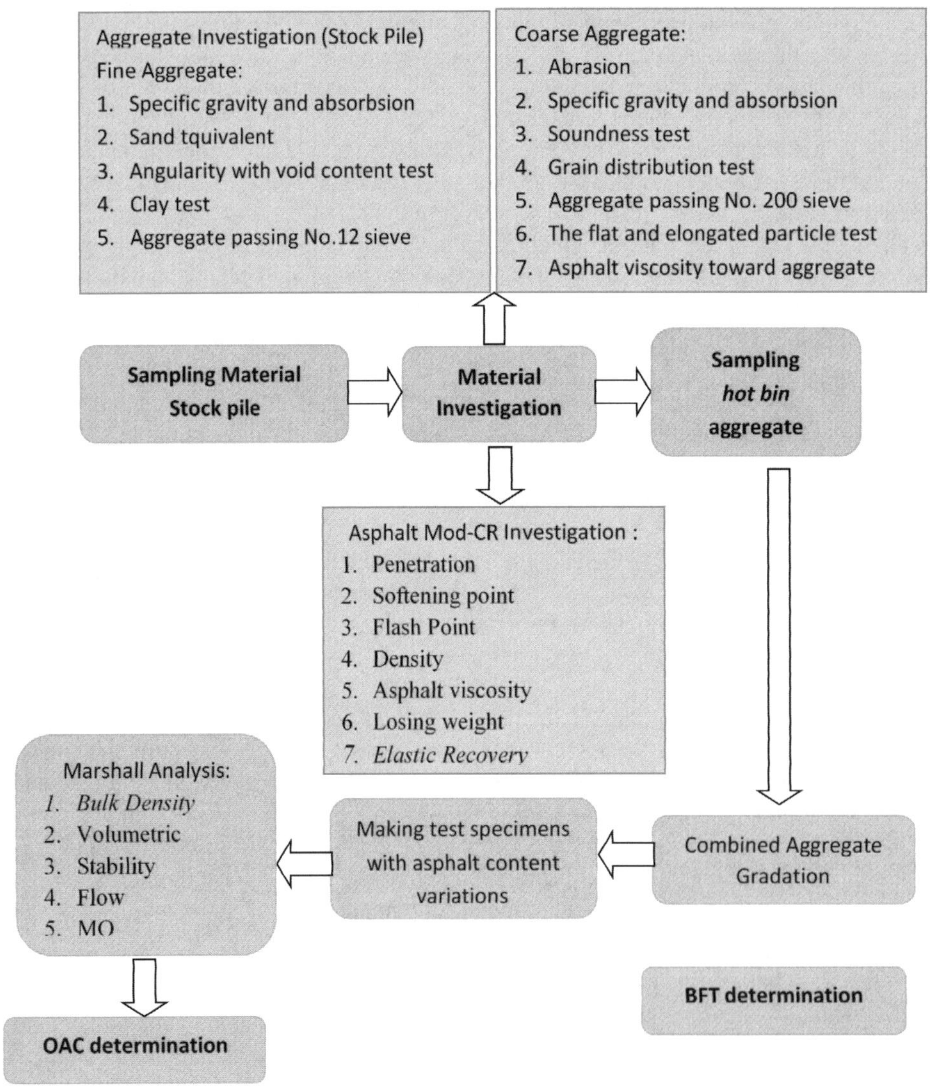

Figure 1. Flowchart of DMF creation

plan. The optimum asphalt content of the plan (Pb) is obtained by the following equation:

$$P = 0.035\ (\%\ CA) + 0.045\ (\%\ FA) + 0.18\ (\%\ Filler) + K \quad (1)$$

where:

P = Middle/ ideal asphalt content, percent by weight of the aggregate mixture plus asphalt
CA = % aggregate retained sieve No. 8
FA = % aggregate passing sieve No. 8 and retained No. 200
Filler = % aggregate minimal 75% passing sieve No. 200
K = Constanta (0.5-1 for Laston; 2-3 for Lataston; and 1-2.5 for another mixture).

The BFT value is associated with the flexibility and resistance to water which of course will produce high pavement durability as well. To calculate the BFT, it is necessary to first find the total aggregate surface area which can be calculated based on combined aggregate gradation data and aggregate surface area factors (Surface Area Factors, SAF). Next, the BFT of the asphalt mixture on DMF is determined by the following formula:

$$T = \frac{b}{100 - b} x \frac{1}{pb} x \frac{1}{SAF} \quad (2)$$

where:

T = Bitumen Film Thickness (mm)
b = Asphalt content (%)
pb = Specific gravity of asphalt
SAF = Surface Area Factors (m2/kg)

3. Analysis and discussion

VTO Mod-CR DMF data processing consists of aggregate gradation analysis, Marshall Test and maximum bulk density (G_{mm}) mixture, which can finally be determined by the OAC of the mixture. From the results of the gradation analysis, it can be determined the aggregate proportion of each faction by graphical and trial error. The aggregate percentages of Hot Bin I, Hot Bin II and Fine Filler (FF) are entered into trial errors in such a way that they are in accordance with specifications. From the calculation results obtained the proportion of each faction is Hot Bin I by 28%, Hot Bin II by 70%, and Fine Filler by 2%. The results of combined aggregate gradations that have been carried out in the form of Figure 2 shows that the gradations have met the required specifications.

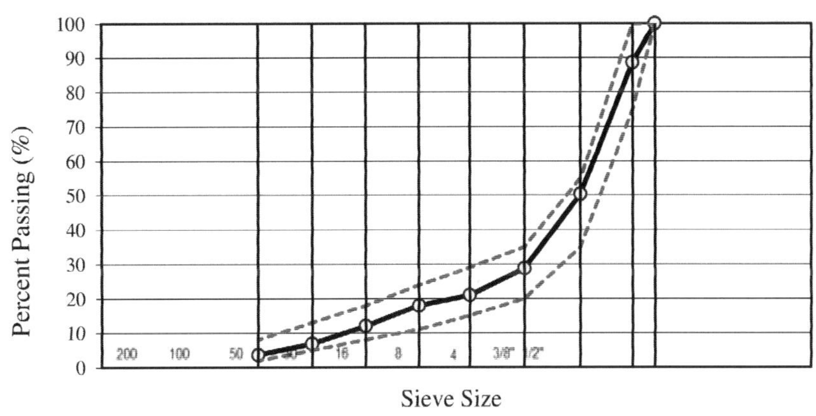

Figure 2. Combined gradation graph

After finding the estimated optimum asphalt content (7.00%), then the maximum bulk d mixture (G_{mm}) can be performed as presented in Table 1.

Table 1. Maximum bulk density mixture calculation results (G_{mm})

		Sample 1	Sample 2
Bottle weight + sample (gr)	a	2126.4	2126.4
Bottle weight (gr)	b	626.4	626.4
Sample weight (gr)	c = a - b	1500.0	1500.0
Bottle weight + water (gr)	d	344.8	344.8
Bottle weight + water + sample (gr)	e	1200.4	1204.4
Fill in the example (cm^3)	f = c + d - e	644.4	643.5
Specific gravity	g = c / f	2.328	2.331
Water temperature (°C)	h	25	25
Temperature correction	i	1.000	1.000
Specific gravity (corrected)	G = g x i	2.328	2.331
Average specific gravity		2.329	

Before conducted the trial mix at AMP, the analysis of the mixture was carried out based on 5 variations of the asphalt content of the Marshall test, to obtain the stability, flow, and Marshall Quotient. As for the volumetric mixture, the results of the examination are bulk density, voids in the mix (VIM), voids in mineral aggregates (VMA), and voids filled with asphalt (VFA). The results of testing the properties are presented in Figure 3 until Figure 9.

Bulk density is the ratio of the weight of the mixture in the air to the volume, including the air cavity, divided by the weight of the water content at the same temperature. The value of the density of the mixed bulk indicates the density of a mixture because this value is obtained from the ratio between the weight and volume of the mixture. From Figure 3, it can be seen that the specific gravity. The highest bulk is at 7% asphalt content of 2.240, which means that mixes with asphalt content below and above 7% have less density.

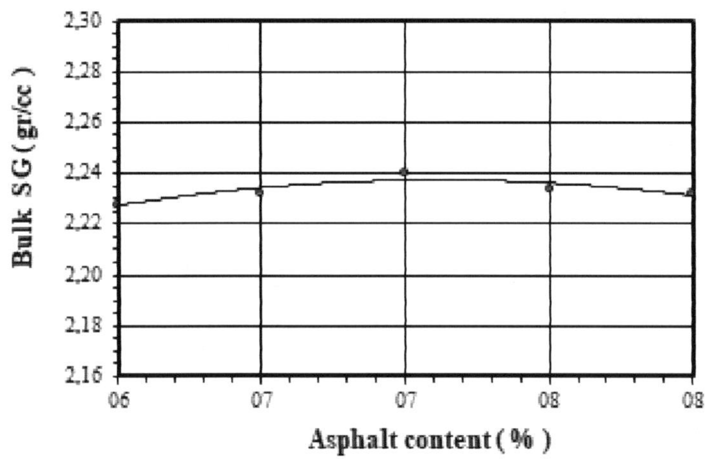

Figure 3. The relationship between asphalt content and bulk density

The range of VIM is 3% - 5% so that the VTO Mod-CR asphalt mixture as a surface layer is more impervious to water and air, which can cause oxidation and asphalt to become brittle. Based on Figure 4 above, the asphalt content that meets the VIM value is 6.2 - 7.7%.

IOP Conf. Series: Materials Science and Engineering **821** (2020) 012016 doi:10.1088/1757-899X/821/1/012016

VMA is the void in mineral aggregate found in the aggregate granules in the asphalt mixture that has been compacted and effective asphalt, expressed in percent of the total volume of the mixture. VMA value will increase along with the increase of asphalt blanket or if the aggregate used is graded open. The test results in Figure 5 shows that the VMA value tends to increase along with the increase in asphalt contents. In this test all mixes with asphalt content of 6% to 8% meet the specified specifications (minimum of 17%).

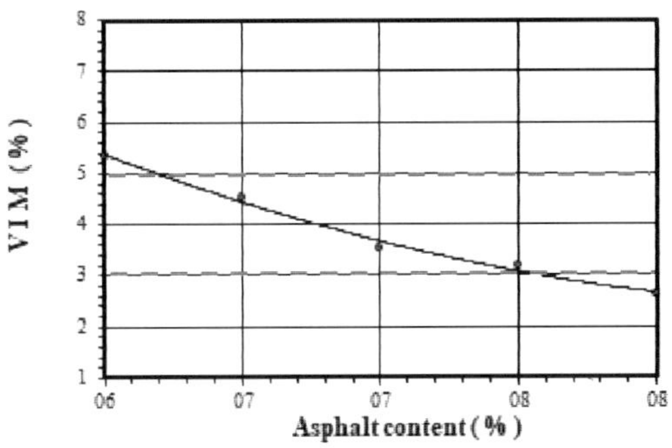

Figure 4. The relationship between asphalt content and VIM.

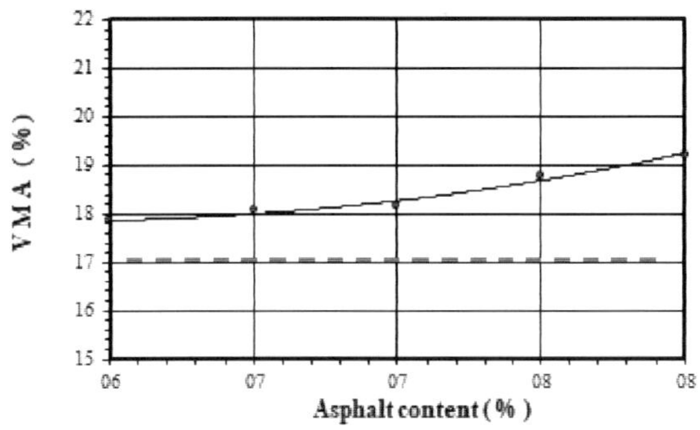

Figure 5. The relationship between asphalt content and VMA

VFA is void filled with asphalt in the mixture after undergoing the compaction process expressed in percent of the cavity between the aggregate grains (VMA) but does not include asphalt absorbed by the aggregate. Factors that can affect VFA include asphalt content, aggregate gradation, compaction energy, and asphalt heating. From the test results shown in Figure 6, the VFA value tends to increase along with the increase in asphalt contents. In this test all mixes with asphalt content of 6% to 8% meet the specified specifications (minimum 66%).

Stability is the ability of the pavement layer to accept the burden of traffic without experiencing permanent deformation such as waves, grooves (rutting), or experiencing bleeding. The stability value is influenced by asphalt cohesion, asphalt content, internal friction, interlocking properties of aggregate particles, surface shape and texture and aggregate gradation. The test results show that the stability value obtained tends to increase to the optimum limit and then decreases (can be seen in Figure 7).

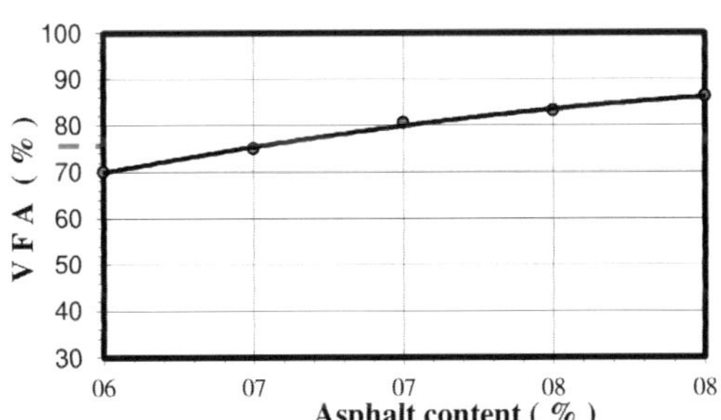

Figure 6. The relationship between asphalt content and VFA.

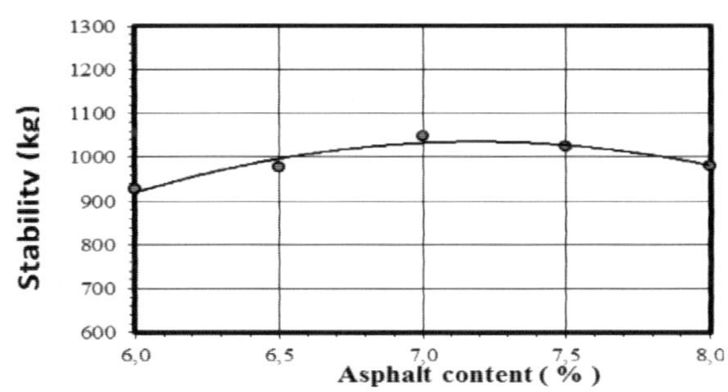

Figure 7. The relationship between asphalt content and stability

The optimum stability value occurs at 7% asphalt content which is equal to 1049.5 kg. Flow is an implementation of the flexibility characteristic of the mixture produced. Flow value is influenced by asphalt content, aggregate distribution, and compaction temperature.

Figure 8 shows the value of flow with variations in asphalt contents. The test results show that the more asphalt content in the asphalt concrete mixture, the greater the value of flow. This is because the addition of asphalt contents makes the mixture more plastic so that the deformation due to load increases. In this test all mixes with asphalt content of 6% to 8% meet the specified Flow value specifications i.e., min. 3 mm.

MQ is the quotient between Stability and Flow. MQ value in Figure 8 shows the flexibility of the mixture that is the greater the MQ value in a mixture, the more rigid (if too rigid tends to crack easily) the mixture, as well as the smaller the MQ value, the degree of flexibility and plasticity (too flexible tends to be less stable) the mixture will get bigger. From the test results obtained the MQ value at 7.00% asphalt content of 251 kg / mm, indicating that the VTO Mod-CR asphalt mixture has fairly good flexibility. Based on the Marshall and volumetric test results of asphalt mixture in Figure 2 through Figure 9, then used as a basis for finding the optimum asphalt content, shown in Figure 10.

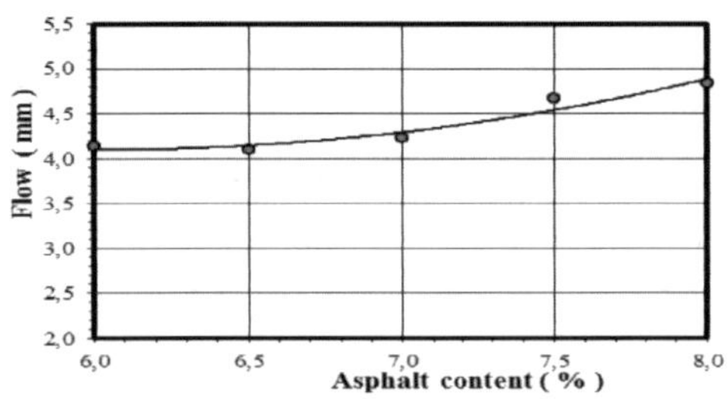

Figure 8. The relationship between asphalt content and flow.

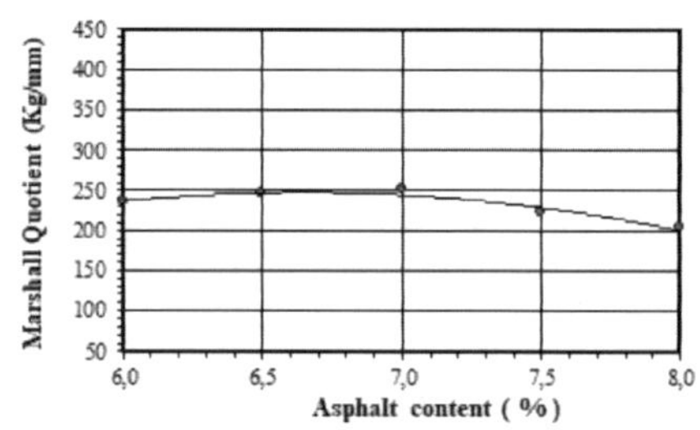

Figure 9. The relationship between asphalt content and Marshall Quotient (MQ)

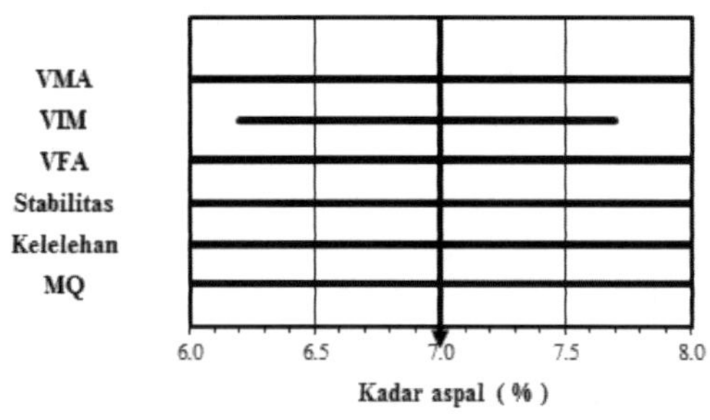

Figure 10. Analysis of optimum asphalt content

Based on the analysis results, the OAC value was obtained at 7%.

The DMF recapitulation results shown in Tables 2 through Table 4 are a mixed draft plan which will then be applied at AMP. Table 2 shows the composition of the mixture, while Table 3 is the total aggregate composition in the mixture before asphalt is added. Table 4 is the result of combined aggregate grading analysis.

IOP Conf. Series: Materials Science and Engineering **821** (2020) 012016 doi:10.1088/1757-899X/821/1/012016

Table 2. DMF mixture composition

	Result	Result
Design of asphalt content	7.00 %	
Hot bin I	26.0 %	28.00 %
Hot bin II	65.1 %	70.00 %
Filler cement	1.90 %	2.00 %
Total	100 %	100 %

Table 3. DMF combined gradation

Sieve size			
Inch	mm	Combined gradation (%)	Specificcation (%)
1/2"	12.7	100.0	100
3/8"	9.5	88.60	75-100
#4	4.8	50.38	35-55
#8	2.4	28.79	20-35
#16	1.2	20.99	-
#30	0.60	17.99	-
#50	0.30	12.01	-
#100	0.15	6.88	-
#200	0.075	3.66	2-8

Table 4. Recapitulation of the properties of the DMT VTO Mod-CR mixture

	Result	Spesification	Units
Optimum asphalt content	7.0	-	%
Bulk density	2.240	-	-
Void in Mix (VIM)	3.54	3.0 – 5.0	%
Void in Mineral Aggregate (VMA)	18.15	> 17	%
Void Filled with Asphalt (VFA)	80.51	> 66	%
Stability	1049.5	Min 600	Kg
Flow	4.23	Min 3	Mm
Marshall Quotient	251	-	kg/mm
Bitumen Film Thickness	1.,08	Min 16	μm

VTO Mod-CR asphalt mixture has advantages in aspects of flexibility, durability, and adhesive compared to conventional asphalt. The analysis shows that the BFT value of asphalt VTO Mod-CR mix results of AMP trial mix of 16.89 μm has met the minimum specification requirements of 16 μm, this shows that the resistance to water and oxidation is higher so pavement becomes durable.

4. Conclusions

Based on the analysis, it can be concluded that the OAC is 7.00%, so that the composition of the mixture with asphalt content is 26.0% for a hot bin I with a screen size of 5 mm and 65.1% hot bin II with a screen size 1/2" and 1.90% cement filler. The DMF asphalt mixture properties were obtained 2.240 of bulk density values, 3.77% of VIM, 18.15% of VMA, 79.23VFA, 1049 kg of Stability, 3.47 mm of Flow, 251 kg /mm of Marshall Quotient, and 16.08 μm BFT. The composition and properties of the mixture have been obtained that meet the specifications so that it can be used as a reference for mixing at AMP. The BFT has fulfilled the minimum specification requirements of 16 μm, this shows that the resistance to water and oxidation is higher so that the pavement becomes durable

Acknowledgement
On this occasion the researchers thanked for the support of publication funding from the Directorate of Research and Community Service Directorate General of Research and Development of the Ministry of Research, Technology, and Higher Education with contract number of 036/B4/PPK-SPK/VII/2019 and support of facilities from Universitas Muhammadiyah Surakarta so that this research could provide the results of scientific development in the field of road materials.

References

[1] Xiao F, Amirkhanian S N, Shen J and Putman B 2009 Influences of crumb rubber size and type on reclaimed asphalt pavement (RAP) mixtures *Construction and Building Materials* vol **23** pp 1028–34.
[2] Yu H, Leng Z, Dong Z, Tan Z, Guo F and Yan J 2018 Workability and mechanical property characterization of asphalt rubber mixtures modified with various warm mix asphalt additives. *Construction and Building Materials* vol **175** pp 392–401.
[3] Kebria, D Y, Moafimadani S R and Goli Y 2015 Laboratory investigation of the effect of crumb rubber on the characteristics and rheological behaviour of asphalt binder *Road Materials and Pavement Design* pp 1-11.
[4] Celauro C, Bernardo C and Gabriele B 2010 Production of innovative, recycled and high-performance asphalt for road pavements *Resources, Conservation and Recycling* vol **54** pp 337–47.
[5] Ghavibazoo A and Abdelrahman M 2014 Effect of crumb rubber modification on short term aging susceptibility *International Journal of Pavement Research and Technology* vol **7** no 4 pp 297-304.
[6] Fakhri M S, Sedghi R, Arzjani D and Zarrinpour Y 2019 Effects of deicing agents on moisture susceptibility of the WMA containing recycled crumb rubber *Construction and Building Materials* vol **227** pp 1-15.
[7] Wahyono D 2014 Karakteristik campuran beraspal panas asphalt concrete wearing course (ac-wc) menggunakan aspal modifikasi crumb rubber dengan variasi tingkat viskositas *Thesis* Yogyakarta: UGM.
[8] MQuillen J L and Hicks R G 1988 Economic analysis of rubber modified asphalt mixed *Journal of Transportation Engineering* vol 11 no 33 pp 259-77.
[9] Zanetti M C, Ruffino B, Dalmazzo D, Lanotte M and Santagata E 2018 Determination of crumb rubber content of asphalt rubber binders *Journal of Materials in Civil Engineering* vol **30**(4) 04018041 pp 1-8.
[10] Fontes L P, Trichês G, Pais J C and Pereira P A 2010 Evaluating permanent deformation in asphalt rubber mixtures *Construction and Building Materials* vol **24** pp1193–200.
[11] Tauste R, Moreno-Navarro F, Sol-Sánchez M and Rubio-Gámez M 2018 Understanding the bitumen ageing phenomenon: A review *Construction and Building Materials* vol 192 pp 593–609.
[12] Rahadian H 2017 *Uji coba karet lokal guna tingkatkan penyerapan nasional* Retrieved Januari 07, 2018, from PU-net : https:/ /www.pu.go.id/
[13] Mashaan N S, Ali A H, Karim M R and Abdeaziz M 2014 A review on using crumb rubber in reinforcement of asphalt pavement *The Scientific Journal Volume 2014* Article ID 2146.
[14] Gong J, Liu Y, Wang Q, Xi Z, Cai, J and Ding G 2019 Performance evaluation of warm mix asphalt additive modified epoxy asphalt rubbers *Construction and Building Materials* vol **204** pp 288–95.
[15] Hajikarimi P, Rahi M and Nejad F M 2015 Comparing different rutting specification parameters using high temperature characteristics of rubber-modified asphalt binders *Road Materials and Pavement Design* pp 1-16.

[16] Indriyati E W 2017 Kajian perbandingan aspal modifiklasi asbuton dan aspal rubber (ar) untuk infrastruktur jalan *Jurnal Teknik SIpil* vol **14** no 2 pp 94 -100.

[17] Dong Y and Tan Y 2011 Mix design and performance of crumb rubber modified asphalt SMA *Geotechnical Special Publication* no 212 pp 78-86.

[18] Jeong K D, Lee S J, Amirkhanian Serji N and Kim K W 2010 Interaction effects of crumb rubber modified asphalt binders *Construction and Building Materials* vol **24** pp 824–31.

[19] Liu X, Li T and Zhang H 2018 Short-term aging resistance investigations of polymers and polyphosphoric acid modified asphalt binders under RTFOT aging process *Construction and Building Materials* vol **191** pp 787–94.

[20] Anwar S 2017 Crumb Rubber Modified Asbuton - Rubber Modified Asphalt *Paparan, Maret 2017* Surakarta.

[21] Attia M and Abdelrahman M 2009 Enhancing the performance of crumb rubber-modified binders through varying the interaction conditions *International Journal of Pavement Engineering* vol **10** no 6 pp 423–34.

[22] Peralta J, Silva H M, Machado A V, Pais J, Pereira P A and Sousa J B.2010 Changes in rubber due to its interaction with bitumen when producing asphalt rubber *Road Materials and Pavement Design* vol **11** no 4 pp 1009-31.

[23] Kirk J V and Holleran G 2000 Reduced thickness asphalt rubber concrete leads to cost effective pavement rehabilitation *1st International Conference World of Pavement* 1-12 Sydney.

[24] Loderer C, Partl M and Poulikakos L 2018 Effect of crumb rubber production technology on performance of modified bitumen *Construction and Building Materials* vol 191 pp 1159–71.

[25] Liu Y, Han S, Zhang Z and Xu O 2012 Design and evaluation of gap-graded asphalt rubber mixtures *Materials and Design* vol **35** pp 873–77.

[26] Caltrans 2010 *Standard Specifications* California State Transportation Agency Department Of Transportation California: Department of transportation

[27] Ibrahim M R, Katman H Y, Karim M R, Koting S and Mashaan N S 2013 A Review on the effect of crumb rubber addition to the rheology of crumb rubber modified bitumen *Advances in Materials Science and Engineering* vol 2013 pp 1-8.

[28] Huang B, Mohammad L N, Graves P S and Abadie C 2002 Louisiana experience with crumb rubber-modified hot-mix asphalt pavement *Transportation Research Record 1789* Paper no 02-2620 pp 1-13.

[29] Ding X, Ma T, Zhang W and Zhang D 2017 Experimental study of stable crumb rubber asphalt and asphalt mixture *Construction and Building Materials* vol **157** pp 975–81.

[30] Xiao F, Amirkhanian S and Juang C H 2007 Rutting resistance of rubberized asphalt concrete pavements containing reclaimed asphalt pavement mixtures *Journal of Materials in Civil Engineering* vol **19** no 6 pp 475-83.

[31] Sangiorgi C, Tataranni P, Simone A, Vignali V, Lantieri C and Dondi G 2018 Stone mastic asphalt (SMA) with crumb rubber according to a new dry-hybrid technology: A laboratory and trial field evaluation *Construction and Building Materials* vol **182** pp 200–9.

[32] Wang, H., You Z, Mills-B J and Hao P 2012 Laboratory evaluation on high temperature viscosity and low temperature stiffness of asphalt binder with high percent scrap tire rubber *Construction and Building Materials* vol **26** pp 583–90.

[33] Sebaaly P E, Gopal V T and Epps J A 2003 Low temperature properties of crumb rubber modified binders *Road Materials and Pavement Design* vol **4** no 1 pp 29-49.

[34] Indriyati, E W and Palupi, K A 2015 Kajian perbandingan perbaikan sifat reologi pada aspal modifikasi asbuton dan aspal modifikasi serbuk ban bekas *Prosiding Konferensi Nasional Teknik Sipil 9 (KoNTekS 9)* pp 633-39.

[35] DgoH 2015 Spesifikasi Khusus Interim-3 Seksi 6.3 Campuran Beraspal Panas dengan Aspal yang Dimodifikasi Crumb Rubber atau Asbuton dengan Crumb Rubber (Skh-3.6.3.2). *Jakarta: Dirjen Bina Marga.*

Laterite Soil Behavior - Geotextile (Study of Laterite Soil, Tanah Laut District)

I Setiawan*, I Muzaidi and M Fitriansyah

Department of Civil Engineering, Faculty of Engineering, Universitas Muhammadiyah Banjarmasin, Indonesia

*Email: ichwansetiawan83@gmail.com

Abstract. Land has different specifications and characters from each type. The land has the main function is as a support of a building construction be it buildings, roads, etc. Seeing from the function is inseparable from the mechanical properties of the soil such as cohesion (c) and soil shear angle (ϕ) where the higher the carrying capacity of the soil the better. The contact force (interface) between the soil surface and the surface of the soil reinforcement material is very dependent on the type of soil. Soil-geotextile contact will influence the carrying capacity of the soil, especially shear strength and soil cohesion. At present research on contact style interface (δ) of laterite-geotextile soil. The method used in this research is to use laboratory-scale experimental methods. This research was conducted with the condition of the soil sample not submerged in water (unsaturated). The test results may be known laterite in Tanah Laut has value ratio δ/ϕ of 0.89 for the type of woven geotextile (HRX250), while the value of the ratio δ/ϕ of 0.86 for the type of nonwoven geotextile (TS600).

1. Introduction

The land is an inseparable part of the construction of civil buildings because all city buildings are more on the ground. Land has different specifications and characteristics of each type. The land has the most important function as supporting the foundation of building construction is its buildings, roads, etc. Judging from the function is inseparable from the mechanical properties of the soil such as cohesion (c) and soil shear angle (ϕ) the higher the carrying capacity of the soil the better.

The contact force (interface) between the soil surface and the surface of the soil reinforcement material is very dependent on the type of soil. Laterite or red soils are known to have diverse soil aggregates. Laterite soils are currently widely used as piles for road construction. As a reinforcement material for road structures geotextiles are usually used. Soil-geotextile contact will influence the carrying capacity of the soil, especially shear strength and soil cohesion. Currently, research on the interface friction (δ) of laterite-geotextile soils.

According to [9], Geotextile is a synthetic geo-material that is impermeable to water. This material can be used as a separator or separator, drainage, filter or filtering, and reinforcement as reinforcement. In addition, geotextiles are also used as filtration or filtering which serves to limit the migration of other particles in fine soil from a soil mass but remain permeable to water movement, and as a reinforcement i.e. stabilization of soil mass by providing tensile strength to the soil system and geotextile fabric.

According to [11] the inclusion of reinforcing geotextile materials in soils improves the CBR and therefore the strength of soils. It implies that geotextile-reinforced soils in unpaved roads will perform better than unreinforced ones and increase load carrying capacity of soils. According to [2] Geotextiles are effective materials for reinforcing soft subgrade ballasted railways. Numerical simulation using finite element analysis provides a potentially useful design method in the construction of reinforced soft subgrade ballasted railways.

Several papers explain the testing of interface friction soils – geotextiles have been shown [1, 4, 6, 9].

The purpose of this study is to review the ratio (δ/ϕ) of the interface friction between the laterite-geotextile soil surfaces. Also, this research is a reference material in research and planning in the field of civil building construction on laterite land that applies geotextiles as reinforcement material.

2. Research methods

Research methods used is to use laboratory-scale experimental methods. The research design as shown in figure 1. This research was conducted with the condition of the soil sample not submerged in water (unsaturated). The tests included original laterite soils, non-woven laterite-geotextile soils, and woven laterite-geotextile soils. The whole test scheme in a condition not submerged in water (unsaturated water).

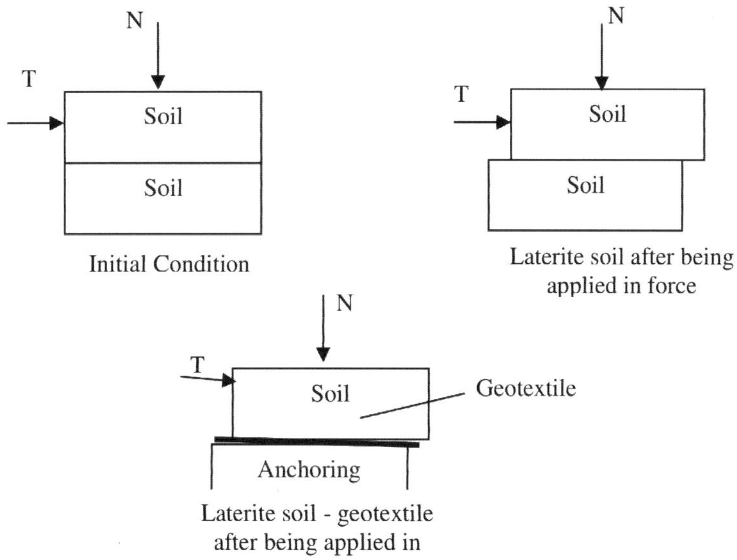

Figure 1. Research Design

3. Tools and materials

A tool used in soil sampling is using a hand drill, while the equipment used in this study is a shear test tool. Geotextile materials used are for woven geotextile (HRX250) and non-woven geotextile (TS600).

4. Theory of soil shear strength

Soil shear strength is the ability of the soil against shear stresses that occur when burdened. Ground shear collapse occurs not because of the destruction of soil grains but because of the relative motion between these soil grains. The shear strength possessed by soil is caused by several factors, namely

a. In fine-grained (cohesive) soils such as clay, the shear strength of a soil is caused by cohesion or attachment between soil grains.
b. In coarse-grained (non-cohesive) soil, shear strength is caused by friction between grains of soil so it is often called the friction angle in the
c. Soil which is a mixture of fine soil and coarse soil (c and ϕ).

Shear strength due to the attachment (cohesion) and friction between grains of land (due to ϕ). Where τ is the shear strength of the soil, c is the soil cohesion, c_a is the geotextile adhesion to the soil, σ is the total stress, and δ is the friction angle of the soil-geotextile interface.

5. Data analysis

Test pieces shaped as many as 3 pieces. The specimens are placed in two rings arranged top and bottom while for geotextile material is placed right in the middle between the lower part of the holder and the laterite soil specimens. Geotextile is glued to the holder, the holder can be made of wood/plywood, and then at the top it is given a normal load (N) of 0.5 kg, 1 kg, and 2 kg. Soil samples are shifted by force (T) whose magnitude is gradually increased. When the ground shifts, the value (T) is recorded and then repeated for the 2nd, 3rd, and so on specimens each using N2, N3, and so on. Calculate τ_2, σ_2, τ_2, σ_2 and so on. Values ϕ and c are searched graphically (test result data) based on the Coulomb formula.

The soil shear strength is expressed by the formula:

$$\tau = c + \sigma \tan\phi \quad \text{(original land)} \tag{1}$$

$$\tau = c_a + \sigma \tan\delta \quad \text{(soil-geotextile)} \tag{2}$$

6. Results and discussion

Based on the results of tests conducted in the laboratory, it can be seen the properties of laterite soils.

6.1. Physical original soils physical

Properties of laterite soils from the results of laboratory tests are shown in Table 1.

Table 1. Physical properties first laterite

No.	Parameter	Results
1	water content(w)	29.61%
2	soil volume weight(γ_b)	1.62 t / m^3
3	Plasticity Index	26.25%

In the Table 1 of laterite soil physical properties above, it can be seen that the laterite soil has a water content of 29.61%, a volume weight of 1.62 t/m^3, and a Soil Plasticity Index of 26.25%. From the results of the laboratory test, it can be seen that the soil is classified as high clay soil with high plasticity, according to USCS is A-7-5, whereas according to AASHTO it is classified as A-7 soil.

6.2. Mechanical of laterite soil

Properties of original soil Mechanical properties of original soil from the results of laboratory tests can be seen in Table 2.

Table 2. Mechanical properties of laterite soil

No.	Parameter	Results
1	Cohesion (*c*)	0.356 kg/ cm^2
2	Shear angle (ϕ)	22.52°

A Table 2 of mechanical properties above can be seen that the cohesion value (*c*) of laterite soil is 0.356 kg / cm^2, while the value of the shear angle of the laterite soil is 22.52^0

6.3. Interface (δ) non-water-geotextile non-water-saturated.

From the results of tests in the laboratory, laterite-geotextile soil in unsaturated water conditions can be seen the value of cohesion (c) and interface friction (δ) can be seen in Table 3.

Table 3. Cohesion and interface unsaturated water non-woven geotextile water (TS600)

No.	Parameter	Results
1	Cohesion (c)	0.325 kg / cm^2
2	Interface Swipe (δ)	19.45°

In the table above it can be seen that the cohesion value of laterite soils with the use of non-woven geotextiles for water unsaturated conditions has a value at 0.325 kg / cm^2. As for the friction value of the interface of laterite soil with non-woven geotextile is 19.45^0

6.4. Interfaces (δ) laterite-geotextile woven not water-saturated.
From the results of tests in the laboratory, laterite-geotextile soils in unsaturated water conditions can be seen the value of cohesion (*c*) and interface friction (δ) can be seen in Table 4.

Table 4. Cohesion and Interface Conditions Unsaturated Water Woven Geotextile (HRX250)

No	Parameter	Results
1	Cohesion (*c*)	0.268 kg / cm^2
2	Friction Interfaces (δ)	20.16°

In Table 4 it can be seen that the value of laterite soil cohesion with the use of woven geotextiles has a value of 0.268 kg/cm^2. As for the value of the interface friction of laterite soils with woven geotextiles is 20.16^0.

Table 3 and Table 4 show that the soil laterite-geotextile after testing in the laboratory has a cohesion value (c) of 0.325 for non-woven type geotextiles, while the cohesion value (c) of 0.268 for Woven type geotextiles, seeing that both values have differences in cohesion values, This is due to the contact surface of laterite-geotextile is influenced by the attachment caused by high levels of clay in the soil. Besides, the friction angle of the laterite-geotextile (δ) has varying values, this is due to the contact between the soil grains and the geotextile surface. Of the two types of Non-Woven and Woven geotextiles, there is a difference in the value of the interface friction angle which is not much different.

6.5. Interfaced angular interface (δ) laterite-geotextile
From the results of testing in the laboratory, it can be seen the shear angle, sliding interface of laterite soil with geotextile. Table 5 shows the relationship between the ratio of the shear angle (ϕ) of the laterite soil to the value of the interface friction (δ) with the use of nonwoven and woven geotextiles. From this table, it can be seen that the ratio (δ/ϕ) for laterite-nonwoven geotextile (TS600) has a value of 0.86. Whereas for laterite - woven geotextile (HRX250) soils, the ratio increased by 0.89.

Table 5. Ratio (δ/ϕ)

No.	Test Scheme	Laboratory Results Testing		
		Shear Angle(ϕ)	Swipe interface(δ)	ratio(δ/ϕ)
1	Laterite soil	22.52^0		
2	Laterite-geotextile nonwoven (TS600)		19.45^0	0.86
3	Laterite-woven geotextile (HRX250)		20.16^0	0, 89

7. Conclusions

From the results of research in the laboratory and data analysis that has been done, it can be concluded that laterite soils originating in the Tanah Laut Regency, South Kalimantan are classified into laterite soils that have high levels of clay with high plasticity. From the test results it can be seen that the interface shear strength that occurs is not only influenced by the type of geotextile, but the grain content in the soil also influences the shear angle value of the laterite-geotextile soil interface. The ratio (δ/ϕ) of woven laterite soil (HRX250) has a ratio of 0.89 and the non-woven geotextile (TS600) has a ratio of 0.86.

Acknowledges
During completing this research, the author has received much direct and indirect assistance from various parties. For that reason, with all humility, the author would like to express my deepest gratitude to all those who helped, especially the ULM Soil mechanics laboratory and all lecturers in the Faculty of Engineering UM Banjarmasin

References
[1] Amu O *et.al* 2011 Geotechnical properties of lateritic soil stabilized with sugarcane straw Ash *American Journal of Scientific and Industrial Research.* http://www.scihub.org/AJSIR ISSN: 2153-649X doi: 10.5251/ajsir.2011.2.2.323.331, pp 323-331
[2] Abebe S M and Qiu H S 2016 Numerical modeling of geotextile reinforcement of soft subgrade ballasted railway under high speed train *Electronic Journal of Geotechnical Engineering* vol **21**.
[3] Bowles J E 1995 *Alih Bahasa Ir. Johan Kelana Putra Edisi Kedua. Sifat-Sifat Fisis dan Geoteknis Tanah* Jakarta: Erlangga.
[4] Datta R and Pal S K 2013 Laboratory investigation on bearing capacity behavior of pond ash reinforced with geotekstil grid. electronic journal of geotechnical engineers **18** pp1631-40.
[5] Das B M 1987 *Advanced Soil Mechanics* Mc Graw Hill Int. Edit New York.
[6] Fourie A B and Fabian K J 1987 Laboratory determination of clay-geotextile interaction *Geotextiles and Geomembranes Journal* vol **6** Issue 4.
[7] Hardiyatmo H C 2012 *Mekanika Tanah I* Edisi 6 Gadjah Mada University Press.
[8] Hardiyatmo H C 2013 *Geosintetik Untuk Rekayasa Jalan Raya Perancangan dan Aplikasi* Edisi 2 Gadjah Mada University Press.
[9] Kumar P S and Rajkumar R 2012 Effect of geotextile on CBR strength of unpaved road with soft subgrade *Electronic Journal of Geotechnical Engineering* vol **17** Bund. J.
[10] Laksono T D 2011 Perbaikan tanah dengan menggunakan geotekstil *Journal Teodolita* **12**(2) pp 19-26.
[11] Naeini S A and Mirzakhanlari M 2008 The effect of geotextile and grading on the bearing ratio of granular soils *Electronic Journal of Geotechnical Engineering* vol **13**.
[12] Saputra N A, Rustam E and Markawie 2013 Kekuatan geser antarmuka laterit palangkaraya dan geotekstil berdasarkan uji geser langsung *Jurnal Teknologi Berkelanjutan* vol **2** no 1 pp 63-72.

IOP Conf. Series: Materials Science and Engineering **821** (2020) 012018 doi:10.1088/1757-899X/821/1/012018

Laboratory study: durability characteristics of Asphalt concrete binder course utilizing Bantak and Clereng as aggregate and Asbuton Lawele as fine aggregate (using Marshall Methods)

E Ruhaidani*, D P Hardiani, E Anggarini, and Y Perdana

Department of Civil Engineering, Faculty of Engineering, Universitas Muhammadiyah Banjarmasin, Banjarmasin, Indonesian

*Email: emma@umbjm.ac.id

Abstract. Construction of roads and will continue and growing, while on the other side of the raw material reserves for the road less. Given this, the researchers tried to use the mining waste material at the bottom of Sabo Dam from Merapi which is bantak to substituted coarse aggregate and asbuton is substituted fine aggregate to tackle the lack of asphalt and to utilise local's natural resources. This research carried out by using a mixture of 40% Bantak and 60% Clereng and substitute fine aggregate (substitution) with asbuton Lawele type 20/25 with the same amount. The asbuton concentrate variation that used in the test are (100%:0), (75%:25%), (50%:50%), (25%:75%), (0%:100%). All the specimens tested and analysed using Marshall Method as Public Work Department specification (2007), determined the optimum asphalt value of each specimen. All specimens with optimum value will immersed in water during half hour, 24 hours and 48 hours before Marshall test. Result of the test is used to know the asbuton substitution variation effects based on stability value, and residual strength index. The result shows that optimum asphalt value of asbuton specimens with percentage asbuton 0%, 25%, 50%, 75% and 100% are 6.2%; 5.6%; 5.25%; 5.1% and 5%. On standardise immersion asbuton variation of 0%, 25%, 50%, 75% and 100% the mix stability value are of 1498 kg, 1629 kg, 1405 kg, 1264 kg and 1258 kg respectively. While on 48 hour immersion, the residual strength indexs are 97.97%, 88.86%; 71.06%; 64.41% and 57%. In the conclusion optimum utilisation asbuton is 25% with optimum asphalt value 56% because has high stability, but has low durability.

1. Introduction

Indonesia is a country with abundant natural resources. One of them is active Merapi volcano with lava eruption releases mining material, known as Bantak that porous material and has a low level of hardness. Research conducts a test between the aggregate mixture of Bantak (porous aggregate from Mountain Merapi) and aggregate from Clereng, resulting the proportion of mixture for the hardness layer are 40% Bantak and 60% clereng [1]. Asbuton Lawele is one type of asbuton that is relatively soft from Lawele. Totomiharjo [2] Asbuton is a type of natural asphalt, found in reef rocks, so that the asphalt is mixed with limestone ($CaCO_3$). The surface of the Bantak is porous, asphalt more absorbed in the mixture to fill the pore.

IOP Conf. Series: Materials Science and Engineering **821** (2020) 012018 doi:10.1088/1757-899X/821/1/012018

Asbuton containing minerals and asphalt is expected to be a pore filler on porous material and to increase strength and durability of the pavement. Pavement Durability is related to the effect of a mixture of water and temperature for a long time. The pavement mixture must be resistant to water due to evaporation and oxidation. Pavement Durability depends on the good mixture, the composition of the mixture in accordance with the requirements, strength according to planning, work implementation and good maintenance. Study state that 8% of granular asbuton growth Marshall Stability and Marshall Immersion and Optimum Asphalt Content values in Asphalt Concrete-Binder Course Mixture [3].

The purpose of this research is to create a mix design asbuton and porous aggregate. Thus obtained optimum level of Asbuton with the required strength. In addition, it also to find out the optimum level of durability from porous aggregate and asbuton mixture through immersion index (IP) for ½ hour, 24 hours and 48 hours at 60 °C. Study state no significant difference for all mixture when BGA replaced petroleum asphalt in ACBC pavement [4].

2. Method and materials

The methodology used in this research is the experimental method, its starts from material testing, design mix and performance testing with Marshall Test (1939).

2.1 Aggregate

In this research, using aggregate Clereng (60%), Bantak (40%), and fillers as can be seen in Table 1.

Table 1. Physical properties of aggregate

Properties	Test Value	Indonesian Standard	
		Lower Limit	Upper Limit
Aggregate			
1. Abration(%)	24.4	-	40
2. Stickiness (%)	98	95	-
3. Absorption (%)	1.79	-	3
4. Specific gravity (gr/cm3)	2.729	2.5	-
5. Soundness test (%)	1.57	-	7
Bantak			
1. Abration (%)	60.21	-	40
2. Absorption (%)	1.523	-	3
3. Specific gravity (gr/cm3)	2.645	2.5	-
40% Bantak + 60% Aggregate			
Abration (%)	-	40	30.55
Fine Aggregate			
1. Absorption (%)	-	3	2.564
2. Specific gravity (gr/cm3)	2.5	-	2.868
Filler			
Berat jenis (gr/cm3)	2.722	2.5	-

2.2 Asphalt

Physical properties of petroleum asphalt 60/70 can be seen in Table 2.

Table 2. Physical properties of petroleum asphalt 60/70

Properties	Test Value	Indonesian Standard
Penetration 25 °C	62.2	60-79
softening point	48	60-79
Flash point	347	48-85
Ductility 25 °C	>100	>200

2.3 Asbuton

The Asbuton used is a varied 25%, 50%, 75%, and 100% (Table 3).

Table 3. Physical properties of Asbuton

Properties	Test Value	Indonesian Standard
Bitumen Content	27.54	25-40
Penetration 25°C	20.6	19-22
Water content	0.83	<2

3. Result

More immersion time, value of Marshall Quotient was decrease because flow was increase and stability index was decrease. After added 25% of Asbuton, Marshall Quotient continues to decline. It is indicate for this condition stiffness on mix desain. This is due to the decreased of stability and the increasing flow value of the test objects. The result of the research on the image shows that the value of MQ in the mix with the degree of Asbuton 25% higher than 0% Asbuton. After adding 25% of the Asbuton MQ value continues to decline because it has reached the optimum value. This indicates in this condition the stiffness of the mixture decreases and is prone to plastic deformation.

The results showed that the strength index of asbuton variation of 50%, 75% and 100% at immersion 24 hours and 48 hours did not meet the specifications required by the Public Works Department which is 80% [5]. This suggests that the mixture has no good enough durability to withstand its stability due to the influence of the weather and the traffic load over the lifetime of the plan.

In the Table 4, it can be seen that the long immersion and added asbuton affects the rise of residual index and stability index values in all variations of Asbuton. That indicate the greater the occurrence of stability decreases which means the durability of the mixture is decreasing.

Table 4. Result of mix design

Marshall Characteristic	Indonesian Standard	Asbuton (%)				
		0	25	50	75	100
Optimum Asphalt (%)		6.20	5.60	5.25	5.10	5.00
Density (gr/cc)		2.33	2.28	2.22	2.13	2.04
VMA (%)	≥ 14	16.29	14.63	13.13	12.82	12.60
VFWA (%)	≥ 63	73.38	71.03	71.83	68.26	66.16
VITM (%)	3,5-5,5	4.34	4.26	3.70	4.08	4.38
Stability (Kg)	≥ 800	1498.48	1629.10	1405.64	1264.39	1257.98
Flow (mm)	≥ 3	3.67	3.70	3.83	4.03	4.27
Marshall Quotient (kg/mm)	≥ 250	409.15	440.28	367.75	312.95	300.37
Residual index (Immersion 24 hours)	≥ 80	99.71	95.87	78.04	73.88	58.29
Residual index (Immersion 48 hours)		97.97	88.86	71.06	64.41	57.00
Stability index (%) (Immersion 48 hours)		0.26	0.41	1.54	1.86	2.68
Loss of strength (Kg) (Immersion 48 hours)		5.82	6.68	21.70	23.53	33.72

4. Conclusion

The mixture of 40% porous agregate and 60% agregate used the optimum Asbuton as a substitute for fine aggregate in the mixture is 25%. The value of decreased stability index (r) as well as the loss of strength/ value of decreased stability (R) of the mixture is increasing as the long of immersion and the variation of Asbuton. This shows that the greater the occurrence of stability decreases which means the durability of the mixture is decreasing.

References

[1] Nugraha W 2007 *Kajian Penggunaan Bantak (Kerikil Merapi) Sebagai Bahan Campuran Lapis Pondasi Atas (Base Course)*, Gadjah Mada University Yogyakarta (Indonesia).

[2] Totomiharjo, S 2000 *Bahan dan Struktur Jalan Raya* Biro Penerbit KMTSJTS FT UGM Yogyakarta.

[3] Setiawan A and Rahmatang R 2011 Studi penggunaan asbuton butir pada campuran beton aspal binder course (AC-BC) *Jurnal Rekayasa dan Manajemen Transportasi* vol **I** no 2.

[4] Gauss A, M Whihardi T, Nur M A and Rudi D 2015 Compressive strength of asphalt concrete binder course (AC-BC) mixture using buton granular asphalt (BGA) *The 5th International Conference of Euro Asia Civil Engineering Forum (EACEF-5) Procedia Engineering* **125**.

[5] Ministry of public works Directorate general of highways 2007 *Specification hot mix asphalt with processing asbuton* Jakarta (in Indonesia).

IOP Conf. Series: Materials Science and Engineering **821** (2020) 012019 doi:10.1088/1757-899X/821/1/012019

Mechanical properties of self-compacting concrete incorporated with high volume fly ash

M Solikin* and T Mulyanto

Study Program of Civil Engineering, Universitas Muhammadiyah Surakarta, Surakarta, Indonesia

*Email: msolikin@ums.ac.id

Abstract. Self-compacting concrete (SCC) is beneficial in concrete casting process as it does not need any tools for consolidation process. In addition, the utilization of High content fly ash in has proven in improving mechanical properties of concrete although the slow strength development is an issue to coop with. This paper presented an experimental program on the mechanical properties test of high strength self-compacting concrete incorporated with high content fly ash. The mix proportion for the concrete was prepared based on ACI high strength concrete with the design strength of 45 MPa. The mechanical properties test consists of slump flow test, compressive strength test development, flexural strength test and water absorption test of concrete. The slump flow test was tested for fresh concrete and the strength development test was tested at the age of 14, 28 and 56 days. In addition, test for flexural strength specimen and absorption specimens were tested at the curing age of 56 days. Moreover, the high strength normal concrete specimen was prepared as control mix design. The result shows, less water was need to reach design flowability for Self-Compacting Concrete High content fly ash concrete and it gives impact to lower water absorption for the type of concrete. In addition, at the curing age of 56 days the compressive strength of high strength high content fly ash-self compacting concrete was comparable to the normal concrete.

1. Introduction

Concrete is globally used construction material to support modern infrastructure. The dramatic increase of modern infrastructure such as skyscraper building needs more reinforcement inside reinforced concrete as element of structure, i.e.: column, beam and shear wall. It leads to narrow spacing of reinforcement and increase the difficulties in compaction of fresh concrete inside the formwork. Therefore, the use of self-compaction concrete become a perfect solution to produce dense, strong and durable concrete.

Self-Compacting Concrete (SCC), was developed for the first time by Japanese researchers in the late 1980's. It is an extremely workable concrete that can flow under its own weight through narrow sections without segregation and bleeding [1]. SCC does not need any vibration equipment in compaction process and its high flowability makes the fresh concrete can easily fill up its formwork to reach its optimum dense. The fresh Self-Compacting Concrete generally have a relatively low yield value to ensure high flowability, a modest viscosity to avoid segregation and bleeding, and should keep its homogeneity during transportation stage, pouring and curing process to ensure the hardened concrete meet with structural performance and long-term durability. The successful casting of Self-Compacting Concrete must guarantee a perfect balance between deformability and stability. Mix

IOP Conf. Series: Materials Science and Engineering **821** (2020) 012019 doi:10.1088/1757-899X/821/1/012019

proportion of SCC always introduce superplasticizer to produce high flowability of fresh concrete and may use appropriate pozzolanic material, such as fly ash, rice husk ash, silica fume, and slag.

Fly ash is the most widely accepted pozzolanic material in concrete, which is a by-product of combustion of ground or powdered coal exhaust fumes in coal-fired power station [2]. The burning process uses burning temperature of 1.300 – 1.400° C. The fly ash is removed from the dust collection system of the exhaust gasses in coal fired power stations before they are discharge in to the atmosphere, as very fine particles, predominantly spherical glassy. The fly ash size of particle ranges from 1 – 150 μm which is generally finer than particle size of Portland cement. In addition, the size of particles is close related to the type of dust collection equipment [3].

There are three major chemical substance in fly ash i.e. Silica (SiO_2) around 5-25%, Alumina (Al_2O_3) about 10-30%, and ferric oxide (Fe_2O_3) about 5-25%. In addition, the other chemical substance in fly ash are calcium oxide (CaO), magnesium oxide (MgO), Sulphur trioxide (SO_3) and alkali oxide (Na_2O) [2]. According to the major chemical contents, type of fly ash can be devided into three i.e. class F, class N, and class C fly ash.

The utilization of fly ash as a binder has demonstrated in improving many concrete properties, both in hardened concrete and fresh concrete [4]. The use of fly ash as partially cement replacement make improvement in fresh concrete for increasing the workability, reducing bleeding, and retarding time set. In addition, for hardened concrete, fly ash contributes in continuing the hardened concrete's pozzolanic activity to gain higher strength for longer period, while the strength contribution rate of Portland cement decreases. Moreover, the most contribution of fly ash in hardened concrete is significantly improve the durability performance [2].

Previous researcher shows the utilization of high content fly ash as partial cement replacement produce a comparable compressive strength in comparison to normal concrete. Despite of its benefit the utilization leads to the slow strength development because fly ash needs $Ca(OH)_2$ a product of cement hydration. As high content fly ash concrete used minimum 50% of fly ash cement replacement, the replacement is beneficially to the environment to reduce significantly the CO_2 emission from cement production.

Based on the discussion this paper will explore the utilization of high content fly ash concrete to produce self-compacting concrete. The discussion in this paper will focus on some properties of self-compacting concrete for fresh concrete (slump flow test) and for hardened concrete (compressive strength, flexural strength and water absorption).

2. Research method

The research programs consist of 5 stages i.e.: material and equipment preparation, concrete specimen casting, curing of the specimens, test of the specimens and analysis and discussion. Material for concrete specimens was selected from local quarry in Surakarta region which is meet the standard as concrete material. Following the material preparation, a 45 MPa mix proportion of concrete was prepared based on ACI mix proportion for high strength concrete and the following table (Table 1) shows the mix proportion.

Table 1. Mix proportion of concrete per m^3

	SCC normal concrete	SCC HVFA concrete
Cement (kg/m^3)	527.0	263.5
Fly ash (kg/m^3)	0	263.5
Fine Aggregate (kg/m^3)	878.0	878.0
Coarse Aggregate (kg/m^3)	756.0	756.0
Water (kg/m^3)	Design: 190.0 Actual: 180.0	Design: 190.0 Actual: 155.8
SP (kg)	7.9	7.9

There are two type of mix design in this experimental work i.e.: high strength self-compacting concrete normal concrete and high content fly ash in. For normal concrete all of binder is Pozzolanic Portland cement (PPC) produce from Portland Cement factory whereas for HVFA concrete half of PPC was replaced by fly ash type F.

In preparation of self-compacting concrete mix proportion, some considerations were taken, i.e.: maximum size of aggregate was 10 mm and the total weight of fine aggregate was higher rather than total weight of coarse aggregate. In addition, 7.9 kg/m^3 superplasticizer type: viscocrete-10 was used to improve flowability of SCC. Although the w/c ratio was designed at 0.36 the actual water used was maintained at minimum amount to reach workability design. It leads to lower water usage rather than the designed w/c ratio.

The lower water usage for HVFA concrete in comparison to normal concrete shows that the utilization of fly ash increases the workability of concrete. The increase of workability is possibly caused by the spherical shapes of the fly ash that reduces the friction between cement and aggregates and results in an increase in the workability of fresh concrete [5]. Previous researcher also reported the higher slump test for high content fly ash was caused by the increase of concrete workability and it is close related to the spherical fly ash particle [6-8].

The specimens for testing consist of cube 15 x 15 x 15 cm for compressive strength test, small beam 15 x 15 x 50 cm for flexural strength test and cylinder diameter 10 x 5 cm height for water absorption test. After 24 hours of casting process the specimens were removed from mold and the curing process is started by immersing them in water until one day of testing date. The number of specimens and the design test date is shown in Table 2.

Table 2. Matrix of concrete specimen

No	Testing program	Specimen	Specimen's age (days)		
			14	28	56
1	Compressive strength	Cube 15 x 15 x 15 cm	8	8	8
2	Flexural strength	beam 15 x 15 x 50 cm			6
3	Water absorption	cylinder dia. 10 x 5 cm height			6

The matrix of specimen table shows, there are 4 compressive strength specimens for each type of mix design and for different curing age. However, there are 3 flexural strength test and 3 water absorption test for each type of mix design. The total number specimens need to be prepared was 36.

3. Result and discussion

The result and discussion of High Strength Self Compacting Concrete High content fly ash concrete starting with the investigation of material to be used in casting the concrete specimen. The investigation on fine aggregate and coarse aggregate shows that all of them meet the standard requirement of SNI as concrete aggregate. Therefore, they can be used to cast the concrete and the experiment can be carried on. The next examination is exploring chemical substances in fly ash. The fly ash used in the research was a by-product of power plant combustion in Java Island and collected from PT. Jaya Ready Mix Sukoharjo. The chemical test was conducted at Laboratorium Pengujian dan Kalibrasi BARISTAND INDUSTRI Surabaya and the following table shows the chemical content of fly ash.

Table 3 The result of fly ash testing

No	**Chemical composition**	**Percentage %**	**ASTM Standard**
1	SiO_2	86.09	$SiO_2 + Al_2O_3 + Fe_2O_3$ min. 50%
2	FeO_3	1.64	
3	Al_2O_3	5.97	
4	CaO	2.31	
5	MgO	0.45	
6	Loss of Ignition	2,16	

The data in Table 3 shows that the content of (SiO_2 + Fe_2O_3 + Al_2O_3) is 93.60%, while the minimum standard of those three major components for fly ash class C is 50%, and the standard of those for Fly ash class F is 70%. Hence, it can be concluded that fly ash from PT. Jaya Ready Mix Sukoharjo belongs to Fly ash class F (ACI Manual of Concrete Practice 1993) [9].

3.1. Slump flow test.

Slump flow test is design to check the flowability of fresh concrete to ensure that the fresh concrete can fill up the formwork to achieve maximum dense. The test carried on by measuring the diameter of fresh concrete on a flat surface after removing from slump cone. The test and the result are shown in Figure 1 and Table 4.

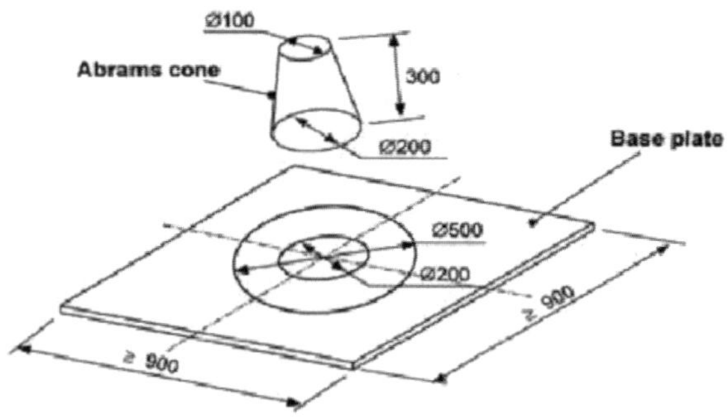

Figure 1. Diameter measurement of slump flow test [10]

Table 4. Slump flow test result

Kode	Diameter		Average
	1 (mm)	2 (mm)	(mm)
SCC normal	580	620	600
SCC HVFA	595	605	600

The diameter of slump flow test result is 600 mm for SCC normal concrete and also for Self-Compacting Concrete High content fly ash concrete. As the requirement of diameter in slump flow test for SCC is 500 – 700 mm, therefore both mix proportion meet the criteria of SCC. In addition, the less water used in HFVA concrete was not reduce the flowability of fresh concrete. This fact shows that the use of HVFA as partial cement replacement increase the workability of fresh concrete due to the spherical shape of fly ash.

3.2. Compressive strength test.

Compressive strength test was conducted at the age of 14 days, 28 days and 56 days to determine the strength development. The test was conducted on 15 x 15 x 15 cm^3 concrete cube and the specimen was prepared by taken out from the curing tank one day before the day of test. As shown in Figure 2 the load was applied on the cube until maximum load and the following formula is used to find out the compressive strength [11].

$$f'c = \frac{P}{A} \tag{1}$$

where:
f'_c = Compressive strength (MPa)
P = Maximum load (N)
A = surface area of cube (mm^2)

Figure 2. Compressive strength test of SCC

As the compressive strength was carried on 15 x 15 x 15 cm cube, it needs 0.83 correction factor to convert the compressive strength to cylinder diameter 15 cm and 30 cm height. Based on compressive strength test result at Figure 3, at early age i.e.: 14 days and 28 days the compressive strength of Self-Compacting Concrete of normal concrete relatively higher than the Self Compacting Concrete High content fly ash concrete. However, at the age of 56 days, differently from Self Compacting Concrete High content fly ash that continuously improve the compressive strength development, the SCC normal concrete did not improve the compressive strength. This result shows a slow strength development of fly ash concrete.

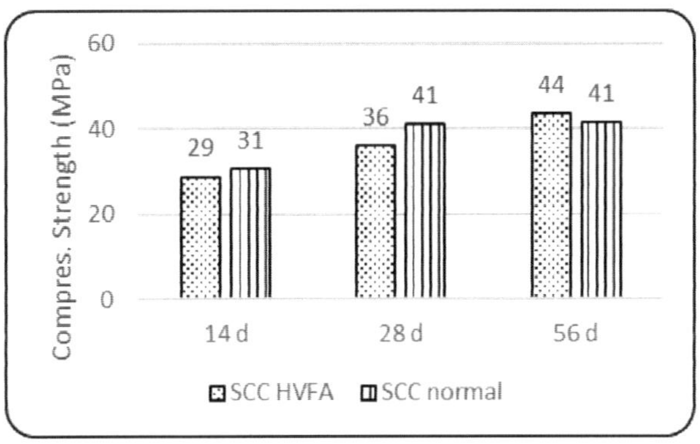

Figure 3. Compressive strength result of SCC

The slow development of concrete compressive strength is possibly caused by the low cement used in high content fly ash, so that the concrete has low C_3A content from cement, the substance which mainly contributes for early age strength. Previous researcher confirmed that phenomena that fly ash concrete usually demonstrates lower strength at early ages especially for high content fly ash concrete in comparison to ordinary portland cement (OPC) concrete, although it shows higher strength at a longer period of time [12, 13].

3.3. Flexural strength test.

The flexural strength was tested on concrete beam with the dimension of 15 x 15 x 50 cm under one-point load (Figure 4). The test was conducted after 56 days of curing age to ensure Self Compacting Concrete High content fly ash concrete reach the comparable strength to normal concrete. The flexural strength test is calculated by using this formula:

$$f_{lt} = \frac{3\,P\,L}{2\,b\,d^2} \quad (2)$$

Where
f_{lt} = Flexural strength (MPa)
P = Maximum load (N)
L = Clear span (mm)
b = average width of beam (mm)
d = average height of beam (mm)

The flexural strength test result in Figure 5 shows at the age of 56 days the SCC normal concrete had higher flexural strength in comparison to Self-Compacting Concrete High content fly ash concrete. It is noted in here that HVFA concrete probably more brittle rather than normal concrete. The lower modulus of rupture for fly ash in comparison to normal concrete was also reported by Duran-Herrera et al [14].

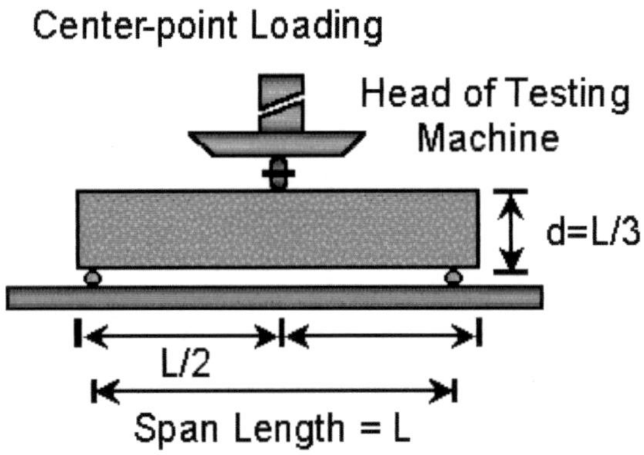

Figure 4. Test set up for flexural strength

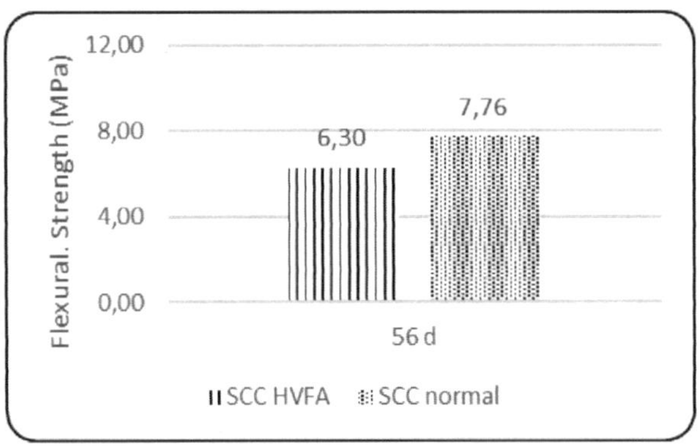

Figure 5. Flexural strength test result of SCC

3.4. Water absorption test.

Water absorption test is aimed to test the amount of water can be absorbed by the concrete and this property closely relate to the concrete durability properties. The test was conducted by finding the differences of water content between saturated specimen and oven dried specimen with diameter 10 cm and 5 cm height. There are two types of absorption tested, i.e. saturated water absorption and immersed water absorption. Saturated absorption measures the total absorption of water when the specimen is in saturated state while Immersed absorption measures the total water that can be absorbed by the concrete after soaking the specimens in water for two days.

This research focus on saturated water absorption and for saturating process of the specimens, the method in ASTM standard of preparing a concrete specimen for rapid chloride penetration testing method was used [15]. The procedure of testing was started by vacuuming the sample to suck up all the entrapped air in concrete voids for three hours following by soaking it into water for 18 hours. The following equation is used to calculate the water absorption of concrete:

$$\text{Water absorption} = \frac{W_2 - W_1}{W_1} \times 100\% \qquad (3)$$

Where:
W_1 = Oven dried specimen (gram)
W_2 = Saturated weight of specimen (gram)

The result of water absorption in Figure 6 at concrete age of 56 days shows the significant decrease of water absorption in HVFA concrete in comparison to normal concrete. The lower the durability in concrete will be beneficial in improving durability of concrete.

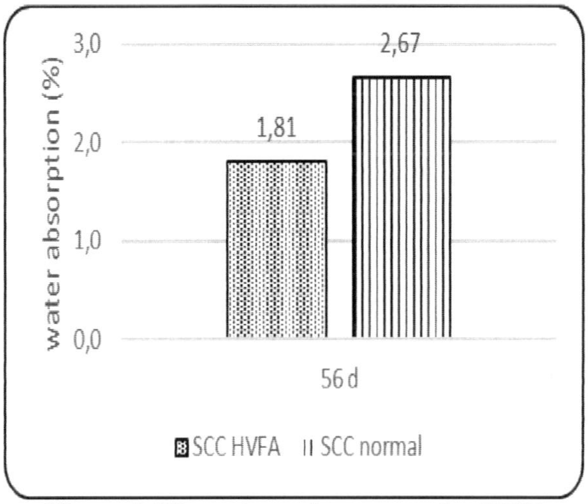

Figure 6. Water absorption test result of SCC

The possible factors affect to lower water absorption of fly ash concrete is the permeability of the transition zone surrounding the aggregate and also the permeability of the cement paste as reported by Haque and Kayali [16]. Moreover, High content fly ash concrete system has discontinuation of pore structure that is also contributes to the lower water absorption [17].

4. Conclusion

Based on the results of the analysis High Strength Self-Compacting Concrete High content fly ash concrete, the research has drawn some conclusions. First, self-compacting concrete with high content fly ash concrete has similar flowability as self-compacting normal concrete, and even better as HVFA concrete needs less water to reach design flowability. Second, the compressive strength of HVFA concrete is better than normal concrete especially after longer period. Third, as flexural strength of HVFA concrete is lower rather than normal, it indicates that HVFA concrete more brittle than normal concrete. Fourth, water absorption of Self-Compacting Concrete High content fly ash concrete is lower rather than SCC normal concrete which is indicate superior Durability properties. Further research needs to be conducted to explore the effect of using fly ash more than 50% to produce high strength high content fly ash Self Compacting Concrete.

Acknowledgement
This research is funded by Ministry of Higher Education through the scheme of *Hibah Bersaing* research grant. The author expresses gratitude to the laboratory of Civil Engineering, Universitas Muhammadiyah Surakarta, which kindly provide the experimental equipment for the research

References

[1] EFNARC 2002 *Specification and Guidelines for Self-Compacting Concrete Surrey* UK.
[2] Nawy E G 1996 *Fundamentals of High Strength High Performance Concrete* Longman Group Limited London UK.
[3] Siddique R 2008 *Waste materials and by-products in concrete* Springer Berlin London.
[4] Oner A, Akyuz S and Yildiz R 2005 An experimental study on strength development of concrete containing fly ash and optimum usage of fly ash in concrete *Cement and Concrete Research* vol **35** pp 1165–71.
[5] Sata V, Jaturapitakkul C and Kiattikomol K 2007 Influence of pozzolan from various by-product materials on mechanical properties of high-strength concrete *Construction and Building Materials* **21**(7) pp 1589-98.
[6] Atis C D and Karahan O 2009 Properties of steel fiber reinforced fly ash concrete *Construction and Building Materials* vol **23** pp 392–99.
[7] Bouzoubaa N, Zhang M H, Malhotra V M and Golden D M 1999 Blended fly ash cements - A review *ACI Materials Journal* November-December pp 641-51.
[8] Idorn G M and Henriksen K R 1984 State of the art for fly ash uses in concrete *Cement and concrete research* vol **14** 463-70.
[9] ACI 226-3R-87 1987 Use of fly ash in concrete: reported by ACI committee 226 *ACI Materials Journal* pp 381-409.
[10] Grdić Z, Despotović I and Ćurčić G T 2008 Properties of self-compacting concrete with different types of additives *FACTA UNIVERSITATIS Series: Architecture and Civil Engineering* vol **6**(2) pp 173 - 77.
[11] Kett I 2000 *Engineered concrete mix design and test methods* CRC Press LLC Boca Raton Florida.
[12] Atis C D 2003 High-volume fly ash concrete with high strength and low drying shrinkage *ASCE Journal of Materials in Civil Engineering* March-April pp 153-56.
[13] Siddique R 2004 Performance characteristics of high-volume Class F fly ash concrete *Cement and Concrete Research* vol **34** 487–93.
[14] Durán-Herrera A, Juárez C A, Valdez P and Bentz D P 2011 Evaluation of sustainable high-volume fly ash concretes *Cement & Concrete Composites* **33** pp 39-45.
[15] ASTM C 1202 -97 2002 Electrical indication of concrete's ability to resist chloride ion penetration *Annual Book of ASTM Standards* Philadelphia United States.
[16] Haque M N and Kayali O 1998 Properties of high-strength concrete using a fine fly ash *Cement and Concrete Research* vol **28** no 10 pp 1445–52.
[17] Malhotra V M and Mehta P K 2005 High Performance, high-volume fly ash concrete: materials, mixture proportioning, properties, construction practice, and case histories 2nd *ed. Suplementary Cementing Materials for Sustainable Development* Inc. Ottawa Canada.

IOP Conf. Series: Materials Science and Engineering **821** (2020) 012020 doi:10.1088/1757-899X/821/1/012020

Feasibility test on the performance of LoRa technology incorporated in the wireless sensor network

R P Astutik* **and H A Winarno**

Department of Electrical, Faculty of Engineering, Universitas Muhammdiyah Gresik, Gresik, Indonesia.

*Email: astutik_rpa@umg.ac.id

Abstract. Wireless Sensor Network (WSN) and Internet of Things (IoT) technologies are developed rapidly in many implementations such as smart cities, smart offices, and smart transportations. They are closely related to communication technology which requires efficient power and covered long distance. To overcome these two problems, this paper proposes the design of a communication system for WSN using LoRa technology that has the ability to receiving lower power and wider area coverage. In this research, the implementation of this communication design is just a prototype that contains two nodes that communicate with each other by using LoRa technology as a communication system. The experiment which was based on measurement had the maximum distance of a point to point communication that reached about 200 m while the lowest received power was accounted at -98 dB. The use of the best spreading factor based on simulation was recorded at 7 using a bandwidth of 250 kbps and 10 using a bandwidth of 125 kbps. Showing the result, it can be concluded that 200 m was the longest distance which can be achieved by this prototype and this was useful for a small system.

1. Introduction

The development of communication system design is currently very rapid and sometimes it is needed to get information quickly also sometimes from a place that is very impossible for humans to retrieve data. One of the designs of the communication system is the incorporated WSN and IoT systems that form a network. This network usually contains an IP (Internet Protocol) network, some devices such as sensors or actuators, telecommunication interface devices, data processing, and storage units [1]. The network model of IP address with many nodes will cost a lot therefore it needs to replace it with technology that uses lower power and wide range.

Various communication devices have been implemented such as [2] discussed communication system model using Zigbee for determining the position of the node. The other implementation is as [3] discussed regarding monitoring air pollution in Gresik, Indonesia using Wi-Fi connection for communication. In this research, the longest distance which each node still could communicate was only 110 m. With this drawback, we try to use LoRa technology as a transceiver in this research. The same discussion which is about temperature monitoring in a data center used LoRa technology that was integrated with WSN [4]. Meanwhile, several papers discuss the performance of LoRa technology in different perspectives such as in [5], it discussed the evaluation of coverage model and channel attenuation for LoRa technology on wide area networks with low power. The analysis of feasibility

WSN using LoRa has discussed in [6] by collecting data obtained from vehicles and submitted to the central node for further processing. However, this paper did not discuss the Doppler Effect. Another paper investigated the limits of LoRa network capacity and the result was LoRa networks could measure fairly well with a note if they used dynamic communication parameter selection and multiple sinks. [7] Beside the limits of LoRa networks capacity, the evaluation performance of LoRa networks also has discussed in [8] with analyzing of total packet collisions that occurred and the result of the maximum total nodes that can communicate on LoRa channel can be determined by considering the duty cycle restrictions.

LoRa technology integrated with WSN has used in many applications either in the rural or urban areas, especially in wide-area since it can cover for wide areas. To explore the performance of LoRa technology, it needs to hold some tests on how far it can cover areas as well as how much cost needed. In this paper, the discussion focuses on the performance evaluation of the integrated LoRa technology in the WSN by testing the feasibility of its system performance. The feasibility of this work is indicated by distance and RSSI (Received Signal Strength Indicator) measurement, packet loss and variations of spreading factor. This paper is organized as follows the first section will discuss the introduction contained in the background of this work, some related work and the aim of this work. The next section will discuss WSN and LoRa technology also link budget analysis. The third section will explain the design of in communication system for WSN using LoRa technology and the following section will share the experiment result.

2. WSN and LoRa technology

In designing communication systems for networks, several components are needed. These components are generally divided into three parts, namely the sensor subsystem, the processor subsystem, and the communication subsystem. The detail discussion for three parts will be explained in the following subsections.

2.1. *Wireless Sensor Network*

WSN is a number of sensors that are interconnected and have a function to control and process data directly in an area. The more sensors that communicate, efficient protocols are needed to support bringing data collection to the center wirelessly. A WSN is not only sensing components but also a board that has a processing unit, storage data and can communicate with each other. A node as part of WSN is divided into three parts, the first part is for the sensing unit, the second part is for processing data and the third part is a communication device. The architecture of a node can be shown in Figure 1 [9].

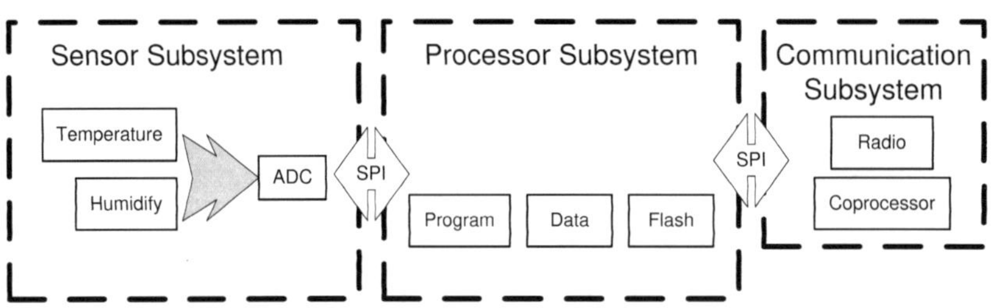

Figure 1. The architecture of a node as part of WSN

The sensor subsystem serves to get data sensing from one or more sensors and integrates it for future processing. It contains a sensor, analog to digital converter (ADC) board and serial-parallel interface (SPI) as connecting to the main processor. The sensor is used to retrieve data from the environment such as temperature, humidity etc. Since the sensing data from the environment is analog data, so an analog to digital converter device is needed for the next process. SPI is needed to connect the sensor subsystem to the processor subsystem, so data from the environment can process on the main processor.

The processor subsystem is the brain of a wireless sensor network which in this part, data processing will be obtained from the sensing subsystem. The main purpose of the processing subsystem is to execute commands related to sensing, communication, and self-organization. It consists of a processor chip, internal flash memory which is used to store command programs, active memory for the temporary storage of sensed data.

The communication subsystem is one of the factors that affect the performance and energy efficiency of wireless sensor networks. This performance is indicated by fast transferring data transmission between all of the nodes in the wireless sensor networks while the energy used for the data transfer process is the lowest.

2.2. LoRa technology

LoRa (Long Range) technology is one of the technologies communication that has long-range coverage area and lower power consumption. LoRa node is divided into three classes, they are A, B and C which each class has a different specification [8]. A class LoRa node has functions to send messages to other nodes for a certain period of time depending on the application used. After sending the message, the other device sends the acknowledgment message (ACK) to let the LoRa node know its message was successfully received while the LoRa device opens a slot of reception. The other node will either continue the process of the data or forward it to other nodes again depends on the needed. B class LoRa node is the extension of class A, by adding the scheduling system to open the acceptance slot so communication between nodes can be organized. C class LoRa node is the prolongation of class A when the message is sent then the receiving slots will always keep open to receive the ACK or other commands. Based on the mechanism of transfer data then A and B class use battery while C class uses an electrical network since it consumes a lot of energy.

LoRa technology uses spread spectrum modulation techniques created by Semtech. This modulation technique provides two-way communication through a special CSS (Chirp Spread Spectrum) mechanism that spreads narrowband signals over a larger frequency band with integrated FEC (Forward Error Correction). The sent message is formed into a packet message with structure as shown in Figure 2 [7].

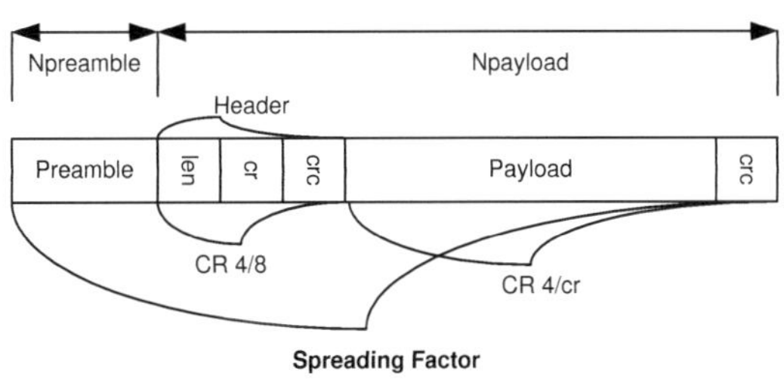

Figure 2. The structure of packet data using LoRa technology

The structure of the packet message consists of preamble, header, payload, and CRC which is each part has different size bytes. The preamble is contained from 6 to 65535 symbols and it can be programmed, meanwhile, the header is just optional and always transmitted with a 4/8 FEC rate and its own CRC. The next header is payload and following by an optional 16-bit CRC. The symbols total, in the packet message, is called Spreading Factor which can be varied between 6 and 12.

There are five parameters in LoRa radio, namely Transmission Power, Carrier Frequency, Spreading Factor, Bandwidth and Coding Rate. To get the best performance by minimizing power consumption and noise resistance then it needs to determine those parameters. The variation of each parameter can be seen in Table 1.

Table 1. LoRa radio parameter

Parameter	**Value**
Transmission Power	-4 dBm to 20 dBm
Carrier Frequency	860 MHz to 1020 MHz with step 61 MHz depending on the chip
Spreading Factor	6 to 12
Bandwidth	7.8 kHz to 500 kHz
Coding Rate	4/5, 4/6, 4/7 or 4/8

LoRa modulation technology uses the chirp spread spectrum, this results in an increase and decrease in frequency during a certain time based on the encoded information. By using this modulation, the signal is resistant to interference since high frequency used and robust to face multipath fading due to the use of broadband chirps. Another effect is the maximum link budget of this technology modulation is 157dB so that the communication range can be longer and save energy. It is as said that LoRa as Long Range communication

3. Design of communication system for WSN using LoRa technology

The design of the communication system on this project focuses on the performance of a communication system where there are some points to be reviewed, they are linked budget calculation based on theory and practice, model modulation used and packet losses. The following paragraphs are explained in detail for the design model and the basis of the system feasibility analysis.

3.1. Design hardware of LoRa-based WSN technology

In this project, two nodes are built where each node consists of a microcontroller, temperature and humidity sensors, LoRa as a transceiver and power bank as a source of electricity. This design can be seen in Figure 3.

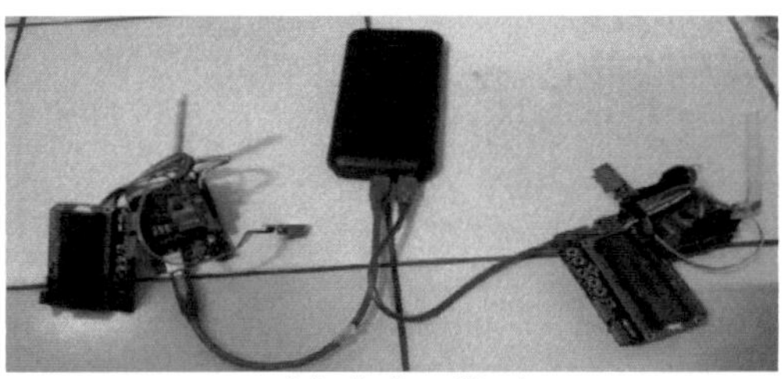

(a) Indoor Design

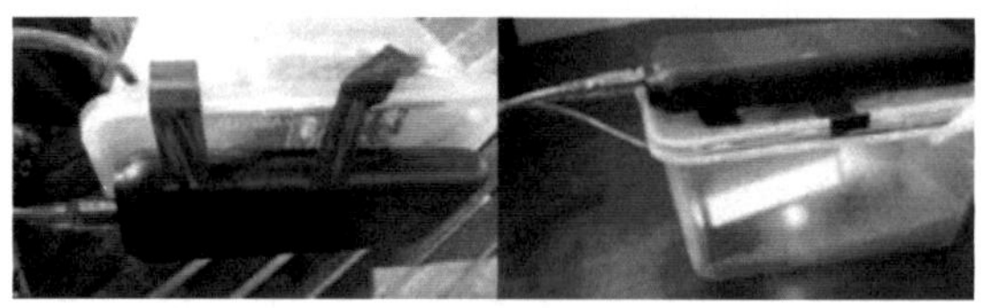

(b) Outdoor Design

Figure 3. Design two node with LoRa technology (a) Indoor Design and (b) Outdoor Design

The hardware of each node consists of a DHT11 sensor, LCD, Arduino, and LoRa also power banks for the electrical resources. Some scripts for sending/receiving temperature and humidity data and also detect the signal strength of each node have been filled into each node. After that, each node is turned on and separated with a certain distance then checked the receiving data. Both data will appear in each of both nodes so it will be known that both nodes have communicated.

3.2. Link budget analysis

One indicator of network feasibility is propagation performance which with this we will know how far the coverage area based on frequency owned by LoRa and the received power was performed. The propagation performance depends on path loss, showing and multipath fading that occurs on the channel of wireless communication. The received power of link budget analysis can be calculated based on Equation (1) where PT is the transmitted power, GT is the transmitted gain, GR is the received gain and LP is the path loss of the link between transmitter node and received node.

$$P_R = P_T + G_T + G_R - L_P \tag{1}$$

Regarding the path loss, based on [6], the LoRa model of propagation has not been determined yet, nevertheless based on the Okumura-Hatta model, it can be calculated in (2) for wireless communication in the urban environment.

$$L_P = 69.55 + 26.16log_{10}(f) - 13.82log_{10}(h_T) - a(h_R) + \left(44.9 - 6.55log_{10}(h_T)\right)log\ _{10}(d) \tag{2}$$

Where the correction parameter an (h_R) is obtained from (3), f is the frequency used and h_T and h_R are the height of transmitter and receiver antenna. Since the area is for urban or suburban even a country then d is for the distance between transmitter and receiver in kilometers.

$$a(h_R) = 3.2\left(log\ _2(11.75h_R)\right)^2 - 4.97 \tag{3}$$

3.3. *Doppler effect*

The Doppler Effect is the change in frequency or wavelength of a wave in relation to an observer who is moving relative to the wave source The direction and the velocity of the source's movement is determined the different of Doppler effect. This is can be used for another indicator of performance in LoRa technology by comparing between coherence time (t_c) and symbol time (t_s). Both times can be determined by (4) and (5) [10].

$$t_c = \frac{2\pi}{\omega_D} \tag{4}$$

where ϖ_D is the angular frequency shift caused by Doppler Effect

$$t_s = \frac{2^{SF}}{BW} \tag{5}$$

Where SF is spreading factor and BW is bandwidth, the relation of t_s and SF is t_s will double when SF is increased by one, in this case, bandwidth does not change. The Doppler Effect that triggers fast fading will cause signal distortion and it will happen when the symbol time more than the coherence time.

4. Results

The first examination is measuring the signal strength of LoRa versus the distance and getting results as the graph shown in Figure. 4. It can be seen that based on measurement, the signal strength plummeted while the distance was extended and the worst signal strength in this experience result fell to -98 dBm when the distance was 200 m. This achievement was almost as two times as previous research [3] which was reached at 110 m.

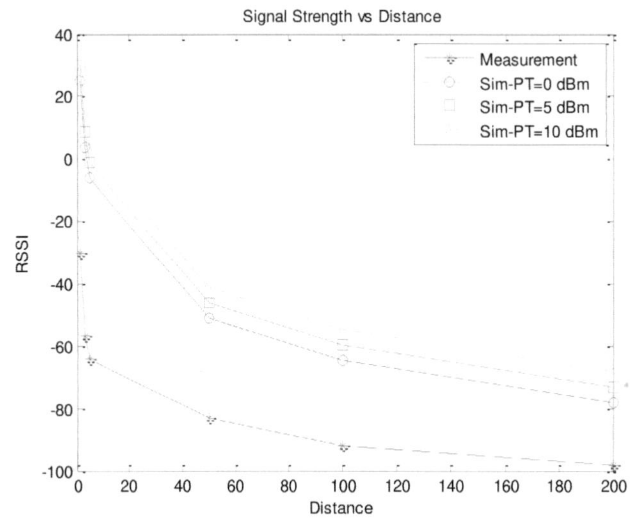

Figure 4. Experience result signal strength versus distance

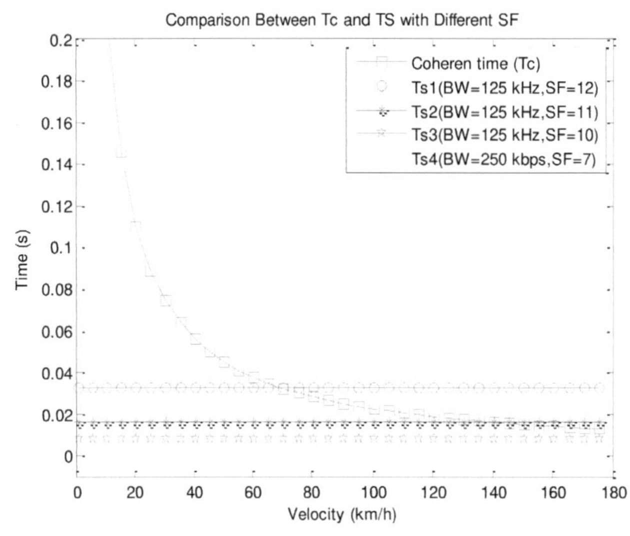

Figure 5. Experiment result comparison coherence time versus symbol time

Besides the measurement both received power and distance between two nodes, experiment also calculated the signal strength based on (1) with selected parameter such as GT and GR was determined as unity, while frequency used was 915 MHz and the transmitted power used in three different values, they were 0 dBm, 5 dBm and 10 dBm. The simulation result shows the highest transmitted power has better signal strength. The best signal strength was achieved at -68.17 dBm with a distance of 200 m when the transmitted power used at 10 dBm.

Another analysis of the feasibility test is the Doppler Effect, in this case, simulate the coherence time and the symbol time has been carried out as in Figure.5. Looking on the details, the three-line graphs of symbol time remained steadily at three different values such as around 30 ms when SF (Spreading Factor) =12 and BW (Bandwidth) =125 kbps, just under 20 ms when SF=11 and BW=125 kbps, around 10 ms when SF=10 and BW=125 kbps and just over 0 ms and BW=250 kbps. Looking at the coherence time, having plummeted to 20 ms when the velocity at 60 km/h, it leveled off at just under 20 ms while the velocity decreased. It can be said that when the SF value is greater than the fast velocity, it will cause the Doppler Effect. When SF=12 and BW= 125 kbps, the Doppler Effect will occur at the velocity of more than 70.5 km/h. However using SF=7 and BW= 250 kbps and SF=10 and BW 125 kbps, the coherence time will always be above the symbol time, it means that the Doppler Effect will not occur during observation.

5. Conclusion

In this paper, we have developed the design of a communication system for WSN using LoRa technology that focused on the communication system. Considering the two issues that should be coped for communication system design such as wider coverage areas and the lower received signal strength. The measurement result of the longest distance in which two nodes can still be communicated to each other reached 200 m when the signal strength measured at -98 dBm. Based on the simulation, by selecting parameters, having the same length distance, the signal strength was accounted at -68.17 dBm. Doppler Effect, based on simulation, occurred when velocity more than 70.5 km/h by using SF=12 and BW=125 kbps. It did not occur when the spreading factor decreased.

Acknowledgments
This research was carried out with financial aid supported by Universitas Muhammadiyah Gresik mainly from the research department.

Reference

[1] Marco C, Lorenzo V, Andrea Z and Michele Z 2015 Long-range communications in unlicensed bands: the rising stars in the IoT and smart city scenarios in *Fabulous 2015 Conference* Ohrid, Republic of Macedonia pp 2-8.
[2] Aries P, Adam S P, Prima K, Amang S, Mike Y and I Gede P A 2017 Skema lokalisasi posisi node terdistribusi pada lingkungan free space path loss *Jurnal Nasional Teknik Elektro dan Teknologi Informasi* vol **6** no 3 pp 352-58.
[3] Misbah, Rini P A, Eliyani, Hendra A W; Andy M, Deny A 2018 Communication system on wireless sensor networks using raspberry pi and Arduino for monitoring gas of air pollution in *International Seminar on Intelligent Technology and its Application* Sanur Bali 2018 pp 39-44.
[4] Tommaso P, Davide B, Andrea B and Luca B 2019 A LoRaWAN wireless sensor network for data center temperature monitoring in *International Conference on Applications in Electronic Pervading Industry, Environment and Society* Switzerland 2019 pp 169-77.
[5] Juha P, Konstantin M, Antti R and Tuomo H 2015 On the coverage of LPWANs: Range evaluation and channel attenuation model for LoRa technology in *2015 14th international conference on ITS Telecommunications* (ITST) Copenhagen Denmark 2015 pp 55-59.
[6] Silvano B, Lorenzo C, Emanuele M, Miryam P, and Marco A 2018 Feasibility analysis of a LoRa-based WSN using public transport *Applied System Innovation* vol **49** no 1 pp 1-12.
[7] Martin B, Utz R, Thiemo V and Juan M A 2016 Do LoRa low-power wide-area networks scale? in *The 19th ACM International Conference on Modelling Analysis and Simulation of Wireless and Mobile Systems* Malta 2016 pp 59-67.
[8] Alexandru L and Velentin P 2018 Performance evaluation of LoRaWAN communication scability in large-scale wireless sensor networks *Wireless Communication and Mobile Computing* vol **2018** pp 1-9.
[9] Dargie W and Poellabauer C 2010 *Fundamentals of Wireless Sensor Networks Theory and Practise* (United Kingdom: John Wiley & Sons Ltd)
[10] Juha P, Konstantin M, Marko P, Janne J and Jari L 2017 Performance of a low-power wide-area network based on LoRa technology: Doppler robustness, scalability, and coverage *International Journal of Distributed Sensor Networks* vol **13** no 3 pp 1-16.

Water content analysis effect on dielectric breakdown voltage in 500 kv transformer oil using coefficient correlation method

M A Auliq* and A B Nugroho

Department of Electrical, Faculty of Engineering, Universitas Muhammadiyah Jember, Jember, Indonesia.

*Email: aan.auliq@unmuhjember.ac.id

Abstract. Transformer is a vital component in distributing electrical power. The use of transformer oil which is continuously for a long time and not monitored can cause fatal damage to the transformer. This condition triggered by high water content in the transformer oil which causes breakdown voltage failure and electrical short circuit on the transformer. Therefore this study aims to analyze the influence of water content and breakdown voltage on 500 kV transformer oil type Kunlun Ki25x using the coefficient correlation method. The coefficient correlation method is statistical analysis technique used to look for relationships between variables that are quantitative. Two variables are said to be correlated if the changes in one variable will be followed by changes in the other variables regularly in the same direction (positive correlation) or opposite direction (negative correlation). For material testing and analysis data researcher using Megger OTS100AF instrument. This experiment conduct in paiton 9 power plant East Java for 36 month. Based on experimental data result researcher obtained a strong negative correlation value on the relationship of water content and breakdown voltage with the coefficient $R = -0.88$ on the function $y = 0.7692x + 85.733$ and it has characteristic that the increase in water content (x) is inversely proportional to increase the dielectric breakdown voltage (y). In addition, it is also known that the increase in transformer oil temperature is directly proportional to the increase in water content with a positive correlation of $R = 0.92$ at the function $y = 0.4208x + 50.155$. From the results above, it can be concluded that the parameters of water content and temperature have an impact on the value of the breakdown voltage and the performance of the transformer

1. Introduction

Transformer is important components in the process of distributing electric power from the generating system to the consumers. This component serves to increase or decrease voltage without changing the power, frequency and phase of the power distribution system. As a result of continuous use in a long time span has potential to cause damage or disturbance to the transformer. Generally this disturbance is caused by overloading. High temperature in overloading triggered hot spots arise in the internal area of the transformer. If this condition left continuously without maintenance it will causes decrease in the transformer oil insulation power capability. Zhao Tao research explains that the characteristics of transformer oil in high temperature conditions will produce water vapor that settles on the ceiling of the transformer [1]. While in low temperature conditions, the water vapor will drip and contaminate the transformer oil content. This condition causes the water content in transformer oil to be high and influences the performance of the transformer oil insulation and encourages the breakdown of dielectric voltage (breakdown voltage) of the transformer. Previous research on transformer oil was carried out by Jonathan Friz [2]. In his research, Jonathan stated that the increase in temperature affects the value of the transformer oil breakdown voltage. Other studies regarding the characteristics of transformer oil were carried out by Firdaus [3]. In his research, Firdaus found that the loading and temperature of the transformer affect the dielectric strength of the transformer oil breakdown voltage. Based on the two theoretical studies, the researcher wants to conduct research with a focus on the causes and effects of water content variables on the transformer oil breakdown voltage characteristics. Research was conducted through testing the water content and breakdown voltage on the kunlun ki25x type transformer oil with a 500 kv transformer test object. Coefficient correlation method is chosen to determine the characteristics influence of the water content on the breakdown voltage. Based on mathematical functions from this research will be obtained that indicates the presence or absence of thermal and electrical failures of the transformer. Furthermore, the results of this research will be analyzed whether the transformer oil insulation tested is still suitable for use, or needs to be purified or replaced with new ones.

2. Research method

The implementation and testing of transformer oil in this research were carried out based on the flowchart shown in Figure 1.

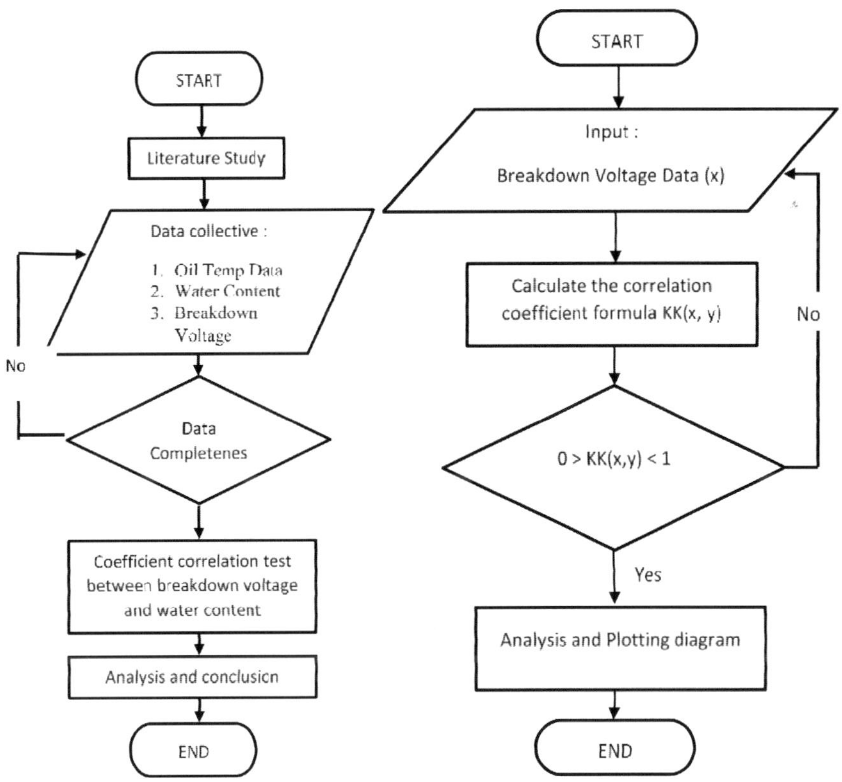

Figure 1. (a) Research flowchart, (b) Transformers oil test flowchart

Transformer oil testing was carried out in the Paiton 9 power plant laboratory using the Kunlun Ki25x oil test material with specifications as shown in Table 1. While the test equipment uses a set of Megger OTS100AF. This test is carried out to determine the relationship between the effect of temperature and water content in oil on the dielectric breakdown voltage of the transformer. Shape and schematic circuit test using instrument Megger OTS100AF are shown in Figure 2. After obtaining the test data the next step is to analyze and calculate using the coefficient correlation method

Table 1. Kunlun Ki25x transformer oil specifications

No.	Properties of Isolation Oil	Unit	Class I / Class II	Test Method	Test Place
1	Clarity	-	clear	IEC 296	Lab
2	Density (20^0) C	g/cm^3	< 0.895	IEC 296	Lab
3	Viscosity (20^0) C	cSt	< 40 < 25	IEC 296	Lab
	Kinematic – (15^0C)	cSt	< 800		
	Kinematic – (30^0C)	cSt	< 1800		
4	Neutrality Number	mgKOH/g	< 0.003	IEC 296 A	
5	Sulfur Corrosion	-	Not corrosive	IEC 296 A	Lab
6	Translucent Voltage	kV/2.5 mm	> 30	IEC 156	Lab
7	Dielectric Leakage Factor	-	< 0.05	IEC 474	Lab
8	Oxidation Resistance **1.** Neutrality number **2.** Impurity	mgKOH/g %	< 0.40 < 0.10	IEC 74	Lab

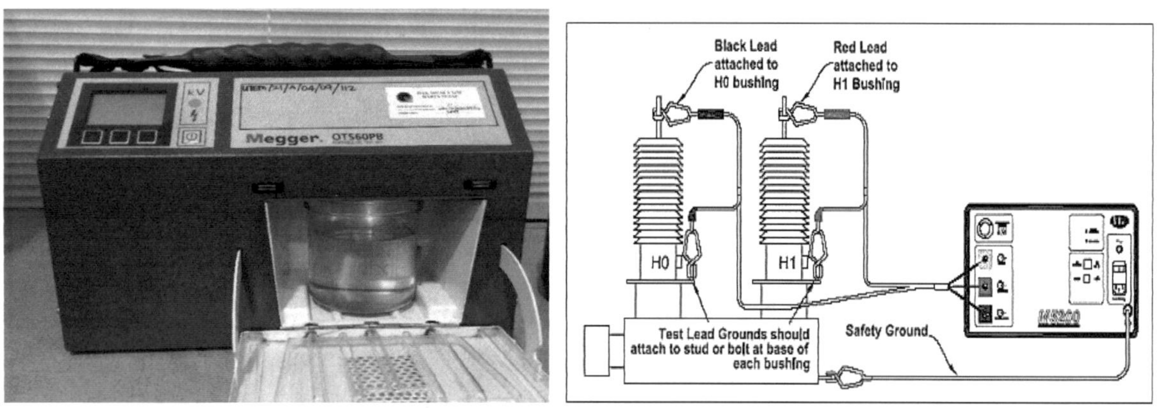

(a) (b)

Figure 2. (a) Megger OTS100AF, (b) Schematic circuit megger OTS100AF

Coefficient correlation method is a statistical analysis technique used to look for relationships between variables that are quantitative. Two variables are said to be correlated if the changes in one variable will be followed by changes in the other variables regularly in the same direction (positive correlation) or opposite direction (negative correlation) [4]. In general, the coefficient correlation method is written in equation 1 to 3 below

$$KK(x,y) = \frac{\sum (xi - x)(yi - y)}{\sqrt{\sum_{i=1}^{N}(xi - x)^2 \sum_{i=1}^{N}(yi - y)^2}} \quad (1)$$

$$a = \frac{n(\sum xy) - (\sum x)(\sum y)^2}{n(\sum x^2) - (\sum x)^2} \tag{2}$$

$$b = \frac{(\sum y)(\sum x^2) - (\sum x)(\sum xy)^2}{n(\sum x^2) - (\sum x)^2} \tag{3}$$

With :

$KK_{(x,y)}$ = coefficient correlationbetween variable x and y

xi = the value of first x variable ; yi = the value of first y variable

x = average value of x variable ; y = average value of y variable

To measure the level of correlation quantitatively, indicators interpretation of the coefficient correlation values are used, which are shown in Table 2 below. The range of coefficient correlation values range from 0 to 1 and is positive or negative

Table 2. Interpretation of coefficient correlation values

No	$KK_{(x,y)}$	Interpretation
1	0	Uncorrelated
2	0.01 – 0.20	Very Low Correlation
3	0.21 – 0.40	Low Correlation
4	0.41 – 0.60	A little low Correlation
5	0.61 – 0.8	Enough Correlation
6	0.81 – 0.99	High Correlation
7	1	Very High Correlation

3. Result and discussion

The 500kV transformer oil test results using two test parameters namely water content testing and breakdown voltage testing are shown in Table 3 below

Table 3. Oil sample trial results based on test parameters

No.	Working Voltage (KV)	Load (R)	Time (Month)		Temp (C)	Water content (ppm)	Breakdown voltage (KV)
			Duration	Date			
1	500	650	1	Apr-16	73	6	80.6
2	500	650	6	Sep-16	73	6	82
3	500	650	9	Des-16	77	7	78.1
4	500	650	12	Mar-17	80	10.09	76.8
5	500	650	15	Jun-17	84	15	76
6	500	650	18	Sep-17	87	18	74.6
7	500	650	21	Des-17	72	6	79.8
8	500	650	24	Mar-18	73	7	81.2
9	500	650	31	Okt-18	80	11	79
10	500	650	33	Des-18	84	15	75.5
11	500	650	35	Feb-19	88	20	65.7
$\sum$=	5500	7150	205		603	121.9	849.3

From the data above shows that initial indication of transformer condition is obtained from each oil sample tested where the results are presented in the form of Figure 3.

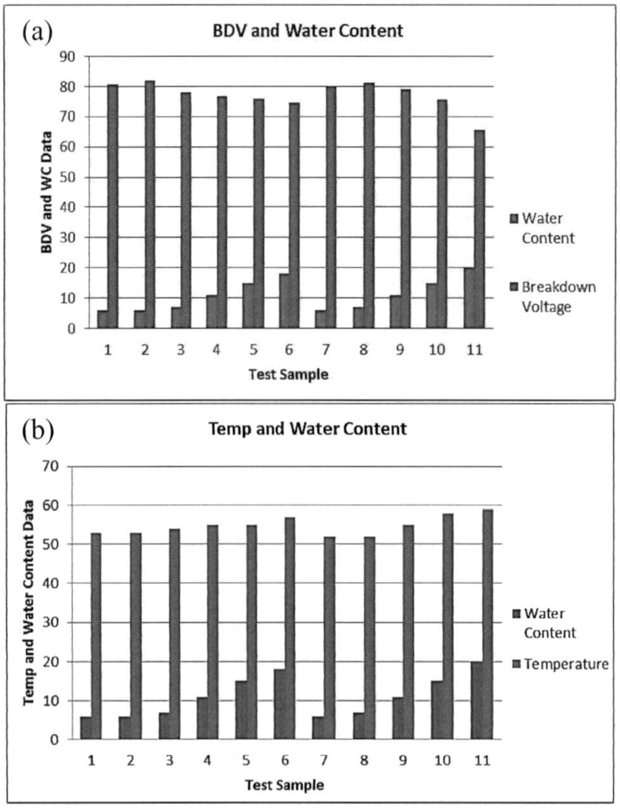

Figure 3. (a) Water content vs BDV graph, (b) water content vs temperature graph

From Figure 3. (a) Researcher found that the increase of water content in transformer oil causes the breakdown voltage value in the transformer oil to decrease. When the lowest water content value is 6 ppm the value of the voltage breakdown is 82 kV, while the highest water content value is 20 ppm the value of the voltage breakdown is 65.7 kV. Seen on Figure 3(a) the value of water content is inversely proportional to the value of the breakdown voltage. An experiment conducted by Zhang L shows when transformer oil is contaminated with water, its performance as insulation and cooling will decrease due to the nature of oil that cannot be fused with water [5]. If more water is contaminated with transformer oil, the dielectric properties of the transformer oil will also decrease as insulation. It will increase voltage insulation failure and short circuit to the transformers.

Characteristic comparison of parameters in water content and temperature can be seen in Figure 3(b). The increase of water content in the transformer oil causes the temperature value in the transformer oil to rise. When the lowest water content value is 6 ppm, the value of the transformer oil temperature is 53 C, while the highest water content value is 20 ppm the value of the transformer oil temperature is 59 C. this

IOP Conf. Series: Materials Science and Engineering **821** (2020) 012021 doi:10.1088/1757-899X/821/1/012021

condition prove that relationship of water content and transformer oil temperature is linear. To analyze the correlation values of breakdown voltage and water content variables, it is necessary to calculate the components of the correlation method shown in Table 4.

Table 4. Coefficient correlation analysis breakdown voltage and water content

No	Water content (ppm)	Breakdown voltage (Kv)	ai - a	bi -b	ai - a)^2	(bi - b)^2	(ai - a)(bi - b)
1	6	80.6	-5.08	3.39	25.82	11.50	-17.23
2	6	82	-5.08	4.79	25.82	22.95	-24.35
3	7	78.1	-4.08	0.89	16.66	0.79	-3.64
4	10.9	76.8	-0.18	-0.41	0.03	0.17	0.07
5	15	76	3.92	-1.21	15.35	1.46	-4.74
6	18	74.6	6.92	-2.61	47.86	6.81	-18.05
7	6	79.8	-5.08	2.59	25.82	6.71	-13.17
8	7	81.2	-4.08	3.99	16.66	15.93	-16.29
9	11	79	-0.08	1.79	0.01	3.21	-0.15
10	15	75.5	3.92	-1.71	15.35	2.92	-6.70
11	20	65.7	8.92	-11.51	79.53	132.46	-102.64
Σ	121.9	849.3			268.94	204.91	-206.87

Table description:
ai = the value of water content to i
a = *water content* average value
bi = the value of *breakdown voltage* to i
b = average value of breakdown *voltage*

By applying equations 1 to 3 we get the value of correlation R = -0.88; a = 0.7692 and b = 85.733 so that we get the linear equation Y = 0.7692x + 85.733. High negative value of R= - 0.88 nearly 1 prove that water content and breakdown voltage have strong inversely relationship. This condition influenced with many factor, one of them is the transfer of wet solid materials such as fiber and other contaminants such as water to the electrically pressure area between the two electrodes [6]. If this continues, a bridge will be formed between the two electrodes. If this continues, bubbles will be formed between the two electrodes [7]. Bubble will form along the bridge and cause conduction, ionization and breakdown voltage to occur. For testing and analysis of water content and temperature parameters using coefficient correlation method are shown in Table 5 below.

Table 5. Test results for water content and temperature parameters

No	Water content (ppm)	Temperature (C)	ai - a	ci - c	$(ai - a)^2$	$(ci - c)^2$	(ai - a)(ci - c)
1	6	73	-5.08	-1.82	25.82	3.31	9.24
2	6	73	-5.08	-1.82	25.82	3.31	9.24
3	7	77	-4.08	-0.82	16.66	0.67	3.34
4	10.9	80	-0.18	0.18	0.03	0.03	-0.03
5	15	84	3.92	0.18	15.35	0.03	0.71
6	18	87	6.92	2.18	47.86	4.76	15.09
7	6	52	-5.08	-2.82	25.82	7.94	14.32
8	7	73	-4.08	-2.82	16.66	7.94	11.50
9	11	80	-0.08	0.18	0.01	0.03	-0.01
10	15	84	3.92	3.18	15.35	10.12	12.47
11	20	88	8.92	4.18	79.53	17.49	37.29
∑	121.9	603			268.94	55.64	113.16

Table description:

a_i = the value of water content to i
a = average value of *water* content
c_i = the value of temperature to i
c = average value of temperature

By applying equations 1 to 3 we get the value of R = 0.92; a = 0.42 and b = 50.155 so we get the function of linear equation Y = 0.42 x + 50.155. The result give high positive value of R = 0.92 nearly 1 this indicate that water content and temperature have strong linear relationship. this condition indicate more water content in the transformer oil causes the condition of the transformer oil to heat up quickly. This is because the oil and water structure cannot be integrated [8]. The graph figure of the coefficient correlation relationship between each parameter test is shown in Figure 4 and Figure 5 .

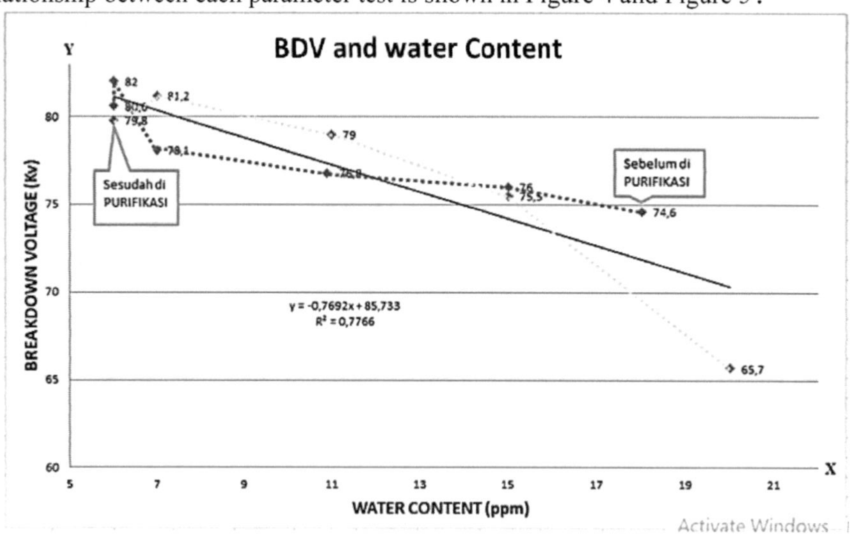

Figure 4 Water Content Vs BDV Value Curve

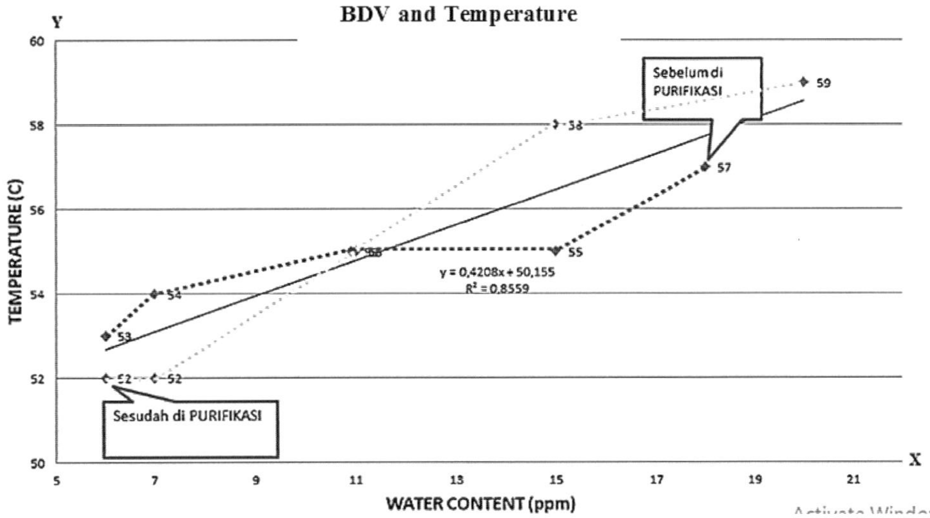

Figure 5. Value of water content vs temperature

The final discussion is the result of testing water content for the time at which these two parameters have an influence in determining the age of transformer oil use. Analysis of the coefficients correlation between two parameters is shown in Table 6 and Figure 6.

Table 6 Water content and time testing results

No	*Water content* (ppm)	Time (Month)	ai - a	di - d	$(ai - a)^2$	$(di - d)^2$	(ai - a)(di - d)
1	6	1	-5.08	-17.64	25.82	311.04	89.62
2	6	6	-5.08	-12.64	25.82	159.68	64.22
3	7	9	-4.08	-9.64	16.66	92.86	39.33
4	10.9	12	-0.18	-6.64	0.03	44.04	1.21
5	15	15	3.92	-3.64	15.35	13.22	-14.25
6	18	18	6.92	-0.64	47.86	0.40	-4.40
7	6	21	-5.08	2.36	25.82	5.59	-12.01
8	7	24	-4.08	5.36	16.66	28.77	-21.89
9	11	31	-0.08	12.36	0.01	152.86	-1.01
10	15	33	3.92	14.36	15.35	206.31	56.28
11	20	35	8.92	16.36	79.53	267.77	145.93
$\sum =$	121.9	205			268.94	1282.55	343.03

Table description:

a_i = The value of *water content* to i
a = average value of *water content*
d_i = The value of time to i
d = average value of time

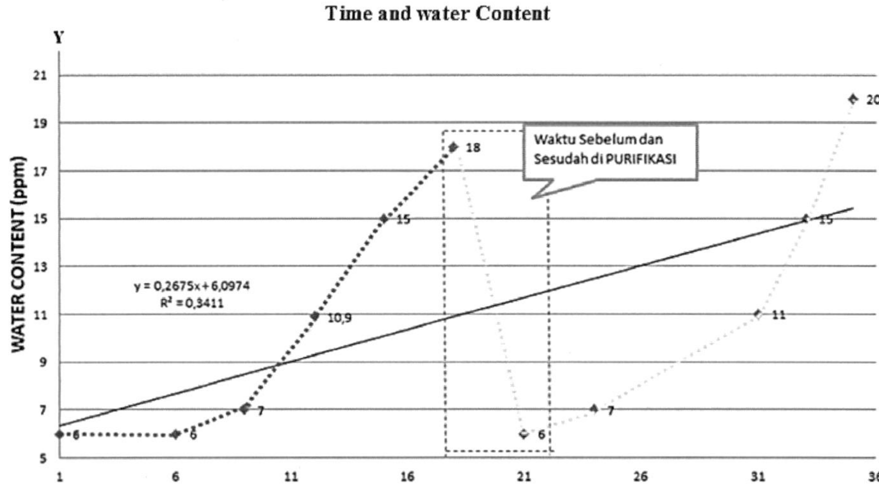

Figure 6. Water content vs time curve value

From the above table using equations 1 to 3 obtained the value of R = 0.58; a = 1.275 and b = 4.5 so we get the function of linear equation Y = 1.275x + 4.5.

From Figure 6, it can be seen data of water content before and after purification, after purification water content value will decrease and enter safe values category. This treatment also make transformer oil structure will not be as new again because there is still a little remaining water content that cannot be removed, this will effect in the performance of the transformer oil is reduced even if only slightly [9, 10].

4. Conclusion

Based on measurements, testing and analysis of water content and dielectric strength of transformer insulation oil in PT Paiton 9 power plant, the authors can conclude that the increase in water content is inversely proportional to the increase in the breakdown voltage of insulating oil with a strong negative correlation r = - 0.88, in other words the greater water content in the transformer, the smaller the breakdown voltage of the insulating oil. The increase in the temperature of the insulating oil is directly proportional to the water content of the insulating oil with a strong positive correlation r = 0.92, in other words the higher the temperature of the insulating oil, the higher the value of the water content in the transformer. From the correlation analysis of the relationship of the increase in water content with the dielectric strength (breakdown voltage) a linear graph is obtained. y = -0.7692x + 85.733 with r = -0.88 and R^2 = 0.7766

Acknowledgments

Researcher would like to thank Mr. Bagus Setya R, M. Kom and Radinal Hadinata for their support during conducting the research.

References

[1]. Zhao T, Liu Y, Fangcheng Lv, Cheng X, Wang Z, Wang F, and Xie 2017 Breakdown characteristics of transformer oil insulation under impulse voltage with different waveform parameters *2017 IEEE Electrical Insulation Conference (EIC).* doi:10.1109/eic.2017.8004709 pp 368-71.

[2]. Jonathan F 2014 Analysis effect of increasing temperature and transformer oil age on the translucent voltage *UMY Journal* pp 75-9.

[3]. Firdaus 2014 analysis the effect of load on dielectric strength of transformer oil isolation 6.6 kv/380*V*, *UNILA Journal* pp 25–30.

[4]. Setiabudy R, Beroual A, Bismo S, and Sitorus H B H 2015 Comparison of statistical breakdown voltages in jatropha curcas methyl ester oil and mineral oil under AC voltage *IEEE 11th International Conference on the Properties and Applications of Dielectric Materials (ICPADM).* doi:10.1109/icpadm.2015.7295325 pp 528-31.

[5]. Li J, Hu D, Zhang L, and Che B 2016 Dielectric breakdown characteristics of the rod-plane electrode system in SF6 gas under oscillating impulse voltage. *2016 IEEE International Power Modulator and High Voltage Conference (IPMHVC)* doi:10.1109/ipmhvc.2016.8012845 pp 308–11.

[6]. Sitorus H B H, Beroual A, Setiabudy R, and Bismo S 2017 Statistical analysis of AC and DC breakdown voltage of JMEO (Jatropha methyl ester oil), mineral oil and their mixtures *2017 IEEE 19th International Conference on Dielectric Liquids (ICDL)* doi:10.1109/icdl.2017.8124647 pp 1-4.

[7]. Foruzan E and Vakilzadian H 2015 The investigation of dielectric barrier impact on the breakdown voltage in high voltage systems by modeling and simulation *2015 IEEE Power & Energy Society General Meeting.* doi:10.1109/pesgm.2015.7285653 pp 978-82.

[8]. Beroual A and Coulibaly M-L 2016 Experimental investigation of breakdown voltage of CO2, N2 and SF6 gases, and CO2-SF6 and N2-SF6 mixtures under different voltage waveforms *2016 IEEE International Power Modulator and High Voltage Conference (IPMHVC).* doi:10.1109/ipmhvc.2016.8012898 pp 292–95.

[9]. Huang C and Ruan J 2016 Prediction study on power frequency breakdown voltage of ring structure short air gaps *2016 IEEE International Conference on High Voltage Engineering and Application (ICHVE)* doi:10.1109/ichve.2016.7800886 pp 1-4.

[10]. Kovacevic U, Vujisic M, Iricanin B, Osmokrovic P, and Milosavljevic S 2014 The correlation and regression between the pre-breakdown current and breakdown voltage of the vacuum switchgear in dependence of switching operation *2014 IEEE International Power Modulator and High Voltage Conference (IPMHVC)* doi:10.1109/ipmhvc.2014.7287290 pp 384-87.

IOP Conf. Series: Materials Science and Engineering **821** (2020) 012022 doi:10.1088/1757-899X/821/1/012022

The first proposed of fiber to the home (FTTH) protection unit for clients scattered placement

H Ramza[*], S A C Aziz, N Z, Mastang, and M S Ab-Rahman

Department of Electrical, Faculty of Engineering, Universitas Muhammadiyah Prof. Dr. Hamka, Jakarta, Indonesia

*Email: hramza@uhamka.ac.id

Abstract. This paper describes the design and analysis of protection schemes that it is not regular customers' location in optical communication system network. Distance and number of customers with the design of optical point protection unit achievements are recognized as the user for a series of rings from CAPU-Ring 1 (Customer Access Protection Unit). CAPU-Ring 1 is made as a safety point of optics that will be placed on the prior user. This system is an optical spot that will be placed in a ring circuit for use in placement of clients scattered. Proposed construction of a series with a protection channel architecture is about designing CAPU-Ring 1 as optical nodes, condition and repair damage to the circuit, the power budget calculations, design in the simulation, and analysis of the node to node. Parameters outlined in this discussion are the relationship of each node optical output power, improving the system of the eye diagram to the distance and also increase the system working against the maximum optical node. Characterization of ring network protection architecture can be identified from the results of a simulation study.

1. Overview of CAPU-RING 1

CAPU-Ring 1 (Customer Access Protection Unit) has been designed with 4 channels of alternative safety that will bring optical signal if there is damage to the channel [1-4]. Two main lines proposed FTTH system circuit architecture includes the main channel and the ring channel [5-9]. Both these channels are combined in the form of this CAPU-Ring 1. As shown at Figure1, CAPU-Ring 1 was used for the filamentous placement position that consists of internal component block system proposed [10-12].

Two main switches are put on the both line, such as primary line and ring line. Each linear line (primary line) will work by duplicating with the protection line. Protection channel will be used in case of damage to the active channels at random. Ring channels are also equipped with two optical channels for each node is an active channel and channel protection will be running in the event of damage to the active channel. Then the optical signal will have four alternative will be chosen to carry the optical signals to each receiver.

The decrement signals will pass through the main switch on the linear channel (primary line) and it will pass through an optical splitter, to distribute power to multiple recipients. Hence under normal conditions without damage, the optical signal goes through an active channel, but if there is damage to the active channel protection channel will be activated. Then, the signal enters the optical coupling, and here the signal is divided according to percentage of power around 1-n% for the signal transmission component and the n% of the signal dropped. The signal drop is a signal that will be dropped directly to the recipient. Therefore, small-signal power ratio can receive at a receiver only. Input signal is a signal to be transmitted to the output optical node, and the output optical signal node will be input signal the next optical node stage. The drop-signal in the optical coupler have n%-ratio on the linear channel then it will enter to the optical switch (2x1) and also, it will be sent to the receiver via an optical circulator. The linear channel of the Transmitted-Signal will be entering the 2x2- optical switch as an input signal at the next optical node stage.

If two channels work and the protection channel on the linear channel is damaged, then the ring will be activated channels. Initially, the ring line will be through the main multiplexer and then enter an optical coupler then the signal will be divided into two percent of the 1-n% (signal out) and n% (signal drop) as the main channel design. The signal will drop into the 2x1 optical switches and optical-circulator. Transmitted-Signal will enter the 2x2 optical switches and there, an optical switch that is controlled by the ACS (Automatic Control System) will select the channel to be activated. The flow of signals from optical node to the next node (for the ring channels) will be equipped with two-channels multiplexer. In addition, all optical switches that serve to transfer the channel in the optical network will be controlled byACS.

Following the network protection architecture of CAPU-Ring 1 is developed; there are two lines of terminals TT(1) and TT(2) that are used in the OLT (Optical Line Terminal) for the signal direction decreased as shown in the Figure 1[13]. Each channel of TT(1) and TT(2) are connecting on the linear line and the ring line. In normal status, TT(1) will transmit the signal through alinear decline, while TT(2) is static and it can be disabled and only works when there is damage to the two lines of the linear path.

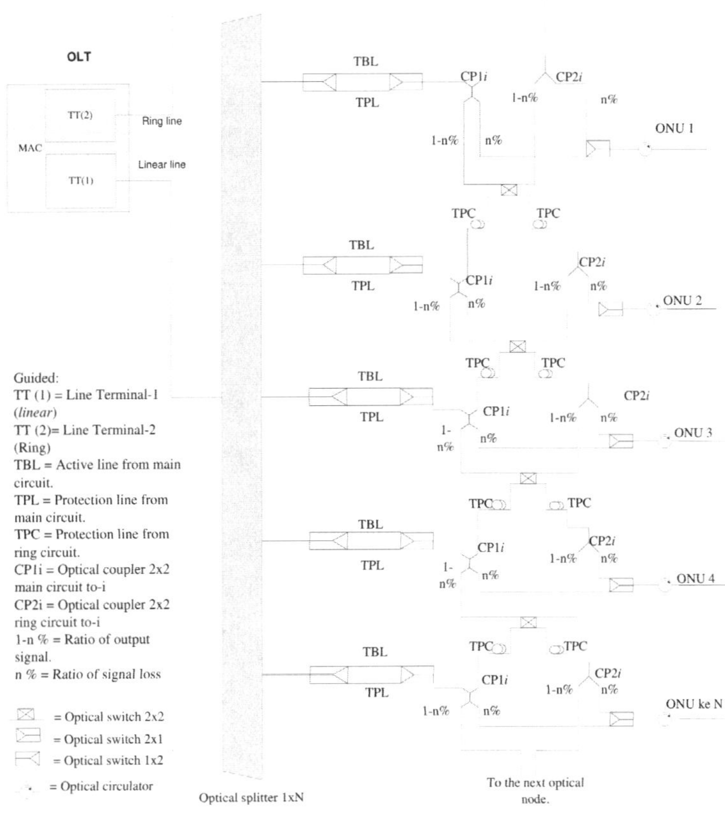

Figure 1. CAPU-Ring 1 with linear protection circuit and ring protection circuit.

2. Fault condition on the CAPU-RING 1 network

This section describes the classification level of damage to the network security. Four of the fault conditions are identified in the proposed recovery system of this network [14-16]. Fault conditions identified involve two types of protection: linear protection and ring protection. Linear protection is categorized as the protection of the main line and ring protection is a ring network.

2. 1. Normal conditions (signalling in\n a linear path)

Normal conditions classified as an optical signal through the line to work in a linear line and after going through an optical coupler, and then the signal will enter the optical circulator that to be sent directly to the recipient. Figure 2 shows the optical signal through the components in the optical node where the line is cut off routes to carry traffic in normal conditions. Then the signal is classified as a signal loss.

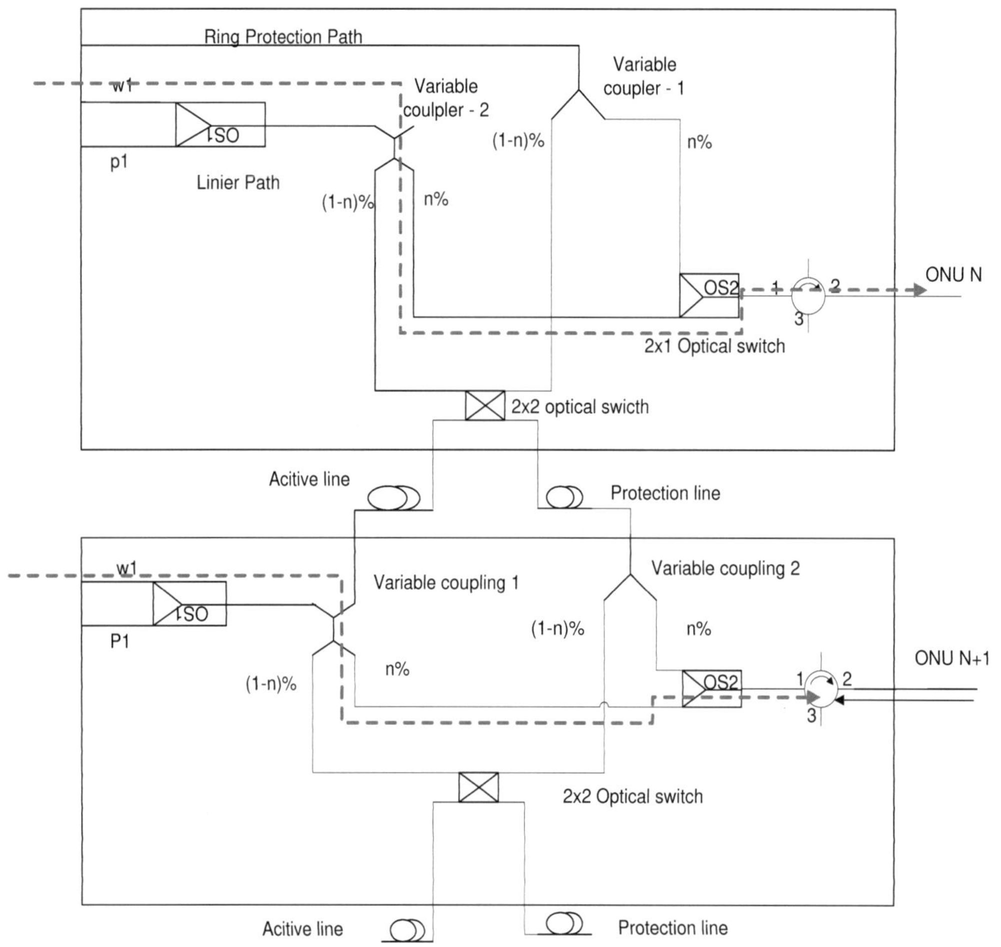

Figure 2. Optical signal flow in normal condition without any damage to the linear channel

2. 2. First-fault condition (linear protection)

First of fault condition is categorized when the optical signalling is switched into the protection line while the active line broken [17, 18]. Figure 3 shows the dotted line that represented optical signal taken into receiver optical node on this protection circuit.

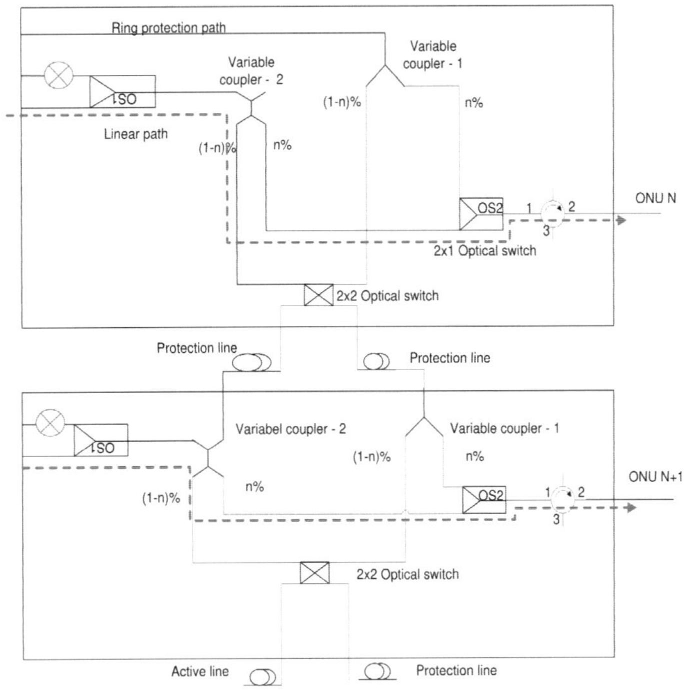

Figure 3. First stage of failure condition is active when protection line is applied for bring the information signal.

2. 3. *Second-fault condition (ring protection)*

The second stage of the fault condition occurs where the channel activated and the protection line broken, then protection line on the ring circuit will be activated [19]. Figure 4 shows the ring line is activated to bring optical signal into the receiver which two-signal type will function as a Dropped-Signal and Transmitted-Signal to next optical node.

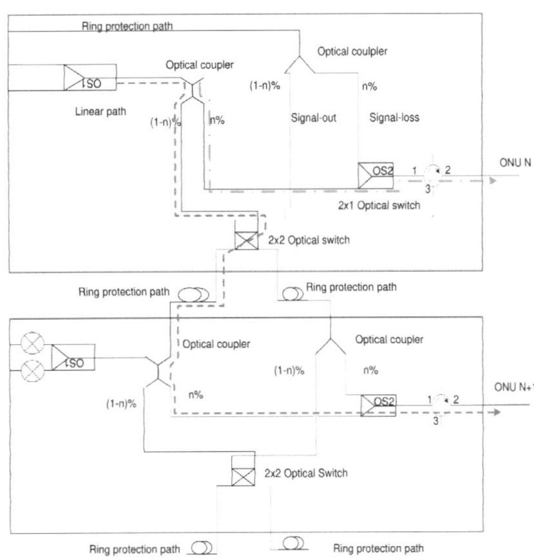

Figure 4. The situations when both channels on a linear circuit are destroyed, then the ring protection channel will be activated.

2. 4. Third-fault condition (ring protection)

Figure 5 shows the third fault condition will active when the ring line broke and both of the line on the linier circuit is off the next series [19]. The Transmitted-Signal from the optical node is switched into the protection line and go into the optical coupler and accepted by the receiver.

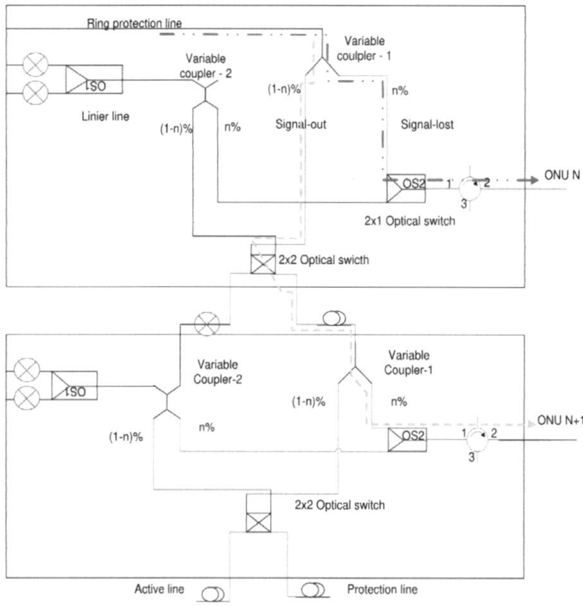

Figure 5. Ring protection line is used as the alternative channel if third condition is occurred

3. Calculation of channel dissipation at CAPU-RING optical node

Calculation analysis in this section aims to quantify the dissipation channel provided by the architecture of this channel protection system [15, 17]. Calculation of dissipation channel is important to know the channel dissipation budget based on the proposed circuit architecture. In this section a calculation of dissipation channel for 1490 nm wavelength for CAPU-RING 1 the first optical node and n-optical node. Calculation of power dissipation channel is divided into two specific channel that consists of a channel through a series of channels through a series of linear and ring.

Optical coupler is a device that can divide the optical signal (fiber) from one fiber to another by following the particular percentage ratio. In addition, the optical coupler can also combine optical signals from two or more lines in to a single fiber.

3. 1. Linear protection scheme.

Based on the design of linear protection, optical signals can be sent directly to the ONU (Optical Network Unit) through the optical fiber (Single Mode Fiber, SMF), Optical Splitter, Optical switch, Optical Coupler, and finally optical circulator. There are two optical signal paths for signal direction that decreased (1490 nm and 1550 nm) in the linear path that consisting of the signal loss and the signal out components as shown in figure 1.2. Both of linear line and ring line using the fiber length of 15 km (cumulative distance).

$$\begin{aligned} &Dropped - signal\ line \\ &\quad = 15\ Km\ (optical\ fiber) + 1\ unit\ (optical\ splitter \\ &\quad + 3\ unit(optical\ switch) + 1\ unit(optical\ coupler) \\ &\quad + 1\ unit(optical\ circulator) \end{aligned} \tag{1}$$

$$= \left(15\,Km \times 0.25\,\frac{dB}{Km}\right) + (1 \times 8\,dB) + (3 \times 1.2\,dB) + (1 \times 1\,dB) + (1 \times 1\,dB)$$
$$= 17.35\,dB$$

$$\begin{aligned} &1 - optical\ node\ of\ transmitted\ signal\ line \\ &= 3\ unit(optical\ swicth) + 1\ unit(optical\ coupler) \\ &= (3 \times 1.2\,dB) + (1 \times 1\,dB) \\ &= 4.6\,dB \end{aligned} \quad (2)$$

$$\begin{aligned} &Transmitted\ Signal\ path\ of\ optical\ node - 1 \\ &= 15\,Km\ (optical\ fiber) + 1\ unit(optical\ splitter) \\ &\quad +(1 \times optical\ node\ of\ signal\ line\ out) \\ &= \left(15\,Km \times 0.25\,\frac{dB}{Km}\right) + (1 \times 8\,dB) + (1 \times 4.6\,dB) \\ &= 16.35\,dB \end{aligned} \quad (3)$$

$$\begin{aligned} &Transmitted\ Signal\ path\ of optical\ node - 5 \\ &= 15\,Km(optical\ fiber) + 1\ unit(optical\ splitter) \\ &\quad +(5 \times optical\ node\ of\ signal\ line\ out) \\ &= \left(15\,Km \times 0.25\,\frac{dB}{Km}\right) + (1 \times 8\,dB) + (5 \times 4.6\,dB) \\ &= 34.75\,dB \end{aligned} \quad (4)$$

3. 2. Ring protection scheme.

Fault lines in the second and third stages involve the activation of the path of a ring network. Thus, the rings protection path can be used as alternative routes to bring optical signals to the ONU. Path analysis of the ring is stressed as a mechanism for protection of optical nodes. In away of the ring network, there are signal loss and the signal out components. The calculation of dissipation channels areas follows:

$$\begin{aligned} &1 - optical\ node\ of\ dropped - signal\ line \\ &= 1\ unit(optical\ coupler) + 1\ unit(optical\ swicth) \\ &\quad +1\ unit(optical\ circulator) \\ &= (1 \times 1\,dB) + (1 \times 1.2\,dB) + (1 \times 1\,dB) \\ &= 3.2\,dB \end{aligned} \quad (5)$$

$$\begin{aligned} &Dropped - signal\ path\ of\ optical\ node - 1 \\ &= 15\,Km(optical\ fiber) + (1 \times optical\ node\ of\ signal\ lossline) \\ &= \left(15\,Km \times 0.25\,\frac{dB}{Km}\right) + (1 \times 3.2\,dB) \\ &= 6.95\,dB \end{aligned} \quad (6)$$

$$\begin{aligned} &Dropped - signal\ path\ of\ optical\ node - 5 \\ &= 15\,Km(optical\ fiber) + 1\ unit(optical\ splitter) \\ &\quad +(5 \times optical\ node\ of\ signal\ loss\ line) \\ &= \left(15\,Km \times 0.25\,\frac{dB}{Km}\right) + (5 \times 3.2\,dB) \\ &= 19.75\,dB \end{aligned} \quad (7)$$

IOP Conf. Series: Materials Science and Engineering **821** (2020) 012022 doi:10.1088/1757-899X/821/1/012022

$$
\begin{aligned}
1 - optical\ node\ of\ Dropped - signal\ out \\
&= 1\ unit(optical\ swicth) + 1\ unit(optical\ coupler) \\
&= (1 \times 1.2\ dB) + (1 \times 1\ dB) \\
&= 2.2\ dB
\end{aligned} \tag{8}
$$

$$
\begin{aligned}
Dropped - signal\ path\ of\ optical\ node - 1 \\
&= 15\ Km(optical\ fiber) \\
&\quad +(1 \times optical\ node\ of\ signal\ line\ out) \\
&= \left(15\ Km \times 0.25\ \frac{dB}{Km}\right) + (1 \times 2.2\ dB) \\
&= 5.95\ dB
\end{aligned} \tag{9}
$$

$$
\begin{aligned}
Dropped - signal\ path\ of\ optical\ node - 5 \\
&= 15\ Km(optical\ fiber) + \\
&\quad (5 \times optical\ node\ of\ signal\ line\ out) \\
&= \left(15\ Km \times 0.25 \frac{dB}{Km}\right) + (5 \times 2.2\ dB) \\
&= 14.75\ dB
\end{aligned} \tag{10}
$$

4. Simulation and research

Network design is applied in scattered settlements of the clients with the linear combination of protection schemes and ring protection schemes [15, 17]. In addition, simulations are carried out to study the effectiveness of the security network architecture so that it can be used to expand and meet the correct specifications. This simulation aims to evaluate the effects of various parameters on the output power and BER performance on each node in the proposed network. Simulation of the entire range of coverage of this type was carried out on two types of protection. Simulations of protection of the network design of linear and ring protection carried out separately. Thus, the results of the simulation can be performed easily and the performance of the network coverage can be seen.

The main fiber distance is fixed at 15 km for the entire length of the optical fiber in the network protection. However, the distance between the ONU not specified. Proposal for a ring network involves five the number of ONU and the coupler is a variable optical coupler where it can be adjusted from 0% to100%. The data rate used to be way down in accordance with the EPON network protocol 1.25 GPS.

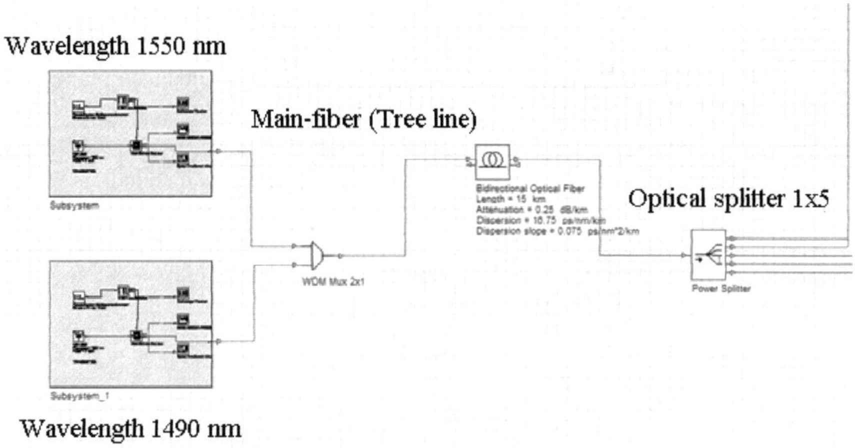

Figure 6. Design of feeder part for linear path

Figure 6 shows the design for the feeder of the linear network (main). 1550 nm and 1490 nm will be multiplexed by the range of protection network. Figure 7 shows the design for the network loss when a signal from the feeder to the five recipients will be divorced and then through the optical nodes. Figure 8 shows the components of the optical nodes that carry the signal to a linear path. For a ring network simulation platform is, Figure 9 shows the signal path for the entire ring network that contains five optical nodes connected together. Figure 10 shows the components through which the optical signal in an optical network node ring. Table 1 provides the parameters for the components - components that are used for the overall network security.

Table 1. Dissipation component parameters that used in the simulation

Devices	Dissipation according to the theoretical value of the product (dB)		
	1550 nm	1490 nm	1310 nm
	0.20 dB/km	0.25 dB/km	0.35 dB/km
15 km Optical Fiber (SMF-28)	4.0	5.0	7.0
Optical Splitter 1x5	8.0	8.0	8.0
Optical Coupler	1.0	1.0	1.0
Optical Circulator	1.0	1.0	1.0
Optical Switch (1x2 and 2x2)	1.2	1.2	1.2
(de) Multiplexing	0.5	0.5	0.5

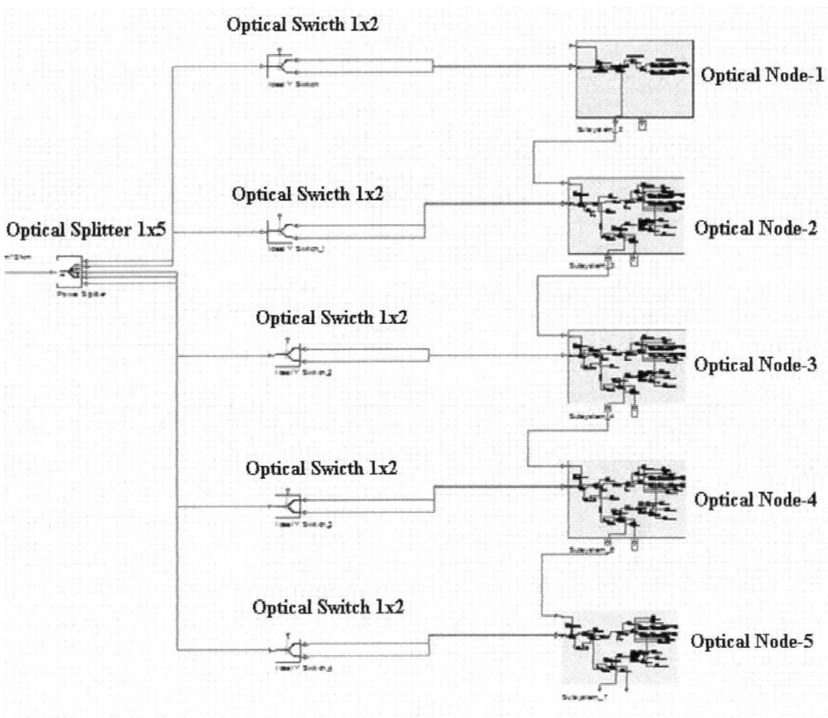

Figure 7. Design of loss part for linier path.

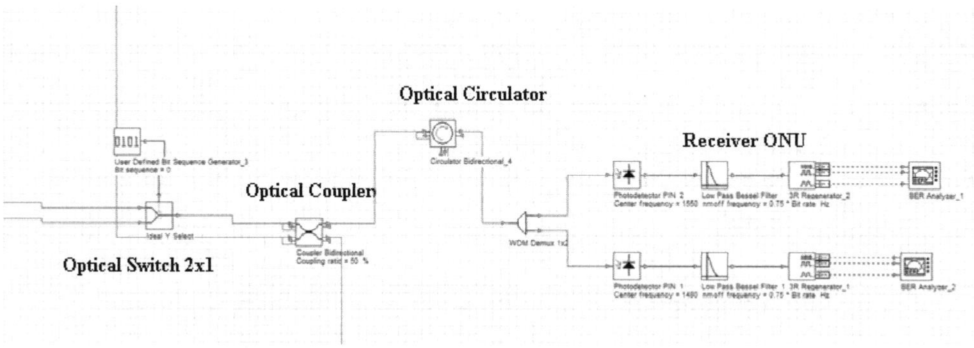

Figure 8. Optical node components to access the ring path.

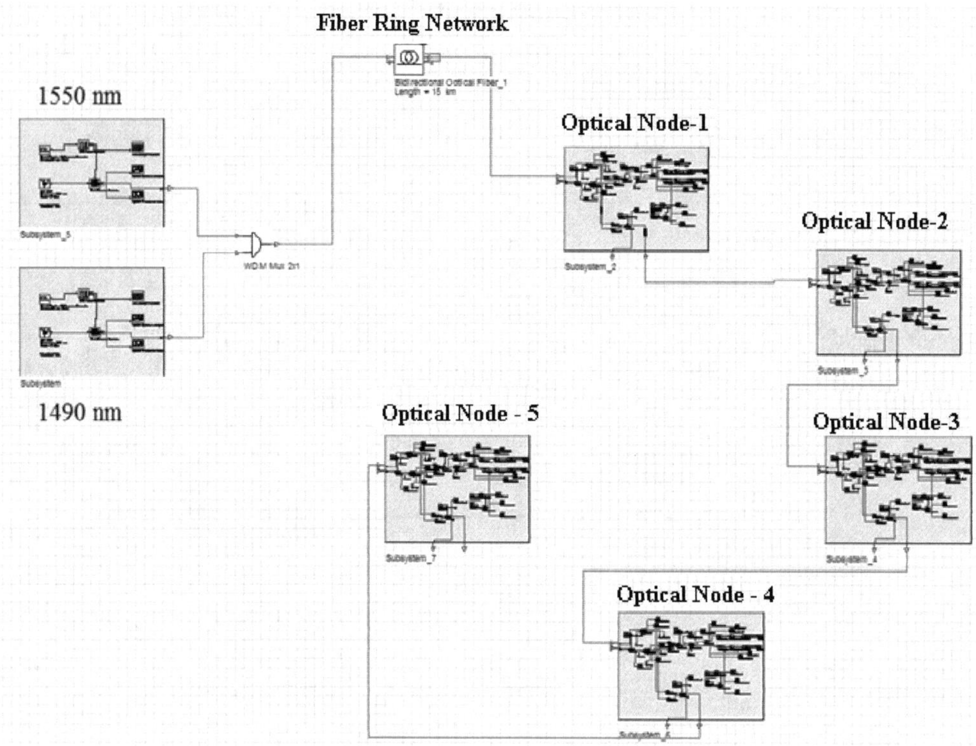

Figure 9. Overall of ring network with five optical node that is used.

IOP Conf. Series: Materials Science and Engineering **821** (2020) 012022 doi:10.1088/1757-899X/821/1/012022

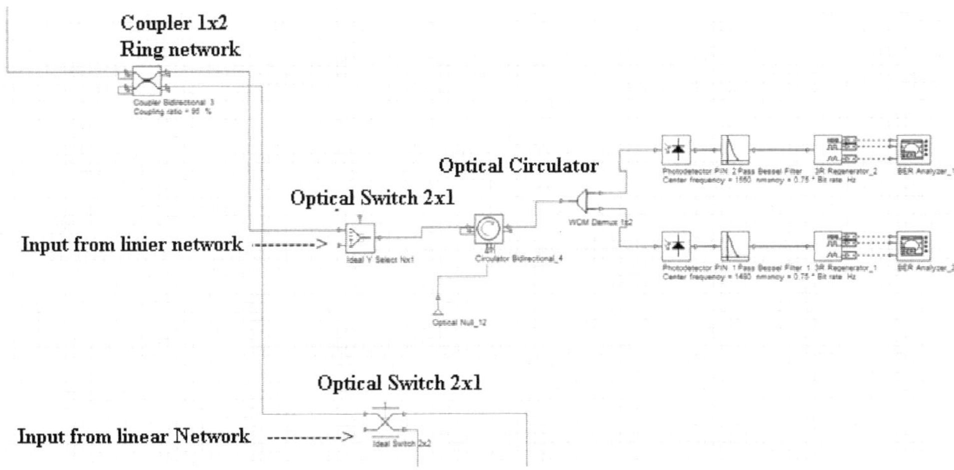

Figure 10. Optical node for signal route decline from ring network.

5. Node to node simulation analysis

Network protection scheme is a combination between the linear protection and the ring protection [15, 17]. Simulation analysis conducted for the node to node is divided into two main parts, namely a linear protection schemes and ring protection schemes. Each part of the protection scheme will investigate and get the power output of each node in the normal state (without damage), the maximum distance that can be achieved per each different protection schemes with the sensitivity and the maximum number of nodes that are -25 dBm, -30 dBm, -32 dBm and -35 dBm.

5. 1. The output power of each node at linear protection schemes

Linier protection path will enter into the optical coupler and it will be divided into two signals of the Transmitted-Signal and the Dropped-Signal. For the analysis of this issue, the first optical coupler is adjusted to a ratio of 50%: 50%. 50% as Dropped-Signal and 50% as Transmitted-Signal will be combined with a new signal from the next optical node. After seeing the performance of the output power of each optical node, the ratio of the optical coupler is varied to obtain optimum power output for each receiver at the optical node. Adjusted Transmitted-Signal goes to the (1-n) % and Dropped-Signal n%.

The design of linear protection, signals through the on-line coverage will not be linear decline upon it through then exit optical node until the end. It is caused by the optical signal power to be donated to the each node as the input power that is transmitted to the OLT (Main network) and it will be divorced by Optical Splitter for entering each optical node. In contrast to the ring network, the power output by each node is expected to shrink by following the optical node position.

Figure 11 shows the power output of the two wavelengths (1550 nm and 1490 nm) respectively to the Transmitted-Signal and the Dropped-Signal when the coupling ratio was adjusted to 50%:50%. Lower output power for the signal loss is the first node of -31.228 dBm (1550 nm) and -31.978 dBm (1490 nm). This may be due on the first node that does not have the power combination of the optical coupler of the previous node and making it the lowest power than other optical node. The lower Transmitted-Signal is the optical node to -29.593 (1550nm) and -30.343 dBm (1490nm). From the output power, the receiver sensitivity of -32 dBm can be carried out to linear protection scheme when the optical coupler is adjusted to 50%:50% for the Transmitted-Signal and the Dropped-Signal.

Subsequently, the percentage of random couplings is adjusted to obtain the lowest power output and the same for each optical node. Thus, the adjusted percentage is shown in Figure 12 where the ratio of the five-optical node will be adjustable individually. Figure 13 shows the results when the ratio of the output coupler is varied as shown in Figure 12. The highest power obtained at node-5 is -26.568 dBm (Dropped-Signal 1490 nm) and the lowest power at node-2 is -30.847 dBm (Dropped-Signal 1490 nm) where the highest Transmitted-Signal goes to the optical node-2 (the Transmitted-

Signal 1490 nm) and the highest Transmitted-Signal goes to the first optical node amount -34.656 dBm (1490 nm Transmitted-Signal). From the output of Dropped-Signal, the sensitivity can be reduced to -31 dBm. So by adjusting the ratio of the coupling is amount of 1 dBm, therefore sensitivity can be reduced.

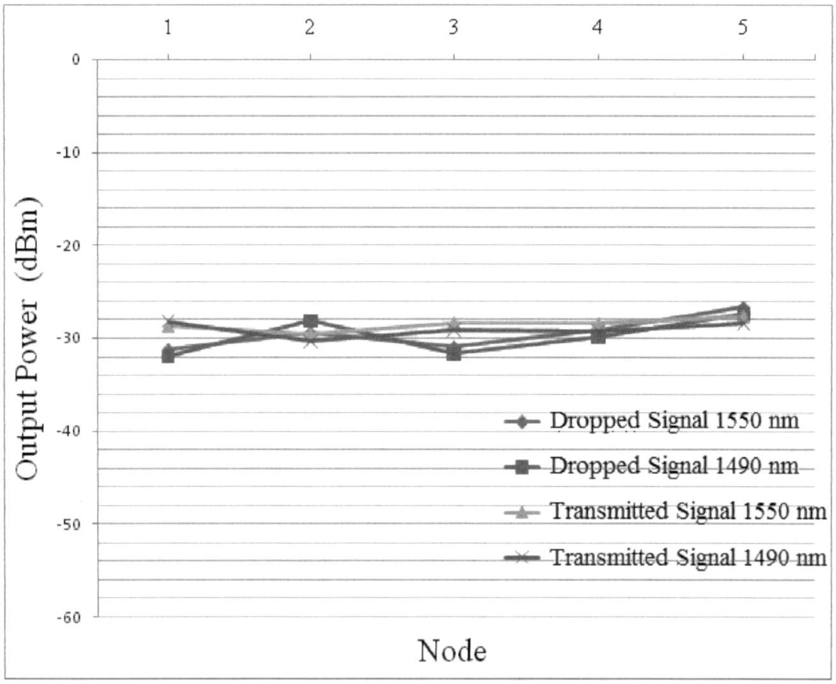

Figure 11. Output power of five optical node for the Transmitted-Signal and Dropped-Signal (1550 nm and 1490 nm) if coupling ration is adjusted by 50%:50%.

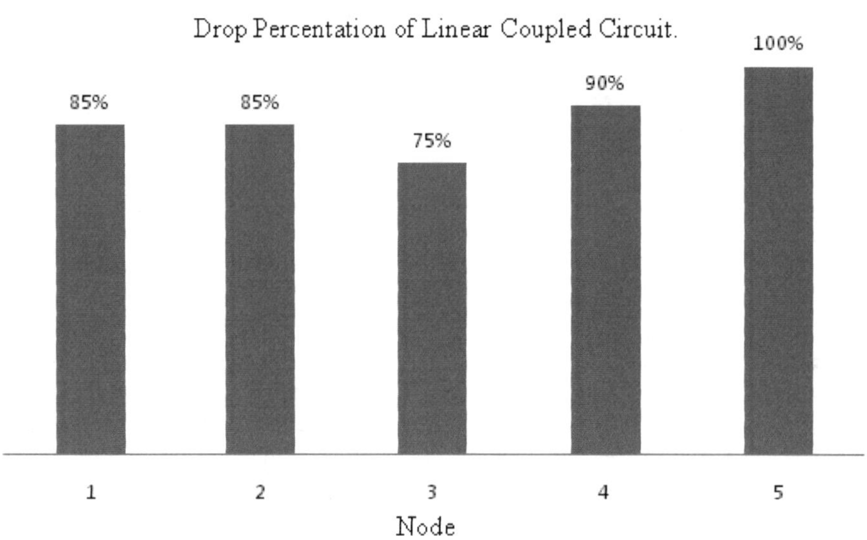

Figure 12. Percent of coupling ratio that is aligned by linier protection path

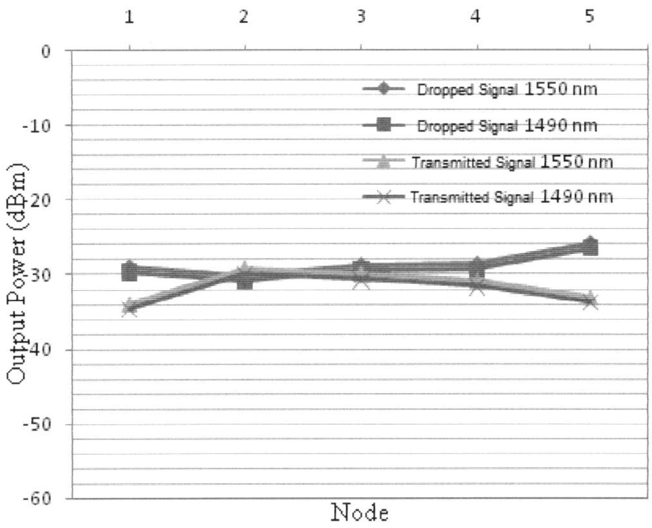

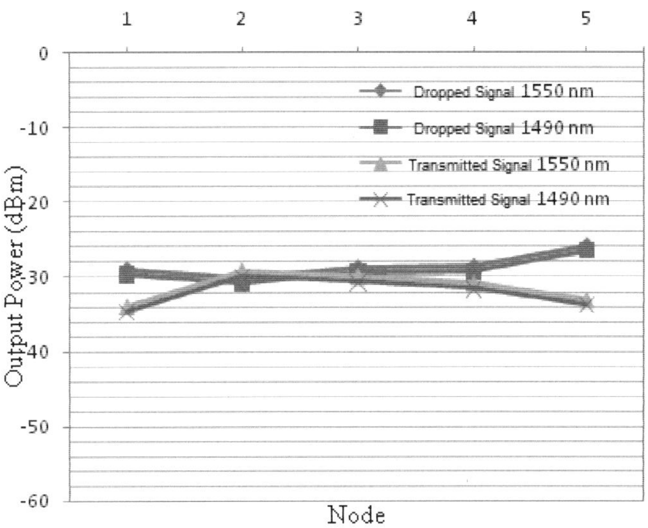

Figure 13. The output power of five optical nodes

5. 2. The output power of each node at ring protection schemes.

Signals through linear protection path will go in to the optical coupler and it will also be divided into the Transmitted-Signal and the Dropped-Signal. For ring protection schemes, the signal will be decreased when overrun the optical nodes. Thus, a study on the distribution of power from the coupler was run to obtain optimum power output. Signals through the coupler is adjusted according to Dropped-Signal (n %) and the Transmitted-Signal (1- n%).

a. (n, 1-n) where n is 5%

In the beginning this power division analysis prescribing at (n,1-n) with n is 5%. Therefore, signal loss is 5% from total power. Figure 14 shows output power per optical node for signal output (1490 nm and 1550 nm) and Dropped-Signal (1490 nm and 1550 nm). To be expected, lowest power for Dropped-Signal found in last optical node namely -32.53 dBm (1550 nm) and -33.28 dBm (1490 nm). Output powers for both kinds signal descend by linier when beyond number optical node. These cases happen when signal that go through the optical node would be experiencing power depreciation.

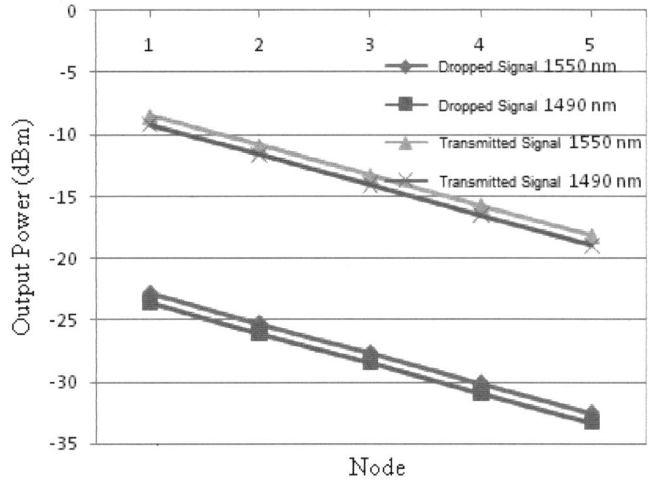

Figure 14. Output power to 5 optical node for Transmitted-Signal and Dropped-Signal (1550 nm and 1490 nm) when coupler ratio adjusted to 95%:5% (95% Transmitted-Signal and 5% Dropped-Signal).

b. (n,1-n) where n is 10%.

Ratio of power division (n,1-n) where n as Dropped-Signal and Transmitted-Signal as much as 10% and 90% respectively. This power division ratio empowering a bit bigger to receiver. There are many primordial nodes that were not appropriate as earliest node and they do not experience power depreciation that so bad. Figure 15 shows output power per optical node for Transmitted-Signal (1490 nm and 1550 nm) and Dropped-Signal (1490nm and 1550nm) that descends by linier when across optical node in the optical network. The output power node-1 to node-5 for wavelength 1550nm are -19.829 dBm, -22.487 dBm, -25.144 dBm, -27.802 dBm and -30.458 dBm respectively. The wide dynamic range is needed to get a good signal.

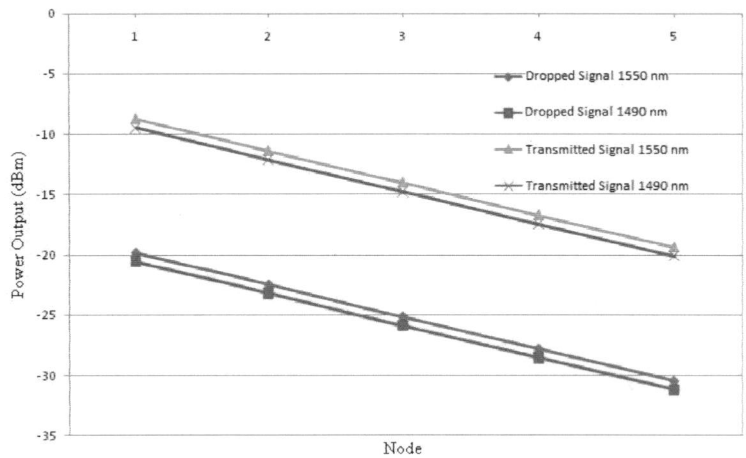

Figure 15. Five optical nodes of output power for the optical transmitted signal and the dropped signal, when the ratio of the coupler is adjusted to 90%: 10%.

c. (n, 1-n) where n is the changing value (1% -100%).

Simulation of the coupling ratio is adjusted by signal changing has been conducted. It seeks to obtain optimum power output to reduce the depreciation of the main applicable to the last optical node. By setting the ratio of these readings, there are couplings in the same ratio to all optical nodes. This is not just looking at a different position in which optical node of the ring structure also contributed to the varying loss and reduces the number of ONU (Yeh&Chi, 2008). Thus, the ratio for each coupler is changed at random so that the output power obtained for optimum optical nodes.

Optimization to obtain the coupling ratio on the power set can be done with the extraction of simulation parameters method. These parameters using optimization method in which the ratio of the optical coupler is extracted to obtain the specific power output. Optimization will find the ratio of the optical coupler to achieve a power output of the power meter as required. As the number of optical nodes exceeds 1, the optimization of various parameters has been carried out. It is found that the coupling ratio of the optical nodes will be varied to obtain the specific power output by using OPTYSYSTEM.

Figure 16 shows reading ratio (percentage) in optical-coupler when output power in -23dBm by using parameter extraction method with simulation result is -23dBm. Output power is chosen as output power that most maximum achievable to supply power at all optical node. If fixed power is less than -23dBm, so power was not enough to supply all optical node as signal, so the optical node have to be divided in each nodes that crossed him. Therefore, based on adjustment that carried out product power average as much as -23dBm to five optical nodes that inserted in the network

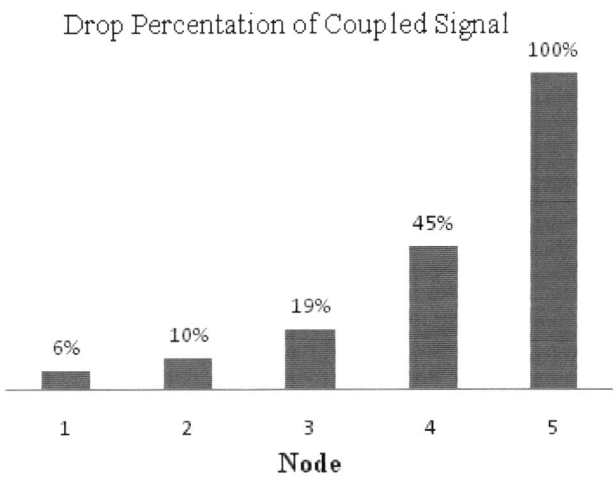

Figure 16. Percentage of optical-coupler ratio for all the optical nodes.

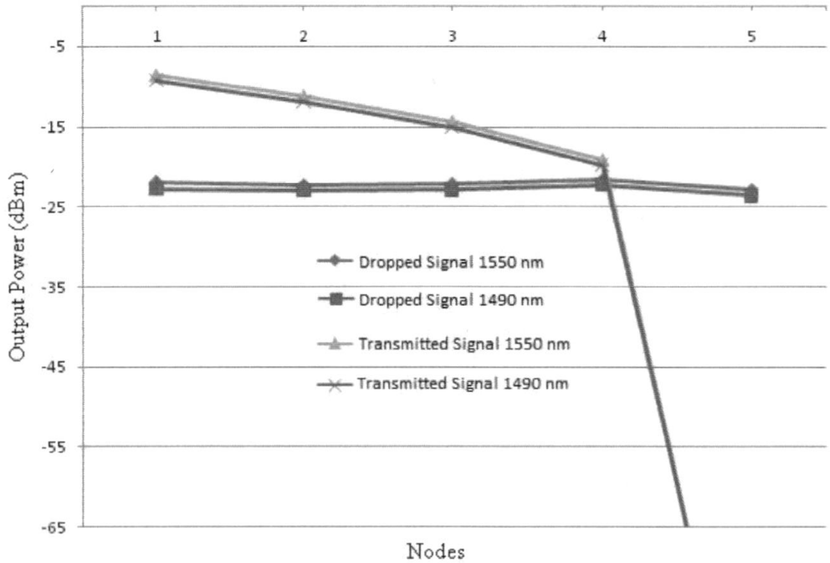

Figure 17. Power output to 5-optical node as a transmitted signal and a dropped signal (1550nm and 1490nm) when coupling ratio adjusted to varied ratio.

Figure 17 shows output power at every optical node for Dropped-Signal and Transmitted-Signal. Power product achieved for dropped signal when optical coupler adjusted to get optimum power output per optical node. The transmitted signal at last node on the other hand adjusted to -100 dBm due to no more optical node that in turn to carry the signal.

6. Discussion

Characterization and study of protection schemes in the placement of order and scattered settlements had been conducted through simulation analysis [16, 17, 19]. The characterization was carried out taking into account the performance and design of the accessibility of the physical layer only. It was found that there were clear differences in terms of dynamic range all the damage, the maximum distance that can be achieved and also the recovery of each design is constructed.

Table 2 shows the comparison of each design of the accessibility of the protection scheme. For proper placement of the protection scheme, the recovery of damages up to the level-3 to the highest level of protection scheme designs from scattered settlements. However, to support the protection scheme until the third recovery phase, the necessary dynamic range is also higher, up to 30.502 dBm for the case as shown (for the cumulative distance of 15km). In addition, the maximum distance that can be achieved at the third level of damage is on the wane for architecture in the placement of proper protection scheme. Accessibility in terms of maximum distance on scattered settlements showed similar values in each case, damage was not different signal path for the route path protection ring and linear (hybrid protection) than the ratio of the optical coupler is adjusted at each node so that the optical output power is same at each node.

Table 2: Comparative performance of the repair, dynamic range, and maximum range for each type of recovery scheme in the network.

Characterisation	Regular placement*	Scaterred placement **	
	Protection *Linear*	Hibrid-Protection	Ring-protection
Repairing Stages :	Stage-3	Stage -3	Stage-1
Dynamic range:			
i) Normal condition	23.80 dBm	28.42 dBm	27.42 dBm
ii) Stage-1 failure	26.05 dBm	28.42 dBm	27.42 dBm
iii) Stage-2 failure	28.10 dBm	29.58 dBm	-
iv) Stage-3 failure	30.50 dBm	29.58 dBm	-
Maximum Distance:			
i) Normal Condition	60.50 km	40.80 km	50.60 km
ii) Stage-1 failure	51.00 km	40.80 km	50.60 km
iii) Stage-2 failure	43.00 km	40.50 km	-
iv) Stage-3 failure	33.80 km	40.50 km	-

Proper placement*: results obtained for the linear network and the sensitivity of -35 dBm at a data rate. 25 Gbps (down path) is taken to the wavelength of 1550 nm. The average readings from the simulation results taken at eight ONU at a distance of 15 km. Reading dynamic range and the maximum distance is achieved in accordance with the results of the simulation.

Placement of Dispersed **: The results obtained for a ring network for the protection scheme and the hybrid ring and the sensitivity of -35 dBm at a data rate. 25 Gbps (halaan down) is taken to the wavelength of 1550 nm. Average of eight readings taken at ONU in the optical node with eight total distance 15 km. Reading dynamic range and the maximum distance is achieved in accordance with the results of the simulation.

7. Conclusion

Among the results discussed in the development of protection schemes to be applied in the proper placement is associated with a dynamic range of each of the damage, range and accessibility of the growing number of consumers and their impact on dynamic range and sensitivity. In addition to the restoration scheme can carry up to 3 the extent of damage are also implications for the network with the highest level of recovery of the resulting dynamic range is also high time to do the recovery is also increased.

Acknowledgments

This project had been carried out in the computer and network security research group laboratory under the department of electrical, electronic and system engineering, Universiti Kebangsaan Malaysia (UKM), Selangor, Malaysia and Department of Electrical Engineering, Universitas Muhammadiyah Prof. Dr. HAMKA, Jakarta, Indonesia. All authors and co-authors are contributed in this project.

References

[1] Yeh C H and Chi S 2008 Self-protection against fiber fault for ring-based power-splitting passive optical networks *Optical Engineering* vol **47** no 2 pp 020501, DOI: 10.1117/1.2841702.

[2] Chuan N B, Ab-Rahman M S and Premadi A 2011 Installation and field testing of video delivery system *Asian Journal of Applied Sciences* vol **4** no 4 pp 423-30 DOI: 10.3923/ajaps.2011.423.430.

[3] Ab-Rahman M S, Chuan N B and Jumari K 2009 Engineering a network management system for FTTH access network *Journal of Applied Sciences* vol 9 no 13 pp 2390-98 DOI: 10.3923/jas.2009.2390.2398

[4] Prat J 2008 Research towards unlimited bandwidth access in *Next-Generation FTTH Passive Optical Networks J Prat* Ed 1st ed Netherlands: Springer Science & Business Media p 188,

ISBN: 978-90-481-7889-6.

[5] Ab-Rahman M S, Chuan N B, Kasimin R and Jumari K 2010 Engineering installation works on FTTH-PON laboratory test bed: Case study *Journal of Applied Sciences* vol 10 no 1 pp 29-36 DOI: 10.3923/jas.2010.29.36.

[6] Chuan N B, Ab-Rahman M S, Premadi A and Jumari K 2010 Optical fault monitoring method in 8-branched PON-based i-FTTH *Research Journal of Information Technology* vol 2 no 4 pp 215-27 DOI: 10.3923/rjit.2010.215.227.

[7] J. Paul E G 2006 *Fiber to the Home: The New Empowerment* vol **4** New Jersey: John Wiley & Sons ISBN: 9780471755630.

[8] Jirachariyakool R, Sraium N and Lerkvaranyu S 2017 Design and implement of GPON-FTTH network for residential condominium in *14th International Joint Conference on Computer Science and Software Engineering (JCSSE)* Nakhon Si Thammarat Thailand 2017 pp 1-5 DOI: 10.1109/jcsse.2017.8025942.

[9] Żotkiewicz M and Mycek M 2017 Reducing the costs of FTTH networks by optimized splitter and OLT card deployment *Journal of Optical Communications and Networking* vol **9** no 5 pp 412-22 DOI: 10.1364/jocn.9.000412.

[10] Chuan N B, Ab-Rahman M S, Premadi A and Jumari K 2010 Optical power budget and cost analysis in PON-based i-FTTH *Research Journal of Information Technology* vol 2 no 3 pp. 215-27 DOI: 10.3923/rjit.2010.127.138.

[11] Ab-Rahman M S, Azizan L A H, Aziz S A C and Jumari K 2011 The eye diagram analysis of restoration scheme in FTTH-PON *Journal of Applied Sciences* vol **11** no 5 pp 840-47 DOI: 10.3923/jas.2011.840.847.

[12] Ab-Rahman M S, Ehsan A A and. Shaari S 2006 Survivability in FTTH PON access network using optical cross add and drop multiplexer switch *Journal of Optical Communications* vol **27** no 5 pp 263-69 DOI: 10.1515/joc.2006.27.5.263.

[13] Ab-Rahman M S, Baharuddin A, Chuan N B, Aziz S A C and Hashim S 2009 Analysis of equipment failure & risk in fiber-to-the home customer access network & reduction techniques proposal *Australian Journal of Basic and Applied Sciences* vol **3** no 4 pp 3692-3702.

[14] Ab-Rahman M S, Hassan S R, Aziz S A C and Jumari K 2009 Implementation of customer protection device (CPD) in FTTH EPON *International Journal of Computer Science and Network Security* vol **9** no 3 pp 125-9.

[15] Aziz S A C, Ab-Rahman M S and Jumari K 2009 Automatic protection unit in application of restoration scheme for FTTH passive optical network in *International Conference on Computer Technology and Development (ICCTD)* Kota Kinabalu pp 256-9 DOI: 10.1109/icctd.2009.244.

[16] Ab-Rahman M S, Hassan S R and Jumari K 2009 Customer access protection unit (CAPU) performing auto-restoration in line with access control system (ACS) for FTTH customer's network in *International Conference on Space Science and Communication* 2009 pp 74-77 DOI: 10.1109/iconspace.2009.5352665.

[17] Aziz S A C, Ab-Rahman M S and Jumari K. 2009 Customer access protection unit for survivable FTTH network," in *International Conference on Space Science and Communication* Port Dickson Neg Sembilan Malaysia pp 71-73 DOI: 10.1109/iconspace.2009.5352668.

[18] Premadi A, Ab-Rahman M S, Aziz S A C and Jumari K 2009 Acces control System: a cost effective protection scheme for fiber fault identification in *International Conference on Signal Acquisition and Processing (ICSAP)* The MINES Reseort city Seri Kembangan Selangor 2009 pp 57 – 60.

[19] Aziz S A C, Ab-Rahman M S and Jumari K 2010 Self restorations through multi line drop mechanism in intelligent FTTH in *Regional Engineering Postgraduate Conference (EPC)* Hotel Equatorial Bangi Malaysia.

Identification parameter system for mathematical modeling BLDC motor using transfer function models

I Anshory[1,2,*], I Robandi[3] and Wirawan[4]

[1] Department of Electrical Engineering, Faculty of Engineering, Universitas Muhammadiyah Sidoarjo, 601271, Indonesia
[2,3,4] Department of Electrical Engineering, Faculty of Engineering, Institut Teknologi Sepuluh Nopember (ITS), 60111, Indonesia

*Email: izzaanshory@umsida.ac.id

Abstract. The parameter system identification of Brushless Direct Current (BLDC) motorbike on electric bicycles is one of the important factors that support success in improving a stable and reliable control system. The purpose of this paper is to explain the mathematical modeling of BLDC motors by system identification parameters. The method of system identification parameters used in this study is experimentation, namely the pulse width modulation (PWM) signal generation process and the collection of input data in the form of current, voltage, and output data such as speed is carried out using sensors connected to the microcontroller. The input and output data are then simulated using the transfer function model found in the system identification toolbox in the MATLAB program. From the results of testing and simulations carried out on the BLDC motor system, a mathematical model was obtained in the form of a transfer function with third-order transfer function is the best fit value than the others. The sequential values of the transfer function 1, 2 and 3 are 69.81, -1945, and 83.34.

1. Introduction

Brushless Direct Current (BLDC) electric motor is one type of Permanent Magnet Synchronous Motor (PMSM) that has many advantages compared to direct current (DC) motors and induction motors such as smooth sound, compact size, large torque, high efficiency [1], long service life, and the most important thing about the advantages of a BLDC motor is the need for setting speed and position, the ability to control speed with changes in voltage [2].

The most important part of analyzing and designing a speed and position control system is knowing the plant model to be controlled [3]. The BLDC motor bike, as one of the most widely used electric motors in the industrial world, required a control system that can work well and produce the desired torque or speed [4]. One effort to improve the efficiency of the BLDC motor is by engineering and optimizing the BLDC motor speed control system [5]. The optimization of the BLDC motor system requires a mathematical model to design the controller, detect failure, and diagnose transient responses [6].

The mathematical model of the BLDC motor system was obtained through the process of input and output signals. The advantage of using this experiment is that it does not require specific knowledge of the system under study, and simple mathematical models that describe the system can be given as a result. The disadvantage of this analysis is that the previous research system must have the fact that the results of the application are limited to the same system. To get the value of the appropriate control parameters, a simulation performed. The value of the control parameter of the simulation results used as a reference for the implementation of the BLDC motor speed control system hardware [7].

The first step of the parameter identification system is hardware design using several main components such as the Arduino Uno, current sensor, voltage sensor, and speed sensor. The Arduino programmed to produce pulse width modulation (PWM) signals [8]. With changes in the PWM value, the motor rotation speed can be monitored and measured to obtain current, voltage and rotation speed. Motor rotation speed, current and voltage data are used as the basis for modeling BLDC motors mathematically using the system identification toolbox contained in the MATLAB program. These parameters are simulated based on input data such as measured voltage and current and output data which are speed and position.

The motor that identified the input and output data in this study is a BLDC motor which is used as a driver on an electric bicycle with 350 watts of power. After the input and output data are obtained, all data transferred from Excel to the Matlab workspace is simulated using the Identification Toolbox System (SIT) which is one of the tools in the MATLAB program so that the transfer function is obtained. This procedure is faster and simpler than using conventional methods.

2. Method

2.1. System identification

The stages for system identification parameters include taking input, output data pairs [9] and the process of determining parameters and validation [10]. The process of retrieving pairs of input and output data is carried out by the microcontroller while the determination of parameters and validation is done on a PC using the MATLAB program. System Identification (SID) is an additional tool from the Matlab software. SID can be used as a means of identification described in Figure 1, simulation and analysis of dynamic systems using graphical user interface (GUI) [11].

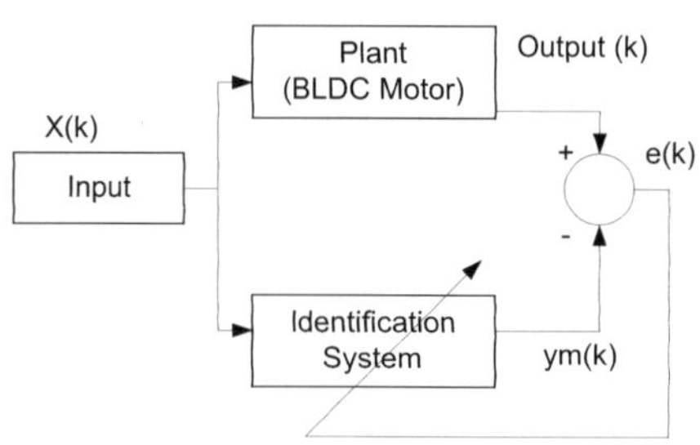

Figure 1. Block system identificaton parameter of BLDC motor

In this study the process of identifying BLDC motors was carried out through four stages, namely [4]: a). Availability of input-output plant data to be identified. b). Model structure selection. c). Estimated model. d). Parameters. Validate the identified model (structure and value of the parameters).

2.2. *Mathematical modeling of BLDC motor*

The design of system identification parameters to obtain a mathematical model in the form of a transfer function [9], the step is to design a BLDC motor speed control system to measure input and output data, through the placement of several sensors connected to the Arduino as shown in Figure 2.

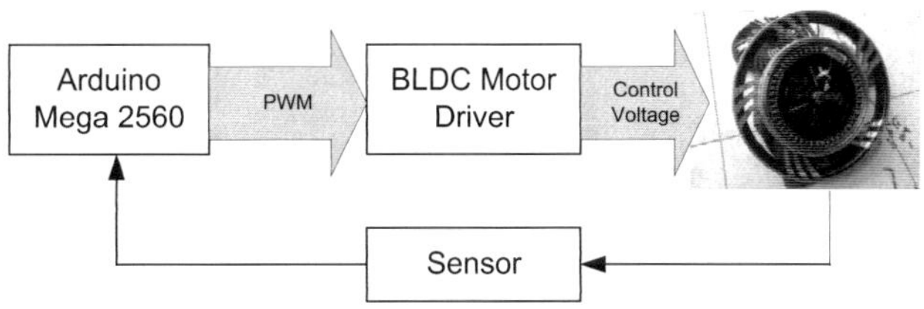

Figure 2. Block design for system identification

The BLDC motor driver circuit is composed of 3 pairs of PNP and NPN transistors. In order to produce a voltage like in Figure 2, each transistor must be given a control signal. The control signal given is a periodic control signal which is divided into 6 conditions. There are two methods for controlling a BLDC motor namely the conventional method or the six step method and the sinusoidal method [12]. The six step method is the method of giving PWM pulses in the form of trapezoidal waves while the second method is sinusoidal method, which is giving PWM pulses in the form of pure sinusoidal waves [13]. This difference is made on the basis of the interconnection of the coil inside the stator winding to provide various types of back emf [14].

In Figure 3. shows a 3 phase inverter circuit that can be arranged by six transistors that can be controlled with the type of MOSFET, Insulated Gate Bipolar Transistor (IGBT). IGBT has work properties that combine the advantages of MOSFET transistors and Bipolar Junction Transistors. Each phase is composed of two switches. Thus the current flowed at each phase can be adjusted. There are two types of inverters commonly used based on the type of conduction, which is 180 degrees and 120 degrees.

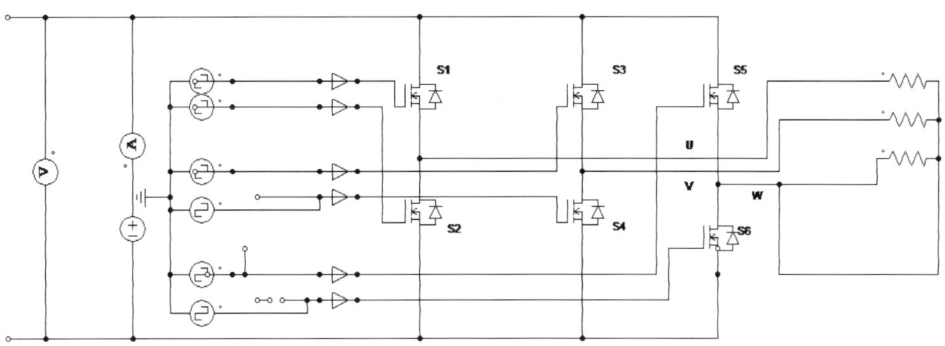

Figure 3. Three phase Inverter circuit

3. Result and analysis

Design testing of BLDC motor system parameter identification is done in several stages, namely taking BLDC motor input and output data, then input and output data simulated and analyzed using the toolbox identification system in the Matlab program.

3.1. Retrieval of input and output data

Collecting 208 data from the input and output of the BLDC motor using the Arduino microcontroller is pulse width modulation, and speed sensors. From input and output data, it is simulated by using the program ident found in MATLAB to get the transfer function. The simulation results are shown in Figure 4.

In Figure 5, we can explain the use of the system identification toolbox found in the MATLAB program by selecting the structure of the transfer function model. With three experiment for first order, second order, and third order. In this study, it has succeeded in realizing an Arduino based BLDC motorbike identification and control module.

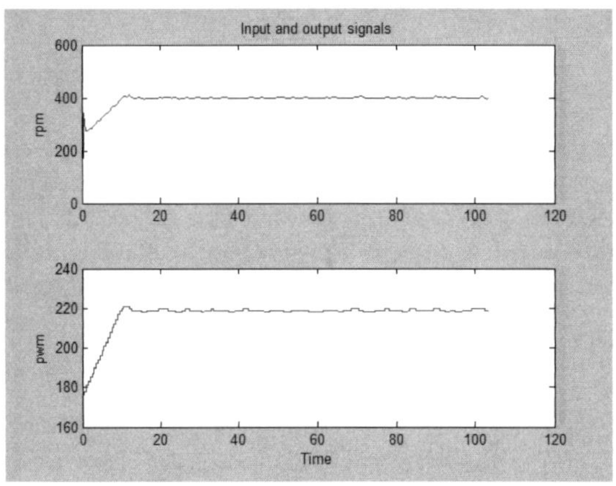

Figure 4. Signal input dan output

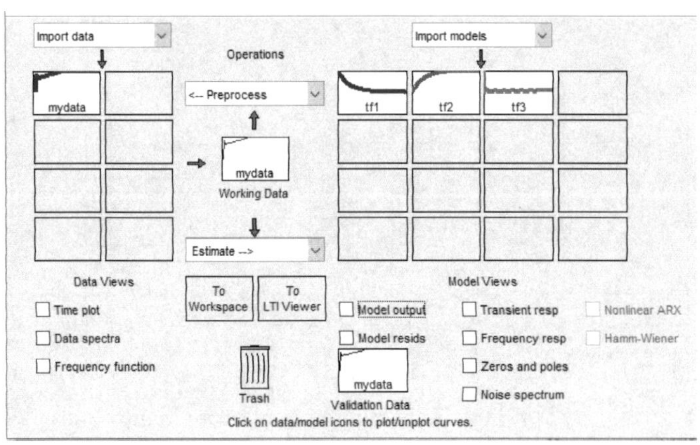

Figure 5. System identification toolbox with transfer function models.

The following Table 1 results in identification with a two type approach.

Table 1. Results of the BLDC motor model approach

1st order approach	$\frac{1589}{s + 875.4}$
2nd order approach	$\frac{-32.53s - 10.51}{s^2 + 26.95s + 6.221}$
3rd order approach	$\frac{12.32s^2 - 7.729s + 4.569}{s^3 + 9.131s^2 + 0.889s + 2.507}$

From Table 1, it was then simulated to find out the output model of first order, second order, and third order. By pressing "Output Model", a comparison between the calculated and measured data will be displayed, and in the right corner of the model accuracy window is calculated because the measured data will appear. When observing the simulation results from experimental data it shows that the calculation for the third order transfer function is the best fit value than the others. The sequential values of the transfer functions 1, 2 and 3 are 69.81, -1945, and 83.34 (Figure 6).

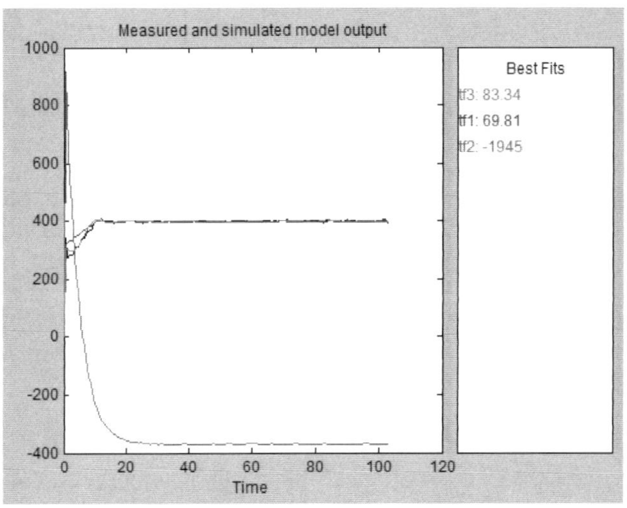

Figure 6. Comparison between the input data and the calculated model

The transfer function of a system is a mathematical model that expresses differential equations that connect output variables to input variables. If the system's transfer function is known, the output or response can be learned from a variety of given inputs. The transfer function provides a comprehensive description of the dynamic characteristics of a system. One-system response analysis uses transfer functions that contain elements termed pole and zero. Pole is a Laplace s variable value which causes the value of the transfer function to become infinite. While zero is the value of the Laplace s variable which causes the value of the transfer function to be zero such as Figure 7. Transfer function is used to determine the stability of the system. Figure 7 shows that the system is stable because the pole is to the left of the plane s. Beside that, this system has two real zeros, marked by o on the plot, and this system also has a complex pair of polishes, with an x.

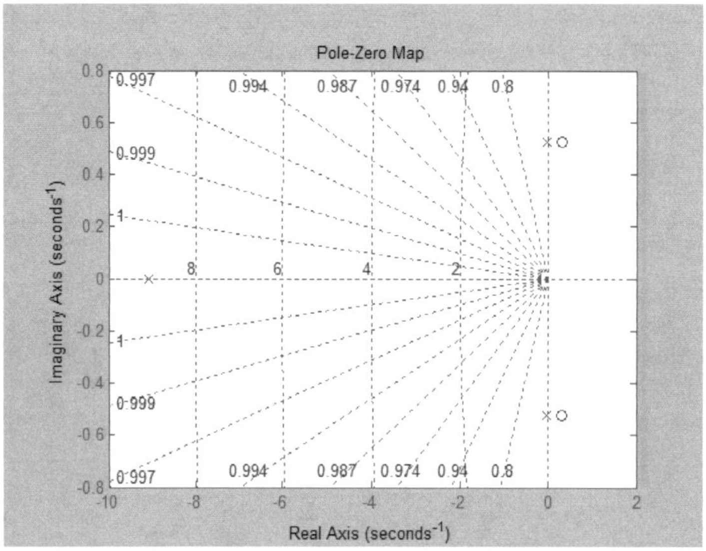

Figure 7. Comparison between the input data and the calculated model

4. Conclusion

This paper explains the approach to identify system parameters with input and one output on the BLDC motor speed control system using the System Identification Toolbox which contains Matlab software. In order to identify the model correctly, it is necessary to record the results of input and output data that use the Arduino microcontroller and some sensors to record results in the form of digital data accurately and precisely. After conducting experiments and simulations by loading data into the Matlab program, the results of mathematical models in the form of transfer functions with third order. The sequential values of the transfer functions 1, 2, and 3 are 69.81, -1945, and 83.34.

Acknowledgment

The author would like to thank the Universitas Muhammadiyah Sidoarjo for providing financial assistance for the International Conference on Engineering Applied Technology 2019.

References

[1] Xiang C, Wang X, Ma Y, Xu B 2015 Practical Modeling and Comprehensive System Identification of a BLDC Motor.

[2] Jigang H, Hui F, Jie W. A 2019 PI controller optimized with modified differential evolution algorithm for speed control of BLDC motor vol **60**(2) pp 135–48.

[3] Alsayid B, Salah WA, Alawneh Y 2019 Modelling of sensored speed control of BLDC motor using MATLAB / SIMULINK vol **9**(5) pp 3333–43.

[4] Trivedi MM, Budhvani MK, Sapovadiya KM, Pansuriya DH, Chirag D 2017 Design & Development of E-Bike - *A Review* vol **1**(5) pp 36–43.

[5] Kumar B, Swain SK, Neogi N 2017 Controller Design for Closed Loop Speed Control of BLDC Motor vol **9**(1).

IOP Conf. Series: Materials Science and Engineering **821** (2020) 012023 doi:10.1088/1757-899X/821/1/012023

[6] Bourogaoui M, Sethom HBA, Belkhodja IS 2016 Speed/ position sensor fault tolerant control in adjustable speed drives – *A Review* ISA Trans [Internet] 64 pp 1–16 Available from: http://dx.doi.org/10.1016/j.isatra.2016.05.003

[7] Anshory I, Robandi I, Wirawan 2018 Identification and optimization speed control of BLDC motor using fuzzy logic controller *Int J Eng Technol* vol **7**(2.14 Special Issue 14).

[8] Xia C 2012 Permanent magnet brushless dc motor drives and controls Singapore: John Wiley & Sons Singapore Pte. Ltd.

[9] Kumar M, Aggarwal A, Kumar T. Bat 2016 Algorithm : Application to Adaptive Infinite Impulse Response System Identification pp 3587–604.

[10] Anshory I, Robandi I 2018 Identification and Optimization Speed Control of BLDC Motor Using Fuzzy Logic Controller vol **7** pp 267–71.

[11] Fathuddin F, Susanto R. 2018 Identifikasi Sistem Motor DC dan Kendali Linear Quadratic Regulator Berbasis Arduino-Simulink Matlab vol **17**(2).

[12] Kumar CS, Kumar NS, Veni KSK 2014 Design and Evaluation of PI Controller for Four Switch BLDC Motor Drive vol **573** pp 136–42.

[13] Salah WA, Ishak D, Abu B, Abu A, Aish A, Shawal M 2015 Implementation of PWM control strategy for torque ripples reduction in brushless DC motors pp 239–50.

[14] Das B, Chakrabarti S, Kasari PR, Chakraborti A 2014 Novel reverse regeneration technique of BLDC motor for capacitor charging. In: 2014 *International Conference on Control, Instrumentation, Energy and Communication* (CIEC) pp 246–53.

IOP Conf. Series: Materials Science and Engineering **821** (2020) 012024 doi:10.1088/1757-899X/821/1/012024

Identification of earthquake types based on seismogram data

D N K Hardani [*], D A Megapiengga, and M T Tamam

Department of Electrical Engineering, Faculty of Engineering and Sciences, Universitas Muhammadiyah Purwokerto. Indonesia

*Email: diannova.kh@ump.ac.id

Abstract. Indonesia is one country that has many active volcanoes, and it can have a high potential for earthquakes due to volcanic activity. Central Java is an area that has an active volcano, the Mount Slamet. Various kinds of earthquake signals, earthquake strength, and frequency are recorded using a seismogram. The process of recognizing seismic signal patterns using short-wave transformations has a higher chance of success. Several types of earthquakes that have occurred on the Mount Slamet. There are three types of earthquake data tested, Shallow Volcanic Earthquakes (VB), Gust, and Tremors. This study aims to identify the type of earthquake vibration signals recorded on the seismograph. Earthquake image processing system consists of several parts. The image is normalized to get the image in the time domain. Then the image is processed with two processes to determine the characteristics of the earthquake. The Fast Fourier Transform process is used to determine the strength of earthquake signals based on the frequency. The quantization process is used to take samples of each data in the time domain. In this study, the method used for identification is pattern recognition and decision trees. The identification system can recognize signals that are approached using the Root Mean Square, Average Power Spectrum, and statistical features. The results of tests carried out obtain 100% accuracy of each method.

1. Introduction

Indonesia is one country that has many active volcanoes, and it can have a high potential for earthquakes due to volcanic activity. Central Java is an area that has an active volcano, Mount Slamet. The area of Mount Slamet starting from the peak to its feet divided into five districts. The west-northwest sector includes the Brebes Regency area, the north sector including the Tegal Regency region, the northeast-southeast sector including the Purbalingga Regency region, and the south-southwest sector including the Banyumas Regency region [1]. The earthquake frequency of an area refers to the type and size of earthquakes experienced over a while. Various kinds of earthquake signals, earthquake strength, and frequency are recorded using a seismogram. Earthquake vibrations recorded on the seismograph can be identified to determine the category of the earthquake that occurred.

Seismogram data in monitoring is raw data, which contains various information about the observed condition. Short-wave switching is a method that can be used to present data and operator functions into different frequency components. The process of recognizing seismic signal patterns using short-wave transformations has a higher chance of success. With the transfer of this short-wave range, it is possible to locate time-frequency. Thresholding method aims to limit and eliminate parts of the signal that are considered not to contain much valuable information. By determining the value of the parameter data, then the restricted parts can be regarded as mixed noise [2].

The automatic classification process of seismic events is fundamental because of the large amount of data that is received continuously. Seismic signal analysis that classifies activities can do by visual inspection and calculation of signal characteristics. This process is subjective and requires significant hard work and time and experience. Reliable automated classification systems can reduce the effort needed and make classification faster and more objective. The approach taken for the classification process by developing an expert classification system based on fuzzy rules can imitate human reasoning and combine analyst knowledge of the classification of seismic events. The classification results on real seismic data show the robustness of the classifier and its ability to operate in on-line classification [3].

Detection and classification of seismic signals can also use an Artificial Neural Network (ANN) methods. A neural network that is trained to recognize actual seismic events makes it possible to save about 70% of computing time in an automatic detector. Automatic associative networks trained in real-time create an internal model of the input signal and errors made in reproducing the signal increase when some modification of the input signal prevents it from matching in the internal model [4]. The unsupervised (neural network) training method consists of a Self-Organizing Map (SMO). The SOM is constructed as a neural classifier and complementary reliability estimator to distinguish seismic events, and was employed for varying map sizes. The SOM achieved a discrimination reliability that could be employed routinely in observatory practice; however, about 6% of all events were classified as ambiguous cases [5]. But in other research prove that the resulting SOM map separated into different areas, each one containing the events of a defined type. It means that the SOM discriminates well the four classes of seismic signals. Moreover, the system will classify a new input pattern depending on its position on the SOM map. This approach can be an efficient instrument for the real-time automatic analysis of seismic data, especially in the case of possible volcanic unrest [6]. A computer-based classifier to automatically identify four seismic event classes of the Llaima volcano are implemented using support vector machines. Results indicate that the features used for recognition of the events of Villarica volcano also provide good recognition results for the Llaima volcano, yielding classification exactitude of over 80% [7].

The new approach combines nonparametric smoothing and classification techniques, which are applied directly to the seismic data, with different graphical representations of the intermediate steps introduced. For each sensor position, potential horizon locations identified along the corresponding time-series traces. Tese candidate locations then examined across all traces and when consistent patterns occur the points are linked together to form coherent horizons [8].

Classifying seismic signals into the appropriate type of volcanic earthquake is one of the most important tasks for monitoring volcanic activity. Such tasks must be carried out routinely. This activity causes a significant workload for personnel. Pattern recognition gives practitioners volcanic seismology theories and methods for designing classification systems with digital signal processing techniques. It will present promising and challenging opportunities for automatic identification of volcanic earthquakes. This study aims to identify the type of earthquake vibration signals recorded on the seismograph by comparing pattern recognition and decision trees analysis model.

2. Method

The morphology of Mount Slamet divided into old volcano morphology and young volcano morphology. Geological structures that develop in the Slamet Mountain and surrounding areas, generally in the form of normal faults that often founded in the Old Slamet group. Traces of these faults in the field based in the form of breccias, scratch lines, fault zones, straightness of hills and valleys, straight and steep escapes and sharp contact between rock units [1]. It is composed of bedrock, old volcano rock, and young volcano rock. It is also cut off by a particular fault line. The geological identification of Mount Slamet is carried out through field observations of the landscape, rocks, and geological alignment [9]. The geological structure of an area can affect the occurrence of active volcanoes and are prone to earthquake disasters.

2.1. Signals database

The data used in this work to evaluate the classification system is data collected at Mount Slamet Monitoring Post, Center for Volcanology and Geological Disaster Mitigation Gambuhan, Pemalang. Signals recorded at three observation stations, namely Jurangmangu station, Blambangan station, and Gunung Cilik station. The type of earthquake recorded on Mount Slamet consists of volcanic A, volcanic B, tectonic, tremor, and gust earthquake. The gust earthquake and the tremor earthquake were earthquakes that dominated the seismograph recordings on Mount Slamet. In this study, based on recorded data, there are three types of earthquake signal data tested, shallow volcanic earthquakes (VB), the gust earthquake, and the tremor earthquake. Figure 1 shows one of the signals recorded by a seismogram from several Mount Slamet monitoring centers during an earthquake. In the spectrogram image, the X-axis shows the time of the earthquake, and the Y-axis represents the frequency.

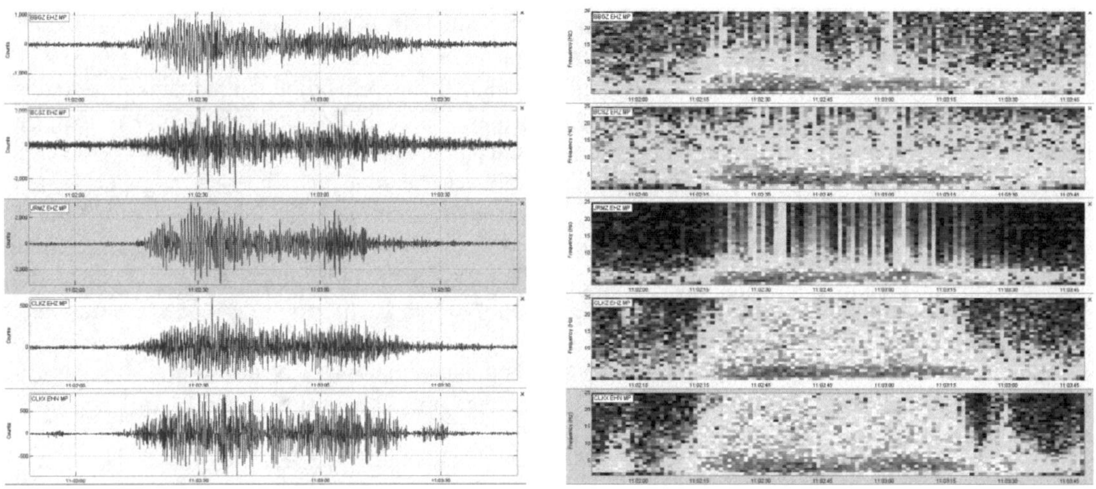

(a) (b)

Figure 1. Example of seismic data (a) One of the signal data when the gust earthquake from several observation centers (b) A spectrogram of the gust earthquake.

2.2. Signal processing

The method adopted in this study can be divided into three main stages, pre-processing, feature extraction, and classification, as shown in Figure 2. This part explains each block in the identification system. Algorithms developed using Matlab, including signals handling, filtering, segmentation, resampling, normalization, and quantization.

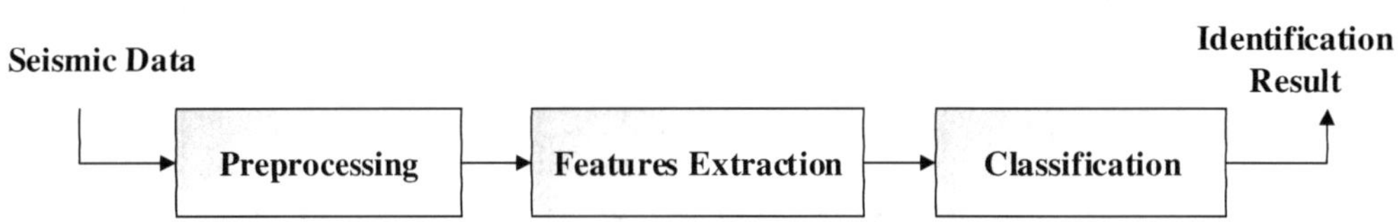

Figure 2. Block diagram of earthquake types identification system

2.2.1. Pre-processing.

Earthquake image processing system consists of several parts. The preprocessing step aims to prepare data that will be used for further processing. The process at this step consists of resampling and normalization. Resampling step is done by resampling existing data by cutting the data based on the time to be processed and eliminating noise contained in the sample data. Then the data is converted into grayscale so that the color data that owned only has a value of 0 and 1. Resampling has the aim of the image cut following the period that can be processed to the next stage.

Normalization is the process of converting two-dimensional image data into one-dimensional signals. Normalization is done by calculating the height and width of the image and the DPI (Dot per Inch) of the image data. The goal of normalization is that the amplitude of the earthquake signal is in the region of 0 to 1, and the earthquake signal will be calculated based on the period.

Then the image is processed with two processes to determine the characteristics of the earthquake. The Fast Fourier Transform process is used to determine the strength of earthquake signals based on the frequency. The quantization process is used to take samples of each data in the time domain. Next, to find out the characteristics of an earthquake, use the Root Mean Square process which is used to calculate the average root FFT and the Average Power Spectrum used to calculate the average.

2.2.2. Feature extraction

Feature extraction aims to get specific characteristics from a signal. Optimal feature extraction will produce different characteristics for each class of data so that it will facilitate the identification process. Feature extraction has methods commonly used for recognition. The extraction method used for earthquake signal processing is Root Mean Square (RMS) and Average Power Spectrum (AVG). Root Mean Square (RMS) is a statistical measure of the size of a variable that varies. RMS is useful when there are positive and negative variations, for example sinusoidal signals. RMS is used in various fields and is most often used in the signal field. RMS is the averaged signal magnitude of each window and its equation is shown in equation (1). In the equation, xi is the value of the ith sample in the processing window and N is the window size [10]. RMS is related to the constant force and non-fatiguing contraction.

$$RMS = \sqrt{\frac{\sum_{i=1}^{n} x_i^2}{N}} \tag{1}$$

In the equation (1), x_i is the value of the ith sample in the processing window, and N is the sample size, that is, the number of observations in the sample and FFT coefficients.

Average Power Spectrum is a process to measure the average power of a deterministic periodic signal. It is a type of continuous signal in the time domain that produces a discrete power spectrum. First of all, the signal filter to be extracted is processed by the Hamming Window method to avoid aliasing. Aliasing is undesirable in signal processing. Aliasing is a new signal where it has a different frequency from the original signal. This effect can occur because of the low number of sampling rates. To reduce the possibility of aliasing, it must go through a windowing process. The window function used in this system is the Hamming window. This process also serves to calculate the average power spectrum, other than that the noise produced is not too large. Equation (*2*) is the formula for the Hamming window method in sample M (α=0.54) [11, 12],

$$w(n) = \{\alpha - (1-\alpha)\}\cos\left(\frac{2\pi\,n}{M-1}\right), 0 \le n \le M-1 \tag{2}$$

where M is number of points in the output window. If zero or less, an empty array is returned. And $w(n)$ is the window, with the maximum value normalized to 1. After the windowing process uses the Hamming window, converting these values into a logarithmic value use equation (3).

$$\text{AVG power} = 10 \times \log 10\left(\frac{w(n)}{2}\right) \tag{3}$$

Another method often used for feature extraction is statistical. This characteristic was chosen based on prior knowledge about the data. The series of features selected contain components: mean, standard deviation, correlation, variance, skewness, and kurtosis [13, 14].

2.3. *Classification*

The identification process used classification method when the whole process has completed. Each earthquake gets the characteristics and produces a value from each earthquake signal image, which arranged into a dataset. There are two approaches used to identify the type of seismogram signal, including the pattern recognition approach and decision tree analysis. The aim is to compare the two classifier's performance.

2.3.1. *Pattern recognition system*

Pattern recognition in the field of engineering study theories and methods for designing machines that can recognize patterns. Many techniques and methods in the field of pattern recognition are borrowed from other basic and applied disciplines such as Digital Signal Processing, statistics, and machine learning. The principle of pattern recognition in this study is a matching process that will state the process ends when dt is equal to the number sought, as equation (4). But, when the value found is not contained in the dataset that created, then dt cannot be the same as the amount requested. The process to be carried out is the approach with the smallest error. It will continue to work until p, dt and q are in one place.

$$dt = \frac{(p+q)}{2} \tag{4}$$

where dt is the number sought, p is the initial value and q is the final value.

2.3.2. *Decision tree models*

In decision analysis, decision trees are used to represent decisions in decision making. The decision tree is one of the most important classification methods. It constructed by data provided, data values, and characters. The results of tree development affected by the number and type of attribute value significantly. The decision tree requires two types of data: training and testing data. Training data is the most significant part of the tree building data and procedures based on it. The more training data, the higher the decision accuracy. Testing data provides the level of accuracy and misclassification of decision trees. There are many decision tree algorithms. The algorithm used in this study is C4.5 and Random Forest.

C4.5 algorithm uses a learning mechanism. The choice of algorithm attributes based on assumptions that depend on the complexity of the decision tree and the amount of information. C4.5 extends the range of classifications to many attributes. This algorithm based on the entropy of data contained by nodal points generated from the decision tree. Entropy is representative of the degree of object disturbance in systematology. Small entropy induces small disorder [15].

Random forest is one of the popular tree-based algorithms that provide the best results for all types of problems. It is part of the ensemble learning method in which two or more weak algorithms are combined to form a robust modeling algorithm. The idea is to make too many trees in the forest. Based on the voting tree output will be done. Some functions generated by Random Forest are used by the "bagging" ensemble to overcome the problem of over-fitting when confronted with small data sets [14].

3. Result

One component that can take from the signal is its frequency spectrum by transforming a signal from the time zone to the frequency region so that a collection of information about the frequency contained in the signal will obtain, the transformation used is the Fourier transform. Getting the frequency spectrum from a seismogram signal can be done using Fourier transform, but directly calculating using FFT is inefficient because it is caused by computational time which will be very large. Therefore a method is used to obtain the signal frequency spectrum, which is the calculation of the estimated power spectrum, as shown in Figure 3. Spectral density has random properties which means it is not periodic, so the signal spectrum is based on Fourier analysis.

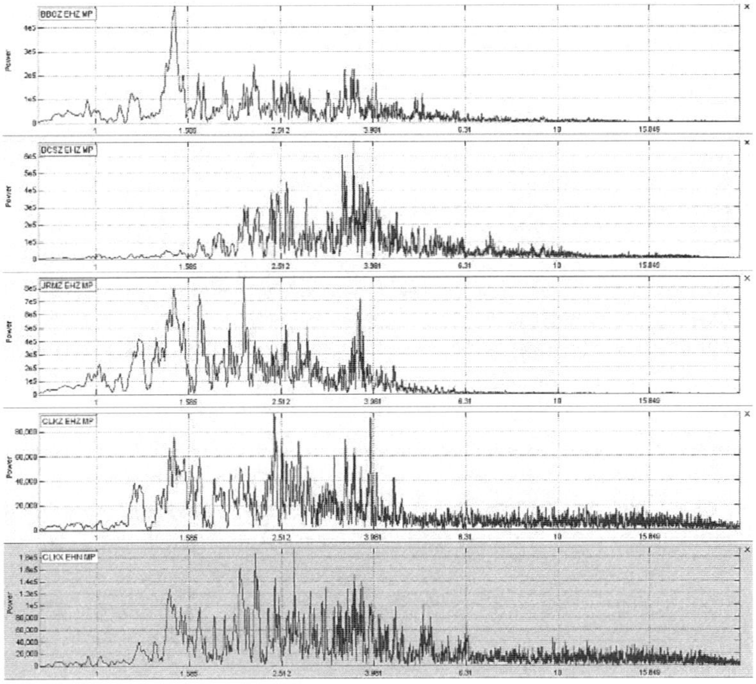

Figure 3. Power spectral when there is an eruption

Data collection aims to make earthquake character references. It is done by taking the value of the feature extraction from each earthquake. Data sampling was carried out at three stations, namely Jurangmangu station, Bambangan station, and Gunung Cilik station. Sample data are grouped into one test data. There are three types of earthquake data tested, shallow volcanic earthquakes, gusts, and tremors. Samples of feature data obtained from the feature extraction are shown in Table 1.

Table 1. Earthquake characteristics data

Data		**Features**	
		RMS	**AVG**
Shallow Volcanic Earth	vb1jrmN.mat	10.4253	-9.162
	vb1bbgN.mat	7.6644	-9.1627
	vb1clkN.mat	6.5649	-9.1627
Gust Earthquake	hb1jrmN.mat	13.9635	-9.1626
	hb1bbgN.mat	14.9888	-9.1626
	hb1clkN.mat	13.283	-9.1626
Tremor Earthquake	lt1jrmN.mat	6.5675	-9.1629
	lt1bbgN.mat	7.0331	-9.1629
	lt1clkN.mat	8.5671	-9.1625

In the pattern recognition process, the extraction results of the *rms* and *avg* feature values used as material for the pattern approach according to the type of earthquake. As an illustration, shown in Figure 4, which shows the value approach process. The approach process is used to determine the outcome of signal processing. For example, for example some data obtained from the *rms* feature extraction are A = (10.4253, 7.5329, 3.658, 13.7372, 17.2639, 14.9888, 15.1844, 8.9468, 13.283). Then the data is sorted first so that searches are faster. Sort *rms* data on A starting from the smallest to the largest value, then obtained B = (3,658, 7.5329, 8.9468, 10.4253, 13.283, 13.7372, 14.9888, 15.1844, 17.2639).

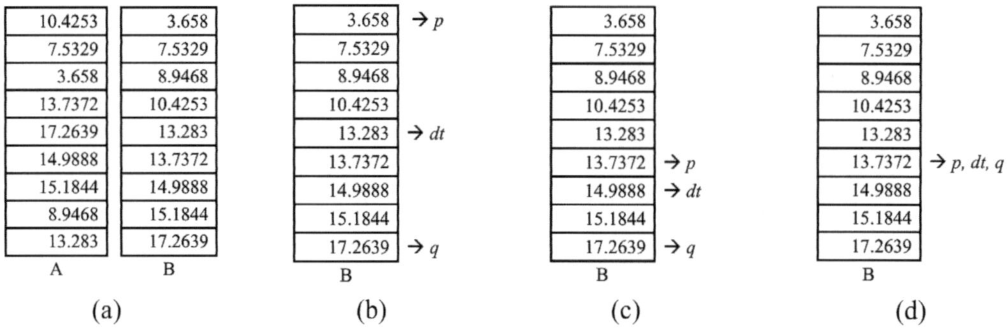

Figure 4. An illustration of the pattern approach according to the type of earthquake. (a) RMS values are sorted from smallest to largest. (b) Determine the initial position, final position, and middle value. (c) Find the middle value based on the new initial position. (d) Find the middle value based on the new end position.

First take the initial position p = 1 and the final position q = N, then find the middle data position with equation (4). If you want the value to match is 13.889, then $dt = \frac{(1+9)}{2} = 5$. The middle value obtained is the 5th data that is 13.283. Because the value is 13.889 > 13.283, then $p = dt + 1 = 5 + 1 = 6$. And then, look for the middle value, so: $dt = \frac{(6+9)}{2} = 7$. The middle value obtained is the 7th data that is 14.9888. Because the value of 13.889 < 14.9888, then $q = dt - 1 = 7 - 1 = 6$. So the value of *dt* must be found again $dt = \frac{(6+6)}{2} = 6$. The final result obtained is 13,889 approaching the 6th data that is 13,737. Likewise, to find the *avg* value approach. Each earthquake has different *rms* and *avg* values. That is what will be used as a reference in the classification process. For example, the results of earthquake signal image processing get *rms* value = 10.4253 and *avg* = 9.1612, the two values obtained can be matched in the dataset.

The next classification method used to identify the type of earthquake is the C4.5 algorithm. The data used as input to this classification process are statistical features consisting of components: mean, standard deviation, correlation, variance, skewness, and kurtosis. The classification process uses the 10-fold cross-validation test mode. Five leaves build the classification system, and the tree size is nine, as shown in Figure 5. The classification results show an accuracy of 100% with the time taken to build model 0.04 seconds.

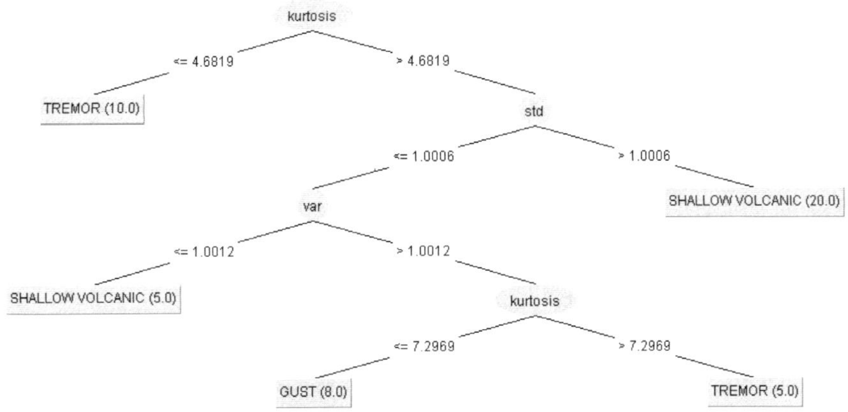

Figure 5. Classification process using C4.5

The following algorithm used for the classification process is Random Forest. Bagging is done with 100 iterations and a basic learning process. Trial mode is carried out with 10-fold cross-validation. The time needed to build the model is only 0.17 seconds. The accuracy of this algorithm is 100%. It shows that the algorithm used is excellent. Based on system testing, the results of identification of the three methods are shown in Table 2. It shows that the method used well with the accuracy of each method by 100%.

Table 2. Identification Results

Classifier		**Accuracy (%)**
Pattern Recognition System		100
Decision Tree Model	C4.5	100
	Random Forest	100

4. Conclusion

Based on the discussion, it can be concluded that signal processing using FFT can be applied to applications and helps in earthquake signal matching. The identification system can recognize signals that are approached using the Root Mean Square and Average Power Spectrum. Also, statistical features such as mean, standard deviation, correlation, variance, skewness, and kurtosis can be used for input data on the classification process. The classification process using the pattern recognition method, C4.5, and Random Forest shows the results of tests conducted with 100% accuracy of each method. It shows that the identification process works excellent.

References

[1] P V. dan M. B. G 2014 Badan Geologi [Online]. Available: http://www.vsi.esdm.go.id/index.php/gunungapi/data-dasar-gunungapi/529-g-slamet?start=2.
[2] Utari E L 2013 Pengenalan pola gelombang seismik dengan menggunakan wavelet pada aktivitas gunung merapi *J Teknol Inf* vol **VIII** no 23 pp 51–6.
[3] Ait Laasri E H, Akhouayri E S, Agliz D, Zonta D and Atmani A 2015 A fuzzy expert system for automatic seismic signal classification *Expert Syst Appl* vol **42** no 3 pp 1013–27.
[4] Romeo G 1994 Seismic signals detection and classification using artiricial neural networks *Annals of Geophysics* vol **37** no 3.
[5] Kuyuk H S, Yildirim E, Dogan E and Horasan G 2011 An unsupervised learning algorithm:

IOP Conf. Series: Materials Science and Engineering **821** (2020) 012024 doi:10.1088/1757-899X/821/1/012024

Application to the discrimination of seismic events and quarry blasts in the vicinity of Istanbul *Nat Hazards Earth Syst Sci* vol **11** no 1 pp. 93–100.
[6] Esposito A M, D'Auria L, Giudicepietro F, Caputo T and Martini M 2013 Neural analysis of seismic data: Applications to the monitoring of Mt. Vesuvius *Ann Geophys* vol **56** no 4.
[7] Curilem M, Fernando H, Daniel B, Caesar S M, Gistavo F, Luis F, Carlos C, Gonzalo A, Max C , Salman K, Nestor B Y 2014 Pattern recognition applied to seismic signals of the Llaima volcano (Chile): An analysis of the events' features *J Volcanol Geotherm Res* vol **282** pp. 134–47.
[8] Aykroyd R G and Hamed F M O 2014 Horizon detection in seismic data: An application of linked feature detection from multiple time series *Adv Stat*.
[9] Widagdo A, Candra A, Iswahyudi S and Abdullah C I 2013 Pengaruh struktur geologi gunung slamet muda dan tua terhadap pola sebaran panas bumi *IRWNS 2013* vol **4** no 2,pp 204–7.
[10] Xiao Z and Menon C 2017 Counting grasping action using force myography: An exploratory study with healthy individuals *JMIR Rehabil Assist Technol* vol **4** p e5.
[11] Islam S, Islam S, Rahman S K and Ferdous N S 2014 Design of FIR filter using hamming window *Int. J Emerg Res Manag Technol* vol **3** no 2 pp 13–6.
[12] Yadav P, Sahu L D, Maravi and Rahi P K 2017 Design of low pass filter using Hanning and Hamming Window techniques *Int J Res Appl Sci Eng Technol* vol **5** no 6 pp 1983–87.
[13] Esmael B, Arnaout A, Fruhwirth R K and Thonhauser G 2013 A statistical feature-based approach for operations recognition in drilling time series *Int J Comput Inf Syst Ind Manag Appl* vol **5** pp 454–61.
[14] Hardani D N K, Wahyunggoro O, Nugroho H A and Faisal N 2014 Analysis of emotional condition based on electrocardiogram signals in *2014 International Conference on Electrical Engineering and Computer Science (ICEECS)* pp 152–57.
[15] Smith M R, Wang X and Rangayyan R M 2009 Evaluation of the sensitivity of a medical data-mining application to the number of elements in small databases *Biomed Signal Process. Control* vol **4** no 3 pp 262–68.

Harmonisa in defibrillator equipment (DC Shock)using simulink Matlab

P Harahap[*] and B Oktrialdi

Departement of Electrical Engineering, Faculty of Engineering, Universitas Muhammadiyah Sumatera Utara, Medan, Indonesia

[*]Email: partaonanharahap@umsu.ac.id

Abstract. The quality of electric power is a combination of voltage quality and current quality. Voltage disturbances often occur from the electricity network which can affect end users. Whereas, current interruption originates from end users which will affect the network. Electrical power quality problems in hospitals often occur due to non-linear loads, harmonic injection, and interaction between medical devices. Because the electrical power quality problem is cumulative, the detected minor event must be taken seriously. The nonlinear load distorts the current waveform and creates a harmonic current to the current system. The nonlinear load distorts the current waveform and creates a harmonic current to the current system. After the simulation, it can be concluded that DC Shock/ Defibrillator equipment has Total Harmonic Current Distortion (THDi) equal to 76.1% and Total Harmonic Voltage (THDv) 3.5% with 330Watt power.

1. Introduction

Electric power quality is a combination of voltage quality and current quality. Voltage disturbances often occur from electrical networks that can affect end users. Meanwhile, current interference comes from end users that will affect the network. The problem of electric power quality in hospitals often occurs due to non-linear loads, harmonic injection, and interactions between medical devices. Because the problem of electric power quality is cumulative, small incidents detected must be taken seriously. Non-linear loads distort current waveforms and create harmonic currents to the current system. For example, from electric lighting. In Indonesia, the lamp must be tested by a medical electric laboratory before being used by the hospital.

Defibrillator (DC Shock) is a tool to provide electrical energy therapy with a certain dose to the patient's heart through an electrode (pedal) placed on the surface of the patient's chest wall. Besides the rapid development of technology from a country, the development is almost comprehensive in all fields, especially in the field of electricity. Since the development of electronic power technology, the device has become the main expectation in almost all electric utilities. But the problems of system power quality caused by the increase in electronic power users by electricity consumers. Basically good electricity quality is electricity which has a stable voltage and frequency. For in Indonesia, electricity sourced from PLN is with a voltage of 220 volts and a frequency of 50 Hz.

There are many aspects that can affect the reduction in electricity quality. One of these aspects is the emergence of harmonics in electric waves rather than the intended equipment is Defibrillator (DC Shock).

A defibrillator is a fibrillation removal device or resets the heart to restore regularity to the heart beat. The defibrillator method/ method is shocked/ electric shock (shock counter) on the heart with a frequency of 60 Hz, and current of 6 A, in 0.25 - 1 second. In this study the author describes previous researchers who are relevant to the problems to be studied about Harmonics in Defibrillator (DC Shock) equipment are anticipated:

- Ariadi Hazmi (2015), in the title Reducing harmonics on Mobile 100ma X-Ray equipment by using a passive filter (Single Tuned And Double Tuned Passive Filter) test results using Single-Tuned Filter Passive Filter and Double-Tuned Passive Filter and merging Single filters -Tuned Passive Filter and Doubel-Tuned Passive Filter on X-rays can reduce 3rd order harmonics to 39th order with each Total Distortion Harmonics (THDi) and Total Voltage Distortion Harmonics (THDv), namely: 57,77 % and 2.79%, 58.48% and 2.75%, 1.74% and 2.77%
- Faisal Irsan Pasaribu (2014), with the title Reducing harmonics on ceragem medical devices with a single tuned passive filter, it can be seen that the result filter designed to function properly, from testing to ceragem loads at 6 C there is a harmonic reduction namely from 133.8% to 17%.

As a result of this harmonics causes current and voltage waves to be deformed and not sinusoidal can be seen, measured, calculated, and analyzed which is expected to cause harmonics by first doing measurements of THDv, THDi, IHDV and IHDi. To answer the approximate harmonics contained in the equipment, the researchers measured THDv, THDi, and IHDi on Defibrillator (DC Shock) equipment [1-3].

2. Preliminary

Harmonics are periodically distorted waves that occur in waves of voltage, current, or power consisting of sine waves whose frequency is a round multiple of the source / fundamental frequency, so that the shape is not sinusoidal. The relationship between harmonic and fundamental frequencies can be written as follows:

$$f_h = n.\, f_i \tag{1}$$

With f_h is the harmonic frequency, n is a multiple of waves (integers), and f_i is the fundamental frequency. This harmonic wave will hitch in the fundamental wave so that a distorted wave will form. This is due to the summing effect of harmonic waves with their fundamental waves.

The role of harmonics in the electric power system is quite large, especially in the tools contained in the power system. Harmonics will cause several effects such as overheating in some devices such as generators and transformers because the tendency for harmonics to flow to places with lower impedances. Some other impacts will be explained in this article. The magnitude parameters of harmonics are stated in Total Harmonic Distortion (THD) which can be written as:

$$THD_V = \frac{\sqrt{V_2^2+V_3^2+V_4^2 \ldots\ldots\ldots\ldots V_n^2}}{V_1} \tag{2}$$

For voltage,

$$THD_I = \frac{\sqrt{I_2^2+I_3^2+I_4^2 \ldots\ldots\ldots\ldots I_n^2}}{I_1} \tag{3}$$

Based on an agreement agreed upon internationally, the THD received is if it is worth under 5% of its fundamental voltage or current. If above this limit, the electronic device must not be used. The harmonic effect cannot be completely eliminated, but can be reduced. There are several methods commonly used in reducing harmonics, namely by using passive filters such as installing capacitors, adding phase quantities, and compensating or injecting negative harmonics..

IOP Conf. Series: Materials Science and Engineering **821** (2020) 012025 doi:10.1088/1757-899X/821/1/012025

As is generally known, harmonic currents arise due to non-linear loads. To eliminate harmonic currents, harmonic filters are commonly used. Harmonic filters can be composed of active or passive circuits. While the combination of the two is commonly called a hybrid filter. Based on the installation on the nets, these filters can be divided into two types: parallel and series. The filter functions to hold the harmonic current and pass the fundamental current Is. The harmonic current is marked with a cross to describe the current that cannot flow because of a series filter.

Passive harmonic filters can be arranged from a series of combinations of inductors, capacitors and resistors. Because it consists of passive components, this type of filter performance is affected by grid impedance. Therefore, the installation of passive filters needs to consider net impedance. Passive filters also inject reactive power because it consists of inductor and capacitor components.

The power electronic circuit is an electrical circuit that can change the electrical power source of certain waveforms such as sinusoidal waveforms into electrical resources with other non-sinusoidal waveforms by using a power semi-conductor device. Power semi-conductors have an important role in power electronic circuits. In harmonic analysis, several indices of Equations (1) and (2) are used to describe the effect of harmonics on electrical power system components.

$$\textit{THD the voltage: } THD_V = \frac{\sqrt{\sum_{h=2}^{\infty} V_h^2}}{V_1} \text{X } 100\% \tag{4}$$

$$\textit{THD current : } THD_i = \frac{\sqrt{\sum_{h=2}^{\infty} I_h^2}}{I_1} \text{X } 100\% \tag{5}$$

Equations (4) and (5) are defined as the ratio of the rms value of the harmonic component to the basic component in (%). This index is used to measure deviation deviations from one period waveforms containing harmonics on one perfect sine wave. For one perfect sine wave at the basic frequency THD is zero. Likewise the measurement of individual harmonic distortion for voltage and current order h is defined as V_h/V_1 and I_h/I_1.

This transformation can decipher a waveform into DC components, cosines and or sines with several different frequencies. The DC component is represented (ao/ 2) and n is the harmonic to n. Therefore, there is a harmonic term of 3, 5 and so on which shows the frequency of multiples of 3, 5 and so on of the fundamental frequency. The figure below shows how a square wave can be decomposed into several sine waves with several frequencies, namely n = 3.5 and 7. If harmonics are present in an electric power system, the fundamental frequency is the frequency of the system e.g. 50Hz or 60Hz. Harmonics in electric power systems can be in the form of harmonic voltages or harmonic currents [4].

2.1. IEC harmonics distortion standard

In this case the standard used as a harmonic limitation is issued by the International Electrotechnical Commission (IEC) which regulates harmonic limits on single phase or three phase small load loads whose current values are smaller than 16 amperes of perfume. For load loads, IEC 61000-3-2 standard is generally used. This is due to the absence of standard standards produced by IEEE. In the IEC 61000-3-2 standard [5, 6] the small load is classified in classes A, B, C, and D, where each class has different harmonic limits which are explained as follows:

Class D includes all types of equipment whose power is below 600 Watt specifically for personal computers, monitors, TVs [7, 8]. The current limits are expressed in the form of mA/ W and are limited to absolute prices whose values are shown in Table 1.

Table 1. Harmonic current limits for class D equipment

Harmonisa To (n)	Maximum Harmonic Flow permitted (mA/W)	Maximum Harmonic Flow permitted (A)
	75< P < 600 W	P > 600 W
3	3.4	2.3
5	1.9	1.14
7	1	0.77
9	0.5	0.4
11	0.35	0.33
13	0.296	0.21
$15 \leq n \leq 39$	3.85/n	2.25/n

2.2. DC Shock/ defibrillator

The Maximum Harmonic Flow that is defibrillated can provide controlled electric shock to the liver that has a life-threatening rhythm, such as ventricular fibrillation (VF). In VF, chaotic activity of the heart prevents blood from pumping to the body and brain. The voltage is stored by a defibrillator conducting an electric current (shock) through the chest by means of electrodes or pads placed on the chest. This short pulse stops the activity when it is chaotic, giving the heart the opportunity to restart normalization rhythm (A) (Figure 1).

External defibrillators can offer a variety of energy choices. So-called "low energy" defibrillators are those that limit their energy choices to 200J or less. Increased energy defibrillators offer a variety of energies, starting with low energy levels with the option to increase energy levels for the next surprise [9, 10].

Many people are confused at this time and energy. This distinction is important in defibrillation, because defibrillators are often explained in terms of energy (for example, 200J) but they are currently - not energy - that defibrillates. Successful defibrillation requires that now enough is sent to the heart muscle during shock.

Defibrillation requires a true middle-of-the-road approach. You must have enough of this time to reach the heart to defibrillate the heart (stop the deadly rhythm), but not so many peak currents that you risk damaging the heart. In fact, the low energy shocks of some defibrillators provide higher current peaks than higher energy shocks than other types of defibrillators.Impedance is the body's resistance to current flow. Some people naturally have a higher impedance than others.

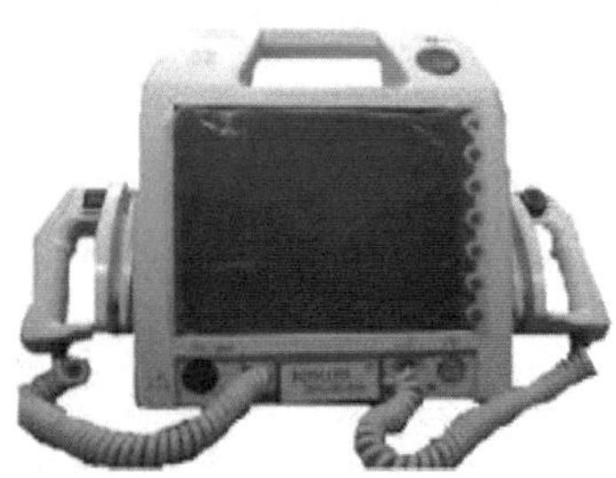

Figure 1. DC Shock/ defibrillator

Effects of Harmonics on Electric Power Systems can cause the appearance of voltage harmonics in the electric power system. This is because the harmonic current flows through the impedance of the system. Impedance can come from channels, transformers etc. [11] Negative effects of the presence of harmonic currents include:

1. Interference with analogue kWh meter.
2. Increased losses and abnormal heating on the transformer

3. Increased losses and abnormal heating on the motor
4. Possible burst of bank capacitors

Even so this effect does not always appear and occur if there are other conditions that support. Therefore, the presence of harmonic currents is not always accompanied by the interference [12, 13].

3. Research methodology

The method used in this study is a simulation method that starts from data collection and then performs harmonic measurements, and filter modelling calculations. Measurements are made on Defibrillator equipment as shown in Figure 2.The measurement results are in the form of voltage harmonics (IHDv) and current harmonics (IHDi). Measurements on Defibrillator equipment to find out the magnitude of the harmonic value contained therein. Measurement and data retrieval was carried out in Medan. Measurement using a Fluke 43B Power Quality Analyzer. Measurement data is displayed in the form of lists and graphs directly, and the data can be stored on the computer.

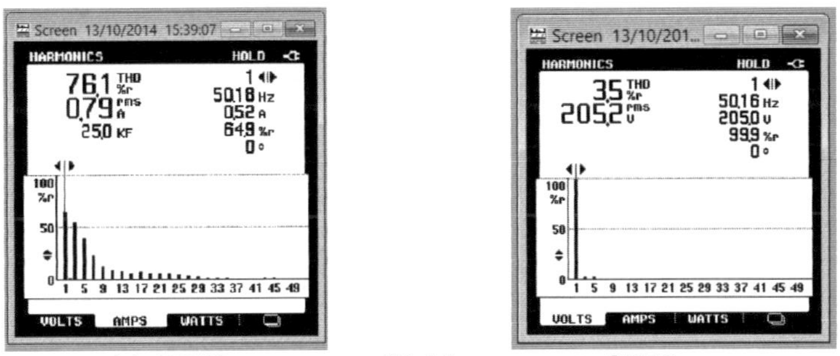

(a). THDi measurement(b). Measurement of THDv

Figure 2. Measurement of defibrillator equipment can result in THDi measurement of 76.1%.

4. Results and discussion

The following is the data that will be taken from the Defibrillator Measurement measured through the Fluke 43B Power Quality Analyzer. The data is attached to appendix B as in Table 2.

Table 2. Defibrillator measurement result

Parameter	Unit	Measurement results
U (Phase Voltage)	Volt	205,2
I (Phase Current)	Ampere	0.79
S (Apparent Power)	KVA	0.59
P (Active Power)	KW	0.33
Q (Reaktife Power)	KVAR	0.49
PF (Power Factor)		0.56
Frekuensi	Hz	50.2
THDv	%	3.5
THDi	%	76.1
IHDi 1	(mA)	2510
IHDi 3	(mA)	2149
IHDi 5	(mA)	1514
IHDi 7	(mA)	931
IHDi 9	(mA)	515
IHDi 11	(mA)	361
IHDi 13	(mA)	306
IHDi 15	(mA)	241
IHDi 17	(mA)	264
IHDi 19	(mA)	241
IHDi 21	(mA)	241

Parameter	Unit	Measurement results
IHDi 23	(mA)	218
IHDi 25	(mA)	186
IHDi 27	(mA)	143
IHDi 29	(mA)	98
IHDi 31	(mA)	55
IHDi 33	(mA)	65
IHDi 35	(mA)	43

In Table 1, we can see the results of individual measurements of current harmonic distortion (IHDi) of each harmonic sequence. The harmonic sequence shown is the harmonic sequence from first to 39th with different values for each harmonic.To make a simulation circuit, measurement and calculation results are used in the form of voltage, harmonic current value (IHDi), capacitance, inductance, and filter resistance (Figure 3).

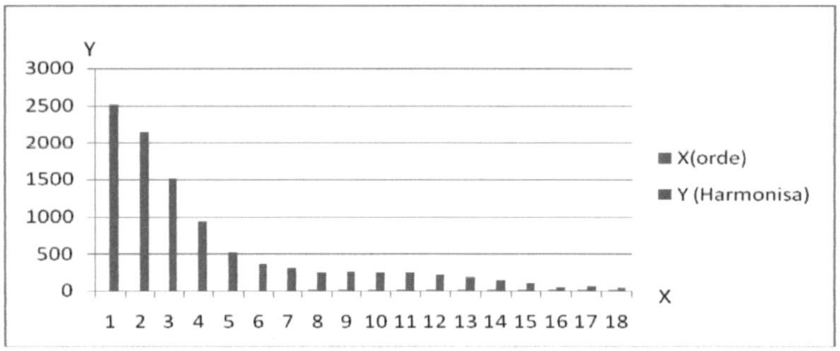

Figure 3.Graph of defibrillator measurement results

Current and voltage waveforms are obtained from the Power GUI Block section of Fast Fourier Transform (FFT) Analysis. (Figure 4 and 5).

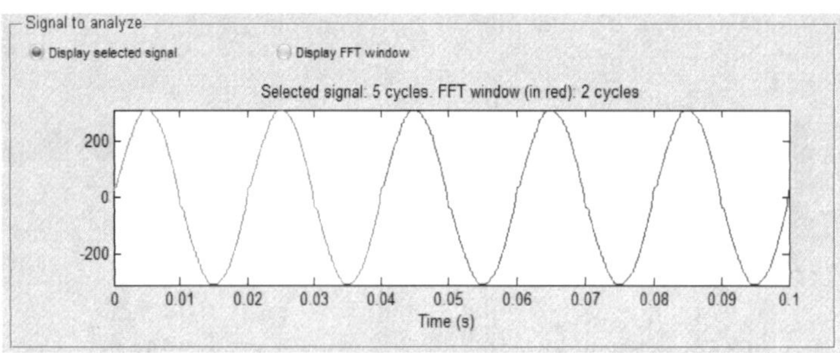

Figure 4. Voltage graphs before installation of filters

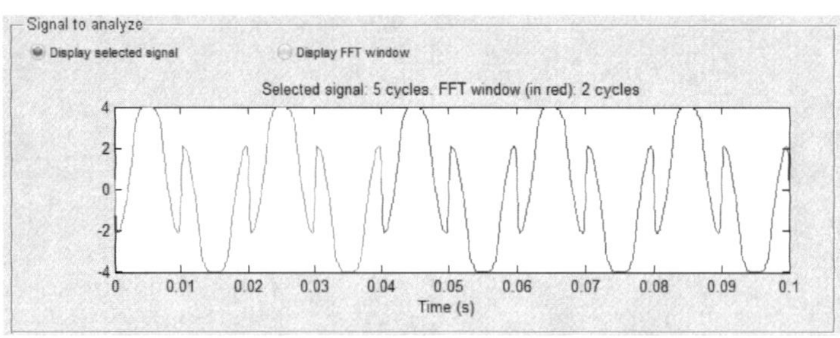

Figure 5. Flow chart before installation of filters

Classification of Harmonics Flow on DC Shock / Defibrillator equipment based on IEC61000-3- 2 Class D.

Standard measurement results obtained from DC Shock/ Defibrillator equipment are indicated harmonic order which is not in accordance with the International Electro technical Standards Commission (IEC) 61000-3-2 Class D from order 3 to 39th order. In Classification of harmonic currents before filter installation is IEC 61000-3-2 Class D Standard. In simulations for DC Shock/ Defibrillation circuit loads with simulation data adjusted for measurement data (Figure 6). Harmonic currents in DC Shock / Defibrator equipment prior to the use of filters are known to not meet the IEC 61000-3-2 Class D standard. For this reason, it is necessary to install a single tuned or double tuned filter.

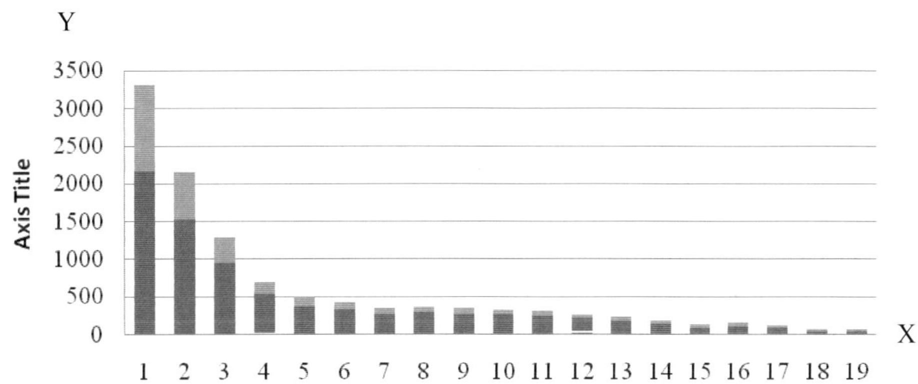

Maximum Harmonic Flow Limit allowed (mA/ W) 75 < P < 600 Watts	1156	646	340	170	119	101	87	77	69	62	57	52	48	45	42	45	37	34	31
Flow Measurement Harmonics HDI (mA)	2149	1514	931	515	361	306	241	264	241	241	218	186	143	98	55	65	43	0	0
Harmonisa To (n)	3	5	7	9	11	13	15	17	19	21	23	25	27	29	31	33	35	37	39

Figure 6. Chartharmonic current on the DC Shock/ defibrator equipment

5. Conclusion

The problem of electric power quality in hospitals often occurs due to non-linear loads, harmonic injection, and interactions between medical devices. Because the problem of electric power quality is cumulative, small incidents detected must be taken seriously. Non-linear loads distort current waveforms and create harmonic currents to the current system. After the simulation, the conclusion is that the DC Shock/ Defibrillator equipment has a Total Harmonic Current Distortion (THDi) of 76.1% and Total Voltage Harmonics (THDv) 3.5% with a power of 330Watt. The measurement of harmonic current in the DC Shock/ Defibrillator equipment before the use of filters is known not to meet the standard IEC 61000-3-2 Class D. For this reason, further research is needed in harmonic reduction in Dc Shock / Defibrillator devices by using tuned and double tune filter types to reduce harmonics in DC Shock/ Defibrillator equipment in accordance with standard harmonic limits International Electronical Commission (IEC) 61000-3-2 Class D.

References

[1] Priyadharsani A, Devarajan N, Umasaranya AR and Anitt R 2012 Survey of Harmonics in Non Linear Loads *International Journal of Recent Technologyand Engineering*.

[2] Cazacu E, Petrescu M-C, Ioniţă V and Petrescu L 2018Nonsinusoidal Load Current Effect on The Electrical and Thermal Operating Parameters of Oil Filled Power Distribution Transformers*18th International Conference on Harmonics and Quality of Power (ICHQP)* Ljubljana Slovenia 13–16 May 2018.

[3] Faisal I P 2016 Reduksi harmonisa menggunakan filter RC pada alat trapi ceragem12 *Jurnal Teknik Elektro dan Komunikasi*.

[4] Emil C, Valentin I and Lucian P 2018 *Thermal Aging of Power Distribution Transformers Operating Under Nonlinear and Balanced Load Conditions*Department of Electro-technics, Faculty of Electrical Engineering, University Politehnica of Bucharest, Splaiul Independentei 313, 060042 Bucharest, Romania Power Engineering and Electrical Engineering vol**16** no 2018 .

[5] De La Rosa F 2006 *Harmonics and Power SystemsDistribution Control Systems* Inc Hazelwood Missouri U.S.A.

[6] Haroon F, Chengke Z and Mohamed E F 2012 Analyzing the Harmonic Distortion in a Distribution System Caused by the Non-Linear Residential Loads*International Journal of Smart Grid and Clean Energy*.

[7] Hotbe H, Panusur SML T 2014 Konsentrasi Teknik Energi Listrik, *Analysis of Harmonic Effect on Heat on Belitan Transformator Distribution* Departemen Teknik Elektro Fakultas Teknik Universitas Sumatera Utara (USU).

[8] Heri S 2013Perancangan filter pasif single tuned filter untuk mereduksi harmonisa pada beban non linier*Jurnal Eltek* (ISSN1693-4024)vol **11**no 1.

[9] IEEE Std 519™ 2014 (Revision of IEEE Std 519-1992) Recommended Practice and Requirements for Harmonic Control in Electric Power Systems*IEEE Power and Energy Society* New York.

[10] IEC 61000-3-2 2014 Electromagnetic compatibility (EMC) - Part 3-2: Limits - Limits for harmonic current emissions (equipment input current≤ 16 A per phase). CENELEC,Bulgarian institute for standardization, Sofia.

[11] J. Mahesh Y 2018 Research Scholar, JNTUH, Hyderabad, INDIA Dr.A. Srinivasula Reddy Professor, Performance study of transformers under non-linear load conditions without and with active filter*International Journal of Electrical Engineering & Technology (IJEET)* vol **9**(2) pp 115–122,

[12] Hernández J L, Castro MA, Carpio J and Colmenar A 2009*Harmonics in Power Systems.International ConferenceRenewable Energies and Power Quality* 15th-17th April 2009.

[13] Kamenka A 2014 Six tough topics about harmonic distortion and power quality indices in electric power systems. *A white paper of the Schaffner Group* 2014.

IOP Conf. Series: Materials Science and Engineering **821** (2020) 012026 doi:10.1088/1757-899X/821/1/012026

River water level forecasting for flood warning system using deep learning long short-term memory network

A Faruq[1,]*, S S Abdullah[2], A Marto[3], M A A Bakar[4], Samin[5] and A Mubin[6]

[1,2,3,4] Malaysia – Japan International Institute of Technology (MJIIT), Universiti Teknologi Malaysia, Kuala Lumpur, Malaysia
[1] Department of Electrical Engineering, Faculty of Engineering, Universitas Muhammadiyah Malang, Malang, Indonesia
[5] Department of Civil Engineering, Faculty of Engineering, Universitas Muhammadiyah Malang, Malang, Indonesia
[6] Department of Industrial Engineering, Faculty of Engineering, Universitas Muhammadiyah Malang, Malang, Indonesia

*Email: faruq@umm.ac.id

Abstract. Flood is considered chaotic, complex, volatile, and dynamics. Undoubtedly, its prediction is one of the most challenging tasks in time-series forecasting. Long short-term memory (LSTM) networks are a state of the art technique for time-series sequence learning. They are less commonly applied to the hydrological engineering area, especially for river water level time-series data for flood warning and forecasting systems. Yet it is inherently suitable for this subject. This paper examines an LSTM network for forecasting the river water level in Klang river basin, Malaysia. The river water level contains of single time series observed data, with time steps corresponding to hourly data and values corresponding to the water level or stage level in meters. In this study, prediction responses for river water level data using a trained recurrent neural network and update the network state function is applied. The result verified that the LSTM network with specified training set options is a promising alternative technique to the solution of flood modelling and forecasting problems. The performance indicates with the root mean square error, RMSE 0.20593 and coefficient of determination value, R^2 0.844 are closely accurate when updating the network state compared with the observed values.

1. Introduction

Urban floods are considered one of the wide-scale devastating natural disasters affected globally, and improved flood forecasting model is essential for better flood warning and management system. It has become an issue of great importance recently due to extensive damage to human life, properties, and lead to large socioeconomic problems caused by floods. Flood forecasting system plays a crucial role in planning and regulating the disaster risk reduction and management (DRRM). The Sendai frameworks 2015-2030, highlighted 'investing in disaster risk reduction for resilience' and 'enhancing disaster risk preparedness for effective response' are respectively for priority number three and four regarding the DRRM [1].

Research on flood forecasting models has a long tradition in the hydrological engineering area. However, the determination of improved flood modelling and forecasting is technically challenging.

As reported in numerous published studies, river water level forecasting for flood management is difficult task and highly dynamic to model. Recently, researchers have shown an increased interest in river flood prediction and modelling, namely, data-driven methods. The types of models use the generalized relationship between input and output variable datasets without requiring the physical mechanism behind the process of the model. In which, the model is constructed based on historical data

Neural network models, especially as data-driven approaches, are developed through training the network to demonstrate the relationships and processes that are inherent within the data. Research on flood water level forecasting has been successfully taken by the authors [2, 3]. The works are growing advance in exploring more suitable flood forecasting model. Reviewing some published research works, by using the Artificial Neural Networks (ANNs) model, Elsafi [4] has demonstrated real-time flood forecasting in the case of River Nile at Dongola Station. Reference in [5] explored the applicability of a deep learning approach for predicting hourly air temperature using real sensor data of the Northwestern Nevada region. The results indicate that stacked denoising auto-encoder (SDAE) as deep learning approach performs better rather than standard neural networks model. Moreover, as a part of machine learning methods, ANNs is the most popular Artificial Intelligent (AI)-based technique in flood forecasting [6].

The continued advancement of AI methods in recent years has seen a growing trend towards deep learning techniques and their application in time-series prediction. Deep learning is being studied in many types of problems such as image processing, speech recognition, and natural language processing. In the field of forecasting, a recent experiment has been reported the successful use of deep learning in various fields, respectively for power load and probability density forecasting [7], traffic flow forecasting [8], and rainfall forecasting [9]. As it developed that deep learning can be quite promising reported better results than the traditional ANN model [10]. Research on the subject in deep learning has the most significant current discussion are advancing computer vision utilizing convolutional neural networks (CNN) and modelling time-series or type of sequential data through recurrent neural network (RNN) [11].

It was until the late 1990s that more advanced RNN architectures have been developed and one of the successful networks is the long short-term memory (LSTM) [11, 12]. Although studies have recognized LSTM in solving time-series prediction [13], to date, few published research works have explored the use of LSTM as deep learning approach in hydrological engineering problems, especially for flood forecasting. Thus, it is a chance to employs the LSTM network to predict river water levels for flood forecasting and warning system. The specific objective of this study is to build real-time data-driven models that enable to simulate and forecast river water levels from historical data using the LSTM network. The long short-term memory network is a type of recurrent neural network used in deep learning due to large architecture can be successfully trained.

2. Methodology

The usefulness of the approached method will be evaluated using a case study for the Klang River, Kuala Lumpur, Malaysia. The Klang River was chosen as a flood forecasting point based on the department of irrigation water level at Sulaiman Bridge. The total catchment area up to the station is approximately 466 km^2 [14]. 19008 hourly time-series observations dataset provided by the department of irrigation and drainage (DID) system Malaysia recorded in November 2011-December 2013, was used as historical input data to the LSTM network. The main feature of the LSTM network is the hidden layer called memory cells. LSTM networks consist of an input layer, one or more memory cells, and an output layer. The number of neurons in the input layer is equal to the number of explicative variables [11, 12]. Each of the memory cells has three gates maintaining and adjusting its cell state S_t: a forget gate (f_t), an input gate (i_t), and output gate (O_t). Figure 1 shows the LSTM architecture memory cell.

The state of the layer consists of the hidden state (also known as the output state) and the cell state. The hidden state at time step t contains the output of the LSTM layer for this step. The cell state

IOP Conf. Series: Materials Science and Engineering **821** (2020) 012026 doi:10.1088/1757-899X/821/1/012026

contains information learned from the previous time steps. At each time step, the layer adds information to or remove information from the cell state. The layer controls these updates using gates. At every time-step t, each of the three gates is presented with the input x_t (one element of the input sequence) as well as the output h_{t-1} of the memory cells at the previous time-step $t-1$. According to the references [11], [12] and [15], hereby, the gates serve as filters, each having different purposes:

- The forget gate determines what information is removed from the cell state
- The input gate specifies what information is applied to the cell state
- The output gate specifies what information from the cell state is used,

and the sequential update formulas are can be referred to as reference [11]. The specified topology of the proposed LSTM networks is according to `predictAndUpdateState` provided by Matlab toolbox function can be seen in Table 1, to predict time steps one at a time and update the network state at each prediction.

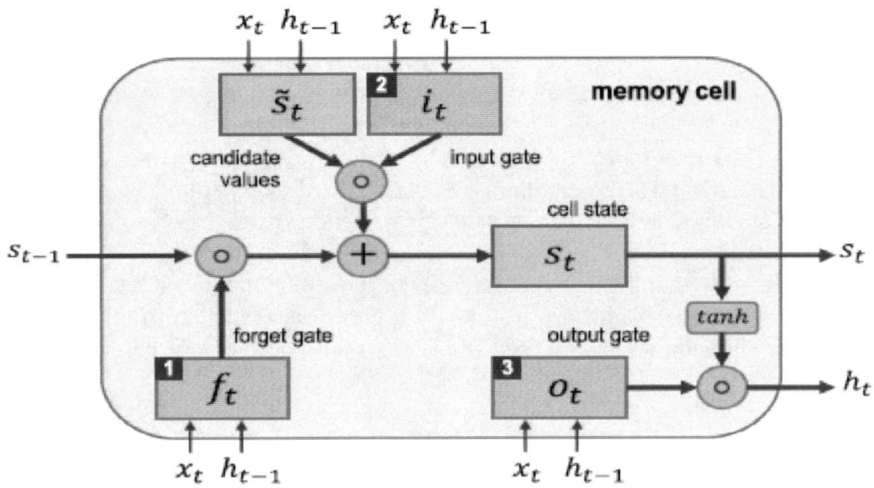

Figure 1. The LSTM network architecture memory cell [15]

Table 1. Parameters setting used in LSTM network

Variables	Value
Number of features and response	1
Time step	1
LSTM layer hidden units	200
Network type	Fully connected
Training dataset	80%
Validation test	20%
Training epochs	300
Gradient Threshold	1
Initial learn rate	0.005
Learn rate drop period	125
Learn rate drop factor	0.2
State and Gate activation function	tanh and sigmoid

In order to demonstrate the efficiency of the LSTM network model, in this study, performance is assessed by using the root mean square error, $RMSE$. It is formulated in Equation (1). While R^2 in Equation (2) is the coefficient of determination, value describes the proportion of the variance in the observed dataset that can be explained by the model.

$$RMSE = \sqrt{\frac{1}{n}\sum_{i=1}^{n}|y - y'|^2} \quad (1)$$

$$R^2 = \frac{\sum_{i=1}^{n}(y' - \bar{y})^2}{\sum_{i=1}^{n}(y - \bar{y})^2} \quad (2)$$

Where n is the number of data points, y' represents the forecasted value of river water level, y is the observed river water level at the time i, and $\bar{y}$ is the average value of the actual or observed records. In this study, Matlab and its environment were used to be the platform for programming and simulation tool in establishing the developed model.

3. Result and discussions

The simulated model was done using `trainNetwork` Matlab function to train an LSTM network for deep learning. For a better fit and to prevent the training from diverging, standardized the training data is applied to have zero mean and unit of variance. The standardized data obtained from the difference between training data and the mean of training data divided by the standard deviation value of training data. At prediction time, the standardized test data using the same parameters as the training data. The standardized water level data can be illustrated in Figure 2. In this study, the network state was initialized at first using 80% training data that is 15207 of the 19008 input data set. Then simulated the next prediction of 20% or 3801 validation test data using the last time step of the training response. Continued by looping over the remaining predictions and input the previous prediction to the LSTM network.

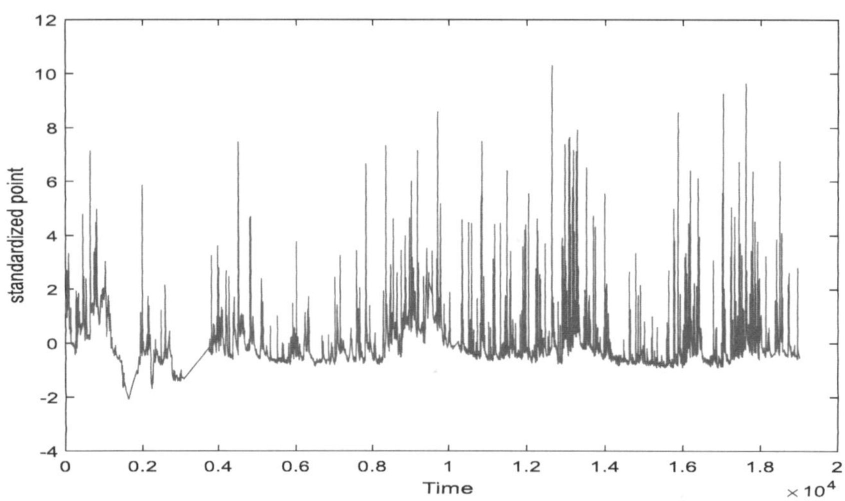

Figure 2. Hourly River Klang water level after data standardized

The experiment result of 3801 validation data sets can be referred to in Figure 3 with the red line distribution values. Quantitatively, with forecasted and observed values booths are unstandardized to their original scale. From Figure 3, it showed that the forecasted red line distribution value quite follows the validated pattern from the actual value after it's unstandardized. The result was shown in the original stage river water level. The result shown in Figure 4 seems that the LSTM network provides a considerable accurate in terms of error criteria. Although LSTM has difficulties in explaining the detailed peak values, the network is adequate to follow the dominant trends. The LSTM network generally reflected observed river water level. The RMSE and the coefficient of determination, R^2 were calculated to evaluate model performance. These two evaluated performances were calculated from the unstandardized prediction results. The results showed that the RMSE and R^2 in Figure 5 were respectively of 0.20593 and 0.844, which showed quit precision.

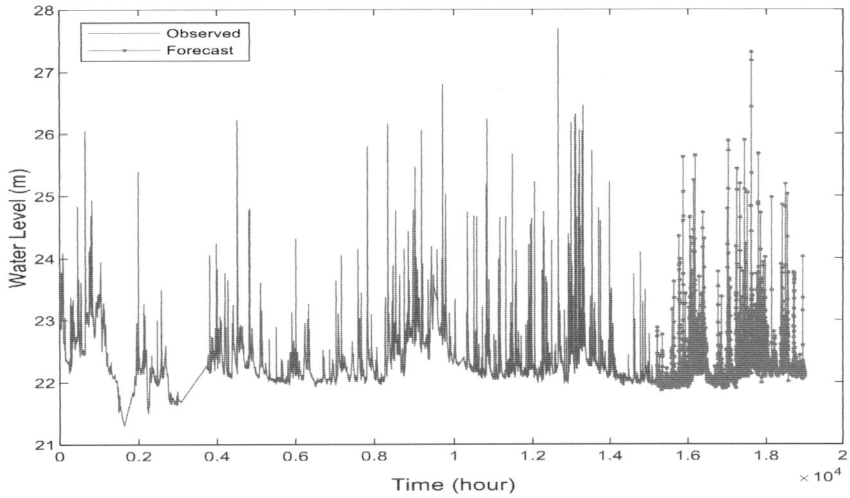

Figure 3. Hourly River Klang water level forecasting and observed data after unstandardized

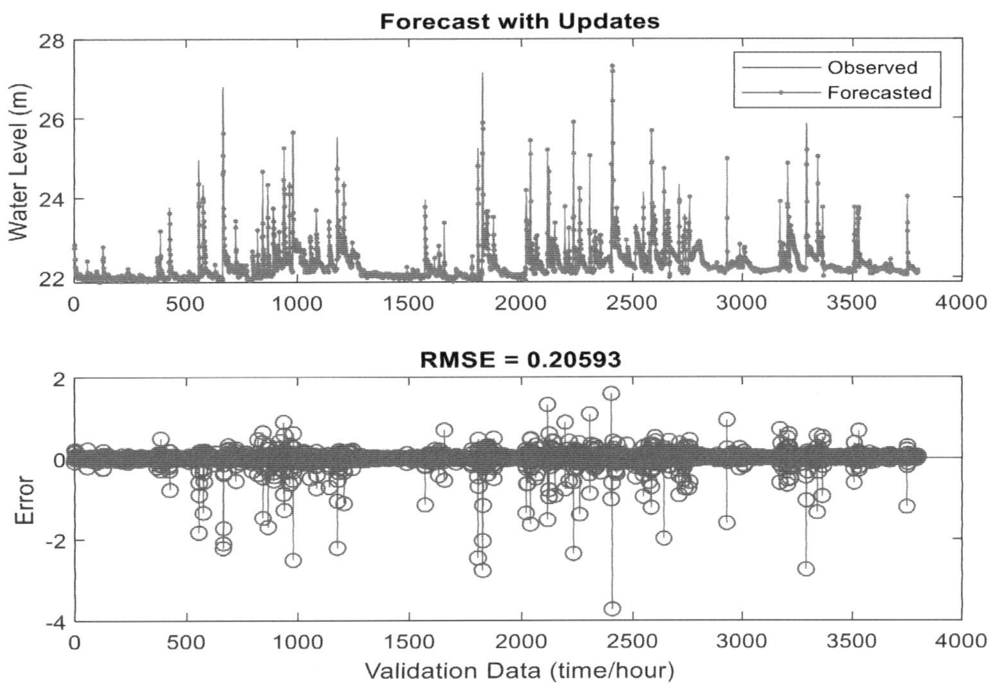

Figure 4. Hourly forecasting based on test data and error distribution

After the quantitative analysis of LSTM models, the observed and simulated water level distribution value also scattered. The LSTM model has a high value showing that the model could well represent the relationship between observed and simulated performance. From Figure 5, the data is scattered relatively closer to the LSTM model. It is clearly shown that LSTM model is good correlation with the observed water level data.

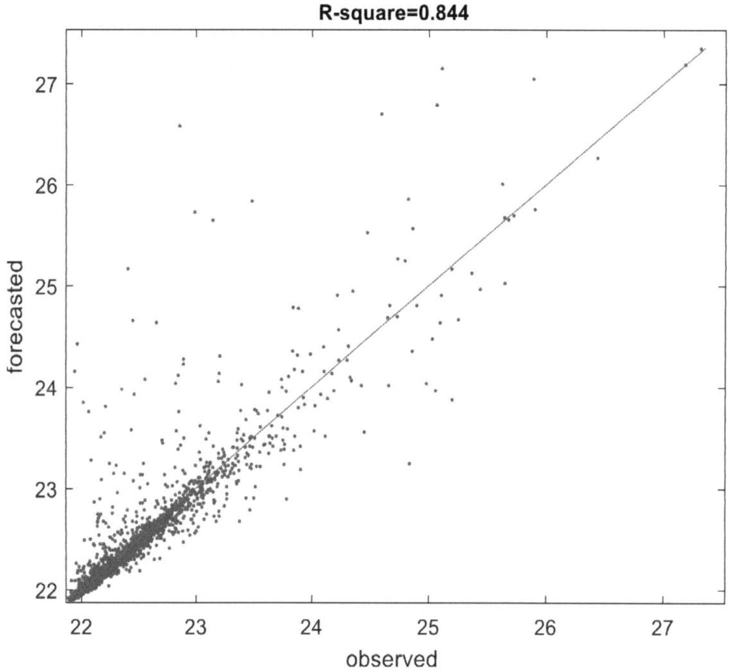

Figure 5. Scatter-plot distribution of the observed and forecasted data

In connection with disaster risk management and flood warning system, according to DID [16, 17] for river level data-above mean sea level, it can be classified as three main categories including "alert level", "warning level", and "danger level". The alert level for Klang River, in Sulaiman Bridge is 27.00 meters, while for warning level is 28.25 meters, and the danger level is 29.50 meters. From the inside of forecasted data, the maximum value is about 27.314 meters compared to the maximum actual value is 27.35 which was categorized as "alert level" at that time. It indicates when the river water level more than the maximum observed value, can be assumed flood will occur in that particular area since it could be reached at a danger level.

4. Conclusions

This paper set out to develop a simulated model for river water levels using the LSTM network. The study has examined the concept of RNN as a deep learning LSTM cell. The LSTM model not only takes full advantage of the current data characteristics but also uses its gate structure to decide whether to remember or forget the previous feature. With the progress of AI techniques, the deep learning method of long short-term memory network model could be better used in the hydrological engineering problems, due to the LSTM is very effective in modelling large time-series data. In this paper, the input data set was considered from one downstream to forecast the river water level of the next hours for flood warning system. The result verified that the LSTM network with specified training set options is a promising alternative technique to the solution of flood modeling and forecasting problems. The performance indicates with the root mean square error, RMSE 0.20593 and R^2 value 0.844 are closely accurate when updating the network state compared with the observed value.

For further research study, employing more input variables including weather forecast, rainfall data, and streamflow data can be addressed in order to get the relationship impact of the improved flood forecasting performance. Furthermore, the enhancement of machine learning algorithm in advance AI-based methods will be more challenging to develop robustness of the flood forecasting system.

Acknowledgement

The authors would like to acknowledge the Malaysia-Japan International Institute of Technology, Universiti Teknologi Malaysia, and Faculty of Engineering, Universitas Muhammadiyah Malang for supporting and facilitating this work. As well as the department of irrigation and drainage (DID) Malaysia for the research materials supplied.

Research materials

The research data material can be accessed through https://osf.io/gh5es/ it contains water level of Klang River at Sulaiman Bridge, Kuala Lumpur, Malaysia, in meters stage.

References

[1] UNISDR 2015 Sendai framework for disaster risk reduction 2015-2030.

[2] Faruq A, Abdullah S S, Marto A, Bakar M A A, Hussein S F M and Razali C M C 2019 The use of radial basis function and non-linear autoregressive exogenous neural networks to forecast multi-step ahead of time flood water level *Int. J. Adv. Intell. Informatics* vol **5** no 1 pp 1–10.

[3] Anuar M, Bakar A, Azarshah F and Aziz A 2017 Flood water level modeling and prediction using radial basis function neural network: Case study Kedah *Model. Des. Simul. Syst. AsiaSim* vol **751** pp 225–34.

[4] Elsafi S H 2014 Artificial Neural Networks (ANNs) for flood forecasting at Dongola Station in the River Nile Sudan *Alexandria Eng. J.* vol **53** no 3 pp 655–62.

[5] Hossain M, Rekabdar B, Louis S J and Dascalu S 2015 Forecasting the weather of Nevada : A deep learning approach *2015 Int. Jt. Conf. Neural Networks* vol **2** pp 1–6.

[6] Suliman A, Nazri N, Othman M, Malek M A and Ku-Mahamud K R 2013 Artificial neural network and support vector machine in flood forecasting: A review *Proc. 4th Int. Conf. Comput. Informatics ICOCI 2013 28-30 August 2013* no 030 pp 327–32.

[7] Guo Z, Zhou K, Zhang X and Yang S 2018 A deep learning model for short-term power load and probability density forecasting *Energy* vol **160** pp 1186–200.

[8] Qu L, Li W, Li W, Ma D and Wang Y 2019 Daily long-term traffic flow forecasting based on a deep neural network *Expert Syst. Appl.* vol **121** pp 304–12.

[9] He X, Luo J, Zuo G and Xie J 2019 Daily runoff forecasting using a hybrid model based on variational mode decomposition and deep neural networks *Water Resour. Manag.*.

[10] Cai M, Pipattanasomporn M and Rahman S 2019 Day-ahead building-level load forecasts using deep learning vs. traditional time-series techniques *Appl. Energy* vol **236** pp 1078–88.

[11] Hu C, Wu Q, Li H, Jian S, Li N and Lou Z 2018 Deep learning with a long short-term memory networks approach for rainfall-runoff simulation *Water* vol **10** no 11 pp 1–16.

[12] Hochreiter S and Schmidhuber J 1997 Long Short-Term Memory *Neural Comput.* vol **9** no 8 pp 1735–80.

[13] Raabe N 2016 Deep learning for solar power forecasting – An approach using autoencoder and LSTM neural networks in *IEEE International Conference on Systems, Man, and Cybernetics • SMC 2016* pp 2858–65.

[14] Goh Y C, Zainol Z and Mat Amin M Z 2016 Assessment of future water availability under the changing climate: Case study of Klang River Basin, Malaysia *Int. J. River Basin Manag.* vol **14** no 1 pp 65–73.

[15] Fischer T and Krauss C 2018 Deep learning with long short-term memory networks for financial market predictions *Eur. J. Oper. Res.* vol **270** no 2 pp 654–69.

[16] Azad, Fauzi F and Ghazali M 2019 National flood forecasting and warning system of Malaysia:

Automated Forecasting for The East," *Hydrol. Water Resour. Div. Dep. Irrig. Drain. (JPS), Malaysia.*
[17] Bin Sulaiman A H 2009 *Flood Management in Malaysia* Department of Irrigation and Drainage Systems.

IOP Conf. Series: Materials Science and Engineering **821** (2020) 012027 doi:10.1088/1757-899X/821/1/012027

Stator slotted design of axial flux permanent magnet generator for low-speed turbine

M Irfan*, R F Ariyanto, L Syafaah, A Faruq, Nurhadi, and N Subeki

Department of Electrical Engineering, Faculty of Engineering, Universitas Muhammadiyah Malang, Malang, Indonesia

*Email: irfan@umm.ac.id

Abstract. The wind power plant has an important component, namely a generator which is an electrical engine that converts mechanical energy into electrical energy. This paper discusses the design of the permanent rotor axial flux generator (Single Rotor Single Stator) using a slotted stator and using a slotted stator reference design. To find out more in detail, this work was conducted by comparing four different generator designs, the number of slots and tooth widths. Testing the width of the teeth and the number of slots varied, including 12 slots width of teeth 23 mm, 15 slots width of teeth 18 mm, and 18 slots width of teeth 15 mm. The permanent magnet used in this study is neodymium iron boron, NdFeB 48/11 with a total of 10 poles. The desired reference power is 100 Watt with a speed of 500 Rpm. The outputs of each design in the form of voltage, current, power, and efficiency are discussed. The conclusion obtained that the additional teeth to the stator and increasing the number of axial flux slots affect the specified output parameters. The result of using teeth is that the voltage, current, and power are getting bigger and the more number of slots the smaller the voltage, current, and power.

1. Introduction

Increasing energy needs encourage us to look for alternative energy sources. At present most of the energy produced uses fossil fuels that can cause global warming. In addition, the amount of fossil fuels available is limited. To overcome the problem of global warming and to meet energy demand, renewable energy sources are targeted as alternative energy. Wind energy is a very productive renewable energy source. Abundant wind to exploit the potential of this renewable source.

An electric machine is a device that can convert mechanical energy into electrical energy or electrical energy into mechanics. When such a device is used to condition mechanical energy to electrical energy, it is called a generator. When converting electrical energy into mechanical energy, it is called a motor. Because every electric engine supplied can convert power in one direction, each machine can be used as both a generator and a motor. Almost all motors and generators practically convert energy from one form to another through the action of the magnetic field.

Min-Fu Hsieh et al [1] conducted a study on reducing cogging torque in axial flux machines for wind turbines by changing the direction of the slot and changing the slope of the magnet. In this study using the finite element model. From the results of this study that by changing the direction of the slot and the slope of the magnet can reduce peak cogging torque by 88% compared to not changing the magnetic slope. This shows the method to be effective and able to reduce cogging torque to a much lower level.

J.G. Wanjiku et al [2] conducted research on minimizing torque cogging in PMSG flux axial with the parallel stator. Permanent magnets with slot-less stators can eliminate hysteresis and cogging torque losses, but greater eddy current losses. In this study using finite element analysis. The results of this study effectively reduce cogging torque by around 70% while reducing skewing by around 50%. Both methods show minimal effects on machine work. Gyeong-Chan Lee et al [3] conducted research on the design of dual structural permanent magnet generators for wind turbines. In this study to reduce cogging torque slotted flux axial permanent magnet generators, where the rotor is clamped with two stators. From the results of this study, the permanent magnet is tilted and the design of the slot transfer can be as high as 1.3 kW and the cogging torque is below 1.61%.

Ionut Bogdan Stoenescu et al [4] conducted a study of simulations of a flux axial magnet synchronous candy generator under no-load conditions. From the results of this study that the simulation results from phase to phase are far from reality, due to the distribution of magnetic flux in the coil and magnetic flux produced by magnets is not sinusoidal. Wearing teeth is very important because the heat dissipation from stator friction with a magnet that causes heat can be flowed to the teeth and divided into all slots that are on the stator, but when not using a slot then the heat caused by friction will collect on the stator. From the accumulation of heat, there will be a very high eddy current.

From some of the mentioned studies it can be concluded that the slotted stator on the axial flux generator is needed for excessive heat dissipation when the generator is operating at its maximum. In small scale axial flux generators do not use slots, but for larger-scale will definitely use. Therefore teeth need to be made on the stator so that the generator design obtained is the optimal design. In this work the teeth will be made on the axial flux generator using slots and poles totaling 12 10-pole slots, 15 10-pole slots, and 18 10-pole slots. The analysis is done by using the formula calculation using finite element. The software used for the finite element method (FEM) analysis simulation is MagNet Infolytica software [5, 6].

2. Methodology

Modeling and simulation are the methods used to model and simulate the model to see whether the model gives results that are in accordance with what is desired. With modeling and simulation, research larger-scale can be done more effectively. This work is about modeling 4 axial flux generator designs namely, permanent magnet generator axial flux 15 slots-10 poles single rotor-single stator (SR-SS) model with resin, 12 slots-10 poles single rotor-single stator (SR-SS), 15 slots-10 single rotor-single stator poles (SR-SS) and 18 slots-10 single rotor-single stator poles (SR-SS). This generator model is made with the same specifications and material then simulated at different rotational speeds. The output of the model is compared with the magnetic flux, voltage, current, power and efficiency so that the most optimal design can be produced.

This paper discusses the analysis of the maximum magnetic field strength with MagNet Infolytica software [5, 6] and compares the results of the calculations and the output voltage will be calculated on the MPAF generator. The type of core laminated material chosen must be classified as a type of soft magnetic material, which means that the material chosen will only be magnetic when there is a magnetic field about it. The type chosen in this research is the type of silicon steel. The content of silicon in silicon steel materials ranges from 0.5% - 3.25%. The core laminate used in the generator design is M250-50A steel type silicon iron with a thickness of 0.5 mm. The types of magnetic properties of the material can be seen in Table 1. The flow of magnetic flux through the material must be maintained so as not to exceed the saturation point.

The permanent magnet material used is NdFeb type NdFeb 48/11 [7, 8]. The selection of magnetic material is based on previous research. The standard used is the Chinese standard with the magnetic characteristics shown in Table 2. The type of coil material used is Copper: 5.77e7 Siemens / m with characteristics as shown in Table 3. This type of material is very often used because it has a high conductivity value. In this axial flux permanent magnet generator design, there are several parameters that have been previously known. These parameters refer to references [9, 10], as shown in Table 4.

Table 1. Magnetic propertis type M250-50A material

Parameters and Symbols	Value	Unit
1,5 T	2,38	W/kg
1,0 T	1,02	W/kg
Anisotropy of loss	10	%
2.500 A/m	1,55	Tesla
5.000 A/m	1,64	Tesla
10.000 A/m	1,77	Tesla
Coactivity (DC)	30	A/m
Relative permeability at 1,5 T	740	-

Table 2. Characteristics of NdFeb Magnets 48/11

Parameters and Symbols	Value	Unit
Remanence flux density	1,39	T
Coercive force, H_{cb}	1060650	Oe (A/m)
Intrinsic Coercive Force, H_{cj}	11,0 (876)	kOe (kA/m)
Energy Product, BH_{max}	50-53 (398-422)	MGO (KJ/m^3)
Max. Operating Temp.	20	^{o}C

Table 3. Copper Characteristics: Siemens 5.77e7 / m

Parameters	Value	Unit
Conductivity	57700000	M
Thermal Conductivity	386	W/m.C
Thermal Specific	383.1	J/Kg.C
Density	8954	Kg/m^3
Max. Operating temp	20	^{o}C

Table 4. Mathematical Equation Parameters [9]

Parameters	Dimension
Daya	100 watt
Cosφ	0,8
Efficiency (η)	90%
$V_1 = 380/\sqrt{3}$	220
$\varepsilon = E_f/V_1$	2
B_{mg}	0,65 T
A_m	22000 A/m
G_t	0.55
G_c	0.55
K_{w1}	0,96
K_D	0,131
J_a	4,5 x 10^6 A/m^2

To get simulation data in the form of no-load voltage, voltage with load, current, and power, it is necessary to make a circuit so that the data can be generated. In this series is a series of three phases, each phase is connected with Wye-Connection with each phase composed of 4, 5, and 6 coils arranged in series. As for simulations with loads, an additional circuit is needed in the form of a diode that serves to rectify waves at each phase, and obstacles in the form of resistors that function as loads. The value of the resistor used varies from 5 *Ohms*, 10 *Ohms*, and 15 *Ohms*. Display a simulation circuit with a load as shown in Figure 1 and Figure 2.

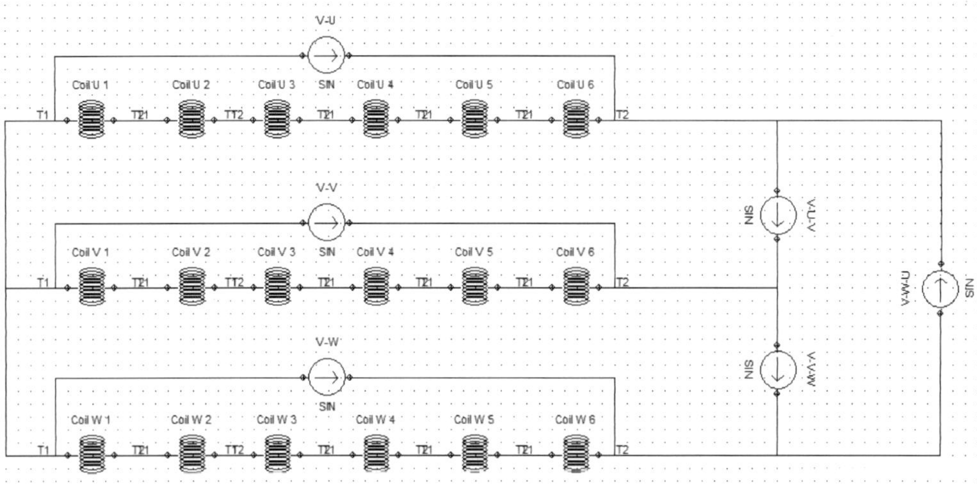

Figure 1. Schematic simulation design without load for 18 slots

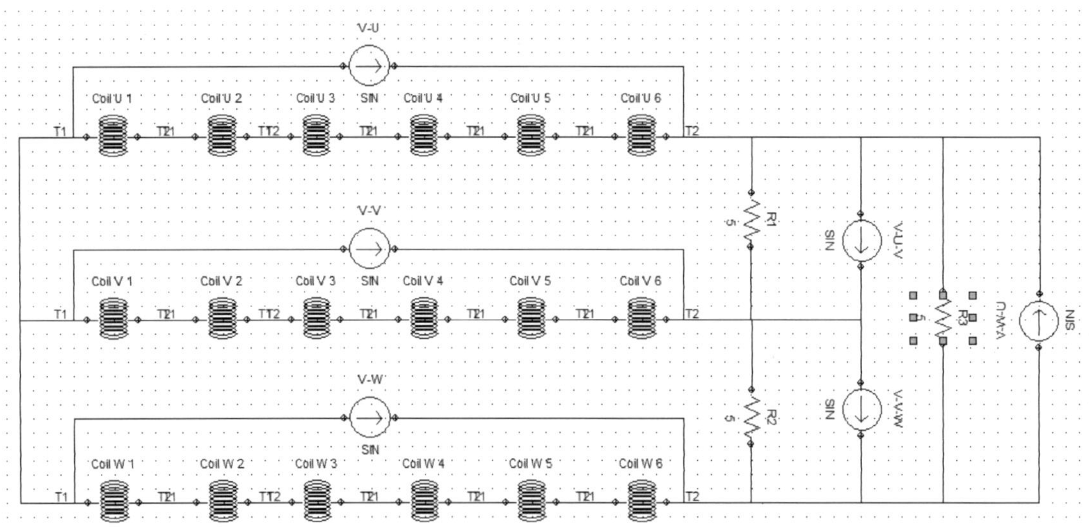

Figure 2. Schematic simulation design 18 slots with load of 15 *Ohm*

3. Result and discussions

This section discusses data analysis from simulation results carried out using MagNet Infolytica software. Also in this section discusses the comparison of the results of 4 model simulations. Data obtained from the MagNet Infolytica software is analyzed. The following parameters obtained from the calculation results are outlined in Table 5.

Table 5. Calculation result of dimension generator

Variable Components	Setting value
Stator diameter outside (D_{out})	220 mm
Stator diameter inside (D_{in})	100 mm
Airgap width (lg)	3 mm
Rotor outer diameter (Di)	220 mm
Rotor inside diameter (Da)	100 mm
Magnetic Thickness (Lm)	3 mm
Slotted width (b_t)	23 mm (12 slot) / 18 mm (15 slot) / 15 mm (18 slot)
Slotted length (h_s)	20 mm
Stator thick yoke (h_y)	8 mm
Total of Slots (Ns)	12 slot / 15 slot /18 slot
Total of Pole (p)	10 *pole*
Number of coil each teeth	170 *turns* (12 *slot*) / 134 *turns* (15 slot) / 112 *turns* (18 *slot*)
Stator and Rotor Core Materials	M250 – 50A
Permanent Magnet Material	NdFeb 48./11
Coil materials	Copper : 5.77e7 Siemens/m

All models ware simulated and compared to determine the effect of changes in the slotted stator and the number of slots on the output of the AFPM design. After comparing each model. Output compared is in the form of no-load current and output power. As for the selection of the model parameters used are the current limit and output power, for consideration. The data being compared is the *Rms* voltage data for each model, in this section the data shown is only comparative data at a speed of 700 Rpm. To facilitate analysis, you can see in Figure 3.

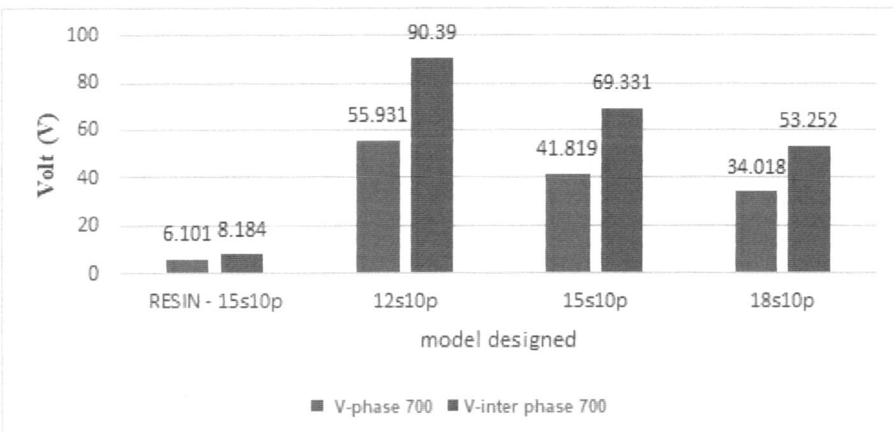

Figure 3. Inter Phase Phase *Rms* Voltage and phase *Rms* Voltage each model at 700 *Rpm* speed

From the table and graph above, it can be noted that the difference between the slotted 15S10P model and the slotless 15S10P model produced by the output voltage has increased very significantly, it can be seen the results of the 15S10P shoeless with a voltage of 6,101 V for the 15S10P slotted with a voltage of 41,819 V. , and 18 slots reduce the large output voltage of the generator design, the pattern is seen in the models with the number of slots 12, 15 and 18. If observed thoroughly, the 12S10P model at 700 Rpm speeds produces the largest output voltage compared to models 15 and 18 slots.

3.1. Comparison of output power for each model

The data being compared is the simulation data with a speed of 700 Rpm at a load of 15 *Ohm*s. Data of each model is presented in graphical form to facilitate analysis. Comparison data can be seen in Figure 4.

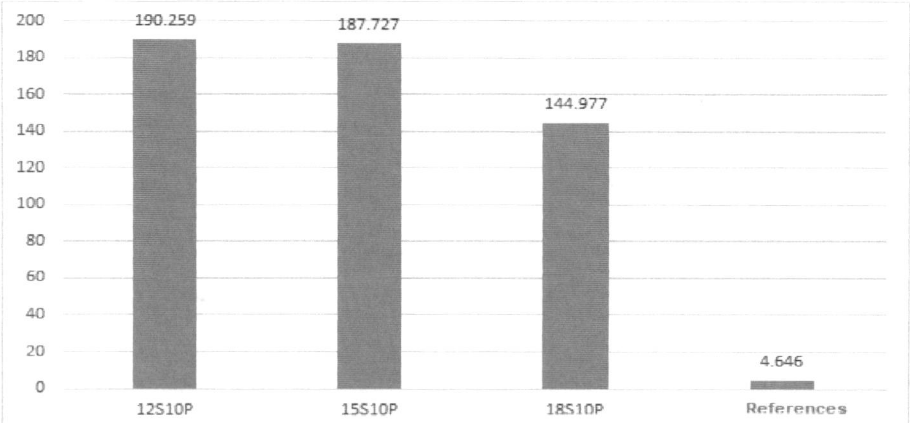

Figure 4. The output power of each model at a speed of 700 *Rpm* and a load of 15 *Ohms*

From the Figure 4, it can be seen that the largest output power is owned by the 12S10P model with an output power of 190,259 Watt, while the smallest output power is generated by the 15S10P model with resin with an output power of 4,646 Watt. The comparison of 15S10P slotted and 15S10P slot-less output power increased significantly for the 15S10P slotted model. For comparison between the number of slots can be seen from the results of the above simulation that the number of slots with the same number of poles, the less power generated will be generated. Overall designed slotted stator in this study as it demonstrated in Figure 5.

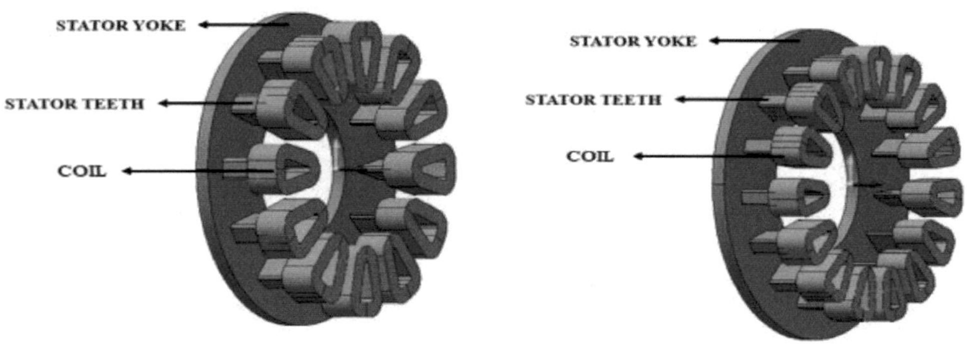

(a) (b)

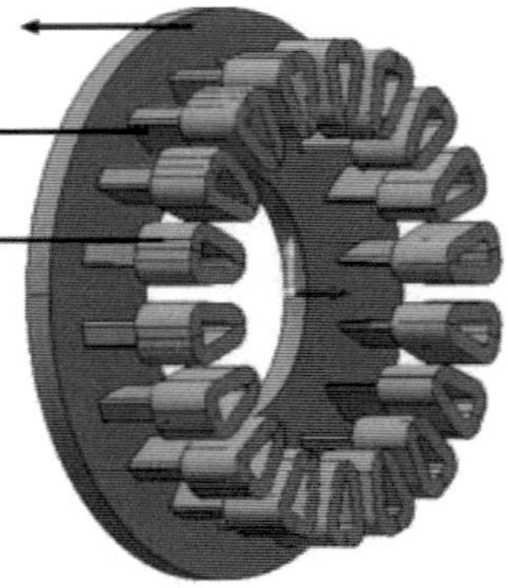

(c)

Figure 5. Designed permanent magnet generator. (a). Model 12S10P Stator *Slotted*. (b). Model 15S10P Stator *Slotted*. (c). Model 18S10P Stator *Slotted*

4. Conclusions

The design of axial flux permanent magnet generators at different speeds and loads, it can be concluded that the increase in speed in each model has an effect perpendicular to the generated voltage. The faster the rotation the greater the voltage generated. The number of slots has effect to determine the width of the slotted stator. The more the number of slots, the smaller the width of the slotted. Seen from the calculations for each slot, the difference in size is obtained when the 12 model slots measure 23 mm, for the 15 slot model the size is 18 mm, and for 18 slots the size is 15 mm.

Comparison of the slot-less 15S10P and 15S10P reference models where the output voltage, current and power are very different, the results of the 15S10P slotted are greatly increased compared to the slot-less 15S10P. Due to the losses in the slot-less 15S10P model so the results are not too large as the slotted 15S10P model. Whereas the efficiency of slot-less 15S10P is greater than the slotted model. For comparison of the number of different slots and the same number of poles the results obtained from the simulation for voltage, current, and power decrease but for increased efficiency.

Further developments that can be carried out include: Analyze the effect of the stator slotted on cogging torque and torque ripple. Developing a slotted stator with multi-stator and multi-rotor. And varying with different materials on cores and magnets.

Acknowledgement

The authors wish to express their gratitude to the Faculty of Engineering, Universitas Muhanmmadiyah Malang, and Direktorat Penelitian dan Pengabdian Masyarakat (DPPM) for supporting this work.

References

[1] Hsieh M, Dorrell D G, Yeh Y and Ekram S 2009 Cogging torque reduction in axial flux machines for small wind turbines in *2009 35th Annual Conference of IEEE Industrial Electronics* pp 4435-39.

[2] Wanjiku J G, Member S, Jagau H, Member S, Khan M A and Barendse P S 2011 Minimization of cogging torque in a small axial- Flux PMSG with a parallel-teeth stator in *2011 IEEE Energy Conversion Congress and Exposition* pp 3687–93.

[3] Lee G 2013 Design of dual structural axial flux permanent magnet generator for small wind turbine in *IEEE 2013 Tencon - Spring* pp 90–4.

[4] Stoenescu I B, Costinas S and Chicco G 2014 The simulation of an axial-flux permanent magnet synchronous generator in no-load condition in *2014 16th International Conference on Harmonics and Quality of Power (ICHQP)* pp 375–9.

[5] Anam M C, Irfan M and Nurhadi 2017 Perancangan generator 100 watt menggunakan software elektromagnetik infolityca *Kinet. Game Technol. Inf. Syst. Comput. Network, Comput.*

Electron. Control vol **2**(1) pp 27–36.

[6] Yusuf A A and Irfan M 2018 A design of coreless permanent magnet axial flux generator for low speed wind turbine in *2018 5th International Conference on Electrical Engineering, Computer Science and Informatics (EECSI)* pp 637–41.

[7] Irfan M, Effendy M, Suhardi S and Pakaya I 2018 A design of electrical permanent magnet generator for rural area wind a design of electrical permanent magnet generator for rural area *Int. J. Power Electron. Drive Syst.* vol **9**(1) pp 269–75.

[8] Irfan M, Hakim E, Suhardi D, Kasan N, Faruq A and Anam M 2018 Simplified design of low speed electric permanent magnet generator for small wind power plant *IOP Conf. Ser. Mater. Sci. Eng.* vol **403** pp 1–6.

[9] Gieras J F, Rong-Jie W and Kamper M J 2005 *Axial Flux Permanent Magnet Brushless Machines* (New York: Kluwer Academic Publishers).

[10] Messina G, De Bella E T, and Morici L 2019 HTS axial flux permanent magnets electrical machine prototype: Design and test results *IEEE Trans. Appl. Supercond.* vol **29**(2) pp 1–5.

Immunostimulating effect of sweet potato fiber extract on IgM production by HB4C5 cells

I D Kumalasari[1,*], T Sugahara[2], K Nishi[2]

[1]Department of Food Technology, Faculty of Industrial Technology, Universitas Ahmad Dahlan, Yogyakarta, Indonesia
[2]Department of Bioscience, Graduate School of Agriculture, Ehime University, Ehime, Japan

*Email: ika.kumalasari@tp.uad.ac.id

Abstract. Sweet Potato (*Ipomea batatas*) is a local tuber potentially to be developed as functional food. The aims of this research is to evaluate the immunomodulatory effect of the sweet potato fibre extract (SFE) in vitro by animal cell culture techniques and identification the active compounds of SFE. SFE were prepared by autoclaving the sweet potato fiber powder in distilled water with room temperature 25 °C for 2 hours in distillated water and dialyzed with membrane with molecular weight cut off 14 kDa and 30 kDa. The IgM production of SFE against human hybridoma cell line HB4C5 cells was evaluated. The results of this research indicated that SFE stimulated IgM production by HB4C5 cells with heat and dialysis treatments in dose dependent manner. IgM production was increased by SFE with molecular weight more than 14 kDa and less than 30 kDa. The conclusion of this evaluation, SFE facilitated IgM production by HB4C5 cells and revealed that the SFE has positive effects on immunostimulatory activity *in vitro*.

1. Introduction

There are several various of Indonesian origin tubers that have dietary fiber sources. Those tubers are sweet potatos (Ipomoea *batatas*), arrowroot (*Marantha arundinaceae*), canna (*Canna edulis*), etc. The tubers contains high amount of dietary fiber. The SFE extracted from sweet potato has potency as prebiotic source [1]. Sweet potato dietary fiber could be isolated from sweet potato pulp. Sweet potato pulp contains of 49,7% dietary fiber, which is rich in pectin (39.5%), cellulose, hemi cellulose and lignin [2]. Sweet potato pectin obtained by sonication- induced modification showed lower molecular weight, strong antioxidant capacity and could induce apoptosis-like cell death in colon cancer [3, 4]

The immune function is very important for prevention and recovery from infectious diseases. There are many study about activation of the immune function by functional food. Many foodstuffs that can activate the immune system have been evaluated [5-7]. The immune response can be divided into two categories such as innate immune response and adaptive immune response. Although the adaptive immune response takes a long time after infection, it gives high specific prevention against pathogens. B cells give an important contribution to the humoral immune response by production of immunoglobulins (Igs). Our study focused on the foods that stimulate the immune function for development of functional foods [8-10].

2. Materials and methods

2.1. A preparation of SFE

Sweet potato was obtained from Karanganyar, Central Java, Indonesia. SFE was prepared as previously described [11]. Briefly, sweet potato tubers were peeled, grated, and suspended in distilled water and settled down overnight. The supernatant was collected as the sweet potato crude fiber after removing the precipitate. Then fresh of sweet potato fiber were steamed for 30 min, soaked in 80% ethanol at 60 °C for 20 min, filtrated, squeezed, oven-dried, and ground into powder. A SFE solution was prepared by suspending the sweet potato fiber powder in distilled water at 10 g/100 mL. The suspension was heated at 121 °C for 20 min or stirred at room temperatur for 2 h. Both suspensions were centrifuged at 15,600 × g for 20 min to remove insoluble materials and collect the concentrate. The supernatant was dialyzed against distilled water using a dialysis membrane with molecular weight cut off of 14 kDa and 30 kDa (Wako Pure Chemical Industries) and then sterilized by filtration.

2.2. HB4C5 cell and cell culture

The IgM production for immune system of SFE was evaluated using the human–human hybridoma HB4C5 cells. HB4C5 cell is a fusion product of a human B lymphocyte from a lung cancer cells with a NAT-30 [12]. HB4C5 cells producing monoclonal IgM were used for the assay of the Ig production stimulating activity [13].

Briefly, HB4C5 cells were cultured in ERDF medium (Kyokuto Pharmaceutical, Tokyo, Japan) containing 10 lg/mL of insulin (Sigma, St. Louis, MO, USA), 20 lg/mL of transferrin (Sigma), 20 lM ethanolamine (Sigma), and 25 nM sodium selenite (Sigma) (ITES-ERDF medium) at 37 °C under humidified 5% CO2 –95% air for 6 hours [14]. The amount of IgM production in the culture medium was determined by enzyme-linked immunosorbent assay (ELISA)

2.3. Ig production stimulating effect of the sweet potato extracts on HB4C5 cells.

The stimulating activity and Ig production *in vitro* was evaluated by measuring the amount of IgM production by HB4C5 cells. The amount of IgM in culture medium 6 h after inoculation was measured by ELISA as described previously [11]. Briefly the SFE were solubilized by heat-treatment at 121 °C for 20 min and evaluated for the Ig production stimulating activity on human hybridoma HB4C5 cells. HB4C5 cells were inoculated in ITES-ERDF medium containing various concentrations of the SFE at 5.0×10^4 cells mL^{-1}, then performed in a 96-well culture plate. After incubation in a CO2 incubator at 37°C for 6 h, the amount of IgM production in the culture medium was determined by ELISA using anti-human IgM [13]. The absorbance was measured at 415 nm after the addition of 1.5% oxalic acid to terminate the coloring reaction. The assays were carried out in triplicate

2.4. Statistical analysis

Each result is expressed as the mean ± standard deviation (SD). One way ANOVA followed by Tukey's test was used to assess the statistical significance of the difference

3. Results and discussion

In this research we prepared two kind of SFE were obtained from different extraction methods. One SFE was prepared by heat treatment of the Sweet Potato fiber powder at 121 °C for 20 min in distilled water. The other SFE was prepared by stirring the powder at 25 °C for 2 h in distilled water and the stimulatory activity was evaluated. The IgM production-stimulatory activities of both SFE were compared to determine the suitable extraction method. The HB4C5 cells were inoculated in ITES-ERDF medium supplemented with various concentrations of either SFE and cultured for 6 h. Immunoglobulin production stimulatory effect of SFE was evaluated. As indicated in Figure 1, the IgM production by HB4C5 cells was enhanced by both SFE preparations in a dose-dependent manner. Sweet potato heat-extracts stimulated IgM production by HB4C5 cells in a dose-dependent manner. Interestingly, the sweet potato extracts prepared at room temperature slightly enhance IgM production.

These results suggest that the active substance in sweet potato tuber is insoluble in water. In addition, the active substance in the sweet potato tuber was solubilized by heat treatment. It is supposed from this result that heating process solubilized the active substance in the sweet potato tuber powder.

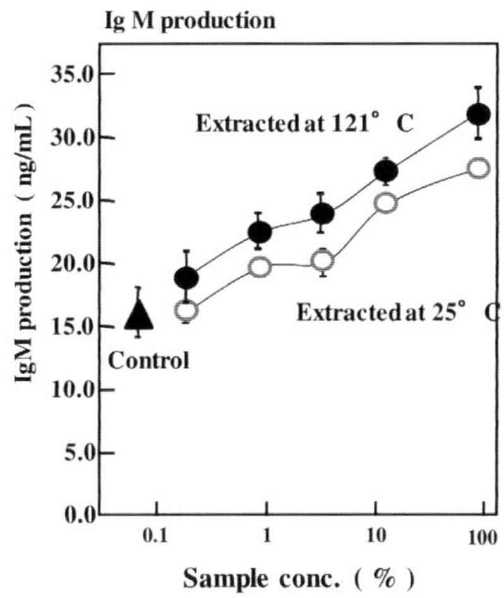

Figure 1. Supplementation effect of the sweet potato fibre extracts (SFE) on IgM production by HB4C5 cells. Sweet potato fibre powder was suspended in distilled water and extracted by heated at 121 °C for 20 min (black circle) and no heated at 25 oC for 2 h p (White circle). HB4C5 cells were inoculated in ITES-ERDF medium with various concentrations of each sweet potato fiber extracts and cultured for 6 h. distilled water as control (black triangle) was supplemented distillated water. Each result is represented as the mean ± SD of three independent measurements

Then, both of the SFE were dialyzed against distillated water with a dialysis membrane that cuts off molecules under the molecular weight of 14 kDa. As shown in Figure 2 the specific activity of heat-extracted SFE was much higher than that of non-heated SFE, indicating that a heating process during extraction from the Sweet Potato fiber powder would effectively solubilize the active substance. This result also suggests that the active substance in SFE is heat-stable. Because SFE was dialyzed using a dialysis membrane with molecular weight cut off of 14 kDa, the active substance seems to be macromolecules. It is supposed from these findings that the active substance in SFE may be a complex of carbohydrates and proteins.

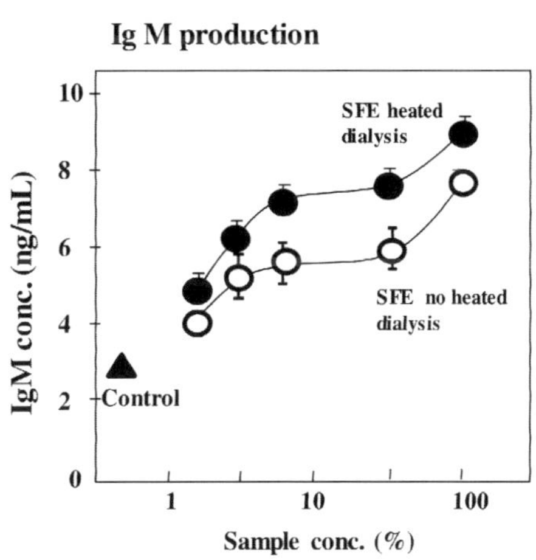

Figure 2. Supplementation effect of the sweet potato fibre extracts (SFE) on IgM production by HB4C5 cells. Sweet potato fiber powder was suspended in distilled water and extracted by heated at 121 °C for 20 min with dialysis treatment (black circle)and unheated at 25 °C for 2 h with dialysis treatment (White circle). HB4C5 cells were inoculated in ITES-ERDF medium with various concentrations of each Sweet Potato fibre extracts and cultured for 6 h. distilled water as control (black triangle) was supplemented distillate water. Each result is represented as the mean ± SD of three independent measurements

Effect of the Sweet potato extracts production on IgM production by HB4C5 cells with heat extraction and heat extraction dialysis. SFE with heat treatment extraction were dialyzed with using a dialysis membrane with molecular weight cut off of 14 kDa and compared with heat treatment but not dialyzed. In another experiment, SFE with room temperature treatments also compared between dialysis and no dialysis using a dialysis membrane with molecular weight cut off 14 kDa. The result showed, IgM production were enhanced 2.3 fold and 2 fold with both SFE prepared by heat treatment and dialysis and no dialysis compared with control (Figure 3A). Sweet potato extracts prepared at room temperature with dialysis and no dialysis did not enhance IgM production sharply (Figure 3B).

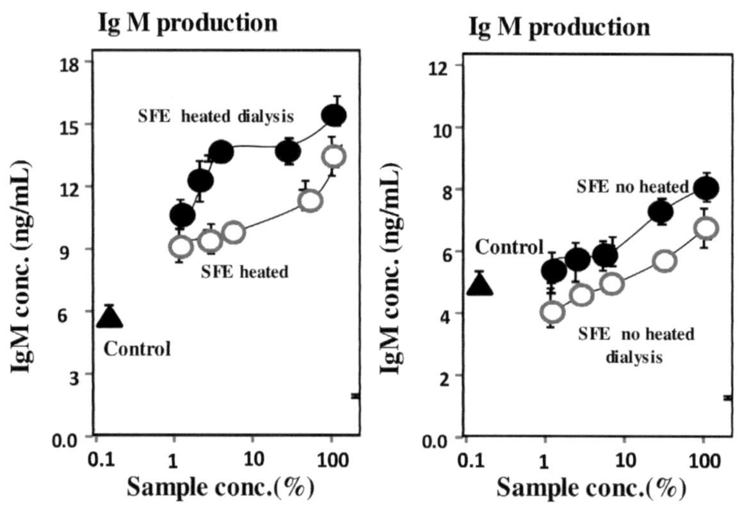

Figure 3. (A) Effect of the sweet potato fibre extracts (SFE) on IgM production by HB4C5 cells. Sweet potato fibre powder was suspended in distilled water and extracted by heated at 121 °C for 20 min with dialysis treatment (black circle) and without dialysis treatment (white circle). HB4C5 cells were inoculated in ITES-ERDF medium with various concentrations of each SFE and incubated for 6 h. Control (black triangle) was supplemented distillated water. (B) Effect of SFE on IgM production by HB4C5 cells. Sweet potato fibre powder was suspended in distilled water and extracted at 25 °C for 2 h without dialysis treatment (black circle) and extracted at 25 °C for 2 h with dialysis treatment (white circle). HB4C5 cells were inoculated in ITES-ERDF medium with various concentrations of each Sweet Potato fibre extracts and cultured for 6 h. distilled water as control (black triangle) was supplemented distillated water. Each result is represented as the mean ± SD of three independent measurements.

Stimulating effect of SFE on IgM production by HB4C5 cells were analyzed. Sweet potato fiber powder was suspended in distilled water and extracted by heated at 121 °C for 20 min with filtration treatment using membrane with molecular weight cut off < 30 kDa and heated at 121 °C for 20 min with filtration treatment using membrane with molecular weight cut off > 30 kDa. The results showed that SFE were prepared with heating treatment at 121 °C for 20 min then filtrated using membrane with molecular weight cut off > 30 kDa and < 30 kDa has stimulating activity on HB4C5 cells (Figure. 4). SFE with membrane with molecular weight cut off > 30 kDa has lower activity compared with molecular weight cut off < 30 kDa membrane. The result suggested that the dominantly active substance of SFE are substance with molecular weight less than 30 kDa.

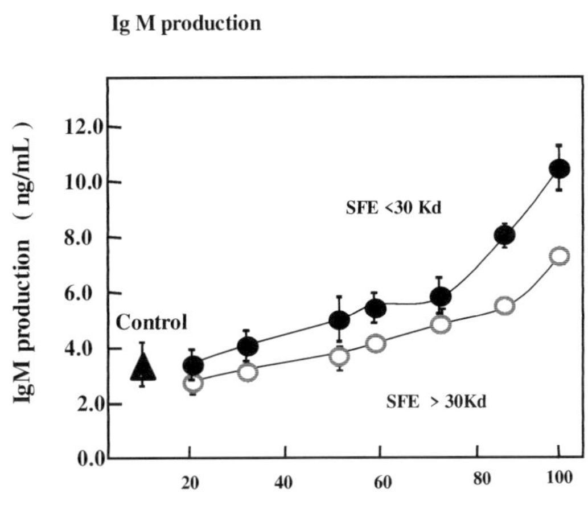

Figure 4. Supplementation effect of the sweet potato fibre extracts (SFE) on IgM production by HB4C5 cells. Sweet potato fiber powder was suspended in distilled water and extracted by heated at 121 °C for 20 min with filtration treatment using membrane with molecular weight cut off < 30 kDa (black circle) and heated at 121 °C for 20 min with filtration treatment using membrane with molecular weight cut off > 30 kDa (White circle). HB4C5 cells were inoculated in ITES-ERDF medium with various concentrations of each Sweet Potato fiber extracts and cultured for 6 h. distilled water as control (black triangle) was supplemented distillated water. Each result is represented as the mean ± SD of three independent measurements.

Because BFE was dialyzed using a dialysis membrane with molecular weight cut off of 14 kDa and under 30 kDa, the active substance seems to be macromolecules. As the previous results shown the active substance from bengkoang fiber extracts with heat treatment that has immunomodulatory effects is pectin [15]. Sweet potato also contains pectin, which has many important functions. Pectin has wide applications in both the pharmaceutical and food industries, where it acts as thickening and gelling agents, regulates the thickness and mouth-feel of fruit drink powder when the powder is dissolved in cold water [16]. It is reported that sweet potato has a good resource of pectin, contain about 15% pectin on dry matter basis [17]. In additon, pectin has proven to have beneficial effects on human health [18-20].

4. Conclusions

SFE stimulated the Ig production by HB4C5 cells with heat and dialysis treatments. The active substance in SFE is heat-stable. Hence, it would be possible to consider that sweet potato fiber extracts has a potentially beneficial effect on human health and on prevention of the immune system-related diseases.

References

[1] Lestari L A, Soesatyo M H N E, Iravati S and Harmayani E 2013 Characterization of Bestak sweet potato (Ipomoea batatas) variety from Indonesian origin as prebiotic *International Food Research Journal* vol **20**(5) pp 2241-45
[2] Takamine K, Abe J, Iwaya A, Maseda S and Hizukuri S 2000 A new manufacturing process for dietary fiber from sweet potato residue and its physical characteritics *Journal of Applied*

Glycoscience vol **47**(1) pp 67-72.

[3] Ogutu F O and Mu Sonochem T H 2017 Ultrasonic degradation of sweet potato pectin and its antioxidant activity *Ultrason Sonochem* vol **38** pp 726-34.

[4] Ogutu F O, Mu T H, H Sun and Zhang M 2018 Ultrasonic modified sweet potato pectin induces apoptosis like cell death in colon cancer (HT-29) cell line *Nutr Cancer* vol **70**(1) pp 136-45.

[5] Miyazaki Y, Kusano S, Doi H and Aki O 2005 Effects on immune response of antidiabetic ingredients from white-skinned sweet potato (Ipomoea batatas L) *Nutrition* vol **21** pp 358–62

[6] Cherng JM, Chiang W, Chiang LC 2007 Immunomodulatory activities of edible beans and related constituents from soybean *J Food Chem* vol **104** pp 613–18

[7] Chen X, Nie W, Yu G, Li Y, Hu Y, Lu J and Jin L 2012 Antitumor and immunomodulatory activity of polysaccharides from Sargassum fusiforme *Food Chem Toxycol* vol **50** pp 695–700.

[8] Nishi K, Muranaka A, Nishimoto S, Kadota A and T. Sugahara 2012 Immunostimulatory effect of b-cryptoxanthin in vitro and in vivo *J Func Food* vol **4** pp 618–25.

[9] Morishige H, Sugahara T, Nishimoto S, Muranaka A, Ohno F, Shiraishi R and Doi M 2011 Immunostimulatory effects of collagen from jellyfish in vivo *Cytotechnology* vol **63** pp 481–92.

[10] Daifuku M, Nishi K, Okamoto T, Nakano H, Nishimoto S, Sugahara T 2012 Immunostimulatory effects of water extract from bulbus arteriosus in tuna in vitro *J Func Food* vol **4** pp 263–70.

[11] Kumalasari I D, Sugahara T, Nishi K, Harmayani E and Raharjo S 2012 Immunomodulatory activity of Bengkoang (Pachyrhizus erosus) fiber extract in vitro and in vivo *Cytotechnology Journal* vol **66**(1) 75-85.

[12] Murakami H, Hashizume S, Ohashi H, Shinohara K, Yasumoto K, Nomoto K and Omura H 1985 Human-human hybridomas secreting antibodies specific to human lung carcinoma *In Vitro Cell Dev Biol* vol **21** pp 593–96.

[13] Sugahara T. Onda H, Shinohara Y, Horii M, Akiyama K, Nakamoto K and Hara K 2005 Immunostimulation effects of proteose-peptone component 3 fragment on human hybridomas and peripheral blood lymphocytes *Biochim Biophys Acta* vo1 **725** pp 233–40.

[14] Murakami H, Masui H, Sato GH, Sueoka N, Chow TP and Kono-Sueoka T 1982 Growth of hybridoma cells in serum-free medium: Ethanolamine is an essential component *Proc Natl Acad Sci* USA vol **79** pp 1158–62.

[15] Kumalasari I D, Sugahara T, Nishi K, Putra A B N 2014 Activation of macrophage stimulated by bengkoang fiber extract through toll-like receptor 4 *Food and Function Foods Journal* RSC vol **5**(7) pp 14-23.

[16] Yokoi H, Hirose J, Hayashi S and Takasaki Y 2002 Flocculation properties of pectin in various suspensions *Bioresource Technology* vol **84** (3) pp 287-90.

[17] Mei X, Mu T H and Han J J 2010 Composition and physicochemical properties of dietary fiber extracted from residues of 10 varieties of sweet potato by a sieving method *Journal of Agricultural and Food Chemistry* vol **58**(12) pp 7305-10.

[18] Yamaguchi F, Shimizu N and Hatanaka C 1994 Preparation and physiological effect of low-molecular weight pectin *Bioscience, Biotechnology and Biochemistry* vol **58** pp 679-82.

[19] Tian Cand Wang G 2008 Study on the anti-tumor effect of polysacharides from sweet potato *Journal of Biotechnology* 136S S351.

[20] Ingjerdingen M, Ingjerdingen K T, Patel T R, Alen S, Chen XY and Rolstad B 2008 Pectic polysacharides from Biophytum petersianum Klotzsch, and their activation of macrophages dendritic cells *Glycobiology* vol **18** pp 1074–84.

IOP Conf. Series: Materials Science and Engineering **821** (2020) 012029 doi:10.1088/1757-899X/821/1/012029

The comparison of ergonomic risk assessment results using job strain index and OCRA methods

D P Restuputri*, I Masudin and A R C Putri

University of Muhammadiyah Malang, Jln. Raya Tlogomas 246 Malang, Indonesia

*Email: restuputri@umm.ac.id

Abstract. Musculoskeletal Disorder (MDSs) is a real problem in terms of ergonomics. MSD is a muscle/pain disorder which is frequently undergone by operators after doing work, e.g. in the press line of a printing company. Some activities in press line are lifting books, reaching out the engine levers, and turning the engine levers. Many MSDs are undergone by operators, such as aches in the upper arms, hands, feet, and back. This is caused by excessive burdens and wrong posture when doing work. This study is aimed to compare the assessment results of operators' body posture by using OCRA and JSI methods. OCRA (Occupational Repetitive Action) Index is a method used to evaluate and analyze a job. Meanwhile, Job Strain Index (JSI) is a method used to estimate the risk of pain in the upper body. Evaluation is carried out on repetitive activities. Based on the calculation results, the OCRA values obtained are 3.35 for the right hand and 1.13 for the left hand. Meanwhile, JSI value for the right hand is 9 and the left hand is 4.5. The results of the assessment indicate that these activities are categorized as high-risk so and need to be corrected. The author's recommendation is by adding the number of operators on the press line.

1. Introduction

The Indonesian printing industry has been increasing since 2012 [1]. Printing is a mass production process of writing and images, particularly ink printed on paper using a printing machine. However, various complaints have been encountered in printing companies as almost 75% of the activities are carried out manually. The press line is a section with the highest number of complaints. Some activities carried out on these sections are lifting books, reaching out the levers, as well as turning the levers. There are many complaints from operators, ranging from aches in legs, upper arms, hands, and back.

Humans play a very important role in the production process as many locations still use human power, especially in Manual Material Handling (MMH) activities. The unergonomic Manual Material Handling activities will cause accidents during production process. Any inappropriate manual work can also result in the risk factors for musculoskeletal disorders (MSD) [2]. The Musculoskeletal disorder complaint is a kind of skeletal muscle pain, ranging from mild complaints to very severe one [3, 4]. Several observation techniques have been developed to assess MSD risk factors, namely Bulduk, et al. [5], Chander and Cavatorta [6], Kee and Karwowski [7], and Hellig et al. [8]. Chiasson, et al. [9] divide the methods of assessing MSD risk factors exposure into three categories: subjective assessment, systematic observation, and direct observation.

Occhipinti and Colombini [10] developed the occupational repetitive action (OCRA) methods to analyze workers' exposure to tasks featuring various upper-limb injury risk factors (repetitiveness, force, awkward postures and movements, lack of recovery periods, and others, defined as "additionals"). The OCRA methods are largely based on a consensus document of the International Ergonomics Association (IEA) technical committee on musculoskeletal disorders (Colombini et al., 2001), and they generate synthetic indicators that also consider worker rotation among different tasks. The OCRA index can be predictive of the risk of upper extremity (UE) work-related musculoskeletal disorders (WMSDs) in exposed populations. The OCRA index was the first, most analytical, and most reliable method developed. It is generally used for the (re)design or in-depth analysis of workstations and tasks (Colombini et al., 1998, 2002). The OCRA checklist, based on the OCRA index, is simpler to apply and is generally recommended for the initial screening of workstations featuring repetitive tasks (Colombini et al., 2001, 2002).

It has to be added that the current OCRA method forms the basis for two technical standards currently being developed by CEN (pr EN 1005-5) and ISO (ISO DIS 11128-3). In EN 1005-5 OCRA is the method of choice. ISO 11228-3 chose OCRA as the preferred method for detailed risk assessment because it considers all the relevant risk factors, is also applicable to "multitask jobs", and provides criteria — based on extensive epidemiological data — for forecasting the occurrence of UL-WMSD (upper limb work-related musculoskeletal disorders) in exposed working populations.

Both OCRA methods are observational and are largely designed to be used by corporate technical specialists (occupational safety and health [OSH] operators, ergonomists, time and methods analysts, production engineers), who have proven in practice to be best suited to learning and applying the methods for prevention and also for improving production processes in general.

The methods have been applied in a wide cross-section of industries and workplaces. They target any jobs in manufacturing and the service sector that involve repetitive movements and/or efforts of the upper limbs (manufacture of mechanical components, electrical appliances, automobiles, textiles and clothing, ceramics, jewelry, meat and food processing). In Europe, it is estimated that these methods are currently used in more than 5,000 tasks that fit these categories, involving over 20,000 employees. The methods are not suitable for assessing jobs that use a keyboard and mouse, or other computerized data-entry tools.

In this study, the work posture analysis of printing operators was conducted by using direct observation. This was done to find out the specific painful part as well as give the best suggestion to reduce the complaint. This research was conducted using the Job Strain Index and Occupation Repetitive Index methods. Job Strain Index serves to evaluate the work toward the risk of musculoskeletal disorders in the upper body or Distal Upper Extremity (DUE), which is on elbows, forearms, wrists, and hands [2]. While the Occupation Repetitive Index is a quantitative method to identify the ways of workings on repetitive workers, especially those raising the risk of pain during upper body movements and repetitive activities [11, 12]. As in the journal of Hudiantoko, et al. [13] OCRA is used to improve the ways of workings in the logistics department, which has many complaints of muscles and nerves. In risk analysis, there are several methods such as OWAS, RULA, REBA, OCRA, and Checklist [14]. Some incorporations of this methods has been carried out in various studies such as Enez and Nalbantoğlu [15] who compared two results from OWAS and REBA methods for harvesting forest products, Paulsen et al. (2015) who used OCRA and strain index methods for processing cheese, and Restuputri [16] had carried out work risk assessments using the strain index method to batik workers.

2. Method

The two assessment methods evaluate four main collective risk factors based on their respective duration, Repetitiveness, Force, Awkward posture and movements, and Lack of proper recovery periods. Other "additional factors" are also considered, such as mechanical, environmental, and organizational factors for which there is evidence of causal relationship with UE WMSDs. Each identified risk factor is properly described and classified to help identify possible requirements and preliminary preventive interventions. All factors contributing to the overall "exposure" are considered in a general and mutually integrated framework. This research was conducted in a printing company located in Malang area.

This study observed the whole work activities during printing process. The data needed include work posture, work cycle, work time data, and work duration per day. The data retrieval was done by observation and interview. Work posture data was obtained by recording, in the form of photos and videos. For analyzing, a method which assesses the work focusing on the upper body was used. Occupational Repetitive Action (OCRA) and Job Strain Index (JSI) are selected for analyzing the ergonomic risk levels in printing activities because it is considered as the most appropriate semi-quantitative comprehension method in measuring MSD risk factors for the upper body [17, 18].

2.1. Determining the highest risk occupation type of RSI

There are two stages in this research, namely Nordic Body Map questionnaire of pain complaints on entire body and Nordic Body Map questionnaire of complaints to the highest risk occupation causing Repetitive Strain Injury. The Nordic Body Map questionnaire of pain complaints on entire body is given to 14 employees of the production section, whose results are processed by accumulating the selected highest percentages of employees' body parts. After that, the Nordic Body Map questionnaire of complaints to the highest risk occupation causing Repetitive Strain Injury will be conducted.

The questions given to respondents in Nordic Body Map questionnaire of highest risk occupation type has already had scores to be chosen. The scores used in this questionnaire is the Likert scale, in which letter A shows pain complaints, letter B shows a-few-pain complaints, and C shows no-pain complaints. The Nordic Body Map questionnaire of highest risk occupation is given to 14 employees of the production section.

2.2. Determining ATA (Actual Technical Actions)

To determine the number of technical actions, the first thing to do is replaying the research videotape on work in slow motion, in order to identify the technical actions as well as counting the number of it. The second is determining the frequency per minutes by dividing the total number of technical actions with the time used in one cycle. The third is determining the duration of repetitive work, namely the interval of repetitive work carried out by the operator during a work shift in units of minutes. The fourth is determining ATA by multiplying frequency per minute and duration of repetitive work. Therefore, the value of ATA is determined by the following formula:

$$\text{Frequency} = \frac{number\ of\ technical\ actions\ x\ 60\ seconds}{cycle\ time\ (in\ seconds)} \quad (1)$$

$$\text{ATA} = \text{Frequency x the time total of } \textit{repetitive} \text{ action} \quad (2)$$

2.3. Determining RTA (Recommended Technical Actions)

To determine the value of RTA (Recommended Technical Actions), these steps are needed:

a. Determining the strength factor based on the interview with the operator using the CR-10 Borg scale.

$$\text{Time proportion in one cycle} = \frac{Work\ duration\ in\ one\ cycle}{Cycle\ time} \quad (3)$$

$$\text{Average score} = \text{Borg scale score x time proportion in one cycle} \quad (4)$$

b. Determining posture factor and operator movement by paying attention to hand segment movement, namely shoulder, wrist, elbow and hand grip. Then select the smallest multiplier score of the four-hand segments as the posture and movement factor score.

c. Determining additional risk factors based on direct observation. This factor is not always present in certain work situations. If additional factors do not exist, then the additional risk factor is 1.

$$\text{Cycle time proportion} = \frac{Total\ Time}{Cycle\ TIme} \text{ x100 \%} \quad (5)$$

d. Determining recovery period factor by classifying each working hour as 'risky' or 'no risk' hour, with a reference that each 'no risk' hour have comparison between work time (with repetitive movements) and minimum recovery time of 5:1.
e. Determining the duration factor. The time lapse during repetitive work in one work shift matched with the criteria of duration factor determination.

$$\text{Cycle time - Total time of doing technical action} \quad (6)$$

$$\text{Total Micro Break Time} = \frac{Work\ Time}{Cycle\ Time}\ x\ Micro\ Break\ Time \quad (7)$$

$$\text{Actual Break Time} =$$

$$\text{Macro Break Time + Total Micro Break Time} \quad (8)$$

$$\text{Actual Working Time} = \text{60 minutes} - \text{Actual Break Time} \quad (9)$$

f. Determining RTA by multiplying constant frequency (30 actions per minute), strength factor, posture factor, additional factor, total duration, recovery period factor, and duration factor.
g. Therefore, the value of RTA is searched by the following formula:

$$\text{RTA} = \sum_{i=I}^{n} \quad [CF\ x\ (Ffi\ x\ Fpi\ x\ Fci)\ x\ Di]\ x\ Fr\ x\ Fd \quad (10)$$

2.4. *Calculating OCRA Index*

According to Stanton [12], OCRA Index is a result of comparison between the number of technical actions during the work shift and the number of recommended technical actions. Calculating OCRA Index is carried out by dividing ATA (Actual Technical Actions) by RTA (Recommended Technical Actions).

$$\text{OCRA} = \frac{Number\ of\ technical\ actions\ carried\ out\ in\ one\ shift\ (\Sigma\ ATA)}{Number\ of\ recommended\ technical\ actions\ in\ one\ shift\ (\Sigma\ RTA)} \quad (11)$$

2.5. *Classifying OCRA index calculation results based on OCRA index provision*

After calculating the OCRA Index, according to Stanton [12], the meaning of OCRA Index calculation results can be classified as Table 1:

Table 1. The classification of OCRA index result

OCRA Index	Area	Note
≤ 1.5	Green	Optimal
1.6 – 2.2	Green	Condition can be accepted
2.3 – 3.5	Yellow	Condition needs to be checked or improved
3.6 – 4.5	Red-Low	Low-risk condition
4.6 – 9.0	Red-Medium	Medium risk condition
> 9	Red-High	High-risk condition

IOP Conf. Series: Materials Science and Engineering **821** (2020) 012029 doi:10.1088/1757-899X/821/1/012029

3. Discussions

3.1. OCRA Methods

From the problem of risk factors for repetitive strain injury, the next step is to determine Actual Technical Action (ATA). Based on the results of recorded video that is played back in slow motion (slow motion). As an example of how to fill Table 2 in the technical action right hand, the number of technical actions is 1 because it requires one basic operation to complete the activity. The time to complete this technical action is 2 seconds.

Table 2. The details of every hand movement activities

Left Hand	Action	Time	Right Hand	Action	Time
Reaching the books (A1)	1	2	Reaching the books	1	2
Holding the books (A2)	1	1	Holding the books	1	1
Lifting the books (A3)	1	2	Lifting the books	1	2
Putting the books (A4)	1	1	Putting the books	1	1
Arranging the books (A5)	1	13	Arranging the books	1	13
Taking the engine control (A6)	1	8	Organizing the books	1	8
Holding the control handle (A7)	1	3	Turning the engine	1	5
Turning the engine (A8)	1	2	Delay	0	0
Delay (A9)	0	0	Reaching the levers	1	5
Taking the cut paper (A10)	1	2	Taking the cut paper	1	2
Reaching the engine control (A11)	1	1	Reaching the engine control	1	1
Holding the engine control (A12)	1	2	Turning the engine control	1	2
Reaching the books (A13)	1	1	Delay	0	0
Taking the results (A14)	1	5	Taking the results	1	5
Arranging the stacks of books (A15)	1	13	Arranging the stacks	1	13
Cycle time	14	56	Cycle time	13	60

$$Freq_{Right\ Hand} = \frac{13\ x\ 60\ seconds}{60} = 13\ actions/minutes$$

Therefore, the frequency in performing operators' duties for the right hand is 13 actions/minutes and the left hand is 14 actions/minutes (using Equation (1)). After calculating the frequency, the next step is to calculate the ATA by multiplying the results of the frequency calculation with the total work time of the repetitive work duration so that the number of actual technical actions can be calculated using Equation (2) as follows:

$ATA_{Right\ Hand}$ = 13 actions/minutes x 460 minutes = 5980 actions

Therefore, the ATA value in performing operators' duties for the right hand is 5980 actions and the left hand is 6440 actions in one working shift (Table 3). After that, we search the value of force factor (Ff).

Table 3. The details of right and left hand activities based on strength factor weighting classification

Activities	Right Hand				Left Hand			
	Time (seconds)	Score of Borg Scale	Time proportion in one cycle	Average Score	Time (seconds)	Score of Borg Scale	Time proportion in one cycle	Average Score
A1	2	0.3	0.0333	0.01	2	0.3	0.0333	0.01
A2	1	0.3	0.0167	0.005	1	0.3	0.0167	0.005
A3	2	3	0.0333	0.1	2	3	0.0333	0.1
A4	1	1	0.0167	0.0167	1	1	0.0167	0.0167
A5	13	0.5	0.2167	0.1083	13	0.5	0.2167	0.1083
A6	8	0.5	0.1333	0.0667	8	0.3	0.1333	0.04
A7	5	4	0.0833	0.3333	3	0.3	0.05	0.015
A8	0	0	0	0	2	4	0.0333	0.1333
A9	5	0.3	0.0833	0.025	0	0	0	0
A10	2	0.3	0.0333	0.01	2	0.3	0.0333	0.01
A11	1	0.3	0.0167	0.005	1	0.3	0.0167	0.005
A12	2	4	0.0333	0.1333	2	0.3	0.0333	0.01
A13	0	0	0	0	1	0.3	0.0167	0.005
A14	5	0.7	0.0833	0.0583	5	0.7	0.0833	0.0583
A15	13	0.5	0.2167	0.1083	13	0.5	0.2167	0.1083
TOTAL	**60**			**0.8133**	**56**			**0.625**
	Strength Factor (Ff)			**0.906**				**0.9625**

The released power is estimated by a scale proposed by Borg (scale CR-10 Borg) as in the table above. The example of calculations using Equation (3) and (4) in arranging the stacks is:

$$Time\ proportion\ in\ one\ cycle\ = \frac{13}{60} = 0.2167$$
$$Average\ score = 0.5\ x\ 0.2167 = 0.1083$$

The next step is calculating Posture Factor (Fp). In the right hand, there are 15 steps to work. At action number 3, there is 2 point for shoulder movement abduction 45° - 80°, 2 point for elbow movement flexion-extension more than 60°, 2 point for wrist movement radial deviation more than 15°, 2 point for kind of grips. Action number 7 spend 5 second, there are 5 point for shoulder movement, 5 point for elbow movement, 5 point for wrist movement extension more than 45° and 5 point for kind of grips. Action number 12 spend 2 second. The point of A12 are 2 point for shoulder movement abduction 45°-80°, elbow movement flexion-extension more than 60°, wrist movement extension more than 45°, and the last is 2 point for kind of grips. After that we calculate the percentage of that posture. After percentage, we score from segment risk and all of that posture's score is 1 (for right hand left hand)

The next is Additional Risk Factor (Fc). Additional factors are important to take into account but they are not always present. The existed additional risk factors is vibration. The risk from vibration for right hand is 9 and left hand is 4. So the proportion of cycle time for right hand is 15 % and left hand is 6.67 %. As the proportion of cycle time is between 0 % - 50 %, and then the score of additional factors (Fc) is 0.95. The next is Recovery Period Factor (Fr). Below is the example of how to determine the actual working time and break time in the first hour (Table 4):

Table 4. Determination of the actual working and break time.

Hour	Left Hand			Right Hand		
	Actual Break Time (minute)	Actual Working Time (minute)	Risk Value	Actual Break Time (minute)	Actual Working Time (minute)	Risk Value
I	8.67	51.33	0	5.00	55.00	1
II	4.00	56.00	1	0.00	60.00	1
III	4.00	56.00	1	0.00	60.00	1
IV	4.00	56.00	1	0.00	60.00	1
V	60.00	0.00	0	60.00	0.00	0
VI	4.00	56.00	1	0.00	60.00	1
VII	4.00	56.00	1	0.00	60.00	1
VIII	13.33	46.67	0	10.00	50.00	0
IX	8.67	51.33	0	5.00	55.00	1

If the comparison between the actual working time and the actual break time is ranged from 5:1 to 6:1 or below 5:1, then the risk value is 0. If the ratio is between 7:1 and 11:1, then the risk value is 0.5. If the ratio is greater than 11:1, then the risk value is 1. For example, in the first hour of the right hand, the ratio is 55 minutes : 5 minutes or 11:1. Therefore, the risk value in the first hour is 1. After that, each risk is added up to know the value of Fr. For the left hand, the risk number is 5 so that the value of Fr is 0.45, while for the right hand, the number of the risk is 7 so the value of Fr is 0.15.

After that, we calculate Duration Factor (Fd). The duration of repetitive work is 460 minutes. As the duration of repetitive work is between 421 to 480 minutes, then the duration factor (Fd) is 0.5 for the right and left hand.

After we know all the indicator value, the next step is recommended technical Action (RTA). The RTA values are obtained by multiplying all indicators (CF, Ff, Fp, Fc, D, Fr, and Fd). CF value is 30 actions/minutes. The RTA Value for right hand is 1781.6 actions and left hand is 5678.3 actions. After that we calculate Occupational Repetitive Action (OCRA). The value of OCRA Index from ATA value divided by RTA value. OCRA Index for right hand is 3.356 and left hand is 1.13

3.2. *JSI Method*

The first step is calculate the Intensity of Exertion (IE). This is an estimated strength needed to do a task at a time. Based on the analysis, researchers find that this activities are considered quite heavy as operators have to move books and pull the lever manually. Therefore, the IE value is 3.

The second steps is Duration of Exertion (DE). The percentage of DE is obtained from amount of energy use multiplied by 100 divided by total time of observation. The percentage of DE is 85.18 % its mean the DE Multiplier value is 3.

The next step is Effort per Minutes (EM). EM is the amount of energy expended by workers in one minute. EM is obtained amount of energy use divided by total time of observation. The value amount of energy use for the right hand is 429.33.total time of observation is 540. So the value of EM for right hand is 0.851. Therefore, the EM Multiplier value is 0.5

The next step is Hand/Wrist Posture (HWP). Hand/wrist posture is an estimation of relative hand and wrist position to the neutral position. Based on the analysis, the operator's left hand is straight which means that the position is very good, with HWP value of 1. Meanwhile, the right hand forms an angle of 50° which means the position is bad, with the HWP value of 2.

After that is Speed of Work (SW). The speed of work is an estimation of how fast a worker is doing the job. Based on the observation, the operators working normally, which is categorized as Fair, the value is 1 in SW table. This factor shows the duration of the task per day as well as the total time it was done in one day. As these activities are carried out for 460 minutes or equivalent to 7 hours 40

IOP Conf. Series: Materials Science and Engineering **821** (2020) 012029 doi:10.1088/1757-899X/821/1/012029

minutes, so the DD value is 1. So Job Strain Index (JSI) is obtained by multiplying IE, DE, EM, HWP, SW, and DD. The value of JSI for right hand is 9 and the left hand is 4.5.

4. Conclusion

In this case study of the printing field, researchers take objects in the process of pressing paper and cutting. At this work station, the operators work manually by standing within 8 working hours. Based on the interview, there were several complaints undergone after work, e.g. in back, shoulders, neck, and hands. The OCRA and JSI methods are carried out to analyze the workload of operators and to find out whether it is dangerous.

Based on the OCRA method, the right hand gets a value of 3.3564 which means it is in the yellow area. This situation needs to be checked or improved again. While the left hand got an OCRA value of 1.1341 which means it is in the green area so that the activities are considered optimal. The OCRA index for right hand is bigger than left hand. Because the value of Ff and Fr right hand is smaller than left hand. This is due to the lowest recovery time for right hand compared to let hand. This cause the RTA value for right hand lower than left hand. Where, RTA is inversely proportional to the OCRA Index.

The results of the analysis using JSI method show that the value of the right hand is 9. It means that the right hand has entered Hazardous area which needs to be analyzed as soon as possible. Whereas the JSI value of the left hand is 4.5. It entered uncertain area which means that the activities do are not at risk of injury. This is due to the value of HWP for right hand bigger than left hand. HWP is an indicator that explains hand and wrist posture. Its mean that posture for right hand is worse than left hand. Right hand forms a 50° position which means the position is bad. That position can cause work accidents.

Based on the comparison of the two analyzed methods, it is found that the right hand does need to be repaired immediately because of its great risk. It is particularly the activity of turning the engine lever to cut the paper together as well as the distance between the lever and machine table which is considered to be too high. Whereas the left hand activities are considered good and optimal.

List of Notations

1, n = Tasks displaying repetitive movements of upper limbs which carried out during a shift.
CF = Constant frequency = 30 action/minute
Ff = Strength factor
Fp = Posture factor
Fc = Additional factor
D = Total duration of each work with repetitive movement
Fr = Lack of recovery time factor
Fd = Duration factor

References

[1] Wiranata B 2013 Pengelolaan dan pengembangan bisnis percetakan pada PT. Ubital Offset Printing *Agora* vol **1** pp 554-58.

[2] Yunita D R and Indah P S 2017 Analisis Postur Kerja Pengrajin Batik Menggunakan Metode Job Strain Index (JSI) dan Loading on The Upper Body Assessment (LUBA) Universitas Muhammadiyah Surakarta.

[3] Faqihudin M H 2014 Evaluasi Penanganan Material Secara Manual Pada Operator Mesin Small Press Dan Big Press Dengan Metode Occupational Repetitive Action (OCRA) Index (Studi Kasus: PT New Armada) Universitas Gadjah Mada.

[4] Tarwaka P and Bakri L S 2010 Ergonomi Industri Dasar-dasar pengetahuan ergonomi dan aplikasi di tempat kerja ed Surakarta Harapan Press.

[5] Bulduk E Ö, Bulduk S, Süren T and Ovalı F 2014 Assessing exposure to risk factors for work-related musculoskeletal disorders using Quick Exposure Check (QEC) in taxi drivers *International Journal of Industrial Ergonomics* vol **44** pp 817-20.

[6] Chander D S and Cavatorta M P 2017 An observational method for postural ergonomic risk assessment (PERA) *International Journal of Industrial Ergonomics* vol **57** pp 32-41.

[7] Kee D and Karwowski W 2001 LUBA: an assessment technique for postural loading on the upper body based on joint motion discomfort and maximum holding time *Applied Ergonomics* vol **32** pp 357-66.

[8] Hellig T, Mertens A and Brandl C 2018 The interaction effect of working postures on muscle activity and subjective discomfort during static working postures and its correlation with OWAS *International Journal of Industrial Ergonomics* vol **68** pp 25-33.

[9] Chiasson M È, Imbeau D, Aubry K and Delisle A 2012 Comparing the results of eight methods used to evaluate risk factors associated with musculoskeletal disorders *International Journal of Industrial Ergonomics* vol **42** pp 478-88.

[10] Occhipinti E and Colombini D 1996 Proposal of a concise index for the evaluation of the exposure to repetitive movements of the upper extremity (OCRA index) *La Medicina del lavoro,* vol **87** pp 526-48.

[11] Andriani D P and Sugiono S 2017 Penjadwalan waktu istirahat optimal untuk mengurangi risiko musculoskeletal disorders berdasarkan OCRA Index *Jurnal Ilmiah Teknik Industri* vol **15** pp 157-67.

[12] Stanton N A, Hedge A, Brookhuis K, Salas E and Hendrick H W 2004 *Handbook of human factors and ergonomics methods*: CRC press.

[13] Hudiantoko M N, Anugraha R A and Nugroho Y 2015 Mengurangi risiko Ue Msds yang disebabkan oleh awkward posture di operator instalasi PT Xyz dengan menggunakan metode ocra index *JRSI (Jurnal Rekayasa Sistem dan Industri)* vol **2** pp 47-52.

[14] Davids D 2019 Recovery effects in binary aluminum alloys *Ph. D. thesis* Harvard University 1998.

[15] Enez K and Nalbantoğlu S S 2019 Comparison of ergonomic risk assessment outputs from OWAS and REBA in forestry timber harvesting *International Journal of Industrial Ergonomics* vol **70** pp 51-57.

[16] Restuputri D P 2018 Penilaian risiko gangguan musculoskeletal disorder pekerja batik dengan menggunakan metode strain index *Jurnal Teknik Industri* vol **19** pp 97-106.

[17] Dale A M, Harris A C, Rempel D, Gerr F, Hegmann K and Silverstein B 2013 Prevalence and incidence of carpal tunnel syndrome in US working populations: pooled analysis of six prospective studies *Scandinavian journal of work, environment & health* vol **39** p 495.

[18] Colombini D, Occhipinti E and Fanti M 2005 Il metodo ocra per l'analisi e la prevenzione del rischio da movimenti ripetuti *Manuale per la valutazione e la gestione del rischio* vol **15** FrancoAngeli.

IOP Conf. Series: Materials Science and Engineering **821** (2020) 012030 doi:10.1088/1757-899X/821/1/012030

Predictions of new students interest in entering the XXX University with statistic modeling

E D Priyana

Department of Industrial Engineering, Faculty of Engineering, Universitas Muhammadiyah Gresik, Gresik, Indonesia

Email: eftadhartikasari@umg.ac.id

Abstract. XXX is the establishment of one of Muhammadiyah's charitable businesses in the field of education. From the beginning of its formation on May 25, 1980 until now XXX has been able to open 18 study programs plus other study programs at the Faculty of Health and the Faculty of Engineering which are planned to be opened. The addition of the study program is expected to filter the number of new student interests to become XXX residents. But unfortunately, the many universities in Gresik were opened, making XXX have to work extra in attracting new students. This research will only be focused on single forecasting on the number of new student interests in several study programs and one university. In the descriptive statistical results, it was obtained the results that Management Study Program contributed the most interest in entering students and as Study Programs which were not new, Fisheries Cultivation Study Program contributed the least interest of students entering. The results of cluster analysis were formed by 8 groups. Group 5 consisting of Management Study Program, Industrial Engineering and PGSD contributed the most interest in terms of the average count per study program. Forecast with Moving Average also proves that this group has a tendency to rise. ARIMA results (1 0 0) for one University produces a value that falls smoothly with a less significant value, but if left unchecked it can also harm the campus.

1. Introduction

XXX is one of Muhammadiyah's charitable endeavors in the field of education. XXX began to be inaugurated by the regent of KDH Level II in Gresik Regency, Mr. Colonel Wasiadji with a location on Jalan Kh.Kholil No.90 Gresik, on May 25, 1980 under Number: E.1 / 017-V / 1980 [1, 2]. From the beginning only opened 4 Faculties (Faculty of Agriculture, Faculty of Fisheries, Faculty of Economics, Faculty of Teacher Training and Education) at the beginning of its formation, XXX has now been able to add to its Study Program by grouping several Faculties (Faculty of Agriculture, Faculty of Economics, Faculty of Teacher Training and Education (FKIP), Faculty of Islamic Religion, Faculty of Engineering, Faculty of Psychology, Diploma Program) [2, 3].

Predictions of new students 'interest in entering the XXX university with statistic modeling is a title carried by the author in the hope of getting a forecast with minimal error value, which means the predicted value obtained will get closer to perfect. Statistics itself is a tool that has long been known by the general public. Prediction with the smallest error is the best model that can be used for many things. The first is to analyze whether the number of new students who enter XXX is in line with the targets management wishes to achieve. The second is for further research, is opening a new study program felt necessary for student worship. The results of this study are also expected to be an alternative reference for other researchers or writers as further research. [4].

The statistical method used to predict interest in entering new students in XXX University is the Autoregressive Integrated Moving Average (ARIMA) forecasting model which is known to have high accuracy in predicting univariate data. This study will use data on the number of new students who entered from the beginning of the XXX University standing until the last available data. Not only predicting the whole student entering XXX university, the researcher also presents descriptive domain analysis of several study programs, with grouping using cluster analysis (the least group and the most group) so that XXX university management can consider appropriate sustainability management to strengthen the university [5, 6].

The ARIMA method studied was developed by George Box and Gwilyn Jenkins in 1976 (Makridakis et al, 1999) [7]. This method has the concept of ignoring the predictor variables in preparing the forecast model. The ARIMA approach is believed to be able to improve the accuracy of the predictions produced. This trust in research on ARIMA has made ARIMA's concepts contribute a lot in various fields. One of the studies was conducted by Bagoes Rahmat Widiarso in 2012, which used ARIMA as the basic concept of forecasting research in rainfall in Ngawi District with the result that the Post Mantingan showed poor performance, so special handling was needed [8].

2. ARIMAZ

Autoregressive Integrated Moving Average (ARIMA) is one of the models used for forecasting time series data. ARIMA is a non-stationary ARMA model that has been differentiated so that it becomes a stationary model. The assumptions that must be met are the stationarity of the data and residuals which are white noise. White noise which means residuals do not autocorrelate and have a normal distribution of $N(0,\sigma_a^2)$. The models in the non-seasonal ARIMA are as follows [9, 10].

a. Autoregressive Model (AR (p))
The Autoregressive (AR) model with order p denoted by AR (p) with the general form of the AR (p) model:

$$Z_t = \emptyset_1 Z_{t-1} + \emptyset_2 Z_{t-2} + \cdots + \emptyset_p Z_{t-p} + a_t \quad (1)$$

b. Model *Moving Average* (MA(q))
The Autoregressive Model (AR (p)) with the order q is denoted by MA (q) with the general form of the MA (q) model:

$$Z_t = a_t - \theta_1 a_{t-1} - \theta_2 a_{t-2} - \cdots - \theta_q a_{t-q} \quad (2)$$

c. Model *Autoregressive Moving Average* (ARMA(p,q))
ARMA model with order p and q is a mixed model between autoregressive order p and moving average order q. The general models:

$$Z_t = \emptyset_1 Z_{t-1} + \emptyset_2 Z_{t-2} + \cdots + \emptyset_p Z_{t-p} + a_t - \theta_1 a_{t-1} - \theta_2 a_{t-2} - \cdots - \theta_q a_{t-q} \quad (3)$$

Model *Autoregressive Integrated Moving Average* (ARIMA(p,d,q))
The general model autoregressive order p, Integrate order d, and moving average order q (ARIMA (p, d, q)) is the result of a merger between a stationary process with a stationary non-stationary product. In this case, d is the order of differentiation. The forms of the ARIMA model (p, d, q) are:

$$\emptyset_p(B)(1-B)^d Z_t = \theta_q(B)a_t \quad (4)$$

$$\emptyset(B) = 1 - \emptyset_1 B - \emptyset_2 B^2 - \cdots - \emptyset_p B^p \quad (5)$$

$$\theta(B) = 1 - \theta_1 B - \theta_2 B^2 - \cdots - \theta_q B^q \quad (6)$$

3. Descriptive statistics

In this study, the beginning of things done to find out the data description is to make descriptive statistics. Descriptive statistics used are by making bar charts, to determine the prospect of study programs that have the most amount of interest to the smallest. Figure 1 shows the results that Management Program study is a study program that is able to absorb the most students. While the study program that stood for a long time but did not absorb many students was the Fisheries Cultivation Study Program (Prodi Budidaya Perikanan).

Figure 2 shows an overview of the number of incoming interests per period. In 2007/2008 it was seen that Informatics Engineering Study Program was the most popular study program. But in the 2012/2013 period the amount of entry interest tended to drop significantly and began to stabilize in 2015. In contrast to the Management Study Program, which used to be less attractive in early 2007/2008, its presence in 2018/2019 actually contributed to many interested people.

Figure 3 shows prodi groups are less desirable. It can be seen that the Study Program for Aquaculture is less desirable, although it continues to move up smoothly, the numbers are very small or even can be said to be not very meaningful. This is not in line with the environment of Gresik which is famous for the cultivation of ponds that this study program should be able to become a doctrine for the Gresik region.

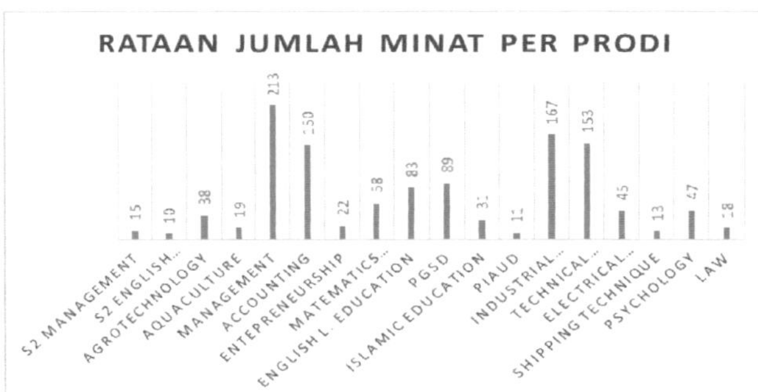

Figure 1. The average number of students entering interest per study program

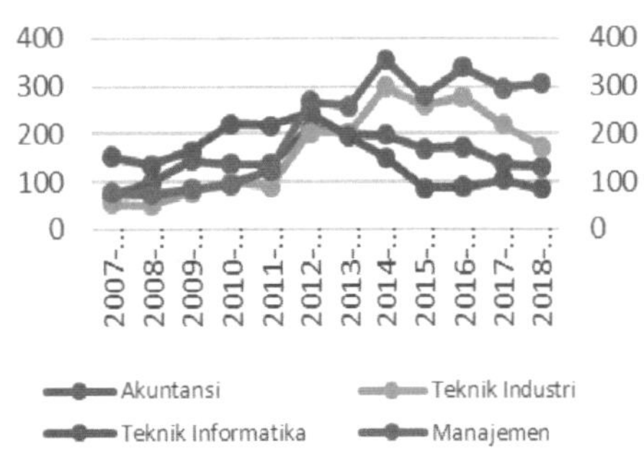

Figure 2. Time guide of the most interested study programs

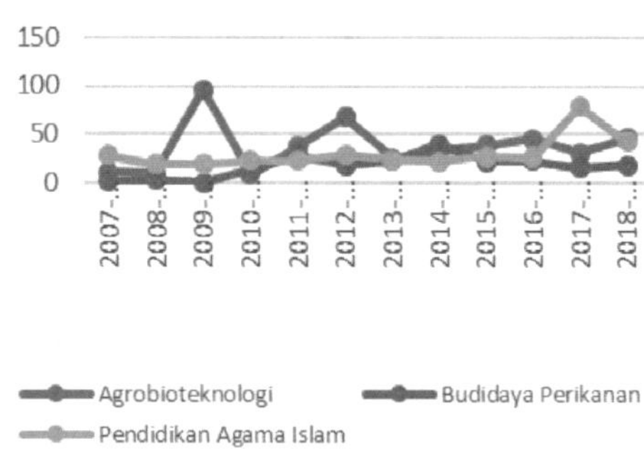

Figure 3. Time-consuming study program not interested

Figure 4 show the output of cluster analysis. 8 groups will be formed. The number of groups is prioritized to find out the tendency of homogeneity of more significant characteristics [11-14]. Table 1 is the result of cluster analysis. It was found that Islamic Religious Education program study and Agrobiotechnology program study formed a single group, which means that the two study programs did not have homogeneous characteristics in filtering the amount of student interest to enter with any study program.

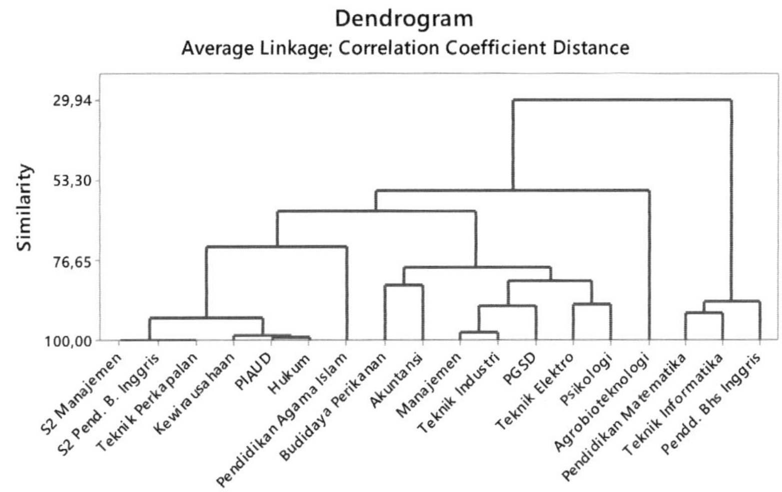

Figure 4. Dendogram cluster analysis

Table 1. Grouping the similarities of study program characteristics

Program Study	Group	Program Study	Group
S2 Management S2 English Language Education Shipping Technique	I	Management Industrial Engineering PGSD	V
Entrepreneurship PIAUD	II	Electrical Engineering Psycology	VI
Law		Agrobiotechnology	VII
Islamic Education	III	Mathematics Education	VIII
Aquaculture Accounting	IV	Technical Information English Language Education	

Figure 5 is an actual data plot without groups 1 and 2, because both groups are new study programs that do not have enough data to be analyzed using time series analysis. It can be seen that group V is the most absorbing group of new students. It is inversely related to group VII which is a less desirable study group, and group VIII which has an interest rate that tends to go down in a certain period [15].

Figure 6. shows the results of forecasting with the Moving Average (MA) method without including groups I and II because of the limitations of existing data. The figure shows the plot of group V which tends to increase significantly compared to other groups. while group VI moved slightly up but not significantly [16, 17].

Figure 7 is a time series of data plots of actual student interest as a whole. The graph shows fluctuating but less irregular values. In the 2007/2008 period, the interest in entering new students tended to increase significantly and in 2012 the plot tended to be irregular and even decreased [18-20].

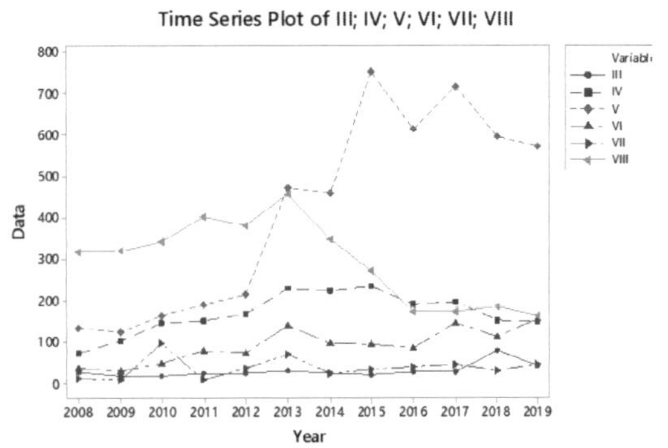

Figure 5. Time series plot cluster analysis

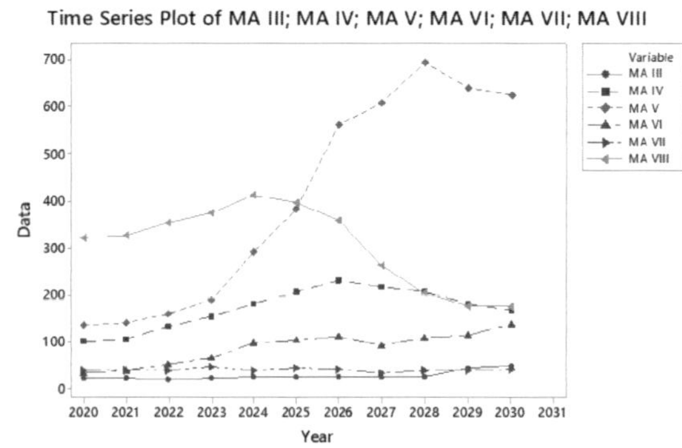

Figure 6. Forecasting 10 periods ahead

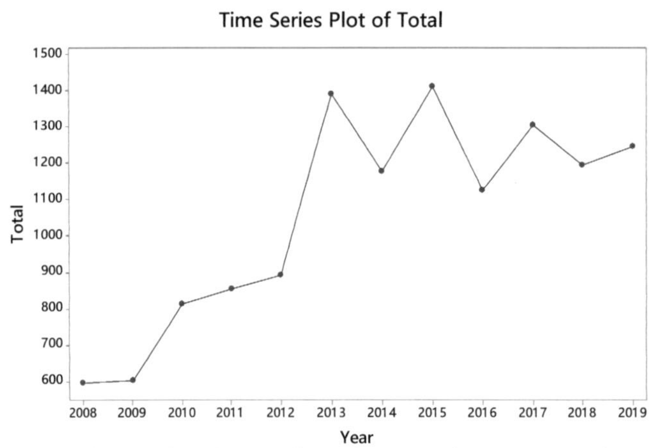

Figure 7. Time series plot number of new students XXX University

4. ARIMA

The assumptions that must be fulfilled in ARIMA are stationary in mean and stationary in variance. This is not found in traditional forecasting models such as Moving Average. It is this consideration of stationarity that makes ARIMA more accurate.

Figure 8 the transformation results show that the rounded value is 1 which means the data is stationary in the variance. Next is the analysis of stationarity in the mean by looking at the ACF value (Figure 9). The way to determine stationarity in the mean is to look at the lag that is out of line. If there are more than 3 lags out of line then the data is considered not stationary in the mean. Since there is no data coming out, it can be ensured that the data is stationary in the mean.

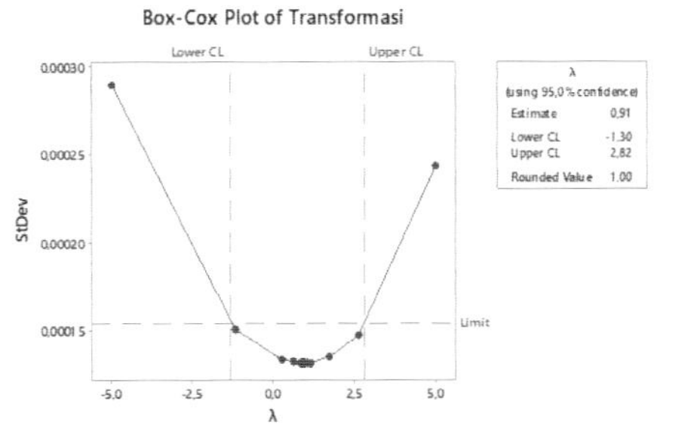

Figure 8. Box cox plot on transformation of number of interest new students in XXX University

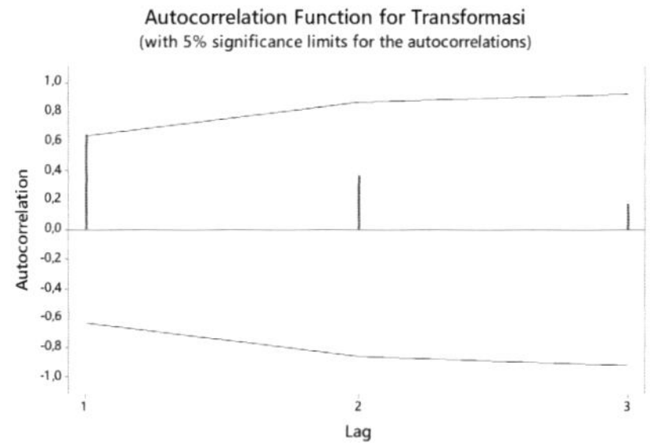

Figure 9. ACF plot data transformation interest in entering new students XXX University

Table 2 is the result of the ACF test with a 95% confidence interval. The comparison of the decision area used by T_{count} adalah T_{table} (α=0.05; db=11) with value of 1,796. T test statistics are used to detect lags to z_t with z_{t-k} (k is lag).

Hypothesis:
H_0: $\rho_k = 0$
H_1: $\rho_k \neq 0$

Table 2 is output T value in lag 1 with value 2.25> 1.796 which means reject H_0, so it can be concluded that there is a significant correlation (between z_t and z_{t-1}) in this lag. Besides lag 1, the lag value is less than 1,796, which can be concluded that there is no significant correlation in the other two lags.

Table 2. Autocorrelation function on the amount of interest in entering new students of XXX University

Lag	ACF	T	LBQ
1	0.649539	2.25	6.44
2	0.370223	0.94	8.75
3	0.172266	0.41	9.30

Figure 10 visually, the PACF plot shows that the data on the amount of interest in entering new University XXX students is stationary in the mean. Seen from no lag coming out of the line. Table 4 is the value that shows the T test for hypothesis testing with a 95% confidence interval. Similar to the ACF test, the decision area used as a comparison for T_{value} is T_{table} (α = 0.05; db = 11) with a value of 1.796. T test statistics are used to detect lags to z_t with z_{t-k} (k is lag).

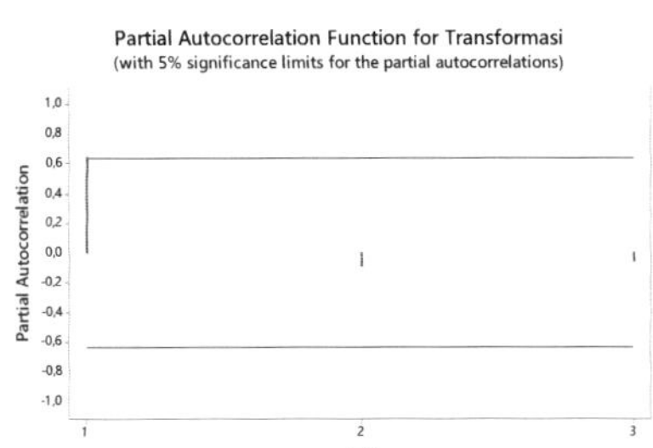

Figure 10. ACF plot data transformation of incoming student interest

Table 3 is the Output T value in lag 1 with a value of 2.25> 1.796 which means rejecting H_0, so it can be concluded that there is a significant correlation (between z_t and z_{t-1}) in this lag. Besides lag 1, the lag value is less than 1,796, which can be concluded that there is no significant correlation in the other two lags.

Table 4 shows the output results from the ARIMA model (10 0 0). The model is the best result that has the smallest P-value value from several experiments that have been carried out. There is a limitation of the data obtained so that in ARIMA forecasting (1 0 0) the data is doubled twice, so that it has the number of n as many as 24, which was previously only 12.

Table 3. Partial autocorrelation function on the interest of new students entering the XXX University

Lag	PACF	T
1	0.649539	2.25
2	-0.089393	-0.31
3	-0.055173	-0.19

Table 4. Output ARIMA (1 0 0)

Type	Coef	SE Coef	T	P
AR 1	0.6166	0.1698	3.63	0.001
Constant	396.50	47.41	8.36	0.000
	Mean	1034.1	123.6	

Hypothesis:

H_0: $\rho_k = 0$

H_1: $\rho_k \neq 0$

ARIMA (1 0 0) will produce AR (1) values that involve constant values, so we get 2 parameters in the model. In general, the significance of a constant does not need to be tested so only the autoregressive parameter (∅).

Hypothesis:

H_0: $\emptyset = 0$

H_1: $\emptyset \neq 0$

The same is true in the case of testing the ACF and PACF hypotheses. The comparison used in the T-hitung decision area is T-table. Normal comparison tables cannot be used because the analyzed data falls into the small sample category. Table 4 is the result of interpretation of the T_{value} for the AR parameter (1) or ∅ is 3.63 greater than the value of T_{table} of 1.717 and $P_{value} < \alpha$. So it can be concluded that the AR (1) model, parameter ∅ is quite significant. The resulting model equation is:

$$Z_t = 0{,}6166 Z_{t-1} + 396{,}5$$

The data processing also mentions the results of the Mean Square Error (MSE) model of 53929 with a free degree of 22. Table 5 is the Ljung-Box value for detecting white noise processes. Several tests will be conducted, one of which is the correlation test used to detect residual independence and next is the residual normality test model. Residual independence test is performed to detect residual independence between lags. Time series uses this test in Ljung-Box-Pierce.

Hypothesis

$$H_0: \rho_{a_t, a_{t+k}} = 0 \text{ (There is a correlation between lags)}$$

$$H_1: Minimal\ ada\ 1\ lag\ yang\ \rho_{a_t, a_{t+k}} \neq 0$$

Ljung-Box-Pierce statistical test uses a comparison between χ^2_{value} dengan χ^2_{table}. Reject H_0 if the value $\chi^2_{value} > \chi^2_{table}$. Table 5 shows the Ljung-Box value only at lag 12, this is due to the limited data available so it is not possible to get more data. The value of χ^2_{value} shows the calculation result of 30.5 and the value of χ^2_{table} of 18.3 which means reject H_0. If the decision is taken from the side of the P_{value}, the result is less than the error value (5%) so it can be decided that the decline in H_0 can be interpreted that between the residuals in lag t with the next lag has a relationship. This is in line with the situation that the output produced only shows lag 12.

Table 5. ARIMA white noise identification (1 0 0)

	Modified Box-Pierce (Ljung-Box) Chi-Square statistic			
Lag	12	14	36	48
Chi-square	30.5	*	*	*
DF	10	*	*	*
P-value	0.001	*	*	*

Figure 11 shows the results that no 1 lag was out of bounds, so it can be concluded that the data has been independent. The lack of data variability in the Ljung Box makes the test not yet consistent. The analysis that has been done shows that the best model for the ARIMA test is ARIMA (1 0 0) selected based on the P_{value}. Having succeeded in the variant also meant that it was agreed to analyze ARIMA. A forecast of 10 years is produced in Figure 4 which has been compared with the overall MA results.

Analysis that has been done shows that the best model for the ARIMA test is ARIMA (1 0 0) which is chosen based on the smallest P_{value}. After stationary success in the variance or mean, the ARIMA analysis test will be performed. The forecast for the next 10 years in Figure 4 has been compared with the overall MA results.

Figure 12 shows the results of MA forecasts that tend to be irregular fluctuations. This can be identified because MA has many shortcomings in terms of correlation between series due to the absence of a stationary assumption in variance or mean. While the ARIMA results show a smooth decline in value. ARIMA testing is more reliable because of the close correlation between time series.

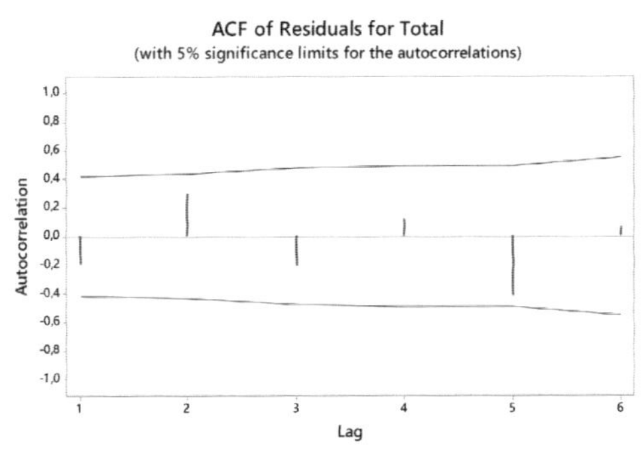

Figure 11. The ACF plot of the AR (1) model

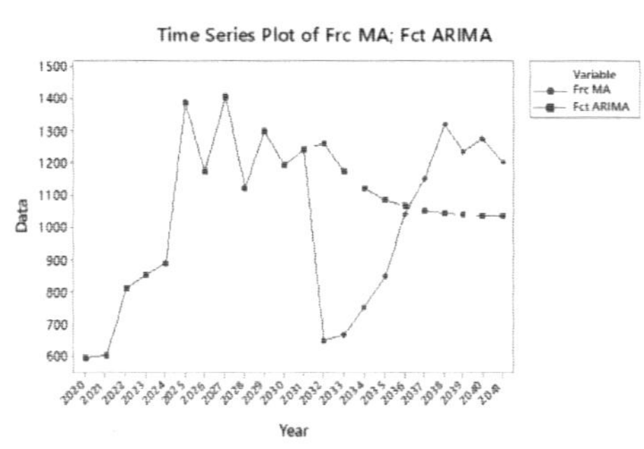

Figure 12. Comparison MA VS ARIMA

IOP Conf. Series: Materials Science and Engineering **821** (2020) 012030 doi:10.1088/1757-899X/821/1/012030

5. Conclusion

The conclusion obtained from this analysis is that in the case of ARIMA forecasting (1 0 0) the resulting value that in the next 10 periods the number of students entering interest in XXX University will decrease smoothly or may be interpreted to have a less significant value but tends to continue to fall. While in group forecasting with MA, it is produced the value that group 5 which consists of Management Study Program, Industrial Engineering and PGSD has the best forecast among the other groups and it can be assumed that the three study programs can contribute the most new students among other study programs.

References

[1]. Muhammadiyah Central Management 2000 *Guidelines for the Islamic Life of Muhammadiyah Citizens* Jakarta.

[2]. Team UMG 2017 *Academic Gudelines* Gresik.

[3]. Asmaria I and Ponirin 2013 Development of Charitable efforts of the Muhammadiyah Organization Medan *Journal of Government Science and Social Politics* UMA.

[4]. Montgomery, Douglas, Jennings and Kalahci. 2008 *Indtroduction To Time Series Analysis and Forecasting* America: A John Wiley & Sons Inc.

[5]. Namin, Sima, Namin and Akbar 2018 *Forecasting Economic And Financial Time Series: ARIMA VS LSTM.* Lubbock USA Texas university.

[6]. Iwok I A and Udoh G M 2018 A Comparative study between the ARIMA – Foutier model and the wavelet model *American Journal Of Scientific And Industrial Research.* ***7-6*** pp 137-44

[7]. Wei W W 2006 *Time Series Analysis Univariateand Multivariate Methods* AmerikaSerikat: Pearson Education Inc.

[8]. Wahyudi 2011 *Identification of Extreme Rainfall in Ngawi District Using Generalized Extreme Value and Generalized Pareto Distribution. Surabaya:* ITS Press. Walpole R E 1996 Pengantar Statistika Jakarta: PT. Gramedia Pustaka Utama.

[9]. Rachman Rizal 2018 *Application of Moving Average and Exponential Smoothing Methods in Garment Industry Production Forecasting* Jakarta: STMIK Nusa Mandiri.

[10]. Satya Febi 2015 *Use of the ARIMA Method for Short-term Electric Consumption Load Assembling.* Semarang: Universitas Negeri Semarang.

[11]. Sukmawati. 2017 *Cluster Analysis with Hierarchy Method for Regency / City Grouping in South Sulawesi Province Based on Macroeconomic Indicators.*

[12]. Hamidi, Seyed. 2018 Consensus clustering algorithm based on the automatic partitioning similary graph *Science Direct Journal. Data & Knowledge Engineering.*

[13]. Khadir and Khedaira. 2019 A multiple clustering combination approach based on interative voting process *Journal of King Sand University-Computer and Information Sciences.*

[14]. Socoro, Joan, Alias and Francesc 2019 Parallel hirearchial architectures for afficient consensus clustering on big multimedia cluster ensembles *Journal Information Sciences* vol **511** pp 212-28.

[15]. Walder, Adam, & Hanks, Ephraim 2019 Bayesian analysis of spatial gneralized linier mixed models with laplace moving average random fields *Journal Pre-Proof.*

[16]. Heinrich C, Mikko S P and Almut E D V 2019 Hybrid simulation scheme for volatility modulated moving average fields *Journal Mathematics and Computers In Simulation* vol **166** pp 224-44.

[17]. Huang, Jing, Huang, Zhijan 2019 Testing moving average trading strategies on ETFs *Journal of Empirical Finance.*

[18]. Zifwen 1999 ENSO forecasting and modeling the relationship of ENSO with Monson rainfall *Skripsi* Jurusan Statistika Bogor: FMIPA IPB.

[19]. Fanoodi B, Behnam M and Farzad F J 2019 Reducing demand uncertainty in the platelet supply chain though artificial neural networks and ARIMA models *Computers In Biology and Medicine* **113**(2019) 103415.

[20]. Wang Q, Shuyu L and Rongrong L 2018 China's dependency on foreign oil will exceed 80% by 2030: Developing a novel NMGM-ARIMA to forecast china's foreign oil dependence from two dimensions *Science Direct Journal. Energy* **163**(2018) pp 151-67.

IOP Conf. Series: Materials Science and Engineering **821** (2020) 012031 doi:10.1088/1757-899X/821/1/012031

Green productivity application for improving productivity and environmental performance through the selection of the best solution scenario in the agroindustry

A Mubin

Department of Industrial Engineering Department, Faculty of Engineering, Universitas Muhammadiyah Malang, Malang, Indonesia

Email: ahmadm@umm.ac.id

Abstract. Improving productivity and environmental performance is very important for agroindustry to improve performance and competitiveness .The application of the green productivity (GP) can help the company to be able to increase productivity and environmental performance in the same time. The purpose of this research is to improve the productivity and the environmental performance through improvement in the system of production by selecting alternative scenario as the best solution from some of alternative solution scenarios. The method of the research is conducted in several stages that are measuring process productivity and early environmental performance, arrangement of the improvement scenario system that can increase productivity and environmental performance simultaneously and provide suggestions for refurbishing and improving productivity and environmental performance. The result of the calculation from early productivity level is obtained the average of the past 7 period is 139.4%. While the value of early environmental performance is 0.69 which means that the quality of waste has met the standard of regulations, although they are still relatively small so they still needs to do restitution and improvement. The measurement value of green productivity index (GPI) is obtained that the existing condition, the scenario 1, the scenario 2 and scenario 3 are 1.39, 1.43, 1.45 and 1.41. Thus, then the selected best scenario is scenario 2, it is expected that the productivity and the environmental performance, as well as the competitiveness of companies can improve.

1. Introduction

Sustainable industry is the industry that has the sustainability economically, environmentally and socially. One indicator of the economic aspect can be shown through level of productivity that is the ratio between output and input in production process in the industry. While the environment aspect can be measured through environmental performance or environmental performance indicator index (EPI). Improving green productivity is an important way to achieve sustainable development [1].

Improving productivity and environmental performance is quite important for agroindustry to improve of performance and competitiveness. The application of green productivity method (GP) may help the company to be able to improve productivity and environmental performance in the same time [2]. GP is a broad strategy for enhancing productivity and environmental performance. Green Productivity's greatest attribute is its potential for integrating environmental protection into the operations of a business as a mean of improving productivity. Towards GP measurement and improvement, Hur et al. [3] developed a measurement tool that indices economic and environmental performance in one index called GP index that is defined as the ratio of productivity of a system to its environmental impacts. This index is intended for estimating the GP performance of an existing product or process and comparing it with other equivalents. Gandhi et al. [4] developed indicators and Green Productivity Index (GPI). A casting case indicates that the GPI can be used as an actionable feedback for leadership to make effective decision. Green Productivity Indicator as a measure the level of Green Productivity companies [5]. Green productivity (GP) index is a measurement tool developed to analyze economic performance and environmental performance in one index [6].

Moharamnejad, et al. [7] did a research regarding to management of GP in Iran Aseman Airline Company current situation are analyzed such as the consumption of energy, water, airplanes fuel and evaluating environmental pollutants. The research result from Marimin, et al. [8] concluded that the best scenario is suggested to improve the productivity of motorcycle tires was a combined treatment of controlling raw material characteristics and reusing water and materials. The research result from Ghahremani, et al. [9] showed that Green Productivity Index depends on factors such as manpower, materials, energy and machinery and environmental factors. Mubin et al. [10] has conducted a study on the application of GP to improve productivity and environmental performance in the leather tanning industry. Liu et al. [11] introduces energy consumption and carbon emission into the analysis framework of the green productivity of tourism.

The purpose of this research is to improve productivity and environmental performance through the improvement of production system in agroindustry by selecting the best alternative solution scenario from some alternative solution scenarios. Improving productivity and environmental performance is expected could increase efficiency in the use of the resources and could reduce the environmental impact, thus it can increase competitiveness and sustainability of the agroindustry.

2. Methods

The GP assessment methodology works within the framework of an Environmental Management Systems (EMS) to help a company concentrate on opportunities to prevent pollution and improve material productivity [12]. The research was conducted in several phases that are the measuring process productivity and early environmental performance, the arrangement of scenario improvement system which is able to increase productivity and environmental performance simultaneously and provide suggestion for refurbishing and increasing productivity and sustainable environmental performance [13].

On the measuring process that is to measure level of productivity and environmental performance has achieved by this company. The result productivity and environmental performance measurement has been done and it will be referred to estimation of increasing productivity and environmental performance. The measurement of productivity is calculated by dividing total output with total input.

Case study research was conducted in an agro industry sugar A. The output is the total income and input is the material cost, labors cost, energy consumption, water cost, and the waste processing cost. The total calculation of level of productivity in company uses this following equation:

$$\text{Productivity} = \frac{Total\ Output}{Total\ Input} \tag{1}$$

The next stage is arranging improvement scenario systems based on the identification result of factor that can increase productivity and environmental performance. The final stage is providing suggestion for improvement and enhancement of productivity and environmental performance based on the selection result of best scenario.

3. Analysis and discussion

3.1. *Productivity calculation*

The result of company productivity in period 1 – 7 can be observed in Figure 1. From the data below, it can be recognized that company productivity is above 100%. In other words, the productivity is good enough and it has increasing trend pattern. The average productivity is 139,4%.

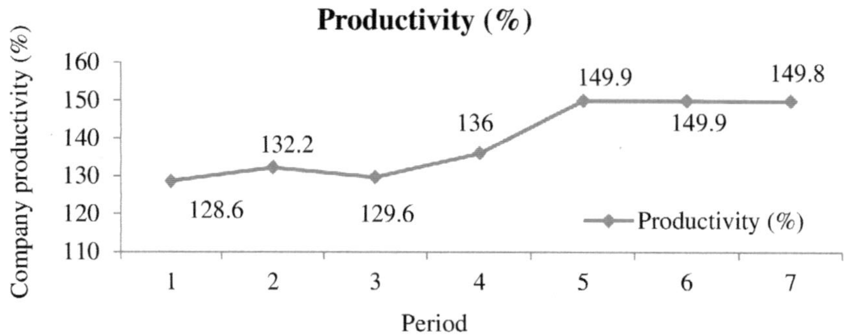

Figure 1. Productivity value period 1 - 7 [14, 15].

3.2. Calculation of environmental performance

The measurement of environmental performance is an important part of environmental management system. A study in environmental performance based on the environmental policy, and environmental target (ISO 14004, ISO 14001). Environmental performance indicator (EPI) can be defined as parameter or the number of measurement based on the number of subjects that are observed or counted. An environment indicator is a thing that is expected to be described various impacts of environment activity and efforts to reduce it. EPI described environment efficiency from the process production by involving numbers input and output. EPI that is used in this study includes: (1) BOD, (2) COD, (3) pH, (4) TSS, (5) sulphide, (6) oil and fat, and (7) the volume of waste.

EPI index can be calculated by using this formula: $$\text{EPI Index} = \sum_{i-1}^{k} W_i P_i \quad (2)$$

K is the number of waste criteria that are lodged and Wi is weights of each criteria. This weight is obtained through the questionnaire to the experts. The value of Pi is deviation percentage between standard quality with the analysis result of company. The measurement result of EPI total index is obtained a positive value that is 0.69 which means that in general the quality of waste has met the standards as specified, although it is still relatively small so that it is still needed refurbishing and improvement.

3.3. Selecting the best scenario

Alternative solution scenario are: (1) scenario 1: Biotray installation, (2) scenario 2: adding outlet capacity from the result of waste water processing, and (3) scenario 3: DAF (*Dissoveled Air Flotation*) installation. Selecting the alternative solution has been done based on calculation of GPI (*Green Productivity Index*) estimation towards the selected alternative solution. The calculation result and comparative GPI value can be observed in Table 1.

Table 1. Ratio of productivity level and environmental impact

No.	Description	Productivity Level (%)	Environmental Impact (EPI Index)	Percentage (%)	GPI
1	Early	139.04	0.69	100.00	1.39
2	Scenario 1	139.05	0.71	97.10	1.43
3	Scenario 2	139.07	0.72	95.65	1.45
4	Scenario 3	139.19	0.70	98.55	1.41

From Table 1, it can be identified that scenario 2 has higher GPI value than scenario 1, 3 and early GPI. Therefore, Scenario 2 is defined as the best scenario and it is suggested as the suggestion to refurbish productivity and environmental performance.

IOP Conf. Series: Materials Science and Engineering **821** (2020) 012031 doi:10.1088/1757-899X/821/1/012031

4. Conclusions

The calculation result of the early productivity level is obtained the average over 7 periods that is 139.4 %. While early environmental performance value with the parameters of EPI, BOD, COD, pH, TSS, sulphide, oils and fats, and the volume of the waste is obtained 0.69 which means that the quality of waste has met the regulation standard although it is still relatively small so it is still needed an improvement and enhancement.

The measurement result of green productivity index (GPI) value that is obtained for the existing condition is 1.39, scenario 1 is 1.43, scenario 2 is 1.45 and scenario 3 is 1.41. Thus, the best scenario that is selected is scenario 2, it is expected that productivity and environmental performance, and the competitiveness of companies can increase.

References

[1]. Li D and Wu R 2018 A dynamic analysis of green productivity growth for cities in Xinjiang *Sustainability* **10** p 515.

[2]. APO 2003 Asian Productivity Organization: *A Measurement Guide to Green Productivity*, Tokyo.

[3]. Hur T, Kim I, and Yamamoto R 2004 Measurement of green productivity and its improvement *Journal of Cleaner Production* vol **12** pp. 673–83.

[4]. Gandhi N D, Selladurai V and Shanti P 2006 Professional practice green productivity indexing: A practical step towards integrating environmental protection into corporate performance *International Journal of Productivity and Performance Management* vol **55** (7) pp 594-606.

[5]. Singgih M L 2010 Waste reduction with green productivity approach for increasing productivity *The 11th Asia Pacific Industrial Engineering and Management Systems Conference* Melaka.

[6]. Findiastuti W, Anityasari M and Singgih M L 2011 Green productivity index: Do different terms measure the same things? *Proceeding of Industrial Eng. and Service Science* pp 20-1.

[7]. Moharamnejad N and Azarkamand S 2007 Implementation of green productivity management in airline industry *International Journal of Environmental Science and Techn* **4** (1) pp 151-58.

[8]. Marimin M, Darmawan M A, Yuliana R P W and Teniwut K 2018 Green productivity improvement and sustainability assessment of the motorcycle tire production process: A case study *Journal of Cleaner Production* vol **191** pp 273-82.

[9]. Ghahremani F T and Omidvari M 2018 Providing an evaluation model of green productivity in paper-making industries *International Journal of Environmental Science and Technology* **15** pp 333–40.

[10]. Mubin A and Alfarisi S 2014 Increasing productivity and environmental performance in leather tanning industry by using method of green productivity *Proceedings of National Seminar of MM-ITS* Surabaya.

[11]. Liu G, Shi P, Hai F, Zhang Y and Li X 2018 Study on measurement of green productivity of tourism in the Yangtze River economic zone China *Sustainability* **10** p 2786

[12]. Balist J, Hoveidi H and Faryadi S 2016 Environmental Management System and Green Productivity (EMS_GP) implementation in Kurdistan Cement Plant *International Journal of Business and Management Invention* vol **5** pp 1-7.

[13]. Hakim M H, and Mubin A 2016 Analysis of environmental performance and productivity by using the concept of green and lean productivity *Journal of Teknik Industri* vol **17** no 1 pp 31-41.

[14]. Pratama H H 2015 Increasing productivity and environmental performance method using green productivity *Jurnal Teknik Industri* vol **16** no 2 pp. 63-73.

[15]. Zahroh A, Mubin A, and Baroto T 2019 *Proposed Enhancement of Productivity and Environmental Performance in the Sugar Industry Using the Green Productivity and AHP Methods* Industrial Engineering Department UMM Malang.

IOP Conf. Series: Materials Science and Engineering **821** (2020) 012032 doi:10.1088/1757-899X/821/1/012032

Computational thinking digital media to improve digital literacy

I Yuliana[1,2,*], L P Octavia[3], and E Sudarmilah[3]

[1] Department of Informatics Education, Universitas Muhammadiyah Surakarta, Surakarta, Indonesia
[2] Departement of Vocational and Technology Education, Universitas Negeri Yogyakarta, Yogyakarta, Indonesia
[3] Department of Informatics, Universitas Muhammadiyah Surakarta, Surakarta, Indonesia

*Email: irma.yuliana@ums.ac.id

Abstract. Computational Thinking (CT) has been defined as an important skill that supports digital literacy in the industrial revolution 4.0. Computatiomal Thinking contained in the informatics curriculum is classified as local content for several formal schools. Four key techniques (cornerstones) to Computational Thinking consist of abstraction, algorithm, decomposition and pattern recognition, are very important in the processes involved in formulating problems and the solutions so that the solutions are represented in a form that can be effectively carried out by an information-processing agent The research started by analysis, design, development, implementation and evaluation in series and give each other feedback to create an interactive learning media. The subjects of this study were students of grade IV and V of elementary school as great starting point for starting digital literacy as early as possible. Data collection techniques in this study are by observation, questionnaire, pre-test and post-test. The data analysis technique used was calculating the percentage of feasibility from the results of the questionnaire. The result is the use of CT learning media able to improve student's understanding of CT in problem solving. The respondents could find the concept by themselves, analogically they can apply to similar problem. This is evidenced by the result of the post-test in similar questions from the pre-test. The respondents finally learnt how computer works, and this is important as basic creativity to improve digital literacy.

1. Introduction

A gadget has variety of diverse application features, it is like a double-edged knife that has many functions, but also potential to endanger the user, especially for the younger. A number of children who access the game and social media features even experience mental disorders due to addiction to devices in the mild to acute level. This is very possible due to the absence of assistance to children in adapting to very intense technological developments. The introduction of technology as a part of digital literacy through the subjects of Information and Communication Technology (ICT) is limited just using the applications. Giving provision of ICT should not only be on how to use it wisely. Computational thinking gives knowledge that builds critical power and creativity by learning the computing process itself. For future reason improving people digital literacy, computational thinking must be taught in formal education, not just for as local content (especially in developing countries). The important thing is that computational thinking acts as the main base of an ICT Learning Building, as shown in figure 1. Computational Thinking is very important as digital literacy as a person's initial capital to interact with digital devices and its applications. One does not need to learn in depth about programming or coding, but it is very important to be able to learn how logic works as a critical power to overcome problems. The use of digital applications will be easier if someone understands how it works and its implications.

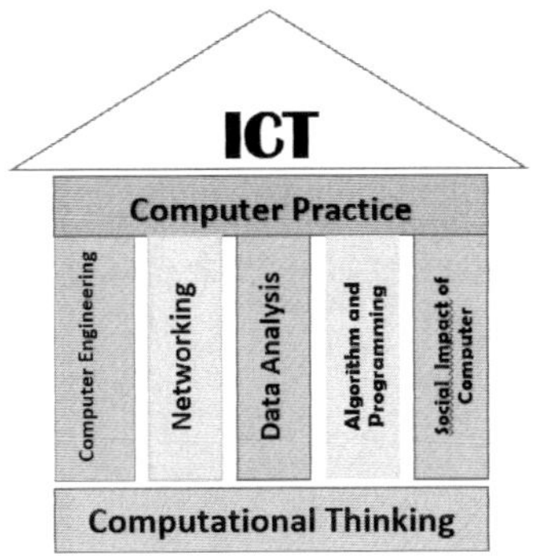

Figure 1. ICT Learning Building

Initially in 1080 Seymour Papert was credited as concretizing Computational Thought, then began to be popularized by Wing In 2006 and brought it to the attention of the international community, more research has been done on CT in education. Breakthrough technology has not fully increase Indonesian human index at the international level (United Nations Development Program, 2018). According to the behavior today, the younger are very addicted in using their gadgets. And so that education, both, devices and applications, hardware – software have not shown interrelated and supportive effort to gain some advantages of the technology, people only become consumers. There needs to be assistance in the use of ICT facilities, through data, technology and digital literacy. In general, learning ICT has to face the real one, the hardware. Due to this specific practical reason, gadget including the smartphone and tablet could be bought in affordable prices and it becomes semi-primary need nowadays. Beside it is a communication device, it contains of a lot of user needs, so it is vary in user age range.

In the previous curriculum, ICT was only about using word processors and graphics, so people have a poor understanding of what a computing career involves [1] and some other future works that will expand their competencies. In developing countries Indonesia digital literacy, is still need to be improved. Not all the people realize that this is very important, including in XXI future skills [2]. Digital literacy was introduced as ICT curriculum known as informatics in formal education. Although the government has rearranged the informatics as part of the curriculum, the implementation in school is still slow in response. Many schools do not have sufficient teacher whom computer science background to teach informatics at the general level including kindergarten, elementary school, junior high school and senior high school, in cities and even in rural area. In people's opinion, ICT is new paradigm called informatics which has complicated sub-subject like programming. They have not realized yet the important of ICT learning through informatics which not only about programming and algorithm. Informatics-computational thinking in computer science is very interesting in some way of teaching it. Even in some countries such as US, UK, Scotland, German [3]–[9], ICT is very crucial in its own curriculum as in STEM-C (Science, Technology, Engineering, Mathematics and Computer). STEM teaching is equipped with computational thinking skills, using ICT. It is very important to provide assistance to introduce digital technology that build the ICT application.

Computational Thinking is the thought processes involved in formulating problems and their solutions so that the solutions are represented in a form that can be effectively carried out by an information-processing agent [10-12]. Computational Thinking (CT) is one method of approach in the learning process. CT does not have an important role in developing ICT applications, but CT can also be used to support problem solving in all disciplines, including the humanities, mathematics and science. Students who study where CT is applied in the curriculum (learning process) can begin to see and compare the relationships between subjects, and between life inside and outside the classroom. Computational thinking is a problem-solving technique that is very broad in the area of its application, so having this ability is a must for someone who lived in the twenty-first century. Like playing music and learning foreign languages, Computational Thinking trains the brain to get used to proceed logically, structured, creative and reasonable. This is be the main thing to build cognitive capacity, skill and smart attitude to be able to survive and sustain in the future.

CT is actually not a new thing and cannot only be applied in an ICT environment. This critical-analytical ability can be applied to various other sciences. ICT education as a form of community digital literacy enrichment is still done partially on the hardware and its use, so it is often hampered by the reason that infrastructure does not support. Though CT can be taught even without having to rely on the availability of infrastructure. Viewed from the minimum number of participants and the results that were not optimal from all elementary school (SD) participants to High School Levels throughout Indonesia who participated in the CT competition (Bebras Challenge 2018). The Bebras Challenge is organized by Bebras Indonesia as one of the community initiatives engaged in the CT field. This competition was attended by 1561 participants with an average value of 32.09 while for the maximum value that could be achieved was 100 (www.bebras.org). This competition is followed by 3 categories, namely Standby for Elementary (Primary) equivalent, Raisers for Middle School (Junior High School) equivalent and Enforcement for High School (High School) and the equivalent. Based on this matter, the socialization of CT to educational institutions is considered still far from expectations.

From the description above, this study aims to create a CT learning media through digital games that are expected to become media that can introduce CT to students who are also expected to improve digital literacy in students. Providing interesting learning media alternatives in the form of theory and problem to train the critical thinking and mapping problems with CT concepts for the best solution.

2. Methodology

This learning media development model uses the ADDIE development model. According to Benny A. Pribadi (2009: 125) explains there are 5 stages in the ADDIE development model, Analysis (Analysis), Design (Design), Development (Development), Implementation (Implementation), and Evaluation (Evaluation). The analysis start from the user requirements, as shown in Figure 2 as Use Case. The

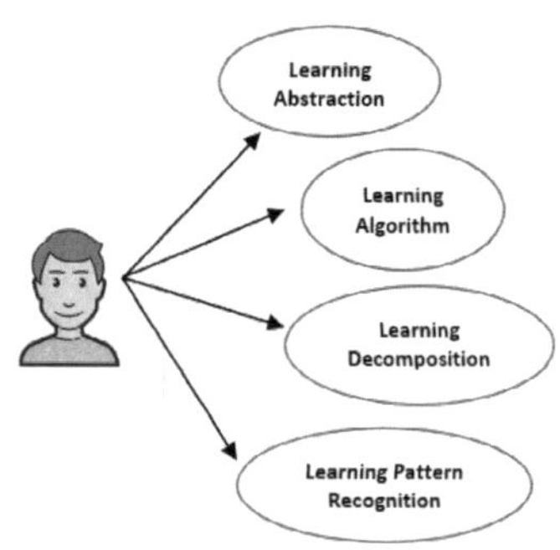

Figure 2. Use case

computational thinking learning media provides learning content about abstraction, algorithm, decomposition and pattern recognition.

Learning material that is compiled consists of simple learning materials that are equipped with examples of problems and solutions. Storyboard and Activity Diagrams are designed as a benchmark in designing using Construct2. After all before being tested on users, the media must go through validation tests by material experts and media experts the feasibility of media by the media and material expert. The media expert test has given overall result on the average on 81.25% which is included in the very decent category, meaning that this learning media is an efficient appearance, where technical errors in the media are very few and easy to use. So it permitted to the next test. While the score of the material test are 75% and 80.05% for learning and content aspect in order.

3. Result and discussion

The respondent of this study were student of K-10 of elementary school. Some instruments were used to gain the data including questionnaires, pre-test and post-test. The questionnaires, given to students, was the SUS (System Usability Scale) questionnaire developed by John Brooke [13], that consisting of 10 items of questions that use a 5 point Likert scale in the form of strongly disagree to strongly agree.

Figure 3, 4, 5 and 6 represent the content of the learning media, which can be accessed randomly by the user, do not have to be sequential. Abstraction takes the example of the name of a famous tourist destination in Indonesia, namely Wakatobi, which turned out to be a series of 4 islands (Wangi-Wangi, Kaledupa, Tomia and Binongko).

Figure 3. (a) Theory of abstraction; (b) Problem of abstraction

Algorithm elements are introduced through sequential processes in serving typical food in the form of fried rice. In the production process a cook must prepare ingredients and equipment, process them in sequence so as to produce the ordered food. The order from the initial state, process and final state is an algorithm which developed a step-by-step solution to the problem, or the rules to follow to solve the problem. While on the decomposition, the rule is breaking down a complex problem or system into smaller, more manageable parts, as shown on Figure 3 which mapping some schedules into well organized.

Figure 4. Theory of algorithm of fried rice receipt

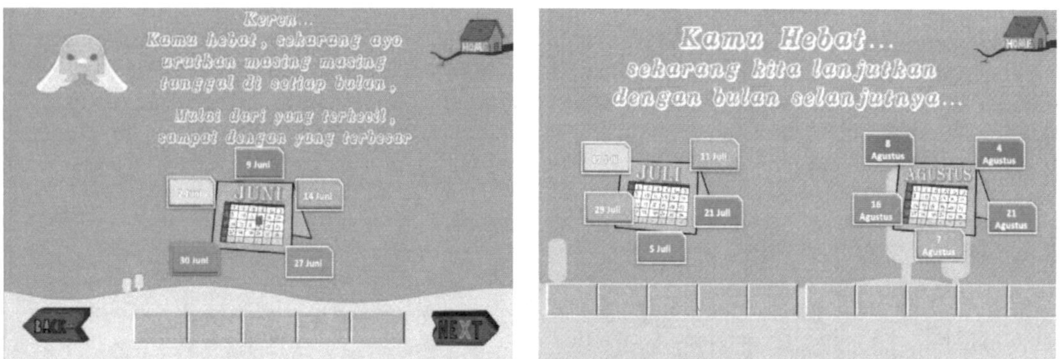

Figure 5. (a) Theory of decomposition; (b) Problem of decomposition

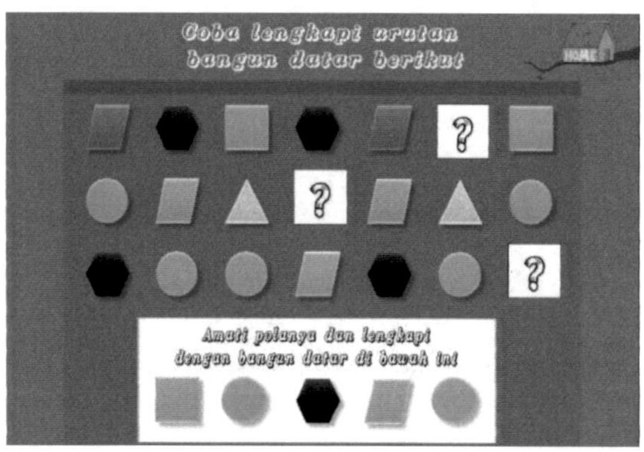

Figure 6. Theory of algorithm of fried rice receipt

Talking about pattern recognition is about looking for similarities among and within problems that might be drive to the alike solution.

Comparison of Learning Outcomes taken from the value of the pre test and post test given to respondents from elementary school students with processing time of 20 minutes each. The results of the pre test was on the an average on of 74.28 and the post test scored is 83.57. It means that student learning outcomes use computational thinking learning media higher than before. While from the SUS test gained the average score of 70.0 that it can be said that the SUS score is in the Marginal Hight or Good category.

4. Conclusion

Based on the results of media testing and material testing conducted, computational thinking learning media is appropriate to be tested delivered to respondents. The results of the pre-test and post-test students showed an increase in the value of the test results which indicated the learning media was able to improve students 'understanding on CT which was expected to be able to increase student's digital literacy levels.

Acknowledgements

The authors would like to express my very great appreciation to Universitas Muhammadiyah Surakarta for allowing the authors to conduct the research, and special thanks for everyone who contributed in this study especially to the Bebras Community, students of Al Azhar Syifa Budi Elementary School in Surakarta and the editors for the constructive criticism and suggestions.

References

[1] Bell T, Alexander J, Freeman I, and Grimley M 2009 Computer science unplugged: school students doing real computing without computers *J. Appl. Comput. Inf. Technol.* vol **13** no 1 pp 20–9.
[2] Rassol N 1999 *Literacy for Sustainable Development In The Age of Information* WBC Book Manuacturers.
[3] Cicirello V A, King V, and Drive F 2013 A cs unplugged activity for the online classroom *J. Comput. Sci. Coll.* vol **28** no 6 pp 162–68.
[4] Jun S J, Han S G, Kim H C, and Lee W G 2014 Assessing the computational literacy of elementary students on a national level in Korea *Educ. Assessment, Eval. Account.*, vol **26** no 4 pp 319–32.
[5] McInerney C, Hinchey M, and McQuade E 2009 Investment in information and communication technologies in the Irish education sector *IFIP Adv. Inf. Commun. Technol.* vol **302** AICT pp 83–91.
[6] Nanjundeswaraswamy T S and Swamy D R 1990 *Review of Literature on Computational Thinking* vol **7** no 2 pp 201–14.
[7] Zhang X and Zou H On computational tools, computational thinking and computational methods *Proc. 1st Int. Work. Educ. Technol. Comput. Sci. ETCS 2009* vol **1** pp 503–7.
[8] Ribeiro L, Nunes D J, Da Cruz M K, and De Souza Matos E 2013 Computational thinking: Possibilities and challenges *Proc. - 2013 Work. Theor. Comput. Sci. WEIT 2013* pp 22–5.
[9] T B B and Vahrenhold J 2018 *Adventures Between Lower Bounds and Higher Altitudes* vol **11011** Springer International Publishing.
[10] Wang L 2011 Computational thinking and computer fundamental education *2011 Int. Conf. Comput. Sci. Serv. Syst. CSSS 2011-Proc.*pp 1158–59.
[11] Selby C 2013 Computational thinking: The developing definition *ITiCSE Conf. 2013* pp 5–8.
[12] Wing J M 2006 Computational thinking-The begening *Commun. ACM* vol **24** no 3 p 33.
[13] Brooke J 1996 SUS—a quick and dirty usability scale 1996 *Usability Eval. Ind.* vol **189** no 194 pp 4–7.

IOP Conf. Series: Materials Science and Engineering **821** (2020) 012033 doi:10.1088/1757-899X/821/1/012033

IT balanced scorecard (IT BSC) based strategic framework for assessing the impacts of Business Strategic-IT alignment

S Balafif [*] and T Haryanti

Department of Computer Science, Universitas Muhammadiyah Surabaya, Surabaya, Indonesia

*Email: sabri.balafif@ft.um-surabaya.ac.id

Abstract. Alignment of IT strategy with business strategy to optimize the achievement of organizational goals. In general, there is a paradox between the application of information technology to the productivity directly. This paper aims to determine the suitability of the IT strategy with business strategy on company X by measuring IT Balanced Scorecard. The strategy implemented over a period of five (5) years. The study was conducted in the year to three (3). The results showed a low maturity level of application of IT for IT investment has not been implemented in accordance with the IT strategy. Anticipated funding has gradually become the cause of IT investments cannot be done thoroughly. The study presents recommendations for improvements every perspective ITBSC to achieve the expected level of maturity in the fifth year.

1. Introduction

Many studies have shown discrepancies in Strategic Business-IT alignment is the main reason for the failure to realize the full potential of IT investments [1], Organization alignment with Information Technology Strategy needs to be done to improve the organization's productivity [2], In general, there is a paradox between the application of information technology with the productivity directly [3] .Ward and Peppard [4] explained that the problem occurred because the productivity paradox of IT investment / IS still has not managed to make a real contribution expected by the organization. In some cases, not all of the work program ICT department supports the business processes and aligned with the organization's vision and mission. IT usage and sometimes impede the operations in an effort to trap binding.

There is a very important system in some parts of the business, but difficult for another section [5], Example Usability and Security in User Interface Design, obstacles often encountered is the application of multiple layers of security, such as the Multi-Factor Authentication, which affects the user-friendly experience [6], The gap occurs between the use of IT and finance. In terms of the financial department, the contribution of ICT adoption and utilization sometimes only judged Tangible value, which represents the only measured the benefits from Increased Sales, Reduced Operating Costs, and Reduced Interest Costs, whereas the IT investment which is necessary not directly visible in the form of profits [7].

This study is held in one of private manufacturing industry in Indonesia. Measurement of the weight of the fit between IT strategy IT organizations use a combination of Balanced Score Card and SWOT analysis. The study was conducted in Mining Equipment and Machinery Company with the goal of increasing productivity and optimizing production between consistency and reliability in the life circle of lead time. Shorter lead-time have been the key challenges for a manufacturing industry Findings Based on study investigates by Woosang Hwang in the role of ERP in the supply chain [8], The results of the research will provide insights for companies to see the maturity of IT implementation in harmony and synergy with the organizational strategy.

2. Literature review

The Mining equipment manufacturing company in the field of a company that sells products that are directly usable by the mining company. Utilization of IT becomes imperative for companies to be able to achieve optimal organizational goals. IT Investment financing used by companies where the study was conducted using the method of financing gradually. Management of IT resources in the company, faced with an alternative choice to develop its own infrastructure with considerable investment value (in-house resource) or services contract Information Systems (IS) with a third party (outsourcing). Overview Re-Engineering as in Figure 1.

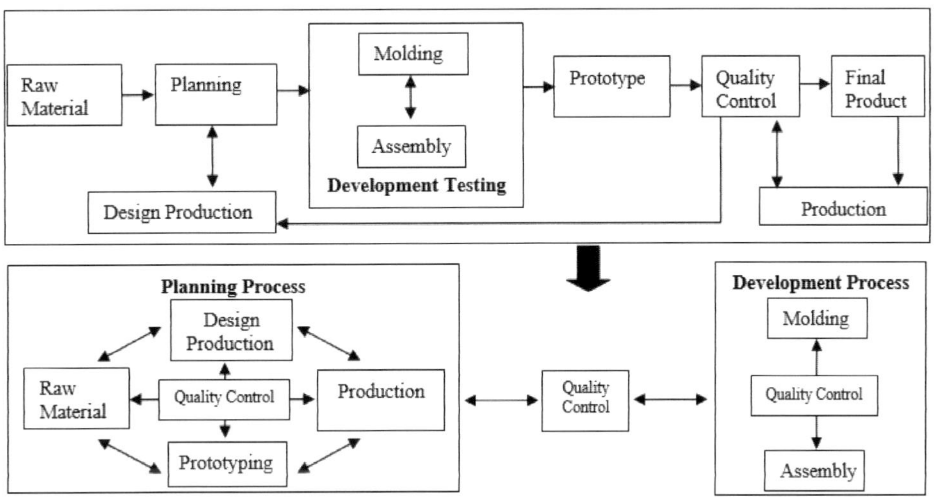

Figure 1. Re-engineering IT

2.1. *IT Balanced Score Card (BSC-IT)*

Measurement of the specific IT needs to be done to develop into the IT-BSC BSC [9]. Perspective ITBSC in this study was supported objectives specifically match the needs of the organization in which to do research.

2.1.1. *Corporate objective perspective*

- Strategic Business-IT alignment (CSF-1), IT utilization is used in accordance with the technology and market development needs of users [10],The Open Group Architecture Framework (TOGAF) [11] a framework the use of IT to meet the development needs of the user. Business success depends on the relationship business strategy, information technology strategy, infrastructure and enterpriseal processes, and IT infrastructure and processes [12],
- IT Service Management (CSF-2) is the craft of implementing, managing, and delivering IT services to meet the needs of an enterprise. In this paper, the main focus in IT management is to optimize IT initiatives to achieve maturity process in the manufacturing automation systems.

2.1.2. *Customer perspective objective*

- Satisfaction Service Performance (CSF-3). The relationship between employee satisfaction, customer satisfaction and organisational performance very tightly. Competition is getting tougher between service companies because many aspects support accelerated change, but consumers are getting smarter. so that progressive improvement and development must periodically be explored in order to achieve the enterprise's vision

- Interactive Support Service (CSF-4) is an important initiative to maintain the perception gap of quality between the services provided and user expectations.

2.1.3. Operational excellence objective

- Reduce Complexity (CSF-5) , Complexity Reduction helps reveal hidden costs and allows companies to determine which products are making money, what customers really value and which enterpriseal or process bottlenecks are getting in the way of effective actions, setting the stage for greater growth and increased profits.
- Technology capability (CSF-6) is the ability of the enterprise to execute any relevant technical function, including the ability of technological knowledge in order to Reduce Complexity and develop new processes for efficiency [13]
- Maturity process (CSF-7) In order to continuously improve performance, enterprises need to control their processes. To do this it is assumed that enterprises need a high level of business process management maturity and employees need a high level of knowledge and experience. Proof of this hypothesis has been found in many empirical studies.

2.1.4. Future perspective objective

- Effective Cost Management (CSF-8)
  The best way to achieve effective is using cost control. it mean concerned with measuring variances from the cost baseline and taking necessary actions to achieve minimum costs possible. Because Effective Cost Management is one of the main measures of business accountability
- Employees Job Satisfaction (CSF-9)
  Job satisfaction contributors in this study were built on two aspects, namely opportunities to use skills and abilities, and trust between employees and senior management.

3. Methods

Methodology of this research is the empirical investigation against a particular phenomenon. The phenomenon in question is Convergence and Integration Strategy in Business-IT Alignment, aligned Reviews. These types of companies have increased the efficiency of technology investments and significantly reduced the financial and operational risks associated with business and technical change. Questionnaire with linker scale will be processed in accordance with the weighting matrix and SWOT IT-BSC. Weighting calculation performed using Analytical Hierarchy Process (AHP) and scoring system using the Key Performance Indicators (KPI).

4. Results and discussion

Respondents to the study were 30. Respondents were selected based on length of employment and job-related fields of research. Corporate strategy plan period 2019 - 2021 are presented in Tables 1 and 2.

Table 1. Strength and weakness mapping company

Code	Conditions map	Weight	rating	Thickness x Rating
S1	Good corporate governance and sustainability	.1035	3	.3105
S2	Focus on the tastes and needs of customers	.1035	3	.3105
S3	Strategic business location and spread	.1293	4	.5172
S4	Product quality trustworthy wider community	.1293	4	.5172
S5	*supply chain* Effective (lead time <3 days)	.1035	3	.3105
S6	A healthy financial ratios	.0603	2	.1206
S7	SI Existing web-based secure connection	.0605	2	.1210
S8	Already using ERP	.1118	3	.1118

	Total			**2.3193**
W1	HR productivity is not optimal	.1293	3	.3879
W2	Investments in the field of IS / IT is not managed optimally	.0344	2	.0688
W3	Implementation of the solution / IT not yet have a framework	.0346	4	.1384
	Total			**.5951**
	Grand Total (SW)			**1.7242**

Table 2. Mapping opportunities and threats

Code	Conditions map	Weight	rating	Thickness x Rating
O1	Has a positive image in the community	.1305	4	.5218
O2	Target marketing is more than 25% can be achieved	.0261	3	.0783
O3	The growing potential new lovers	.1044	3	.3131
O4	The products have already become a tradition in the community	.1044	3	.3131
O5	Support rapid technological developments social media	.1044	3	.4175
O6	Utilizing the technology trends of cloud, mobile app	.1044	4	.4175
	Total			**2.0612**
T1	Regulation limiting the number of product circulation	.1305	3	.3914
T2	The issue of the negative impact on the result of product consumption	.1044	4	.4175
T3	Dependency of raw materials	.1044	4	.4175
T4	Consumers easily replace other brands	.0607	3	.1822
T5	Market share is still dominated by a competitor	.0261	2	.0522
	Total			**1.4606**
	Grand Total (OT)			**.6005**

SWOT quantitative data above represents the conversion of an assessment by calculation SWOT Analysis developed by Pearce and Robinson in order to know exactly the position of the actual organization.

Figure 2 presents a SWOT coordinates are at 2, 0625 in progressive quadrant focused on strategic SO, which is a strategy to maximize the organization's internal strengths to seize opportunities. Table 3 and 4 present the strategy of SWOT adapted to IT.

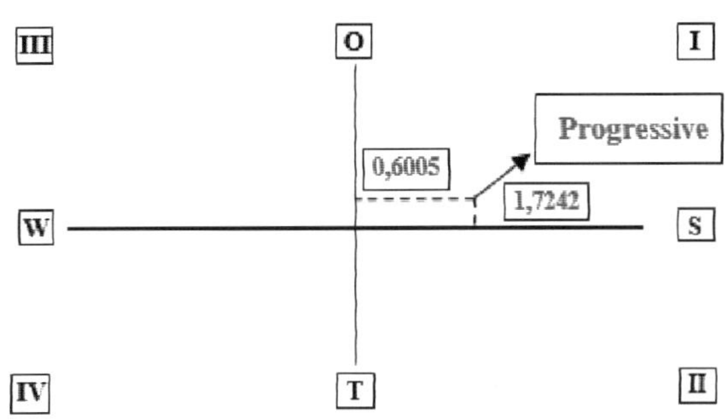

Figure 2. Coordinates SWOT

Table 3. The relationship between IT and business requirements initiative

Code	IT Initiative	Originally code	Business Requirements (BR)	BR code
SO1	Integrating and Online Supply Chain systems	S4, O2, O3	Continuously grow in markets	BR 1
SO2	Support continuous updates directly to the apps without re-downloading	S6, O5, O6	Great user experience	BR 2
SO3	Manufacturing Automation Systems	S6, S8, O5, O6	Cutting lead times in processes	3 BR
SO4	High Availability IT Architecture	S6, S7, O6	Business continuity	BR 4
SO5	Protection of privacy in information systems	S4, S8, O5, O6	Protecting information from falling into the wrong hands	BR 5

Table 4. Results weighting objective KPI measurements of IT-BSC

Perspectives	Objective	KPI Measurements	Weight	Total
Corporate Contribution	Strategic Business-IT alignment	Tangible and intangible benefits in CBA	0154	0.247
	IT Service Management	Service Level Management	0093	
Customer Orientation	Satisfaction Service Performance	User satisfaction and loyalty index	0124	0.248
	Interactive Support Service	Support Availability	0124	
Operational Excellence	Reduce Complexity	Process Velocity	0062	0.255
	Technology capability	IT capability and business performance	0062	
	process maturity	Process Complexity	0124	
Future Orientation	Effective Cost Management	Traceability cost in IT investment	0166	0.250
	Job Satisfaction Employees	Total factor productivity (TFP)	0166	

Figure 3 illustrates the strategy and KPI targets are mapped in the hierarchy priority using AHP. Figure 4 illustrates the relationship of objectivity conduction in ITBSC.

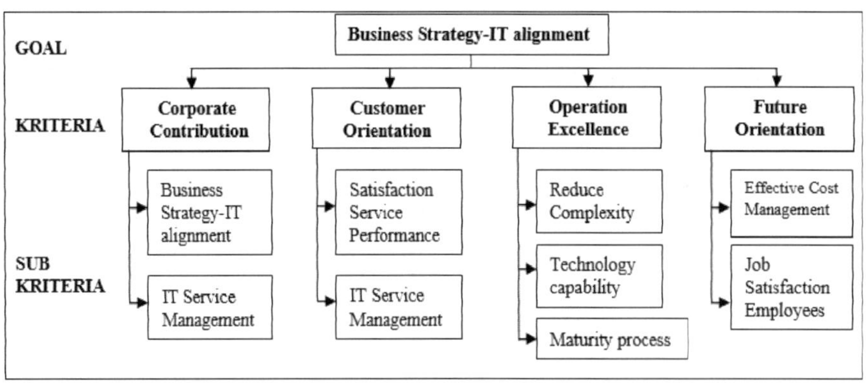

Figure 3. Structure Hierarchy on KPI

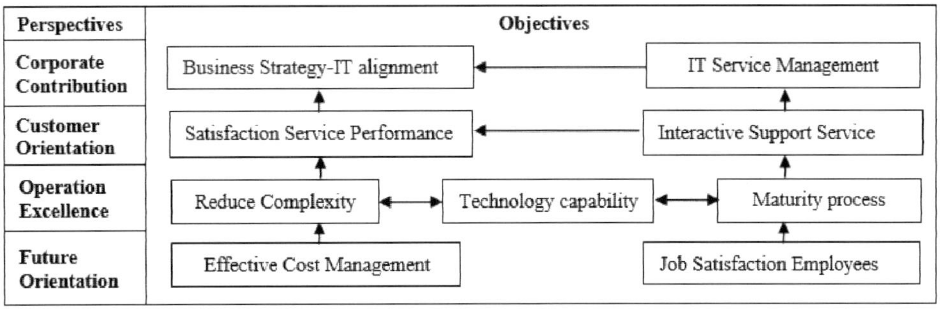

Figure 4. The relationship between Objectives in IT BSC

IT application maturity calculation is based KPIs are presented in Table 5.

Table 5. Results of measurement

Perspectives	Objective	KPI Measurements	Weight	Target in -5th	Target Score	Realization in-3th	Score
Corporate Contribution	Strategic Business-IT alignment	Tangible and intangible benefits in cost benefit analysis	0154	> 80%	3	80%	3
	IT Service Management	Service Level Management	0093	> 95%	5	70%	2
Customer Orientation	Satisfaction Service Performance	User satisfaction and loyalty index	0124	> 90%	4	80%	3
	Interactive Support Service	Support Availability	0124	90%	4	70%	2
Operational Excellence	Reduce Complexity	Process Velocity	0062	> 90%	4	75%	2
	Technology capability	The relationship between IT capability and business performance	0062	> 95%	5	70%	2
	process maturity	Process Complexity Versus Process Capability	0124	> 90%	4	78%	2
Future Orientation	Effective Cost Management	Traceability cost in IT investment	0166	> 95%	5	85%	3
	Job Satisfaction Employees	Total factor productivity	0166	> 90%	4	70%	2

Weighting Procedure for Criteria and Measurement of Success of IT-BSC. Weighting calculations are carried out using the Analytical Hierarchy Process (AHP) method and scoring system using the Objective-Key Performance Indicator Matrix. In these calculations can provide a performance achievement score of each Objective based on the Key Performance Indicator (KPI) and know how to achieve it, so that action can be carried out according to the conditions of achievement. As illustrated in Table 6.

IOP Conf. Series: Materials Science and Engineering **821** (2020) 012033 doi:10.1088/1757-899X/821/1/012033

Table 6. Scoring Table based on traffic light

Score	Progress	Status	Action Recommendation
1	<60%	Warning	Evaluation and control
2	60% - 79%		
3	80% - 89%	Progressive	Evaluation and improvement
4	90% - 94%		
5	≥95%	Excellent	Maintained and followed

Based on the four perspectives that were examined, namely the company's contribution, customer orientation, excellent operations, future orientation each had each criterion with a score of 3 and 2 so that it entered into the Progressive and Warning status. The perspectives activated by Progressive status are Company Contribution with the Strategic Business-IT alignment objective, Customer Orientation with Service Satisfaction Objectives and Future Orientation objectives with an Effective Cost Management objective. As for the Warning status, the Company's Contribution with Objective IT Service Management, Customer Orientation with Interactive Support Service objectives, Operational Excellence objectively Reducing Complexity, Technology capability and process maturity, and Future Orientation with Employee Job Satisfaction objectives.

5. Conclusion

The position of the company based on SWOT mapping obtained on progressive quadrant. This means that the organization is in good shape and healthy. Low achievement of Operational Excellence in the year to -3 allegedly caused by the pattern of corporate financing gradually so that the benefits of IT cannot be measured before the 5th year. Further research needs to be done for proving suitability financing pattern with IT strategy and business strategy of the company. Perspective Operational Excellence has an impact on the other perspective. Further evidence needs to be done to determine the relationship between the degrees of perspective on the IT-BSC. The results of this study are practically is that external and internal ICT capabilities are important drivers of firm performance, while merely having integrated IS do not lead to better firm performance.

In addition, a moderating effect of the IS integration in the relationship between ICT capabilities and business performance is found, although this integration contributes to firm performance only when it is directed to connect with suppliers or customers rather than when integrating the whole supply chain. In general the achievement of KPI based perspective weighting on ITBSC not yet reached a level of maturity that is expected. Score each KPI parameters on ITBSC can be used as reference improvements to achieve the expected level of maturity. Operational Excellence perspective had the lowest scores on the achievement of KPI which at maturity level 2. Lead time is part of the Process Velocity is measured on Operational Excellence Reduce Complexity Perspective. Currently Score 70% of the target of more than 90%, acceleration batch-process lead time is expected to improve Operational Excellence. Future Orientation Perspective is necessary to increase in accordance with the parameters measured in the KPI that is Effective Cost Management and Job Satisfaction Employees

IOP Conf. Series: Materials Science and Engineering **821** (2020) 012033 doi:10.1088/1757-899X/821/1/012033

References

[1]. Cândido C J F and Santos S P 2015 Strategy implementation: What is the failure rate? *J Manag & amp Organ* vol **21** 2 pp 237–62.

[2] Walter J, Kellermanns F W, Floyd S W, Veiga J F and Matherne C 2013 Strategic alignment: A missing link in the relationship between strategic consensus and organizational performance *Strateg Organ* vol **11** no 3 pp 304–28.

[3] Brynjolfsson E 1993 The productivity paradox of information technology *Commun. ACM* vol **36** no 12 pp 66–77.

[4] Ward J and Peppard J O E 2002 *AM Strategic Planning for Information Systems.*

[5] Evans N D 2012 The IT paradox: A diminished role in technology, but greater clout in the business *The IT paradox: A diminished role in technology, but greater clout in the business.*

[6] Nwokedi U O, Onyimbo B A and Rad B B 2016 Usability and security in user interface design: a systematic literature review *Int J Inf Technol Comput Sci* vol **8** no 5 pp 72–80.

[7] Velcu O 2007 Exploring the effects of ERP systems on organizational performance: Evidence from Finnish companies *Ind Manag Data Syst* vol **107** no 9 pp 1316–34.

[8] Hwang W and Min H 2013 Assessing the impact of ERP on supplier performance *Ind Manag Data Syst* vol **113** no 7 pp 1025–47.

[9] Van Grembergen W, Saull R and De Haes S 2011 Linking the IT balanced scorecard to the business objectives at a major canadian financial group in *Strategies for Information Technology Governance*.s

[10]. Van Grembergen W and De Haes S 2005 Measuring and improving IT governance through the balanced scorecard *Inf Syst Control J.*

[11]. Nasrin D, Reza S and Fatemeh S 2011 Developing a combined framework for evaluating IT projects based on IT-BSC and COBIT *Int J Digit Content Technol its Appl.*

[12]. Alfrian C T, Rudianto C and Rahadja Y 2015 *Analisis IT Balanced Scorecard sebagai Pengukuran Capaian Kinerja IT (Studi Kasus: Dinas Kependudukan dan Pencatatan Sipil Kota Ambon).*

[13]. Jairak K and Praneetpolgrang P 2013 Applying IT governance balanced scorecard and importance-performance analysis for providing IT governance strategy in university *Inf Manag Comput Secur.*

Attribute analysis with classification algorithm on election participation

A S Fitrani*, M A Rosid, F Muharram and F L Kodriyah

Department of Informatics Engineering, Faculty of Engineering, Universitas Muhammadiyah Sidoarjo, Indonesia

*Email: asfjim@umsida.ac.id

Abstract. Stages of General Elections of the President, DPD, DPR, Provincial DPRD and Regency / City DPRD in Indonesia are determined by institutions namely the General Election Commission (KPU), where there is a measure of success in holding direct, general, and free. Another component of the implementation of elections is that there are contestants and voters. In the voter factor, this is also a measure for success in the overall process of implementation, namely success if high community participation in the administration of elections. However, vice versa, if community participation is low, one of them is the level of public confidence in the organizers (government) decreases. Data mining classification analysis and modification of attributes in prediction classes "Hadir" and "Tidak Hadir" on the final voter list (DPT). The number of datasets is 4249 instances, and the number of attributes is 11. The percentage results are 89.3417% for the Naive Bayes algorithm for prediction classes in the Presidential Election, DPD, DPR, Provincial DPRD and Regency/ City DPRD in 2019. Further analysis is done on eliminating some attributes to obtain information, whether it has a significant effect on the results of predictions. And in this analysis, for the ten attributes with the removal of "statusKK" (highest rank gain ratio) for the prediction class, the results are the worst. = After nine attributes, the removal of the "rt" and "tps" attribute (second and third rank) for the prediction class is the best result. There is the highest percentage difference for the prediction class on the classification algorithm for modified and or unmodified status attributes, the results of the percentages and classification algorithms are different.

1. Introduction

General Election, abbreviated as PEMILU, is a people's party in determining the representation of both legislative, senator (DPD) and executive (president, governor, and district/city). The implementation of the ELECTION of the President, DPD, DPR, Provincial DPRD, and Regency/ City DPRD is carried out in 2019. ELECTION is an important political event to determine leaders in a democratic country. The Permanent Voters List (DPT), which is then determined to be a DP4, is the right of citizens to be protected by law to choose their representation in the executive, senator and excesses. Residents who have been recorded to channel their aspirations to the General Election are conducted at the polling station (TPS). Requirements for citizens to have the right to vote are regulated by law. The voter segment group is divided into three namely beginner voters, millennial voters and post menial voters. Where in the previous research the focus was on the scope of results, namely, predictions of legislative elections (C4.5 algorithm to predict the results of legislative selection of DKI Jakarta DPRD, Techno Nusa Mandiri Vol. IX No.1, March 2013). Here, we try to take the other side, which is related to voter participation in the democratic party process that takes place.

The involvement of the community for participation in the democratic process is determined by the presence of the community in the booth (TPS), where residents come to be able to vote. The absence of the public in the ELECTION process at polling stations was termed golput (white group) or indeed was not present at the time of the democratic party that was held. Many factors determine the absence of the public at the democratic party, among others, based on the age level that is identical to the millennial voters (ages 17 to 32), beginner voters (aged 17 to 20) who have no experience in a democratic party being held, apathy will representations that have been legally registered and other factors that generally allow the age level to not be able to participate in the democratic party this time.

The study took a sample of data in Wonokasian Village, Wonoayu District, Sidoarjo Regency, East Java Province at the 2019 ELECTION simultaneously in Indonesia. Classification method (Naïve Bayes algorithm, C4.5, and kNN), attribute analysis on attendance data and voter data is the focus of this research so that it can be known whether the determination of the effect of the prediction class is produced. Because there may be many attributes, but they have no influence on the prediction class (results) at all. Many attributes are irrelevant and will place unnecessary computational overhead on the data mining algorithm [1, 2]. At worst, it can cause the algorithm to be a poor outcome.

Classification methods in several algorithms are used in the implementation of research for quality attribute analysis to be tested on several different attributes [3, 4]. The hope is, can it provide a significant percentage change with the same number of instances.

2. Literature review

2.1. Classification

Classification method that is studying a set of data that can be produced by a new classification of data [5]. The classification process in data mining techniques is a set of data that can produce a classification model (target function). So, it requires a dataset on the set for the classification process. The dataset used is attributes and features using training data and testing data [6]. Classification techniques can be grouped into two categories. Namely, the classification technique globally calculates all training data and classification locally taking into account some training data [7].

2.2. C4.5 algorithm

At the learning stage of the data, the C4.5 algorithm constructs decision trees from training data [8, 9], in the form of cases or records (tuples) in the database. The three working principles of the C4.5 algorithm at the learning stage of the data are:

1. Making a decision tree.
2. The decision tree pruning and evaluation (optional).
3. Making rules from decision trees (optional).

The formula for the C4.5 algorithm is:

$$Gain(S,A) = Entropy(S) \sum_{i=1}^{n} \frac{Si}{S} * Entropy(Si) \tag{1}$$

where S is set of cases, A is attribute, n is number of partition attributes A, Si is number of cases on the i-partition, S is number of cases in S.

$$Entropy(S) \sum_{i=1}^{n} \frac{Si}{S} * pi * log2\, pi \quad (2)$$

where S is set of cases, A is feature, n is number of S partitions, pi is proportion of Si to S.

2.3. Naïve Bayes algorithm

Naïve Bayes algorithm is a classification method by calculating probabilities in determining the number of classes and values of a dataset [10]. The advantage of using the Naïve Bayes algorithm is that the small amount of training data can determine the required parameter estimates. Assume a simplification of the Naïve Bayes algorithm with attribute values that are mutually independent if given their output values. Naïve Bayes has a very strong level of accuracy and speed when applied to a database with large data [11]. Based on the Bayes theorem, which can classify the same method as the Decision Tree algorithm and Neural Network [12, 13]. In the Bayes theorem equation, the conditional probability is expressed as:

$$P(H|X) = \frac{P(X|H).P(H)}{P(X)} \quad (3)$$

Where,

X : Data with unknown classes.
H : Data hypothesis with specific classes.
P (H) : Probability of H. hypothesis
P (X) : Probability X.
P (H | X) : Posterior probability H with condition X.
P (X | H) : Posterior probability X with the condition H.

2.4. k nearest neighbor classification

The K-Nearest Neighbor (NN) is the simplest method of machine learning [13, 14]. It is a type of instance-based learning in which object is classified based on the closest training example in the feature space. It implicitly computes the decision boundary; however, it is also possible to compute the decision explicitly. So the computational complexity of K NN is the function of the boundary complexity. The k-NN algorithm is sensitive to the local structure of the data set. The special case when k = 1 is called the nearest neighbor algorithm. The best choice of k depends upon the data set; larger values of k reduce the effect of noise on the classification but make boundaries between classes less distinct. The various heuristic techniques are used to select the optimal value of K. KNN has some strong, consistent results. As the infinity approaches data, the algorithm is guaranteed to yield an error rate less than the Bayes error rate.

3. Methods

Diagram for determining how to select characteristics by determining attendance class (present and absent) in the election process and determine the number of attributes for the best value that results in a comparison of the three conditions for the number of attributes (Figure 1):

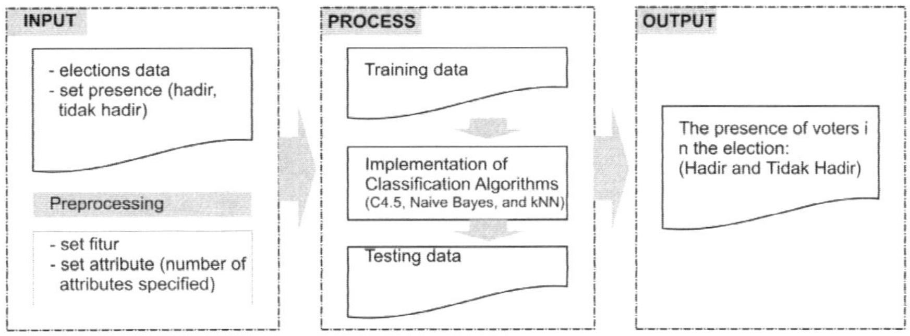

Figure 1. Classification diagram.

For the 'train and test' method, the available data is split into two parts called a training set and a test set (Figure 2). First, the training set is used to construct a classifier (decision tree, neural net, etc.). The classifier is then used to predict the classification for the instances in the test set. If the test set contains N instances of which C are correctly classified the predictive accuracy of the classifier for the test set is p = C/N. This can be used as an estimate of its performance on any unseen dataset. In cases where the dataset is only a single file, we need to divide it into a training set and a test set before using Method 1. This may be done in many ways, but a random division into two parts in proportions such as 1:1, 2:1, 70:30 or 60:40 would be customary.

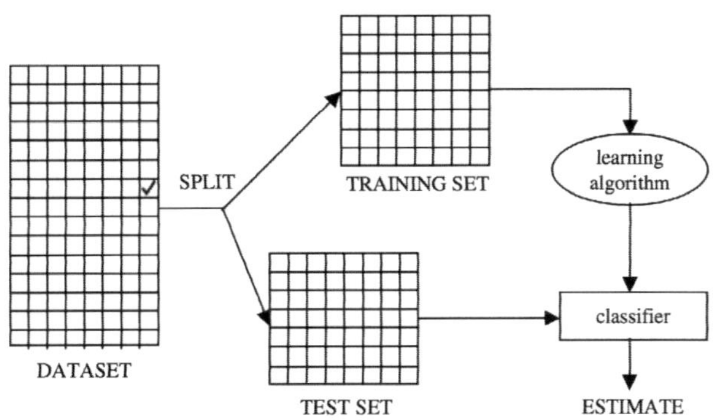

Figure 2. Train and test.

4. Data processing

Data warehouse originating from 2 data sources, namely C7 data (attendance list) and DP4 data (DPT data). From 2 references, a merger process is conducted (C7 data and DP4 data) based on the voter ID, to determine the presence of voters at the polling station. And changes in some attributes to produce new information, to produce better predictive results. Attributes from the results of the changes are the class, origin, category, and status of the KK. For the addition of a new attribute the location_TPS is raised based on the address (rt) of the voter and the name of the tips, so that it can know whether the location of the polling station is in the address "rt" or outside, this determines the distance from the voter to the polling station location.

The data warehouse has been set as many as 11 attributes, and the dataset is 4249 instances, then divided into two parts for further processing. Data train in the capacity of 70% (2973 instance) and test data of 30 (1276 instances). Selecting data that is dismissed information to produce the best value in the testing process.

4.1. Data collection

The dataset is obtained from the final voter list (DPT) in Wonokasian village, Wonoayu sub-district, Sidoarjo, East Java. DPT is used in the General Elections of the President, DPD, DPR, Provincial DPRD and Regency/ City DPRD in 2019. The set datasets are 4249 instances, and 11 attributes are taken from form C7 (voter attendance list). From these datasets scattered in 15 polling stations (TPS), where polling stations are the polling implementation unit Table 1.

Table 1. Dataset description.

Dataset	No. of attribute	No. of instance
Form C7 at 15 polling stations, in the election of the President, DPD, DPR, Provincial DPRD and Regency / City DPRD in Wonokasian village, Wonoayu District, Sidoarjo Regency	11	4249

Form C7 data is sourced from the List of Electoral Potential Population (DP4), where the data summarizes the three conditions, namely probability data, geographic data, and attendance data. Personal and geographical data are found in DP4 data and attendance data in form C7 data. The description of processing 2 data sources (form C7 and DP4) is Table 2:

Table 2. 11 Attribute and future dataset.

No	Attribute	Variable	Information
1	jmlKK	{dua,>dua,singgle}	Number of family members
2	asal	{SIDOARJO,LUAR,'LUAR PROV'}	Origin of voters
3	kategori	{Milenial,PascaMilenial,Pemula}	Type of society
4	kawin	{S,B,P}	Marriage Status (S = Already, B = Not yet and P = Ever)
5	jk	{P,L}	Gender (P = Female and L = Male)
6	alamat	{WONOKASIAN,LENGKONG,NGGODEK,NGEMPLAK,KRAMAT,NDOKOH,KRESAN,KLITIH}	Address Village and or hamlet in one village
7	rt	{rt01,rt02,rt03,rt04,rt05,rt06,rt07,rt08,rt09,rt10,rt11,rt12,rt13,rt14,rt16,rt15,rt17,rt19,rt18,rt20,rt21}	Address RT (Neighborhood) in one village
8	tps	{TPS-1,TPS-2,TPS-3,TPS-4,TPS-5,TPS-6,TPS-7,TPS-8,TPS-9,TPS-10,TPS-11,TPS-12,TPS-13,TPS-14,TPS-15,TPS-16}	Name of TPS (polling station) in one village
9	lokasi_TPS	{dalam,luar}	Distance between polling station locations and voter address
10	statusKK	{TSH,HS}	Attendance at polling stations in one family
11	hadir	{TH,H}	Individual attendance at TPS (class prediction)

4.2. *Ratting attribute*

To see the best weight of 11 attributes, it can be done by calculating the gain ratio, so that the best and worst attribute sequence can be known, namely (Table 3):

Table 3. Ranked attributes

Rangking	Gain Ratio	No	Attribute
1	0.390750	10	statusKK
2	0.054964	8	tps
3	0.036399	7	rt
4	0.018774	6	alamat
5	0.013883	9	lokasiTPS
6	0.004621	1	jmlKK
7	0.004580	4	kawin
8	0.001704	2	asal
9	0.001436	3	kategori
10	0.000126	5	jk

4.3. *Attribute modify*

Modification of attributes that are by the conditions in Table 4, is done to reduce the initial conditions which have several 11 attributes, then from the number 11 on some attributes not included in the testing process. This will be seen here, whether there is a change in the results of the steps for reducing the attribute that has been set. Starting with a crew condition of 11 attributes. Then ten attributes that try to delete one attribute (statusKK). In Table 3 "statusKK" ranks first regarding the quality attribute of the calculation of the gain ratio with a value of 0.390750 and for the condition of 9 attributes, delete two attributes, namely "rt and tps". Information from these two attributes, namely for "rt" is the location of the address of the voter domiciled and "tps" is the name of the place where the election is held. The quality of the two attributes ranks second and third after "statusKK".

Table 4. Three attribute conditions

No	11 attribute
1	jmlKK
2	asal
3	kategori
4	kawin
5	jk
6	alamat
7	rt
8	tps
9	lokasi_TPS
10	statusKK
11	hadir

No	10 attribute
1	jmlKK
2	asal
3	kategori
4	kawin
5	jk
6	alamat
7	rt
8	tps
9	lokasi_TPS
10	hadir

No	9 attribute
1	jmlKK
2	asal
3	kategori
4	kawin
5	jk
6	alamat
7	rt
8	statusKK
9	hadir

4.4. *Condition dataset with deferent attribute*

The dataset with condition class {TH, H} with a total of 4249 instances, training data 2973 instances and testing 1276 instances, displays datasets according to the discussion in Table 4, where changes directly

IOP Conf. Series: Materials Science and Engineering **821** (2020) 012034 doi:10.1088/1757-899X/821/1/012034

occur in each dataset presentation according to the attribute group used. Removal between "statusKK" for ten attribute categories and "rt and tps" for category nine attribute Tables 5, 6 and 7.

Table 5. Dataset, 11 attribute

No.	Data election
1	dua,SIDOARJO,Milenial,S,P,WONOKASIAN,rt01,TPS-1,dalam,TSH,TH
2	dua,SIDOARJO,Milenial,S,L,WONOKASIAN,rt01,TPS-1,dalam,TSH,H
3	>dua,LUAR,Milenial,B,L,WONOKASIAN,rt01,TPS-1,dalam,TSH,TH
4	singgle,SIDOARJO,Milenial,S,L,WONOKASIAN,rt01,TPS-1,dalam,HS,H
…	…

Table 6. Dataset, 10 attribute

No.	Data election
1	dua,SIDOARJO,Milenial,S,P,WONOKASIAN,rt01,TPS-1,dalam,TH
2	dua,SIDOARJO,Milenial,S,L,WONOKASIAN,rt01,TPS-1,dalam,H
3	>dua,LUAR,Milenial,B,L,WONOKASIAN,rt01,TPS-1,dalam,TH
4	>dua,LUAR,PascaMilenial,S,P,WONOKASIAN,rt01,TPS-1,dalam,TH
…	…

Table 7. Dataset, 9 attribute

No.	Data election
1	dua,SIDOARJO,Milenial,S,P,WONOKASIAN,dalam,TSH,TH
2	dua,SIDOARJO,Milenial,S,L,WONOKASIAN,dalam,TSH,H
3	>dua,LUAR,Milenial,B,L,WONOKASIAN,dalam,THS,TH
4	>dua,LUAR,PascaMilenial,S,P,WONOKASIAN,dalam,THS,TH
…	…

5. Testing data results

The results of the classification algorithm (C4.5, Naïve Bayes and kNN) to determine the prediction class "Hadir and Tidak hadir". The results of Table 8, explain the comparison of class predictions in both the number of attributes and algorithm groups. The highest value is in 11 attributes, namely, in the Naïve Bayes algorithm with a percentage value of 89.3417% or 1140 instances. In this analysis, try to see the results in the reduction of attributes. Subtraction is done for attributes that have the first ranking "statusKK" (10 attributes), second "TPS" and third rank "rt" (9 attributes). For the highest ranking of Gini ratio Table 3. For prediction class 9 attributes, where the results of the prediction class show the highest value of 3 algorithms namely C4.5 with a percentage of 99.6364% or 1131 instances, then the highest percentage prediction class for ten attributes is 10.7555 % or 1107 instances with algorithms that have the same nine attributes, namely C4.5 (Figure 3). The best ranking is based on two groups of attributes (9 attributes and ten attributes), namely on nine attributes.

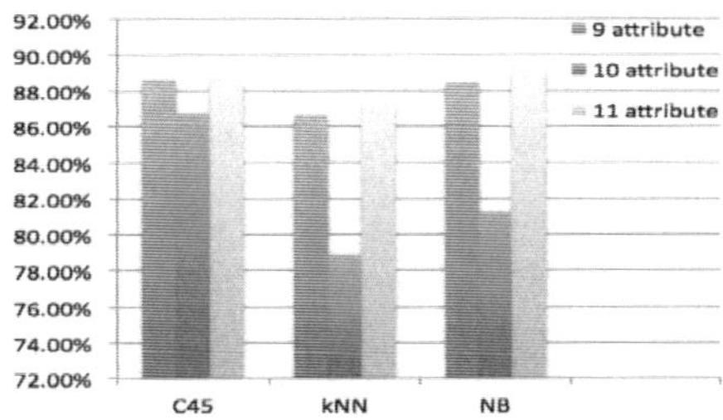

Figure 3. Graph of prediction class results.

Table 8. Result testing data

Attribute		C45		kNN		Naïve Bayes	
9 attribute	True	1131	**88.6364 %**	1106	86.6771 %	1129	88.4796 %
	False	145	11.3636 %	170	13.3229 %	147	11.5204 %
10 attribute	True	1107	**86.7555 %**	1007	78.9185 %	1037	81.2696 %
	False	169	13.2445 %	269	21.0815 %	239	18.7304 %
11 attribute	True	1132	88.7147 %	1114	87.3041 %	1140	**89.3417 %**
	False	144	11.2853 %	162	12.6959 %	136	10.6583 %

6. Conclusion

Analysis of three algorithms (C4.5, Naive Bayes and k-NN) and changes to attributes, which uses a dataset of 4240 instances which are divided into training data 2973 instances and testing 1276 instances. The presentation of the results obtained for the Naïve Bayes algorithm (for 11 attributes) ranks first for the prediction class from the attendance list of the President, DPD, DPR, Provincial DPRD and Regency / City DPRD in 2019, with a percentage of 89.3417%. But here try to look at the analysis for the reduction of the number of attributes, namely the group of 9 attributes (attributes that are reduced by "rt" and "TPS") and ten attributes (attributes that are reduced by "statusKK"). The results show that for nine attributes, it is better with the percentage value for the predictive class, which is 88.6364%. This shows that for the reduction of attributes with the highest gain ratio (statusKK) in the ten attributes, it has a significant effect, so the predictive value of the ten attributes is bad. The results of the attributes that are not deleted (11 attributes) and deleted attributes (10 and 9 attributes) indicate that the results of predictions are different, ie for the best not deleted is Naïve Bayes and the best-deleted algorithm is C4.5. In the end it can be seen that the weight level of an attribute for the value of the gain ratio affects the results of the predictions both on the percentage value of the prediction class and especially on the best value for the algorithm used.

Acknowledgments

We hereby thank you to Universitas Muhammadiyah Sidoarjo for supporting the publication of this research.

IOP Conf. Series: Materials Science and Engineering **821** (2020) 012034 doi:10.1088/1757-899X/821/1/012034

References

[1] Bramer M 2016 *Data for Data Mining:In Principles of Data Mining* London: Springer.
[2] Zhang J, Williams S O and Wang H 2018 Intelligent computing system based on pattern recognition and data mining algorithms *Sustain. Comput. Informatics Syst.*vol **20** pp 192–202.
[3] Xie J and Burstein F 2011 Using machine learning to support resource quality assessment: An adaptive attribute-based approach for health information portals *Lect. Notes Comput. Sci. (including Subser. Lect. Notes Artif. Intell. Lect. Notes Bioinformatics)* vol **6637** LNCS pp 526–37.
[4] Nikam S S 2017 A comparative study of classification techniques in data mining algorithms *Int. J. Mod. Trends Eng. Res.* vol **4** no 7 pp 58–63.
[5] Xia G S, Hu J, Hu F, Shi B G, Bai X, Yanfei Z, Lu X, and Zhang L 2017 AID: A benchmark data set for performance evaluation of aerial scene classification *IEEE Trans. Geosci. Remote Sens.* vol **55** no 7 pp 3965–81.
[6] Novaković J, Strbac P and Bulatović D 2011 Toward optimal feature selection using ranking methods and classification algorithms *Yugosl. J. Oper. Res.* vol **21** no 1 pp 119–35.
[7] Rosid M A, Gunawan G, and Pramana E 2015 Centroid based classifier with TF – IDF – ICF for classfication of student's complaint at appliation e-complaint in Muhammadiyah University of Sidoarjo vol **1** no 1.
[8] Abdallah I, Dertimanis V, Mylonas C, Tatsis K, Chatzi E, Dervilis N, Worden K and Maguire A E 2018 Fault diagnosis of wind turbine structures using decision tree learning algorithms with big data *Saf. Reliab. - Safe Soc. a Chang. World - Proc. 28th Int. Eur. Saf. Reliab. Conf. ESREL 2018* no December pp 3053–62.
[9] Yang Y and Chen W 2016 Taiga: Performance optimization of the C4.5 decision tree construction algorithm *Tsinghua Sci. Technol.*vol **21** no. 4 pp 415–25.
[10] Rashmi G D, Lekha A, and Bawane N 2016 Analysis of efficiency of classification and prediction algorithms (Naïve Bayes) for Breast Cancer dataset *2015 Int. Conf. Emerg. Res. Electron. Comput. Sci. Technol. ICERECT 2015* pp 108–13.
[11] Ginting S L B, Adler J, Ginting Y R, and Kurniadi A H 2018 The development of bank application for debtors selection by using Naïve Bayes Classifier technique *IOP Conf. Ser. Mater. Sci. Eng.*, vol **407** no 1.
[12] Balasaravanan K and Prakash M 2018 Detection of dengue disease using artificial neural network based classification techniquetion *Int. J. Eng. Technol.*vol **7** no 1 pp 13–5.
[13] Jadhav S D and Channe H P 2016 Comparative study of K-NN, Naive Bayes and Decision Tree Classification techniques *Int. J. Sci. Res.* vol **5** no 1 pp 1842–45.
[14] Pouriyeh S, Vahid S, Sannino G, De Pietro G, Arabnia H, and Gutierrez J 2017 A comprehensive investigation and comparison of Machine Learning techniques in the domain of heart disease *Proc. - IEEE Symp. Comput. Commun.* no. Iscc pp. 204–7.

IOP Conf. Series: Materials Science and Engineering **821** (2020) 012035 doi:10.1088/1757-899X/821/1/012035

Application of fuzzy preference relations method in AHP to improve judgment matrix consistency

M Hamka* and Harjono

Department of Informatics, Universitas Muhammadiyah Purwokerto, Purwokerto, Indonesia

*Email: muhammadhamka@ump.ac.id

Abstract. Most issues when using AHP is the difficulty of getting the consistency of a judgment matrix. It can be affected due to decision-making subjectivity. This study proposes the usage of the fuzzy Preference Relations (FPR) method to ensure a judgment in decision-making is acceptable and high objectively due to its obstacle for satisfying consistent judgment matrix from decision-makers in AHP. Improving the consistency of decision matrix can be done through the following steps, namely: 1) Transform pairwise comparison matrix of $A=(a\_ij)$ into FPR $R=(r\_ij)$, 2) construct a new consistency matrix of FRP which satisfies the additive consistency, and 3) compose a new consistent pairwise matrix. As a result, the consistency index value was acceptable (< 0.10). Moreover, the procedure is applied in the determination of agro-industries based SMEs development.

1. Introduction

Most of the land in Banyumas Region, which is part of Central Java Province, Indonesia, is reserved for the agricultural field with an area of about 93.864 Ha or more than 70 percent of the total area in Banyumas Region [1]. In addition, Banyumas Region becomes one of the priorities for an agricultural area in Central Java Province [2]. According to land utility in Banyumas Region, one of the priorities for the development of SMEs in Banyumas Region is the industrial development of agricultural products. [3]. In 2018, The Department of Industry, Trade, and Cooperation of Banyumas region were targeting 300 SMEs would be assisted [3].

The issue of determination of SME's assistance, namely the large of data and the vast number of parameters need to be judged. Regarding this problem, The Decision Support System (DSS) may be performed as a useful tool to assist decision-makers. One of the DSS models for supporting decision making in terms of establishment of agro-industries based SMEs development is the Analytical Hierarchy Process (AHP). AHP is a useful method of decision making, especially when it involves several attributes and preferences [4], which is enable decision-makers to simplify the problem into a hierarchical structure based on human being's perceptions [5]. Furthermore, Ho and Xin (2017) stated that AHP is ease of understanding, clearness, and highly adaptive [6].

In connection with its advantages, AHP is the widely employed decision-making approach that is used in various fields such as education [7] for supporting in the ranking ordering of scholar prospective and deciding the proper teaching process in higher learning institution [8]. Additionally, AHP was utilized for appraising the stage of information technology management that currently used and level of expected information technology management, which involved in the next phase [9], multi-criteria decision making (MCDM) methods was also applied for specifying the most performing information system [10]. On the other hand, Wang et al. (2017) have studied the matching agricultural supply chain data in order to obtain suitable production and marketing models based on AHP [11].

Despite the benefits of AHP, at most times, there is a challenge to satisfy consistent judgment matrix from decision-makers [12] due to lack of information during criteria prioritization or subjective aspects of decision-makers in qualitative assessments are often inappropriate [13]. Moreover, when a consistent state of decision maker's perceptions fails to be achieved, then it has to correct the judgment matrix to obtain acceptable consistency index value, which is less than or equal to 0.1 [14]. It may lead to many problems, such as time-consuming, or requires a fairly large size in the calculations in decision making [12]. Besides that, the judgment of priority criteria will be influenced by the subjectivity of decision-makers [12].

This study aims a method that can be utilized for enhancing properly consistency index (CI) value of a received valuation matrix so that a generated decision matrix represents prioritization based on empirical situations and fulfills highly level of objectivity. According to [15], one approach for improving the consistency of judgment matrix in the AHP is the usage of Fuzzy Preference Relations (FPR). It will transforms pairwise comparison matrix of $\boldsymbol{A} = (\boldsymbol{a_{ij}})$ into FPR matrix of $\boldsymbol{R} = (\boldsymbol{r_{ij}})$ [15]. Based on the matrix of $\boldsymbol{R} = (\boldsymbol{r_{ij}})$, it may construct a new consistency matrix of $G = (g_{ik})$, which satisfies the additive consistency $G_{ik} = g_{ij} + g_{jk} - \frac{1}{2}$ where $1 \leq i \leq n, 1 \leq j \leq n, and\ 1 \leq k \leq n$ [16]. Then, a new consistent pairwise matrix is obtained depending on $G = (g_{ik})$ [15].

2. Literature review

2.1. Analytical Hierarchy Process

The Analytic Hierarchy Process (AHP) is a decision-making technique to determine the best from several alternatives based on several assessment criteria. It's invented to deal with human sensibility and impulse [14]. In using the AHP model, there are several steps as follow [12]:

a. Simplify the purpose of decision making by selecting the proper appraisal attributes and alternatives into relevant hierarchies.
b. Compare the priority of each criterion according to the notion of decision-makers by arranging a pairwise comparison matrix. The priority of the assessment criteria uses a scale, as shown in Table 1 [14].

Table 1. Fundamental scale for pairwise comparison

Intensity of importance	Definition	Explanation
1	Equal importance	Two activities contribute equally to the objective
3	Moderate importance	Experience and judgment slightly favor one activity over another
5	Strong importance	Experience and judgment strongly favor one activity over another
7	Very strong or demonstrated importance	An activity is favored very strongly over another; its dominance demonstrated in practice
9	Extreme importance	The evidence favoring one activity over another is of the highest possible order of affirmation
2,4,6,8	notion between two proximate values	This value is specified when options are adjacent. For instance, four is given in-between 3 and 5
Reciprocals of above	If activity *i* has one of the above nonzero numbers assigned to it when compared with activity *j*, then *j* has the reciprocal value when compared with *i*	

c. Calculate the priority vector of each decision element.
d. Check the consistency of the judgment matrix, which constructed in step 2 above. If the consistency fails to be obtained, refine the relevant pairwise comparison matrix. The consistency index of the judgment matrix can be measured by formula, as shown in Equation (1) [14].

$$CI = \frac{(\lambda \max - n)}{(n-1)} \tag{1}$$

where :
CI: Consistency Index
λ max: Maximum eigenvalue
n: Number of criteria

The next phase is checking consistency, which represented by consistency ratio (C.R). C.R is obtained by calculating the differences between CI and one of the corresponding values from Table 2 Equation (2) [14]. If it is less or equal than 0.10, then the estimate of the matrix is acceptable. Otherwise, attempt to improve consistency.

$$C.R = \frac{CI}{R.I} \tag{2}$$

Table 2. Random Consistency Index (r.i)

N	1	2	3	4	5	6	7	8	9	10
Random consistency index (R.I.)	0	0	0.52	0.89	1.11	1.25	1.35	1.40	1.45	1.49

e. Finally, conduct absolute measurement of each alternative related to criteria to achieve the ranking of alternatives

2.2. Fuzzy preference relations
The Fuzzy Preference Relations (FPR) method is used to induce a pairwise comparison matrix in order to gain adequate consistency measures in AHP [15]. Consistency improvement can be made through three-steps as follow:

a. Transform pairwise comparison matrix of $\boldsymbol{A} = (\boldsymbol{a_{ij}})$ into FPR $\boldsymbol{R} = (\boldsymbol{r_{ij}})$ using the following function (Equation (3)) [15].

$$r_{ij} = g(a_{ij}) = \frac{1}{2}(1 + \log_9 a_{ij}) \tag{3}$$

b. According to Equation (2), we can construct a new consistency matrix of FRP which satisfies the additive consistency [16], where:

$$g_{ik} = \left(\frac{1}{n} \sum_{j=1}^{n} (r_{ij} + r_{jk}) - \frac{1}{2} \right) \tag{4}$$

c. A new consistent pairwise matrix based on additive consistency matrix (Equation (4)) can be acquired by Equation (5) [15]:

$$a_{ij} = 9^{2(g_{ik} - 0.5)} \tag{5}$$

3. Methodology

The aim of this study is utilizing the Analytic Hierarchy Process (AHP) method to work out issues associated with agroindustry-based SMEs in Banyumas Region, Central Java Province, Indonesia. This research is an attempt to apply the AHP technique to offer a recommendation for the empowerment of agroindustry-based SMEs. The second stage is determining the assessment criteria. The set of criteria were adapted from DSS application that has been developed by Ministry of Cooperation and Small and Medium Enterprises of The Republic Indonesia [17], then modified based on the condition of SMEs in Banyumas region namely 1) the level of technology is used in the

IOP Conf. Series: Materials Science and Engineering **821** (2020) 012035 doi:10.1088/1757-899X/821/1/012035

cultivation process, 2) number of cultivation phases, 3) market segmentation, 4) advertisement media, 5) marketing area, and 6) availability of raw materials. Definitions of each criterion are shown in Table 3. The third step is comparing the priority of each criterion based on 1 to 9 scale (Table 1). It will compose a pairwise comparison matrix of $\boldsymbol{A} = (\boldsymbol{a_{ij}})$.

Table 3. Defining each criterion

Criteria	Description
The level of technology is used in the cultivation process (P1)	The level of technology used, whether production is done entirely by humans, production is partly done by humans and partly supported by machines, or the overall production is carried out by modern technology.
Number of cultivation phases (P2)	Stage of the cultivation process, whether it requires a lot of paths and each path takes several days, each path takes less than one day, or a relatively small path and each path takes less than one day.
Market segmentation (P3)	SMEs have plans for consumer segmentation such as age, gender, economic status and geography.
Advertisement media (P4)	Advertisement media are used by SMEs such as newspapers, magazines, radio advertisements, television advertisements on, or advertisements on the Internet (social media).
Marketing area (P5)	Does the sales area covers only in the region, in the province, within the country, or abroad.
Availability of raw materials (P6)	The level of availability of raw materials for cultivation is always rare, available at certain times, quite available, or widely available.

After constructing the pairwise comparison matrix, the next phase is computed priority vector (w) or priority significance of each criterion by counting any column in a row based on normalization of judgment matrix and then divide them with the number of criteria. In order to find the eigenvalue (λ), which is needed inconsistency assessment, is carried out by multiplying between judgment matrix and normalized eigenvector. In addition, to verify the consistency vector (CV), divide each of the eigenvalues with priority vector (w). Based on eigenvalue (λ), the maximum eigenvalue (λ_{max}) can be calculated using Equation (6) [14].

$$\lambda_{\max} = \frac{CV_1 + CV_2 + .. + CV_n}{n} \tag{6}$$

Regarding the maximum eigenvalue (λ_{max}) on Equation (6), it is necessary to perform consistency measurement $(CI\ and\ C.R)$ of matrix $(\boldsymbol{A})$ to make sure that judgment in decision making is acceptable and high objectively. The CI (*Consistency Index*) and the CR (*Consistency Ratio*) are then gained by equation 1, and by comparing CI with corresponding the random matrix (Table 2). The value of the R.I (Random Index) is set as about 1.25 (for n = 6 criteria). When the number of C.R is not adequate (more than 0.1), it will carry out the FPR method for improving consistency. The FPR method in this study is carried out sequentially using Equation (3), (4), and (5) [15, 16]. They both last stages are priority ranking alternatives toward each of the criteria and conducting the addition of all weights to gain the best alternative. The sequences of the research methodology are described in the form of a flowchart diagram, as shown in Figure. 1.

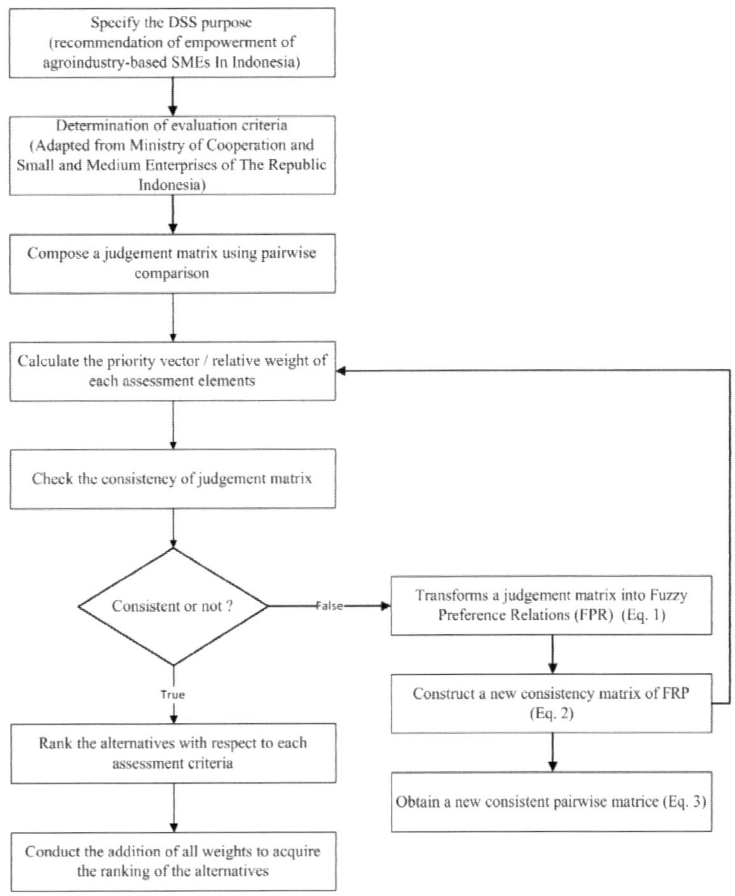

Figure 1. The flowchart of research methodology

4. Results and discussion

4.1. Judgment scales of decision elements

According to the priority determination of each criterion from decision-makers, it can be composed of a judgment matrix presented in Table 4. In addition, Table 5 represents the normalization matrix, which is collected from the decision scale that has been compared by the addition of each column's grades. As a result, the priority significance of each criterion can be achieved.

Table 4. Pairwise comparison

A	P1	P2	P3	P4	P5	P6
P1	1	3	1/5	5	1/5	1/5
P2	1/3	1	3	4	4	1/5
P3	5	1/3	1	2	2	1/5
P4	1/5	1/4	1/2	1	1/3	1/5
P5	5	1/4	1/2	3	1	1/5
P6	5	5	5	5	5	1
Total	16.53	9.83	10.20	11.00	12.53	2.00

Table 5. Normalization and priority vector of each criteria

A	P1	P2	P3	P4	P5	P6	Weight
P1	0.06	0.31	0.02	0.45	0.02	0.10	0.16
P2	0.02	0.10	0.29	0.36	0.32	0.10	0.20
P3	0.30	0.03	0.10	0.18	0.16	0.10	0.15
P4	0.01	0.03	0.05	0.09	0.03	0.10	0.05
P5	0.30	0.03	0.05	0.27	0.08	0.10	0.14
P6	0.30	0.51	0.49	0.45	0.40	0.50	0.44

4.2. Assess consistency of judgment matrix

With regard to fulfilling decision matrix consistency assesment, it is necessary to count the maximum eigen value (Equation (6)) as well as consistency index (CI) (Equation (1)). The marks for both (λ_{max}) and CI are (λ_{max}) = $\mathbf{8.33}$ and CI = 0.467. So, the consitency ratio (CR) can be computed using Equation (2).

$$C.R = \frac{0.467}{1.25} = 0.373$$

Because of CR > 0.10, then the judgment matrix was considered as inconsistent. Hence, it need to change into FPR $\boldsymbol{R} = (\boldsymbol{r}_{ij})$ matrix (Eq. 3). A FPR $\boldsymbol{R} = (\boldsymbol{r}_{ij})$ matrix is as follow :

$$R_{ij} = \begin{bmatrix} 0.50 & 0.75 & 0.13 & 0.87 & 0.13 & 0.13 \\ 0.25 & 0.50 & 0.75 & 0.82 & 0.82 & 0.13 \\ 0.87 & 0.25 & 0.50 & 0.66 & 0.66 & 0.13 \\ 0.13 & 0.18 & 0.34 & 0.50 & 0.25 & 0.13 \\ 0.87 & 0.18 & 0.34 & 0.75 & 0.50 & 0.13 \\ 0.87 & 0.87 & 0.87 & 0{,}87 & 0.87 & 0.50 \end{bmatrix}$$

In order to obtain a new consistency matrix of $\boldsymbol{R} = (\boldsymbol{r}_{ij})$ which satisfies the additive consistency, transform matrik $\boldsymbol{R} = (\boldsymbol{r}_{ij})$into $\boldsymbol{R} = (\boldsymbol{r}_{ik})$ matrik using Equation (4). A $\boldsymbol{R} = (\boldsymbol{r}_{ik})$ matrix shown as follows

$$R_{ik} = \begin{bmatrix} 0.52 & 0.38 & 0.41 & 0.66 & 0.46 & 0.11 \\ 0.62 & 0.50 & 0.53 & 0.79 & 0.58 & 0.24 \\ 0.59 & 0.47 & 0.50 & 0.75 & 0.55 & 0.21 \\ 0.34 & 0.21 & 0.25 & 0.50 & 0.29 & -0.05 \\ 0.54 & 0.42 & 0.45 & 0.71 & 0.50 & 0.16 \\ 0.89 & 0.76 & 0.79 & 1.05 & 0.84 & 0.50 \end{bmatrix}$$

From the new consistency matrix of $\boldsymbol{R} = (\boldsymbol{r}_{ik})$, it can construct A new consistent pairwise matrix $\overline{\boldsymbol{A}} = (\boldsymbol{a}_{ij})$ based on Equation (5). A new consistent pairwise matrix $\overline{\boldsymbol{A}} = (\boldsymbol{a}_{ij})$ appear in Table 6.

Table 6. A new pairwise comparison

$\bar{A}$	P1	P2	P3	P4	P5	P6
P1	1.00	0.58	0.67	2.04	0.83	0.18
P2	1.73	1.00	1.16	3.53	1.43	0.32
P3	1.49	0.86	1.00	3.05	1.24	0.27
P4	0.49	0.28	0.33	1.00	0.41	0.09
P5	1.21	0.70	0.81	2.47	1.00	0.22
P6	5.44	3.15	3.64	11.10	4.50	1.00
Total	11.37	6.58	7.61	23.18	9.40	2.09

With respect to phases in AHP, after pairwise matrix was constructed, the next step is conducting normalization matrix to obtain the priority significance of each criteria and validate consistency measurement ($\boldsymbol{CI\ and\ C.R}$) of matrix ($\bar{\boldsymbol{A}}$). A new normalization matrix ($\bar{\boldsymbol{A}}$) was revealed in Table 7.

Table 7. A new normalization and priority vector of each criteria

($\bar{A}$)	P1	P2	P3	P4	P5	P6	Weight
P1	0.09	0.09	0.09	0.09	0.09	0.09	0.09
P2	0.15	0.15	0.15	0.15	0.15	0.15	0.15
P3	0.13	0.13	0.13	0.13	0.13	0.13	0.13
P4	0.04	0.04	0.04	0.04	0.04	0.04	0.04
P5	0.11	0.11	0.11	0.11	0.11	0.11	0.11
P6	0.48	0.48	0.48	0.48	0.48	0.48	0.48
Total	1.00	1.00	1.00	1.00	1.00	1.00	1.00

Regarding Table 7, the marks for both ($\boldsymbol{\lambda_{max}}$) and CI are ($\boldsymbol{\lambda_{max}}$) = **6** and CI = 00. So, the value of consitency ratio (CR) is 0. Because of CR < 0.10, then the judgment matrix was considered as consistent. With respect to analyses rating (Table 6) and calculating relative weight (Table 7), availability of raw materials (48%) is 3 times more important than number of cultivation phases (15%) and market segmentation (13%). Availability of raw materials was most critical criterion in the sustainability of agro-industrial SMEs, since it was the basic substances that is used in production. Furthermore, it was one of significant acquiring activity in supply chain [18] because naturally it depends on several elements, namely: recycling rate of a raw material, climate change (7.21), agricultural land occupation (0.39), and natural land transformation (1.02) [19]. In addition, its 3 times more important than number of cultivation phases (15%) and market segmentation (13%). The two lowest criteria are the level of technology is used in the cultivation process (9%) and advertisement media (4%). Comparison of the importance degree of criteria are revealed as a pie chart in Figure. 2.

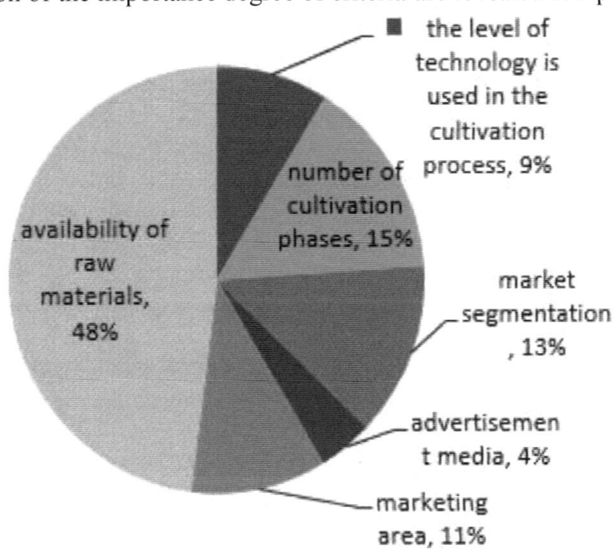

Figure 2. Priority weight scale of 6 criteria

4.3. Compare the alternatives with respect to each criterion

The degree of importance of five alternatives with respect to each associated criteria is gained through pairwise comparison. Similarly, it needs to perform six pairwise comparisons (n = 6). The values in Table 8 are obtained from the weight calculation of each alternative to each criterion. The five alternatives are paired comparisons with each criterion P1, P2, P3, P4, P5, and P6 according to the comparison scale in Table 1. The comparison results are then normalized through the division of each value by the number of columns. Then the normalization results are summed and divided by the number of criteria to get the priority weights of each alternative to each criterion as shown in Table 8. Alternative weights 1 against criteria P1 = 0.32, weights against criteria P2 = 0.33, weights against criteria P3 = 0.51, weights with respect to criterion P4 = 0.21, weight for criterion P5 = 0.29, weight for criterion P5 = 0.27. And so on for alternative 2 to alternative 5. The results of the priorities of the alternatives shown in Table 8.

In accordance with AHP rules (step c and d). The results of pairwise comparison normalization were carried out consistency testing using Equation (1) and (2). Based on the sum of each normalized column, all of them are 1, then it can be ensured that the results of the pairwise comparison of all alternatives to each criterion are consistent.

Table 8. Importance degree of each alternative towards criteria

	Weight					
	P1	**P2**	**P3**	**P4**	**P5**	**P6**
Alternative 1	0.32	0.33	0.51	0.21	0.29	0.27
Alternative 2	0.10	0.25	0.19	0.11	0.16	0.09
Alternative 3	0.29	0.18	0.10	0.37	0.29	0.09
Alternative 4	0.11	0.11	0.10	0.21	0.10	0.27
Alternative 5	0.18	0.12	0.10	0.11	0.14	0.27

4.4. Rank the best alternatives

Finally, the ranking of alternatives is achieved from the sum of the multiplication of each alternative weight (Table 8) with the weight of the corresponding criteria (Table 7), as shown in Table 9. The best alternative is the one that has the most significant weight results, which is alternative 1. On the other hand, it could be inferred that alternative 1 is strongly recommended as a SMEs would be assisted due to higher degree aspects such the level of technology is used in the cultivation process (P1), number of cultivation phases (P2), advertisement media (P3), and marketing area (P5) than other five alternatives. In addition, Whereas the second recommendation is alternative 2, which has a higher weight value on the two criteria with the second and third highest weights, namely number of cultivation phases (P2) and market segmentation (P3) compared to alternatives 3,4, and 5.

Table 9. Ranking of alternatives

Ranking	**Alternatives**	**Overall Weight**
1	Alternative 1	0.20
2	Alternative 5	0.17
3	Alternative 4	0.16
4	Alternative 3	0.12
5	Alternative 2	0.08

5. Conclusions

In this research, AHP has been implemented for suggesting agroindustry-based SMEs that would be encouraged. Moreover, it proposed a consistency enhancement procedure for the pairwise comparison matrix. As a result, there is priority vector refinement (See. Table 7) since the inadequate consistency index. In order to fulfill CR < 0.10, it proposed several procedures, namely 1) transformation pairwise comparison matrix fuzzy preference relations matrix $\boldsymbol{R} = (\boldsymbol{r_{ik}})$ (Eq. (3)), 2) compose a new consistency matrix of FRP which satisfies the additive consistency (Eq. (4)). Further, a matrix $\boldsymbol{R} = (\boldsymbol{r_{ik}})$ satisfies the additive consistency fuzzy preference relations where $\boldsymbol{r_{ik}} = \boldsymbol{r_{ij}} + \boldsymbol{r_{jk}} - \boldsymbol{0,5}$) for $\boldsymbol{i} \in \{\boldsymbol{1,2,\ldots,6}\}, \boldsymbol{j} \in \{\boldsymbol{1,2,\ldots,6}\}$ and $\boldsymbol{k} \in \{\boldsymbol{1,2,\ldots,6}\}$, and 3) obtain a new consistent pairwise matrix $\overline{\boldsymbol{A}} = (\boldsymbol{a_{ij}})$ (Eq. (5)) [16].

The offered procedure above has proven to improve the consistency of judgment matrix as the results of $\boldsymbol{\lambda_{max}} = \boldsymbol{6}$ and $\boldsymbol{C.R} = \boldsymbol{0}$ that related to matrix$(\overline{\boldsymbol{A}})$. From matrix$(\overline{\boldsymbol{A}})$, the availability of raw materials (C1) considered the most significant criterion (48%). Then, an alternative one is the best SMEs would be assisted because it has a larger amount in the level of technology is used in the cultivation process (C1), the number of cultivation phases (C2), advertisement media (C3), and marketing area (C5).

References

[1] Banyumas region 2014 *Development Planning Agency at Sub-National Level of Banyumas Region*.
[2] Local regulation of Central Java Province no. 6 2010 *Concerning The Central Java Province Spatial Plan Year 2009 - 2029* Central Java Province.
[3] Strategic plan of regional work unit department of industry, trade and cooperation of Banyumas region 2013-2018 *Department of Industry, Trade and Cooperation of Banyumas region*.
[4] Muslim E, Riansa I and Komarudin 2017 *International Journal of Technology* vol **8** 1356-60.
[5] Zhu B and Xu Z 2014 *European Journal of Operational Research* vol **239** pp 794-801.
[6] Ho W and Xin M 2017 *European Journal of Operational Research* vol **267** pp 399-414.
[7] Rianto, Setyohadi D B and Suyoto 2017 AHP-TOPSIS on selection of new university students and the prediction of future employment in *1st International Conference on Informatics and Computational Sciences (ICICoS)-2017* IEEE Conference Proceedings pp 126-30.
[8] Sahroni T R and Ariff H 2016 Design of Analytical Hierarchy Process (AHP) for teaching and learning in *11th International Conference on Knowledge, Information and Creativity Support Systems (KICSS)*-2016 IEEE Conference Proceedings pp 1-4.
[9] Yudatama U and Sarno R 2015 Evaluation maturity index and risk management for IT governance using fuzzy AHP and fuzzy TOPSIS (Case study Bank XYZ) in *International Seminar on Intelligent Technology and Its Applications-2015* IEEE Conference Proceedings pp 323-27.
[10] Daghouri A, Mansour K and Qbadou M 2018 *International Journal of Advanced Computer Science and Applications (IJACSA)* vol **9** pp 291-97.
[11] Wang D, Zhang C, Chen T and Tang Y 2017 Research on the production and marketing match model of agricultural products based on Analytic Hierarchy Process in *10th International Symposium on Computational Intelligence and Design (ISCID)*-2017 IEEE Conference Proceedings pp 8 – 12.
[12] Xu Q and Xiong M 2017 A method for improving consistency of judgment matrix in the AHP," in *2nd Asia-Pacific Conference on Intelligent Robot Systems (ACIRS)-2017* IEEE Conference Proceedings pp 83-7.
[13] Ennaceur A, Elouedi Z and Lefevre E 2015 Qualitative AHP models under the Belief Function Framework in *15th International Conference on Intelligent Systems Design and Applications (ISDA)-2015* IEEE Conference Proceedings pp 128-33.
[14] Saaty T L, Vargas L G 2012 *Models, Methods, Concepts & Application of the Analytic Hierarchy Process* New York Springer.

IOP Conf. Series: Materials Science and Engineering **821** (2020) 012035 doi:10.1088/1757-899X/821/1/012035

[15] Fedrizzi M and Brunelli M 2010 *Soft Computing* vol **14** 639-45.

[16] Lee L W 2012 *Expert Systems with Applications* vol **39** 11666-76.

[17] State Minister 2017 *Manual Instructions of Application Decision Support System (DSS) for Feasibility Study SMEs and Cooperative* for Cooperatives Small and Medium Enterprises of Republik Indonesia Retrieved June 6, 2017 (https://smecda.com).

[18] Serin H, Sahin Y, Durgun M, Ceyhan M and Uluca H 2016 A research on raw material problem of small and medium-sized furniture enterprises in Mersin city *14th International Scientific Conference on Economic and Social Development* May 2016 Varazdin Development and Entrepreneurship Agency Varazdin, Croatia pp 373-77.

[19] Bensch S, Kolotzek C, Helbig C, Thorenz A and Tuma A 2015 Decision support system for the sustainability assessment of critical raw materials in SMEs *48th Hawaii International Conference on System Sciences* January 2015 Conference Publishing Services pp 846-55.

Classification of EEG signals uses the coefficient of wavelet transform and K-nearest neighbor

H Hindarto[1,*], A Muntasa[2] , A Efiyanti[3]

[1,3] Department of Informatics Engineering, Faculty of Sains and Technology, Universitas Muhammadiyah Sidoarjo, Indonesia.
[2] Informatics Engineering Department, Faculty of Engineering, Universitas Trunojoyo Madura, Indonesia

*Email: hindarto@umsida.ac.id

Abstract. The research that was built was used to explain the application of Electroencephalography (EEG) signal waves. EEG data is used to move the cursor up and the cursor down. Characterization of each EEG signal uses the Wavelet method taken at each the subband of the wavelet process. Wavelet process by taking the average value and the standard deviation value of the wavelet coefficient. The average value and the standard deviation value is used as an EEG feature. K-Nearest Neighbor is used as an identification whether the cursor will move up or vice versa. This study uses of 100 EEG signal data consisting of 50 test data and 50 testing data. The accuracy of identification uses the 80% K-NN method.

1. Introduction
A keyboard or mouse is usually used by someone to move the cursor on a computer screen. Someone who has physical limitations, especially people who do not have hands and feet will be difficult to run computer applications using a keyboard or mouse. Researchers from both inside and outside the country have the idea of overcoming physical limitations by not being able to run computer applications, including using the signal brain that is coiled on the scalp without opening the brain skull. The head brain that is recorded will produce an electrical signal whose current is very weak. This head brain recording is called the Electroencephalography (EEG) signal. The first researcher who tried to produce this EEG signal was German Hans Berger. one of the tools used to communicate with humans is the Brain Computer Interface (BCI), this tool is used to communicate through brain signals. so with this tool, humans can control other equipment by just thinking. BCI is a system that obtains and analyzes nerve signals to create direct communication channels between the brain (EEG signals) and computers. With the BCI tool, the relationship between humans and other equipment does not require the activity of muscle strength [1]. An order from someone to electronic equipment can be responded to using brain signals through the BCI system [2]. Many uses of BCI include playing games through cellphones [3].

The EEG signal is obtained from recording sensors that are affixed to the head and can be used to communicate directly with other equipment through the BCI system. The usefulness of BCI is very much, by using the mind the human brain can control any object or tool [2-4]. To identify the EEG signal is a separate problem, both the method of finding EEG signal characteristics, the method of EEG signal classification, even the problem of the design of inputs and equipment used for applications of BCI systems. Various attributes are used as input of EEG signals for communication with BCI, including alpha rhythm, beta rhythm or activities related to a BCI device. For example BCI by using P300 [2-4].

Research on identifying epilepsy taking EEG signals and the method used is the Fourier Transform method as feature extraction and Neural Network as a classification has been carried out [5]. Research using a combination of slow cortical potential (SCP) and wavelet packet transformation as a feature extraction and classification using the Artificial Neural Network method, so as to obtain an accuracy of 91.47% [6]. EEG signal classification by obtaining an accuracy of 90.80% using the wavelet packet decomposition method as cirri extraction and nerve tissue as an identification process has been carried out [7].

This research tries to find the feature extraction method and the EEG signal classification method. The method used is the DWT method in the process of finding feature extraction and the K-NN method in the classification process. The EEG signal used is the EEG signal whose recording process imagines the cursor movement up and the cursor movement down.

2. Data and methods

2.1. Material

This research uses data from the 2003 BCI competition, the data from the 2003 BCI competition is an EEG signal recording conducted by Dr. Birbaumer and his team at the University of Tuebingen, Germany (Blankertz 2004). EEG signal data is obtained from recording healthy subjects consisting of six electrode sensors that are affixed to the scalp. In the process of recording the subject is asked to imagine the cursor moving up and down. The recording process lasts for 3.5 seconds and is sampled at a frequency of 256 Hz. This trial data set consisted of 268 trial data and 293 test data [8]. In this study EEG signal data taken in the research process were only taken 50 test data and 50 test data.

2.2. Discrete Wavelet Transformation (DWT)

DWT is one of the feature extractions used in the polo recognition process, which aims to reduce the input dimensions. Input data that is too large can result in a long process, so that the necessary representation of features in this case the representation of these features does not reduce the information needed. The selection of features must be done carefully and accurately and can be used as expected information. Many studies that have tried to find suitable feature extraction for EEG signals, one of them is the use of feature extraction using wavelet transforms can be found in the following studies [9, 10].

The representation of EEG signals in the time domain is still imperfect when using representations using the frequency domain. By using the frequency domain, it provides a good description in non-stationary messages, one of which uses wavelet transforms. Wavelet transform is more suitable for feature extraction for EEG signals [11]. Feature extraction using wavelet transform can represent good resolution for low frequency and high frequency. So the scale owned by wavelet can represent a certain violence from the message to be tried. The formula for multi-axis decomposition of the x [n] signal will be schematically shown in Figure 1.

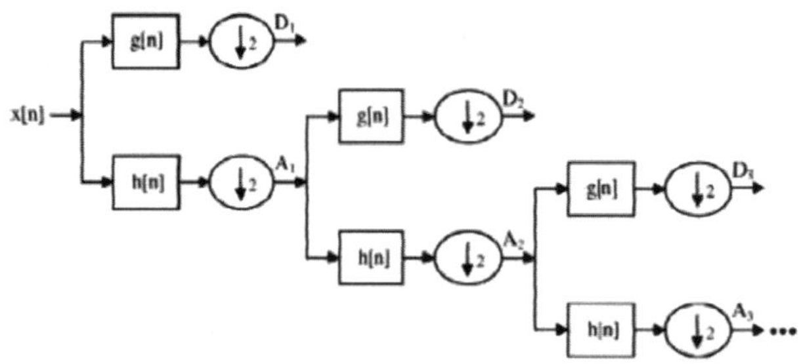

Figure 1: Subband decomposition of discrete wavelet transform

In this study the EEG signal is broken down to breakdown D5. To detect changes in the EEG signal daubechies wavelet sequence 2 is used because it displays the interior more precisely. The feature vector will represent can be obtained from the details of the EEG signal that is calculated and estimated by the wavelet coefficient of the EEG signal. This study takes the statistical value of the wavelet coefficient which can be used to reduce the dimensions of vector features. The statistical characteristics used to represent the frequency-time distribution of EEG signals, namely the average value and the standard deviation value of the wavelet coefficient for each subband.

2.3 K-nearest neighbor classification method

The K-NN algorithm is an easy and simple machine learning algorithm. The essence of the K-NN algorithm is a subject that is close to each other, then the subject has the same characteristics. Thus to find out the characteristics of one object, then we can also predict other objects based on the nearest neighbor. K-NN is the idea that each new instance can be classified based on the most votes of neighbor k, where k is a positive integer, and usually with a small number [12]. The K-NN classification algorithm predicts the category of the test sample according to the training sample k is the closest neighbor to the test sample, and puts it in the category that has the largest probability category [13].

In pattern recognition, the KNN algorithm is a method used to classify objects based on the closest training example in the feature space. KNN is a type of institutional-based learning, or lazy learning where this function is only approached locally and all calculations are deferred to classification [14]. The K-NN classification method has several stages, the first being the k value which is the number of closest neighbors that will determine which new query goes to which class is determined. The second stage, k the nearest neighbor is searched by calculating the distance of the query point with the training point. The third stage, after knowing the distance of each training point with the query point, then see the smallest value. The fourth stage takes the smallest k value, then see the class. The class that is the most is the class of the new query. The distance or distance of a point with its neighbors can be calculated using the Euclidean distance. The euclidean distance is represented as follows [15]:

$$j(a,b) = \sqrt{\sum_{k=1}^{k_n}(a_k - b_k)^2} \tag{1}$$

J (a, b) is the distance between point a which is the point known to its class and b is a new point. The distance between the new point and the training points is calculated and taken by the nearest point. The new point is predicted to enter the class with the highest classification of these points.

3. Results and discussion

This research processes data from EEG signals from 2003 BCI competition data, the data consists of training data and testing data. The initial process is to separate a set of data from 6 channels into data for each channel. Each channel is then processed using the DWT process. From this DWT process, the average value and standard deviation for each subband are taken. Thus the process of finding feature extraction is used to reduce the dimensions of the data, i.e. the initially large amount of data becomes less.

The feature search process via DWT for EEG signals is taken for coefficient values of A5, D5, D4 and D3 (Figure 2). This process is carried out because the coefficients of A5, D5, D4 and D3 enter the frequency of alpha, beta, gamma and tetha waves. Figure 3 is the result of the DWT process which is then taken the average value and the standard deviation of each subband.

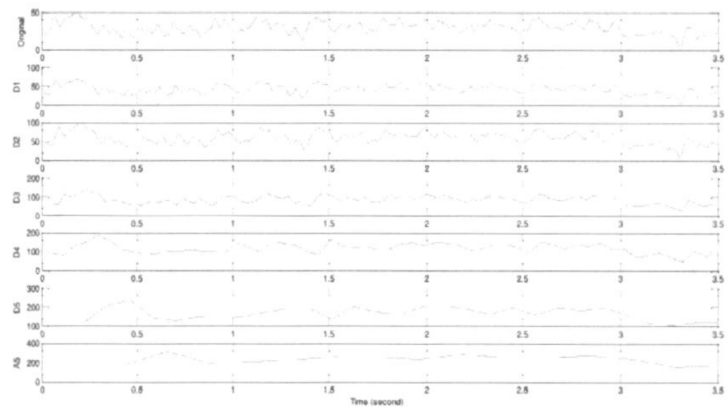

Figure 2. The results of the DWT process

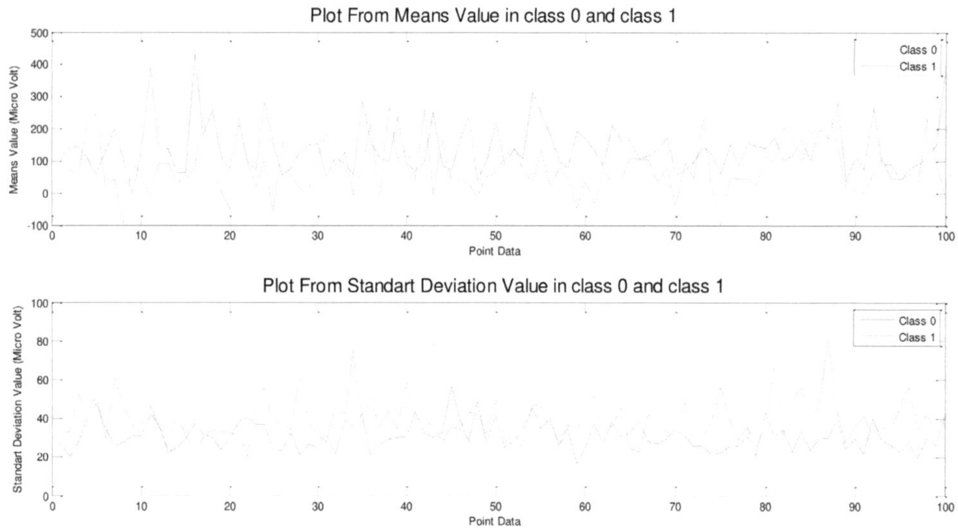

Figure 3. The results of the process of taking the average value and the standard deviation value

To train K-NN, use training data sets, while to verify the accuracy and effectiveness of K-NN test data to detect cursor movements up and down. The test will be carried out with variations in the K value in the KNN system and variations in the training data and test data used. The accuracy in question is the accuracy of the classification results carried out by the program with manual classification results. The level of accuracy can be formulated with Equation 2.

$$Level\ of\ accuracy = \frac{The\ number\ of\ EEG\ signals\ is\ correctly\ classified}{Total\ amount\ of\ test\ data}\ x\ 100\% \quad (2)$$

In this test variations in the number of k are performed on the KNN function. k is the number of closest neighbors. The k values tested are one, three, five, seven, and nine. Odd values are chosen to avoid similarities in proximity to two different points of class. Because KNN will classify based on the most class voting. From this test the average accuracy value for each k value is determined.

- Classification testing with value k = 1, k = 3, k = 5, k = 7, and k = 9
  The classification results with values k = 1 to k = 9 are shown in Table 1. The test results with k = 1 obtain the highest accuracy with a value of 80% and the average value of the KNN classification of 67%. For k = 3 the highest accuracy value is 72% and the lowest value is 48%. For k = 5 the

highest accuracy value is 76% and the lowest is 48%. For k = 7 the highest value is 72% and the lowest is 48%. For k = 9 the highest accuracy value is 76% and the lowest is 48%.

Tabel 1. Results of testing k = 1, k=3, k=5, k=7 dan k=9

Test	training data	test data	Accuracy K=1	Accuracy K=3	Accuracy K=5	Accuracy K=7	Accuracy K=9
1	50	50	80%	72%	76%	72%	76%
2	50	50	68%	60%	60%	60%	56%
3	50	50	64%	52%	40%	52%	48%
4	50	50	60%	60%	60%	60%	64%
5	50	50	60%	48%	48%	48%	52%
Average			67%	59%	57%	58%	59%

- Comparison of the average value of accuracy
  From all tests with variations in the value of k obtained the highest accuracy value is found at the value of k = 1 with an average level of accuracy of 67%. The lowest accuracy value is found in testing with k = 5 with an average level of accuracy of 57%. Overall the value of accuracy has a value close to 60%. Classification of KNN with the description above, the value of k = 1 is the most optimal k value.

4. Conclusion

This study aims to find a cirri extraction method and a good identification method to recognize EEG signals, namely recording the EEG signal of a subject in imagining right cursor movement and imagining cursor movement to the left. The method used for extracting EEG signal characteristics is using the DWT method and the method used for the identification process is the K-NN method. The results of this study are reaching an accuracy value of 80% in the testing process.

Acknowledgement
The authors express our deep gratitude to Umsida and the Directorate of Higher Education for providing time and money in the research process that is being investigated by researchers.

References

[1] Wolpawa J R, Birbaumer N, McFarland D J, Pfurtscheller G and Vaughan T M 2002 Brain-computer interfaces for communication and control *Clin Neurophysiol* vol **113** no 6 pp 767–91.
[2] Vaughan T M 2000 Guest editorial. Brain-Computer interface technology: a review of the second international meeting *IEEE Trans Neural Syst Rehabil Eng* vol **11** no 2 pp 94–109.
[3] Pour P A, Gulrez T, AlZoubi O, Gargiulo G and Calvo R a 2008 Brain-computer interface: Next generation thought controlled distributed video game development platform *2008 IEEE Symp Comput Intell Games* pp 251–257.
[4] Wolpaw J. R., Birbaumer N, McFarland D J, Pfurtscheller G and Vaughan T M 2002 Brain-computer interfaces for communication and control *Clin Neurophysiol* vol **113** no 6 pp 767–91.
[5] Puloma P 2003 Classification of electroencephalogram (EEG) signal based on fourier transform and neural network *A Thesis submitted in partial fulfillment of the requirements for the degree of Bachelor of Technology in Electrical Engineering* Department of Electrical Engineering National Institute of Technology Rourkela-769008 (ODISHA).
[6] Baojun W, Liu J, Jing B, Le P, Yan L, Guang L 2005 EEG recognition based on multiple types of information by using wavelet packet transform and neural networks Engineering in Medicine and Biology Society IEEE-EMBS 2005 *27th Annual International Conference* of the Issue Date 17-18 Jan. 2006 pp 5377 – 5380

[7] Wu T, Yan Guo Z, Yang Bang H, Sun H 2007 EEG feature extraction based on wavelet packet decomposition for the brain-computer interface 0263-2241/$ - see front matter _ 2007 Elsevier Ltd. All rights reserved. doi: 10.1016/ j.measurement.2007.07.007.

[8] http://bbci.de/competition/ii/

[9] Dipti U Classification of EEG signals under different mental tasks using wavelet transform and neural network with one step secant algorithm *International Journal of Scientific Engineering and Technology* vol **2**(4) pp 256-259.

[10] Pravin A, Kharat, Sanjay V, Dudul Daubechies Wavelet Neural Network Classifier for the Diagnosis of Epilepsy was transactions on biology and biomedicine.

[11] Rajya Lakshmi M, Prasad Dr T V, Chandra Prakash Dr V 2014 Survey on EEG signal processing methods *International Journal of Advanced Research in Computer Science and Software Engineering* vol **4**(1).

[12] Khamis, Kipruto H S, W C and Kimani S 2014 Application of k-nearest neighbor classification in medical data mining *International Journal of Information and Communication Technology Research* vol **4**(4).

[13] Suguna N and Thanushkodi K 2010 An improved k-nearest neighbor classification using genetic algorithm *International Journal of Computer Science Issues* vol **7** Issue 4 No 2.

[14] Imandoust S B and Bolandraftar M 2013 Application of K-Nearest Neighbor (KNN) approach for predicting economic events: Theoritical Background *International Journal of Engineering Research and Applications* vol **3** issue 5.

[15] Pramesti R P A 2013 Identifikasi karakter plat nomor kendaraan menggunakan ekstraksi fitur ICZ dan ZCZ dengan metode klasifikasi KNN *Scientific Repository of Bogor Agricultural University.*

IOP Conf. Series: Materials Science and Engineering **821** (2020) 012037 doi:10.1088/1757-899X/821/1/012037

Fire spread simulation using cellular automata in forest fire

H D Bhakti[1, *], H Ibrahim[1], F Fristella[2] and M Faisal[2]

[1] Department of Informatic Engineering, Faculty of Engineering, Universitas Muhammadiyah Gresik , Gresik, Indonesia
[2] Department of Mechanical Engineering, Faculty of Engineering, Institut Teknologi Kalimantan, Indonesia

[*] Email: hennydwi@umg.ac.id

Abstract. Forest fires are adverse events both economically and mentally. Triggers of fires are complex variables so that the spread of fire to date has not been predicted. Modeling is one solution to study the process of spreading fire. The Cellular automota was used to simulate the spread of fire in forest fires. Simulations carried out with simulations carried out by varying the value of burnability. burnProbability is a value that shows the probability of a tree burning. The value of burnability used is 10%, 20%, 30%, 40%, 50%, 60%, 70% and 80%. From each burnProbability value, the percentage of forest that is burned is calculated. The simulation was carried out for 17 iterations for each burnProbability value. The results of the average is the greater the value of burnProbability, the greater the percentage of burned forest. When the burnProbability value of 80% of the burned forest is almost entirely, it is 97.23183%.

1. Introduction

Forest fires are natural events that can occur naturally or intentionally. Natural factors are usually caused by high temperatures and the presence of minerals such as coal, oil or natural gas and triggered by the friction of branches and dry tree branches, while the intentional factor is humans who intentionally burn forests[1]. Until now, the spread of fire during forest fires cannot be predicted with certainty, because there are many variables in the trigger factors of fires. The trigger factors that play a role in the process of fires are fuel, humidity, wind, and slope of the land or slope [2]. Forest fires often occur in Indonesia, especially during the dry season. Fires that have occurred in Indonesia have a detrimental effect if they are not properly anticipated. In 1997 during the Elnino disaster, forest fires burned around 25 million hectares of forest land around the world, and Indonesia in 1997 was the country that experienced the most severe forest fires [1].

Model is a representation of a real system. Complex system studies can be carried out using modeling [3]. Forest fires are one example of a complex problem. The spread of fire and smoke is a complex phenomenon caused by chemical reaction and physical processes [3]. The computational method is progressing, so that the problem of the distribution of fire can be modeled. Cellular Automata is a computational method that can be used to model the phenomenon of forest fires[4]

Basically, there are two types of computational simulation methods used for the spread of fire and smoke: a) the Computational Fluid Dynamics (CFD) modeling method that resolves the basic equations of physical processes and clams [5]; and b) Celullar Automata (CA) which is a modeling method based on probabilitic transition rules between states [6][7].

2. Cellular Automata

Cellular automata are dynamic systems operating discrete in space and time, on a uniform, regular lattice and characterized by "local" interactions. Cellular Automata were invented by the mathematician Stanislaw Ulam and were used by J. von Neumann, followed by A.W. Burks and E. F. Codd, to solve problem of the non-trivial self-reproduction in a logical system [8]. A cellular automata is defined by a grid with start states and set of rules for state transitions.

Generally, cellular automata consist of four elements, which could be considered as a tuple (X, S, N, f):

a) X are cells which are objects in any dimensional space, we can call this cellar space. In cellar space, each cell has the form $x=(x_1, x_2, x_3, .., x_m)$,where m is the dimension of the space. All cells have some forms of neighborhood.
b) S is a nonempty finite set of automaton states. Each cell can take on only one state at any one time from a set of states, $s \in S$. Strict CA also requires state variables to be discrete.
c) Neighborhood Template N - the state of any cell depends on the states and configurations of other cells in the neighborhood n of that cell. In two-dimensional space, there are two well-known templates, von Neumann or 5 cell neighborhood.
d) f is state transition function rule [9]

3. Forest model

The area of spreading fire in the forest is modeled with a 17x17 size matrix. Representation of forest models as shown in Figure 1. Each cell represents the conditions that occur in it. Possible circumstances are having values 0, 1 and 2. Each value indicates the condition of the tree in the cell. A value of 0 (Empty) represents that the cell is empty or there are no trees in it. The value of 1 (Tree) represents that in the cell there is a tree that is not burned. The last possibility is the value 2 (Burning) which represents that the cell contains a burning tree.

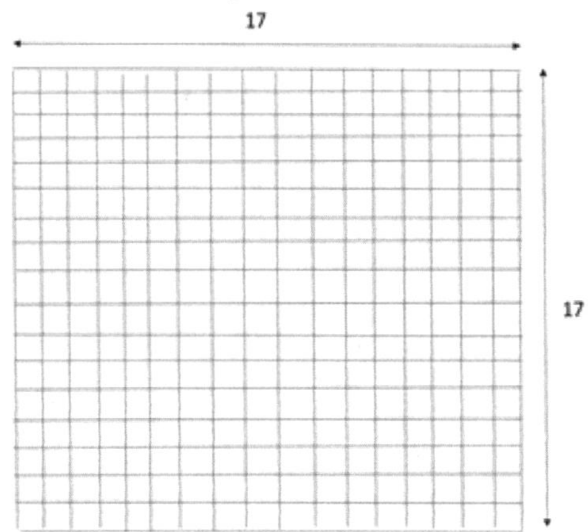

Figure 1. Cell representation in 17x17 size matrices

The initial condition of the system has 2 opportunities, namely probTree and probBurning. The opportunity for a tree to be located in a cell is expressed by probTree, while probBurning states the chance for a tree to burn if there is a tree in the cell. So the presentation of the tree that burned at the beginning of the simulation was stated by probBurning

4. Neighborhood Von Noumann

Neighboring cells in 2-dimensional form are represented using Von Neymann Neighborhood. A neighbor cells are cells that are located in the North, South, West and East and these cells. One cell has four neighbors. As shown in Figure 2. one cell has North, South, West and East neighbors.

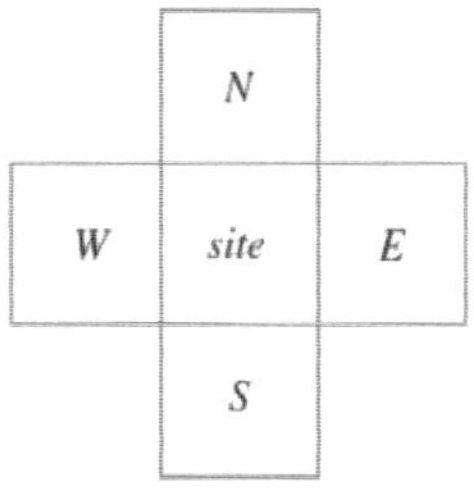

Figure 2. Von Neumann Neighborhood's representation of a cell.

The rule for the spread of fire based on Von Noumann Neighborhood is that trees can burn when their neighbors burn. If a cell is worth 0 (Empty) at time t, then at time t + 1 the cell also remains worth 0 (Empty). If a cell is worth 1 (Tree) at time t, then when t + 1 tree in the cell can burn or not at time t + 1, this depends on the neighboring cell there is fire or not. Cells that have a value of 2 (Burning) represent that in the cell loading the burning tree at time t will always burn out, so that at time t + 1 the cell will be worth 0 (Empty).

5. Periodic Boundary Condition

The spread of fire has rules that can be applied to every cell, including cells whose position is at the matrix boundary. Cells located in the first and last row, the first and last columns must also be burned. This is contrary to Von Neumann Neighborhood's rule because cells located at the matrix boundary have no neighbors. So to overcome the problem of the rules of spread, it is necessary to add one or two cells at each boundary called the boundary expansion.

In Figure 3 shows the geometry built on the model. Area A is the observation area which is a 17x17 sized matrix. Area B is the area used in the simulation. Region C is the area used for Periodic Boundary Condition. The area of the Periodic Boundary Condition as shown in Figure 4. Conditions that occur in cells located at the boundary of the matrix will affect the conditions that are at the boundary of the matrix that is located across. Iteration is done 17 times.

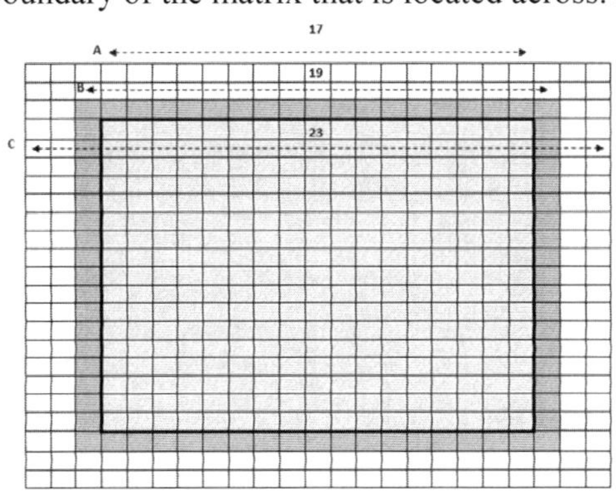

Figure 3. Model Geometry

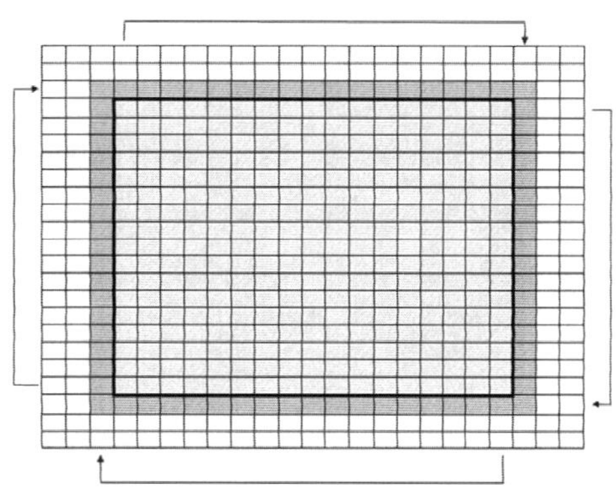

Figure 4. Periodic Boundary Condition

The location of the neighbors of each cell is shown in Figure 5. If a cell with coordinates (i, j) then the cell has 4 for the neighbor. The East neighbor is in the coordinate (i, j + 1), the South neighbor is at the coordinate (i + 1, j), the West neighbor is at the coordinate (i, j-1) and the North neighbor is at the coordinate (i -1, j). The fire spreading algorithm can be shown as in Figure 6.

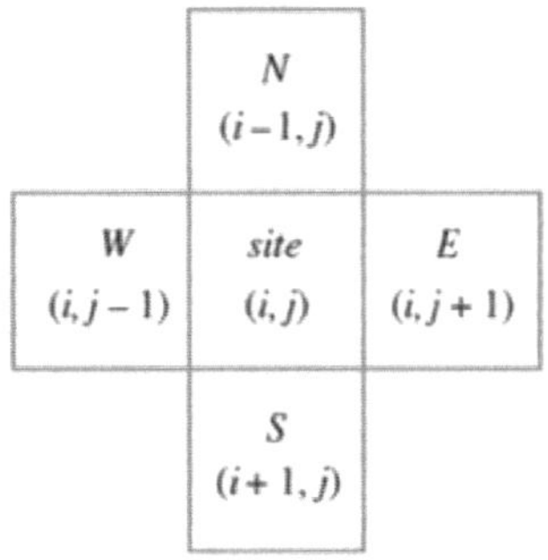

Figure 5. The cell Neighborhood geometry

```
Inisialisasi:
    EMPTY = 1 ; TREE = 2; BURNING = 3;        ← States
    N = 17                                    ← grid length and weight
    T = 17                                    ← total iteration
    forest(N,N);                              ← matriks (forest)
    forest(:,:) = TREE                        ← initial state
    forest(9,9) = BURNING                     ← fire source (9,9)
        burnProbability = 0.8                 ← burn probability (0.1,0.2,…,0.8)
    t = 0                                     ← initial condition t

while t < T

    for j = 2 to N-1
      for i = 2 to N-1
        site = forest(i,j)

        if site == EMPTY
              nextStep(i,j) = EMPTY
        elseif site == BURNING
              nextStep(i,j) = EMPTY
        else
          if NORTH==BURNING or EAST==BURNING or SOUTH==BURNING or WEST==BURNING
              if rand() < burnProbability
                     nextStep(i,j) = BURNING
              else
                     nextStep(i,j) = TREE
              end
          else
              nextStep(i,j) = TREE
          end
        end

      end
    end
    forest = nextStep

    Do Periodic Boundary Conditions           ← Periodic Boundary Conditions
    t = t+1
End

Count the burning area                        ← Percentage the burning area
terbakar
Plot matriks forest                           ← Plot matriks
```

Figure 6. Fire spreading algorithm

6. Results and Discussion

After the model was formed, simulations of the spread of fire in forest fires were conducted using the Cellular Automata method. Simulation is done by varying the value of burnProbability. burnProbability is a value that shows the probability of a tree burning. The value of burnProbability used is 10%, 20%, 30%, 40%, 50%, 60%, 70% and 80%. From each burnProbability value, the percentage of forest that is burned is calculated. The simulation was carried out for 17 iterations for each burnProbability value.

Figure 6(a) is a simulation result for the spread of fire with a 10% burnProbability value. It can be seen that the tree burned a little and only near the source of the fire. The percentage of burning forest with a 10% burnProbability value is 0.34602%. Figure 6(b) is the result of a fire spread simulation with a 20% burnProbability value. There are also not many burning trees, but more than Figure 6(a). The percentage of burning forest with a 20% burnProbability value is 0.69024%.

Figure 7(a) is a simulation result for the spread of fire with a 30% *burnProbability* value. It appears that more trees are burning. The percentage of forest burned with a 30% *burnProbability* value is 1.0381%. Figure 7(b) is the result of simulating the spread of fire with a 40% burnProbability value. Trees that burn more when compared to 30% burnProbability. The percentage of forest burned with 40% burnProbability is 0.3841%.

Figure 8(a) is a simulation result for the spread of fire with a 50% burnProbability value. It is seen that the trees that burn more and more when compared to before, but the location tends to be around the fire source which is located in the center of the matrix. The percentage of burning forest with a 50% burnProbability value is 8.3045%. Figure 8(b) is the result of a fire spread simulation with a 60% burnProbability value. The more burned trees and the direction they spread. The percentage of burned forest with a 60% burnProbability value is 74.0484%.

Figure 9(a) is a simulation result for the spread of fire with a 70% burnProbability value. It is seen that more and more trees are burned and spread throughout the forest. The percentage of forest burned with a 70% burnProbability value is 91.3495%. Figure 9(b) is the result of a fire spread simulation with an 80% burnProbability value. More and more trees are burned and almost all of the forest is burned. The percentage of burned forest with a 20% burnProbability value is 98.6159%. Of all the BurnProbability, which has the largest percentage of burning forest, the value is 80%. The greater the burnProbability, the more the forest will burn. This is because the more likely a tree is burning.

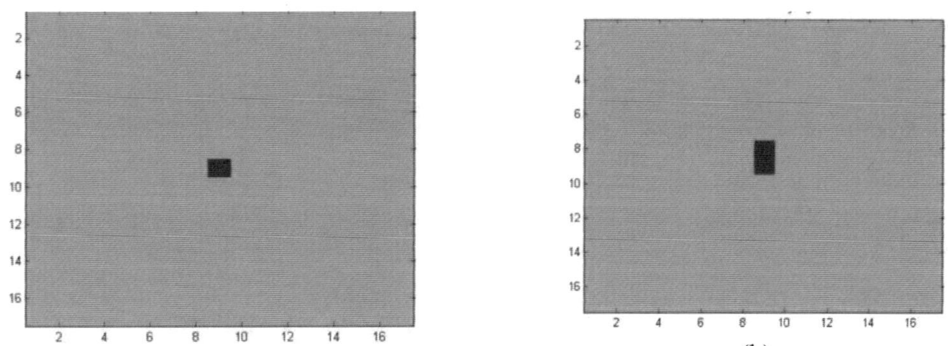

(a) (b)

Figure 6. Fire spread simulation (a) *burnProbability* 10% (b) *burnProbability* 20%

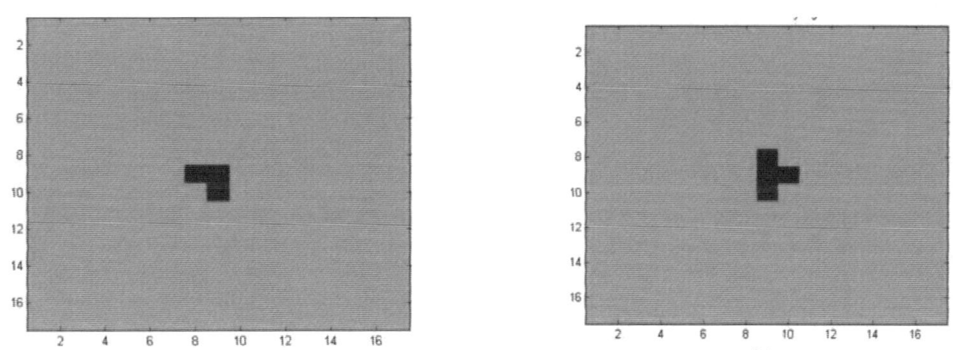

(a) (b)

Figure 7. Fire spread simulation (a) *burnProbability* 30% (b) *burnProbability* 40%

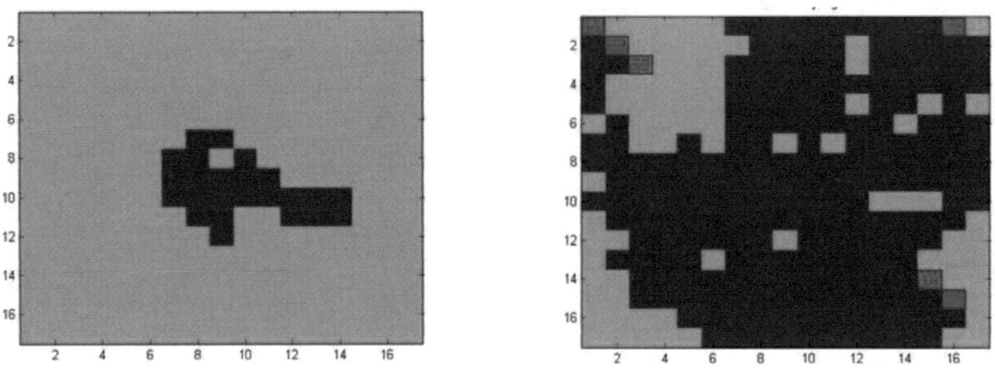

(a) (b)

Figure 8. Fire spread simulation (a) *burnProbability* 50% (b) *burnProbability* 60%

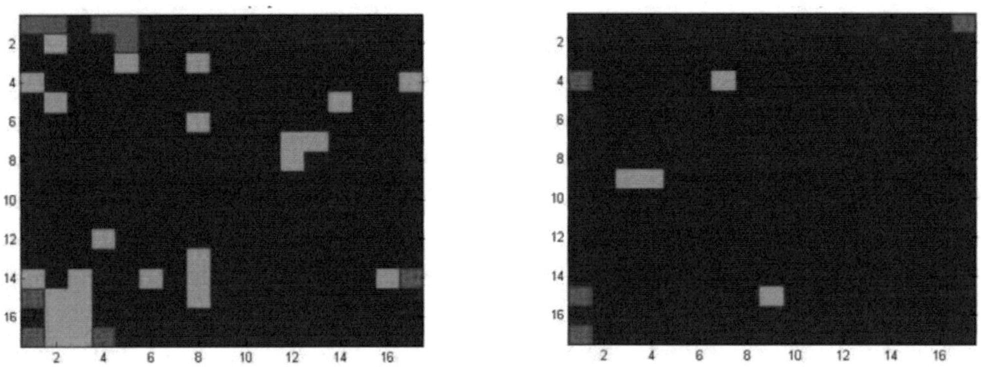

(a) (b)

Figure 9. Fire spread simulation (a) *burnProbability* 70% (b) *burnProbability* 80%

Each *burnProbability* value is simulated 10 times. The simulation results for each BurnProbability are shown in Table 1. Furthermore, the value is calculated on average as shown in Table 2. From the results of the average it can be seen that the greater the value of burnProbability, the greater the percentage of burned forest. When the burnProbability value of 80% of the burned forest is almost entirely, it is 97.23183%.

Table 1. Table of percentage of forest fires

Trial	***burnProbability***							
	10%	20%	30%	40%	50%	60%	70%	80%
1	0.34602	0.69204	0.34602	11.4187	3.1442	51.9031	87.8893	98.2699
2	0.34602	0.34602	0.34602	7.2664	41.5225	82.3529	91.3495	96.1938
3	0.34602	2.4221	6.5744	1.7301	11.0727	65.7439	93.4256	98.2699
4	0.34602	2.4221	2.0761	1.7301	41.5225	64.7059	94.4637	98.9619
5	0.34602	1.7301	1.0381	0.34602	12.8028	73.7024	89.6194	97.2318
6	0.34602	1.7301	0.34602	3.4602	0.346	77.8547	78.8927	96.1938
7	0.34602	2.7682	3.4602	9.3426	53.3633	79.2388	92.3875	95.1557
8	0.34602	1.3841	3.4602	0.96204	20.4152	57.7855	94.4637	98.2699
9	0.34602	0.34602	1.0381	0.96204	17.301	65.0519	83.045	95.1557
10	0.34602	0.34602	0.34602	1.0381	3.4602	68.8581	93.7716	98.6159

Table 2. The average forest burned for each burnProbability value.

burnProbability	**The average percentage (%)**
10%	0.34602
20%	1.41868
30%	1.903118
40%	3.82563
50%	20.49504
60%	68.71972
70%	89.9308
80%	97.23183

7. Conclusion

From this research it can be seen that the higher the BurnProbability, the more trees are burned. That is because the possibility of the tree burning is getting bigger. Cellular Automata can be used to simulate the process of spreading fire in forest fires. For further research real data will be used to validate its performance.

References

[1] A. I. Febriandhika, F. Ramdani, and A. N. Rusydi, “Pengembangan Sistem Simulasi Perkiraan Penyebaran Api Pada Gunung Arjuno Kawasan Tahura R . Soerjo Menggunakan Tangible Landscape,” *J. Pengemb. Teknol. Inf. dan Ilmu Komput. Univ. Brawijaya*, vol. 3, no. 2, 2019.

[2] R. C. Rothermel, “How to Predict the Spread and Intensity of Forest and Range Fires,” *USDA For. Serv. Gen. Tech. Rep.*, p. 161, 1983.

[3] A. Muzy, E. Innocenti, A. Aiello, J.-F. Santucci, and G. Wainer, “Cell-DEVS quantization techniques in a fire spreading application,” pp. 542–549, 2003.

[4] I. K. A. Thanailakis, “A model for predicting forest fire spreading using celullar automata,” *Ecol. Modell.*, vol. 99, no. 1, pp. 87–97.

[5] Dr N. Gobeau;Dr x.x. Zhou, “Evaluation if CFD to Predict Smoke Movement in Complex Enclosed Spaces,” .

[6] A. Ohgai, Y. Gohnai, S. Ikaruga, M. Murakami, and K. Watanabe, “Cellular Automata Modeling For Fire Spreading As a Tool to Aid Community-Based Planning for Disaster Mitigation,” *Recent Adv. Des. Decis. Support Syst. Archit. Urban Plan.*, pp. 193–209, 2005.

[7] D. I. Curiac, O. Banias, C. Volosencu, and D. Pescaru, “Cellular Automata Based Simulation for Smoke and Fire Spreading in Large Buildings,” no. 0.

[8] X. Li and W. Magill, “Modeling fire spread under environmental influence using a cellular automaton approach,” *Complex. Int.*, vol. 8, pp. 1–14, 2000.

[9] L. Bodrožić, D. Stipaničev, and M. Šerić, “Forest fires spread modeling using cellular automata approach,” ... *Split. Dep. Model.* ..., 2006.

Local irregularity chromatic number of vertex shackle product of graphs

R Umilasari[1,*], L Susilowati[2] and Slamin[3]

[1] Department of Informatics Engineering, Universitas Muhammadiyah Jember, Jember, Indonesia
[2] Department of Mathematics, Faculty of Sciences and Technology, Universitas Airlangga, Surabaya, Indonesia
[3] Department of Informatics, Faculty of Computer Science, Universitas Jember, Jember, Indonesia

*Email: reni.umilasari@unmuhjember.ac.id

Abstract. Local irregular vertex labeling is one of graph labeling type that can be used as a tool for graph coloring. A mapping l is called local irregular vertex labeling if there are: (i) a mapping $l: V(G) \rightarrow \{1,2,\cdots,k\}$ as vertex irregular *k*-labeling and a weight $w: V(G) \rightarrow N$, for every $uv \in E(G), w(u) \neq w(v)$ where $w(v) = \Sigma_{\{u\in N(v)\}} l(u)$; and (ii) $opt(l) = min\{max\{l_i\}$. Thus, the labeling l induces a proper vertex coloring of G where the vertex v is assigned the color *w(v)*. The local irregular chromatic number of G, denoted by $\chi_{lis}(G)$ is the minimum cardinality of the largest label over all such local irregular vertex labeling. In this paper, we determine the local irregular chromatic number of a vertex shackle product of graphs. The vertex shackle products of graphs, denoted by $Shack\ (G, v, k)$, is the graph constructed from k copies of connected graph G and v as the linkage vertex.

1. Introduction

A graph labeling is an assignment of integers to the vertices or edges, or both that first introduced in 1964. That is why labelling is one of interest topic in graph theory because until now there are some kinds of graph labelling. Let $\boldsymbol{G}$ be a simple connected graph, finite and undirected graph with vertex set $\boldsymbol{V}$ and edge set $\boldsymbol{E}$. A labeling in $\boldsymbol{G}$ is a mapping from the set of elements in a graph (vertices, edges, or both) to a set of numbers (usually positive integers) [1]. There are many types of labelings that have been studied such as magic and anti-magic labelling, distance irregular labelling, and many others [2].

A vertex coloring is an assignment of labels or colors to each vertex of a graph such that no edge connects by two identically colored vertices. The most common type of vertex coloring seeks to minimize the number of colors for a given graph. A vertex coloring also called as the chromatic number of a graph, it is the minimum number of colors needed to produce a proper coloring of a graph [3]. In formal definition, the chromatic number denoted by $\boldsymbol{\chi(G)}$ is the smallest positive integer $\boldsymbol{k}$ such that $\boldsymbol{G}$ is vertex *k-colorable* (and thus G has *k-coloring*) [4]. One of topic advance of chromatic number is chromatic number local irregular which be introduced by Kristiana *et al.* [5] and written as $\boldsymbol{\chi_{lis}(G)}$. If $\boldsymbol{G = (V, E)}$ be a connected grap, then local irregularity vertex coloring of $\boldsymbol{G}$ is $\boldsymbol{l: V(G) \rightarrow \{1, 2, \cdots, k\}}$ and $\boldsymbol{w: V(G) \rightarrow N}$ where $\boldsymbol{w(u) = \Sigma_{\{v\backslash in\ N(u)\}} l(v)}$ such that $\boldsymbol{opt(l) = min\{max\{l_i\}}$ and for every $\boldsymbol{uv \in E(G), w(u) \neq w(v)}$. The result that have been found before about chromatic number local irregular are chromatic number local irregular of path, cycle, complete, bipartite, star, friendship, and corona product graphs [6]. Figure 1 shows a graph with local chromatic number $\chi_{lis}(G) = 3$.

A shackle product graph is divided into two type. That is vertex shackle and edge shackle product graphs. Both of them are formed from k copies of graph G. A vertex shackle product of graph, denoted by $Shack(G, v, k)$ is the graph constructed from k copies of connected graph G and v as the linkage vertex. Whereas edge shackle products graphs is denoted by $Shack(G, e, k)$ means that the graph is constructed from any G graph of k copies and e as the linkage edge [7]. In this paper we show the local irregular chromatic number of vertex shackle products of graphs. There are some definitions and prepositions that are used in this paper as presented below:

Definition 1. [1] l is called local irregularity vertex coloring if there is a function $l: V(G) \to \{1, 2, \cdots, k\}$ as vertex irregular k-labeling, $w: V(G) \to N$ where $w(u) = \Sigma_{\{v \in N(u)\}} l(v)$ and

i. $opt(l) = min\{max\{l_i\};\ l_i, \textbf{vertex irregular labeling}\}$

ii. For every $uv \in E(G), w(u) \neq w(v)$.

Definition 2. [1] The minimum cardinality of local irregularity vertex labeling of G is called local irregular chromatic number, written by $\chi_{lis}(G)$.

Preposition 1. [1] $opt(l_i) = 1$ if each pair of two adjacent vertices in G have different degrees.

Preposition 2. [1] $opt(l_i) \geq 2$ if two adjacent vertices in G have the same degree.

Lemma 1. [1] For every G be simple and connected graph, $\chi_{lis}(G) \geq \chi(G)$

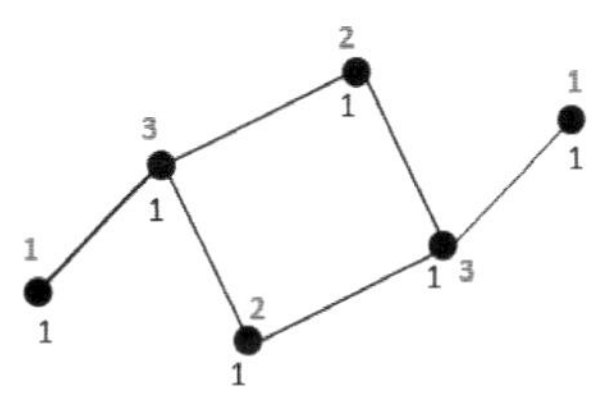

Figure 1. Graph G with $\chi_{lis}(G) = 3$

2. Result

We begin this section with some observations regarding to the chromatic number and local irregularity chromatic number of vertex shackle product of graphs such as path, cycle, star, complete and wheel graph.

Observation 1. The chromatic number of vertex shackle products of path

$$\chi_{lis}(Shack(P_n, v, m)) = \chi_{lis}(P_n)$$

This is due to the definition of vertex shackle graphs that the vertex shackle of k paths with n vertices is also a path with kn vertices. Furthermore, Kristiana *et al.* [1] showed that

$$\chi_{lis}\big(Shack(P_n, v, m)\big) = \chi_{lis}(P_n) = \begin{cases} 2, \text{for } m \times n = 2,3 \\ 3 \text{ for } m \times n \geq 4,5, \dots \end{cases}$$

Observation 2. If $Shack(C_n, v, m)$ is the vertex shackle of m cycles C_n for $m \geq 2$ and $n \geq 3$ then

$$\chi\big(Shack(C_n, v, m)\big) = \begin{cases} 2, \text{for } n \text{ even} \\ 3, \text{for } n \text{ odd} \end{cases}$$

Observation 3. If $Shack(S_n, v, m)$ is the vertex shackle of m stars S_n for $m \geq 2$ and $n \geq 3$, then $\chi(Shack(S_n, v, m)) = 2$

Observation 4. If $Shack(K_n, v, m)$ is the vertex shackle of m complete graphs K_n for $m \geq 2$ and $n \geq 3$, then $\chi(Shack(K_n, v, m)) = n$.

Observation 5. If $Shack(W_n, v, m)$ is the vertex shackle of m wheel W_n for $m \geq 2$ and $n \geq 3$ then

$$\chi\big(Shack(W_n, v, m)\big) = \begin{cases} 3, \text{for } n \text{ even} \\ 4, \text{for } n \text{ odd} \end{cases}$$

We now ready to determine the local irregularity chromatic number of vertex shackle of m cycles as presented in the following theorem.

Theorem 1. If $Shack(C_n, v, m)$ is the vertex shackle of m cycles C_n for $m \geq 2$ and $n \geq 3$ then

$$\chi_{lis}\big(Shack(C_n, v, m)\big) = \begin{cases} 3, \text{for } n \text{ even} \\ 4, \text{for } n \text{ odd} \end{cases}$$

Proof. Let C_n be a cycle on $n \geq 3$ vertices and $Shack(C_n, v, m)$ be the vertex shackle of m cycles C_n for $m \geq 2$. Then the vertex set of $Shack(C_n, v, m)$ is $V(Shack(C_n, v, m)) = \{a_{i,j}, 1 \leq i \leq n - 1, 1 \leq j \leq m, a_{n,m}\}$ and the edge set of $Shack(C_n, v, m)$ is $E(Shack(C_n, v, m) = \{a_{i,j}a_{i+1,j}, 1 \leq i \leq n - 1, 1 \leq j \leq m\} \cup \{a_{n-1,m}a_{n,m}\}$.

By **Preposition 2** we have $opt(l_i(Shack(C_n, v, m)) = 2$. To show the proof we can divide it into two cases below.

Case 1: for n even.

The vertex shackle of cycles $Shack(C_n, v, m)$ has chromatic number $\chi(Shack(C_n, v, m)) = 2$ for n even and $\chi(Shack(C_n, v, m)) = 3$ for n odd. Based on **Lemma 1** the lower bound is $\chi_{lis}\big(Shack(C_n, v, m)\big) = 3 \geq \chi((Shack(C_n, v, m))) = 2$. The upper bound shall be proved by defining $l: V\big(Shack(C_n, v, m)\big) \rightarrow \{1,2\}$ using the following formula.

$$l(a_{i,j}) = \begin{cases} 1, \text{for } 1 \leq i \leq n - 1, i \text{ odd} \\ 2, \text{for } 2 \leq i \leq n - 2, i \text{ even} \end{cases}$$
$$l(a_{n,m}) = 1$$

Therefore, $opt(l) = 2$ and the labeling gives vertex-weight as follows:

$$w(a\_\{i,j)\} = \begin{cases} 2, \text{for } 2 \leq i \leq n - 2, i \text{ even}\} \\ 4, \text{for } 1 \leq i \leq n - 1, i \text{ odd}\} \\ 8, \text{for } i = 1 \text{ linkage vertex}, j = 2,3, \ldots \end{cases}$$
$$w(a_{n,m}) = 2$$

We conclude that $|w(V(Shack(C\_n, v, m)))| = 3$. Based on **Definition 1** the upper bound is $\chi_{lis}\big(Shack(C_n, v, m)\big) \leq 3$. Hence, $\chi_{lis}(Shack(C_n, v, m)) = 3$ for n even.

Case 1: for n odd.

Based on **Lemma 1** the lower bound is $\chi_{lis}\big(Shack(C_n, v, m)\big) = 4 \geq \chi(Shack(C_n, v, m)) = 3$ for n odd. The upper bound shall be proved by defining $l: V\big(Shack(C_n, v, m)\big) \rightarrow \{1,2\}$ using some formula below.

$$l(a_{i,j}) = \begin{cases} 1, & \text{for } 1 \leq i \leq n - 2, i \text{ odd} \\ 2, & \text{for } 2 \leq i \leq n - 1, i \text{ even} \end{cases}$$

$$l(a_{n,m}) = 1$$

Therefore, $opt(l) = 2$ and the labeling gives vertex-weight as follows:

$$w(a_{i,j}) = \begin{cases} 2, & \text{for } 2 \le i \le n-3, i \text{ even} \\ 4, & \text{for } 1 \le i \le n-2, i \text{ odd} \\ 3, & \text{for } i = n-1, 1 \le j \le m \\ 6, \text{for } i = 1 \text{ linkage vertex}, j = 2,3,\ldots \end{cases}$$
$$w(a_{n,m}) = 3$$

Therefore $|w(V(Shack(C_n, v, m)))| = 4$. Based on **Definition 1**, the upper bound is $\chi_{lis}(Shack(C_n, v, m))) \le 4$. Hence, $\chi_{lis}(Shack(C_n, v, m)) = 4$ for n odd. This case completes the proof. □

The following theorem presents the local irregularity chromatic number of vertex shackle of m stars for $m \ge 2$.

Theorem 2. If $Shack(S_n, v, m)$ is the vertex shackle of m star S_n for $m \ge 2$ and $n \ge 3$ then $\chi_{lis}(Shack(S_n, v, m)) = 3$

Proof. Let S_n be a star on $n \ge 3$ vertices and $Shack(S_n, v, m)$ be the vertex shackle of m stars S_n for $m \ge 2$. Then the vertex set of $Shack(S_n, v, m)$ is $V(Shack(S_n, v, m)) = \{a_{i,j}, b_j;\ 1 \le i \le n-1, 1 \le j \le m\} \cup \{a_{n,m}\}$ and the edge set of $Shack(S_n, v, m)$ is $E(Shack(S_n, v, m)) = \{a_{i,j}b_j;\ 1 \le i \le n-1, 1 \le j \le m\} \cup \{a_{n,m}b_m\}$. Based on **Observation 3,** vertex shackle of stars $Shack(S_n, v, m)$ has $\chi\ (Shack(S_n\ , v, m)) = 2$ and by **Lemma 1,** the lower bound is $\chi_{lis}(Shack(S_n, v, m)) = 2 \ge \chi(Shack(S_n, v, m)) = 2$. However, the lower bound cannot be reached. The upper bound shall be proved by defining $l: V(Shack(S_n, v, m)) \to \{1\}$. By using **Proposition 1**, we know that $opt(l_i)(Shack(S_n, v, m)) = 1$ and the labeling gives vertex-weight as follows:

$$w(a_{i,j}) = 1, \text{for } 1 \le i \le n-1, 1 \le j \le m$$
$$w(a_{i,j}) = 2, \text{for } i = n, 1 \le j \le m-1, a_{i,j} \text{ is linkage vertex}$$
$$w(a_{n,m}) = 1$$
$$w(b_j) = n$$

Therefore $|w(V(Shack(S_n, v, m)))| = 3$. By **Definition 1** the upper bound is $\chi_{lis}((Shack(S_n, v, m)) \le 3$. Then it proven that $\chi_{lis}((Shack(S_n, v, m)) = 3$. □

The local irregularity chromatic number of vertex shackle of m complete graphs for $m \ge 2$ is presented in the following theorem.

Theorem 3. If $Shack(K_n, v, m)$ is the vertex shackle of m complete graphs K_n , for $m \ge 2$ and $n \ge 3$, then $\chi_{lis}(Shack(K_n, v, m)) = n$.

Proof. Let K_n be a complete graph on $n \ge 3$ vertices and $Shack(K_n, v, m)$ be the vertex shackle of m complete graphs K_n for $m \ge 2$. Then the vertex set of $Shack(K_n, v, m)$ is $V(Shack(K_n, v, m)) = \{a_{i,j};\ 1 \le i \le n-1, 1 \le j \le m\} \cup \{a_{n,m}\}$ and the edge set of $Shack(K_n, v, m)$ is $E(Shack(K_n, v, m)) = \{a_{i,j}a_{i,j};\ 1 \le i \le n-1, 1 \le j \le m\} \cup \{a_{i,j}a_{n,m}\}$.

The vertex shackle of m complete graphs $Shack(K_n, v, m)$ has $\chi(Shack(K_n, v, m)) = n$ as **Observation** 4. Based on **Lemma** 1, the lower bound is $\chi_{lis}(Shack(K_n, v, m)) = n \geq \chi(Shack(K_n, v, m)) = n$. The lower bound cannot be reached. The upper bound will be proved by defining $l: V(Shack(K_n, v, j)) \rightarrow \{1,2,3, \ldots n;\ 1 \leq j \leq m\}$. By **Preposition 1**, we can conclude that $opt(l_i)(Shack(K_n, v, m)) = n$ and the labeling gives vertex-weight as much as n different vertex weight. Therefore$|w(V(Shack(K_n, v, m)))| = n$. By **Definition 1**, the upper bound is $\chi_{lis}((Shack(K_n, v, m)) \leq n$. Hence, $\chi_{lis}((Shack(K_n, v, m)) = n$. □

Finally, we close this section with the local irregularity chromatic number of vertex shackle of m wheels for $m \geq 2$ as shown below.

Theorem 4. If $Shack(W_n, v, m)$ is the vertex shackle of m wheels for $m \geq 2$ and $n \geq 3$, then

$$\chi_{lis}(Shack(W_n, v, m)) = \begin{cases} 4, & \text{for } n \text{ even} \\ 5, & \text{for } n \text{ odd} \end{cases}$$

Proof. Let W_n be a wheel on $n \geq 3$ rim vertices and $Shack\ (W_n, v, m)$ be the vertex shackle of m wheels W_n for $m \geq 2$. Then the vertex set of $Shack(W_n, v, m)$ is $V(Shack(W_n, v, m)) = \{a_{i,j}, b_j, a_{n,m}, 1 \leq i \leq n-1, 1 \leq j \leq m\}$ and thc edge set of $Shack(W_n, v, m)$ is $E(Shack(W_n, v, m) = \{a_{i,j}a_{i+1,j}, a_{i,j}, b_j: 1 \leq i \leq n-1, 1 \leq j \leq m\} \cup \{a_{n-1,m}a_{n,m}\} \cup\ b_m a_{n,m}\ \}$. By **Proposition 2**, we have $opt(l_i(Shack(W_n, v, m)) = 2$. We divide the proof into two cases below.

Case 1: for *n* even.

The vertex shackle of wheels $Shack(W_n, v, m)$ has chromatic number $\chi(Shack(W_n, v, m)) = 3$ for *n even* and $\chi(Shack(W_n, v, m)) = 4$ for *n odd.* Based on **Lemma 1**, the lower bound is $\chi_{lis}(Shack(W_n, v, m)) = 4 \geq \chi((Shack(W_n, v, m))) = 3$. The upper bound will be proved by defining $l: V(Shack(W_n, v, m)) \rightarrow \{1,2\}$ using the following formula.

$$l(a_{i,j}) = \begin{cases} 1, \text{for } 1 \leq i \leq n-1, i \text{ odd}, 1 \leq j \leq m \\ 2, \text{for } 2 \leq i \leq n-2, i \text{ even}, 1 \leq j \leq m \end{cases}$$
$$l(a_{n,m}) = 1$$
$$l(b_j) = 1, \text{for } 1 \leq j \leq m$$

Therefore, $opt(l) = 2$ and the labeling gives vertex-weight as follows:

$$w(a\_\{i,j\})\} = \begin{cases} 3, & \text{for } 2 \leq i \leq n, i \text{ even}, 1 \leq j \leq m \\ 5, & \text{for } 1 \leq i \leq n-1, i \text{ odd}, 1 \leq j \leq m \\ 10, & \text{for } i = 1 \text{ linkage vertex}, j = 2,3, \ldots \end{cases}$$
$$w(b_j) = \frac{n}{2}(1+2), 1 \leq j \leq m$$

We conclude that $|w(V(Shack(W_n, v, m)))| = 4$. By **Definition 1**, the upper bound is $\chi_{lis}(Shack(W_n, v, m))) \leq 4$. It means $\chi_{lis}(Shack(W\_n, v, m)) = 4$ for *n* even.

Case 2: for *n* odd.

Based on Lemma 1 the lower bound is $\chi_{lis}(Shack(W_n, v, m)) = 5 \geq \chi((Shack(C_n, v, m))) = 4$ for *n* odd. Then the upper bound will be proved by defining $l: V(Shack(W_n, v, m)) \rightarrow \{1,2\}$ using the following formula.

$$l(a_{i,j}) = \begin{cases} 1, \text{for } 1 \le i \le n-2, i \text{ odd}, 1 \le j \le m \\ 2, \text{for } 2 \le i \le n-1, i \text{ even}, 1 \le j \le m \end{cases}$$
$$l(a_{n,m}) = 1$$
$$l(b_j) = 1, \text{for } 1 \le j \le m$$

Therefore, $opt(l) = 2$ and the labeling gives vertex-weight as follows:

$$w(a\_\{i,j)\} = \begin{cases} 2, & \text{for } 2 \le i \le n-1, i \text{ even}, 1 \le j \le m \\ 4, & \text{for } 1 \le i \le n-2, i \text{ odd}, 1 \le j \le m \\ 3, & \text{for } i = n, i \text{ odd}, 1 \le j \le m \\ 10, & \text{for } i = 1 \text{ linkage vertex}, j = 2,3,\ldots \end{cases}$$
$$w(b_j) = \frac{n-1}{2}(1+2) + 12, \text{for } 1 \le i \le n-1, 1 \le j \le m$$

We also can conclude that $|w(V(Shack(W_n, v, m)))| = 5$. By **Definition 1**, the upper bound is $\chi_{lis}(Shack(W_n, v, m))) \le 5$ prove that $\chi_{lis}(Shack(W_n, v, m)) = 5$ for n odd. Therefore, this case completes the proof. Figure 2 shows the example of local irregular chromatic number of the vertex shackle product of W_4.

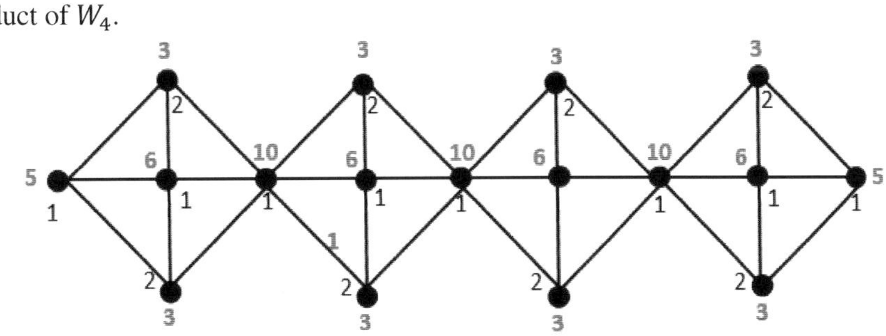

Figure 2. Graph G with $\chi_{lis}(Shack(W_4, v, 4)) = 4$

3. Conclusion

We conclude this paper with an open problem regarding to the local irregular chromatic number of the vertex shackle product of graphs as follows. **Problem 1.** For any connected graph G with n vertices, find the local irregular chromatic number of vertex shackle and edge shackle product of k copies of graph G.

Acknowledgments

We gratefully acknowledge the support from Universitas Muhammadiyah Jember and DRPM Kemristekdikti through World Class Research grant year 2019.

References

[1] Slamin 2017 On distance irregular labeling of graphs *Far East Journal Mathematical Sciences* **102**:05 pp 919-32.
[2] Gallian J A 2016 A dynamic survey of graph labelling *Electron J. Combin.* #DS6.
[3] Amurugam S, Premalatha K, and Martin B 2017 Local anti-magic vertex coloring of a graph *Graphs and Combinatorics* **33** pp 275-85.
[4] Chartrand G and Lesniak L 2000 *Graph and Digraph Third Edition* Chapman & Hall/First CRC Press Reprint (Florida) p 220.
[5] Kristiana A I, Utoyo M I, Dafik, Slamin, Alfarisi R, and Agustin I H 2019 Local irregularity vertex coloring of graph *International Journal of Civil Engineering and Technology* **10**:4 pp 451-61.

[6] A.I. Kristiana, M.I. Utoyo, Dafik, R. Alfarisi, I.H. Agustin and E. Waluyo 2019 On the chromatic number local irregularity of related wheel graph *Journal of Physics: Conf. Series* **1211** (2019) 012003.

[7] Inayah N, Simanjuntak R, Salman A N M, and Syuhada K I A 2013 Super *(a,d)-H* Antimagic Total Labellings for Shackle of a Connected Graph H *Australasian Journal of Combinatorics* vol **57** pp 127-28.

Consumable Fish Classification Using k-Nearest Neighbor

S Winiarti*, F I Indikawati, A Oktaviana, H Yuliansyah

Department of Informatics Engineering, Faculty of Industrial Technology, Universitas Ahmad Dahlan, Yogyakarta, Indonesia

*Email: sri.winiarti@tif.uad.ac.id

Abstract. Fish is beneficial for the human body because it has high protein content. Consuming fish is necessary and expert knowledge is needed to identify fresh fish that are suitable for consumption. In this study, we developed a classification system to identify four classes of consumable fish by grouping fish images based on texture extraction and color features. We use fish meat and fish scale as identification parameters. Fish meat image is measured using the HSV colors model (Hue, Saturation, and Value) and GLCM (Gray Level Co-occurrence Matrix) method. We use these values for texture feature extraction of scales. Then we use k-Nearest Neighbor (kNN) as the classifier. The test results from 320 sample images show that the identification accuracy of tilapia meat is 90% and 97.5% for mackerel meat. Meanwhile for the scales, the accuracy up to 87.5% for tilapia scales and 95% for mackerel scales.

1. Introduction

Fish has high protein content and has a lot of benefits for the human body [1]. The freshness of fish usually determines the quality of consumable fish. Due to the high protein and water content, fish is a highly perishable commodity. Consumers must be careful and have the expertise to identify fresh fish that are suitable for consumption.

Several kinds of research have been done to determine fish freshness using hardware and or software approach. Metallic Potentiometric Electrodes have been proposed to analyse fresh fish. Information from the electrodes is combined and analysed with fuzzy logic [2]. Electrode sensors are used to measure dielectric properties of the skin of fish and muscles [3].

The Torrymeter sensor (a tool to measure the freshness of fish) and RGB (Red, Green, Blue) color indices are compared to detect fresh fish [4, 5]. The result of the comparison shows that Torrymeter produces fast analysis, friendly usage, and shows the exact measurement when compared to the RGB color index [4]. The electronic nose, which consists of 8 metal oxide sensors, are used to evaluate the freshness of mackerel. Information from sensors is classified and combined with Support Vector Machine (SVM) and k-Nearest Neighbour (k-NN) method [5].

Image processing is the basis of detection techniques that can be applied to identify an object or its condition. It can be used to identify fresh fish [6-8], another animal such as birds[9], fruit[10], or food product [11]. The combination of changes in the value of the grey iris and the surface texture of the fish body can be combined to achieve detection in image processing [12]. Other variables to detect the freshness of fish are fish gills [3, 6-8, 13], muscle [3], and fish skin [3].

Data from those variables are classified into several algorithms to detect fish freshness, such as Artificial Neural Networks [14], ensemble learning using SVM and NN classifiers [15], Support Vector

Machine [5, 13], k-Nearest Neighbor [5, 15]. Classification is also carried out with the relationship between electrical parameters and the stage of fish decay [17]. The statistical data-fitting model is a general and effective analysis method for fish freshness classification.

In previous studies, the freshness of fish can be classified into several types. Artificial Neural Networks was used to classify seven fish freshness classes by the storage time: 1, 3, 5, 7, 9, 11, and 13 days [14]. A Matlab Program utilizing Support Vector Machine (SVM) classifier has been developed to determine the freshness of fish and shelf life, but only classified into two classes: fresh and not fresh fish [13].

Another study also classified fish into two classes: fresh and not fresh fish [16]. The parameters that are used to measure the quality of fish are general appearance, eyes, and gills. The types of fish studied were Giant Gourami, Red Snapper and Tilapia. Fish image is taken, cropped and summarized into RGB values. Then, the results of RGB values are classified using the kNN algorithm with the help of WEKA tools.

In this study, the key contributions are summarized as follows. First, we develop a classification system to classify four classes of fish freshness: very suitable for consumption, suitable for consumption, less suitable for consumption, and not suitable for consumption. We identify fish freshness in mackerel and tilapia fish by utilizing 320 meat and scale image samples. Fish meat image is converted from RGB into HSV (Hue, Saturation,Value) color model for meat identification. Fish scale image is converted into a grayscale image and use GLCM (Gray Level Co-occurrence Matrix) for scale texture feature extraction. Then, we classify the converted image using kNN algorithm and develop the algorithm using MATLAB.

Finally, we present the performance evaluation of our fish freshness classification system using mackerel and tilapia fish image. We measure the effectiveness of our classification system by comparing our classification result with expert judgment and calculate the classification accuracy. The remainder of this paper is organized as follows. Section 2 discusses our method for classifying fish freshness based on fish meat and fish scale image. Then, we discuss the demonstration and performance evaluation of our classification system in Section 3. Finally, we conclude our work in Section 4.

Image processing is the basis of detection techniques that can be applied to identify an object or its condition. It can be used to identify fresh fish [6-8] , another animal such as birds [9], fruit[10], or food product [11].

The combination of changes in the value of the grey iris and the surface texture of the fish body can be combined to achieve detection in image processing [12]. Other variables to detect the freshness of fish are fish gills [3] [6-8, 13], muscle [3], and fish skin [3].

Data from those variables are classified into several algorithms to detect fish freshness, such as Artificial Neural Networks [14], ensemble learning using SVM and NN classifiers [15], Support Vector Machine [5, 13], k-Nearest Neighbor [5, 16]. Classification is also carried out with the relationship between electrical parameters and the stage of fish decay [17]. The statistical data-fitting model is a general and effective analysis method for fish freshness classification.

In previous studies, the freshness of fish can be classified into several types. Artificial Neural Networks was used to classify seven fish freshness classes by the storage time: 1, 3, 5, 7, 9, 11, and 13 days [14]. A Matlab Program utilizing Support Vector Machine (SVM) classifier has been developed to determine the freshness of fish and shelf life, but only classified into two classes: fresh and not fresh fish [13].

Another study also classified fish into two classes: fresh and not fresh fish [16]. The parameters that are used to measure the quality of fish are general appearance, eyes, and gills. The types of fish studied were Giant Gourami, Red Snapper and Tilapia. Fish image is taken, cropped and summarized into RGB values. Then, the results of RGB values are classified using the kNN algorithm with the help of WEKA tools. In this study, the key contributions are summarized as follows. First, we develop a classification system to classify four classes of fish freshness: very suitable for consumption, suitable for consumption, less suitable for consumption, and not suitable for consumption. We identify fish freshness in mackerel and tilapia fish by utilizing 320 meat and scale image samples. Fish meat image is converted from RGB into HSV (Hue, Saturation,Value) color model for meat identification. Fish scale image is converted into a grayscale image and use GLCM (Gray Level Co-occurrence Matrix) for scale texture feature extraction. Then, we classify the converted image using kNN algorithm and develop the algorithm using MATLAB.

Finally, we present the performance evaluation of our fish freshness classification system using mackerel and tilapia fish image. We measure the effectiveness of our classification system by comparing our classification result with expert judgment and calculate the classification accuracy. The remainder of this paper is organized as follows. Section 2 discusses our method for classifying fish freshness based on fish meat and fish scale image. Then, we discuss the demonstration and performance evaluation of our classification system in Section 3. Finally, we conclude our work in Section 4.

2. Research Method

In this section, we explain how to identify fish quality based on fish meat and fish scale image. We first present the overall process of classification from fish meat and fish scale image. We then provide a detailed description of each process. Figure 1 depicts the general process in our classification system using the k-NN classifier. There are two kinds of inputs for our system: fish meat and fish scale image. Both images should be in RGB mode.

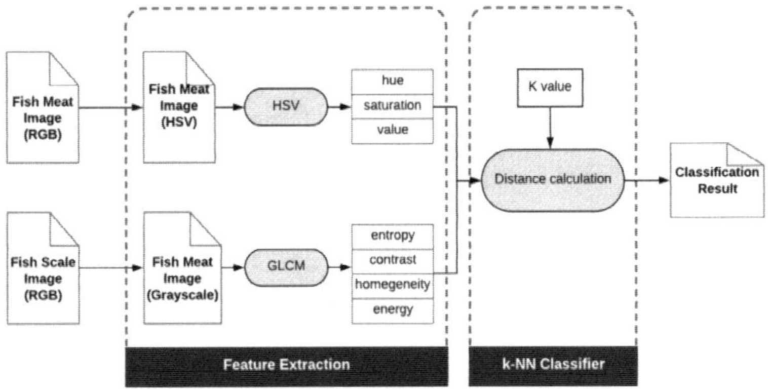

Figure 1. Classification System

There are two main processes in our system, which is the feature extraction process and classification process. Feature extraction process involves converting RGB value into another color mode. Meat image is converted into HSV color model and the extraction result is hue, saturation, and value. Scale image is converted into grayscale and then processed using GLCM method, which resulted in four values: entropy, contrast, homogeneity, and energy.

In the classification process, the k-NN classifier uses those values along with K value from the user to calculate the distance and then display the classification result.

2.1. Fish Meat Measurement

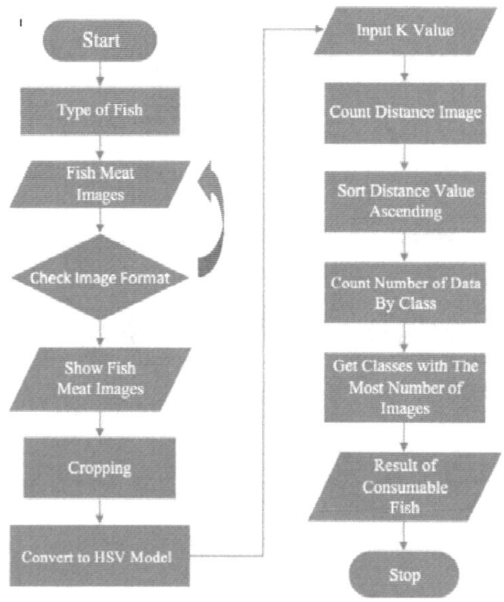

Figure 2. Flowchart for measuring fish freshness using fish meat image

Figure 2 shows the flowchart for fish meat measurement in our system. The first step is to select the type of fish to be identified. Then, the image of fish meat in *.jpg or *.bmp format is loaded and displayed. If the image format is correct, then image will be cropped to focus on the image. The system will perform a conversion from RGB into HSV color model into hue, saturation, and value. These values are used as input for classifier.

To perform classification, the user must fill in the K value to find the amount of closest fish meat image data to the image of the tested fish meat. The system will calculate the distance between the testing image and the training image using kNN. The calculation results will determine the image of fish meat that follows class K (fish class results are suitable for consumption). The final result of the system is the conclusion of the fish freshness class.

2.2. Fish Scale Measurement

Figure 3 shows the process for fish scale measurement. The differences between Figure 2 and Figure 3 are the addition of RGB to grayscale image conversion step. After converting into grayscale image, then the system calculates feature extraction values using GLCM method that produces entropy, contrast, homogeneity, and energy values.

Similar to the process in fish meat measurement, the classification process uses values from the feature extraction process and K value from the user to calculate the distance and display the classification result.

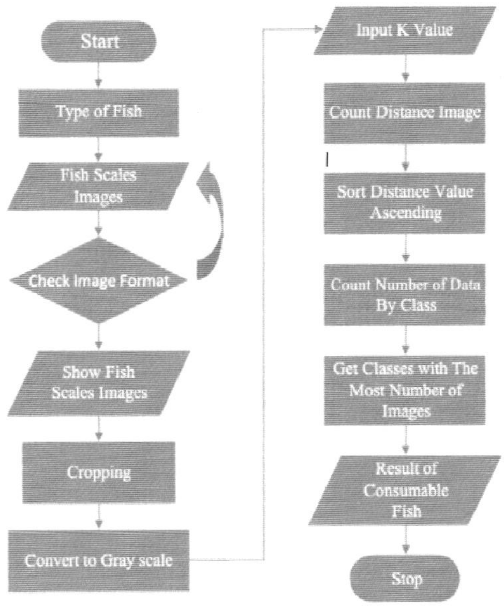

Figure 3. Flowchart for measuring fish freshness

3. Experimental Result

Our fish classification system is developed with Matlab and has two main features, namely identification based on fish meat and fish scales. Users can choose the identification process according to fish image data. We captured the fish meat and scale images using the digital camera and all sample data are recorded into 320 image samples. These image samples are tested using developed Matlab program to get the classification result.

Using the Matlab program that has been created, the image data is analyzed. A total of 320 images were analyzed by grouping 160 images of mackerel and tilapia. Each fish image group consists of two types of images: the image of fish meat and the image of fish scales.

This image data consists of:

40 images of mackerel meat as training data.
40 images of mackerel meat as testing data.
40 images of tilapia meat as training data.
40 images of tilapia meat as testing data.
40 images of mackerel scales as training data.
40 images of mackerel scales as testing data.
40 images of tilapia scales as training data.
40 images of tilapia scales as testing data.

3.1. Fish Meat Measurement

An example result of the identification of fish meat is shown in **Error! Reference source not found.**. A user can demonstrate classification process based on the flowchart in Figure 2. A user only needs to click the button in the Matlab program. Button Open Images to load the fish image, button Crop Images to crop the fish images, button Process to perform fish classification. After the identification process is finished, a user can get the result of the classification directly in the program GUI.

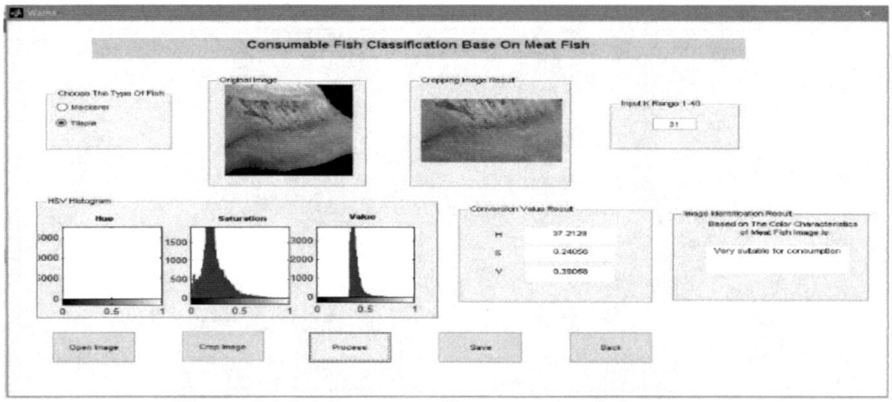

Figure 4. Matlab program for measuring fish meat

Table 1 and
Table are HSV values from mackerel and tilapia, respectively. These values from each of 40 sample data that have been analyzed in testing are used to determine classification results.

Table 1. HSV Value for mackerel meat testing image

#	Image	Cropping Image	HSV Value			Result
			H	S	V	
1	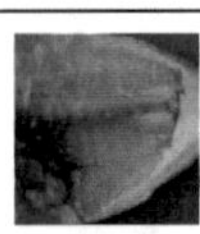		31,6225	0,72686	0,42503	Very suitable for consumption
2		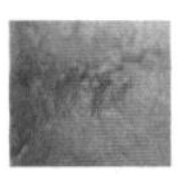	31,477	0,73413	0,43282	Suitable for consumption
3	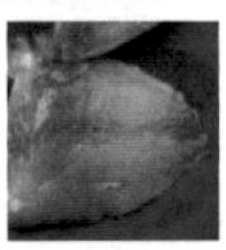	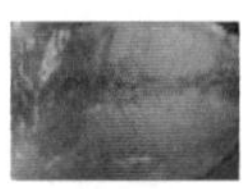	32,0623	0,73078	0,42179	Less suitable for consumption
4	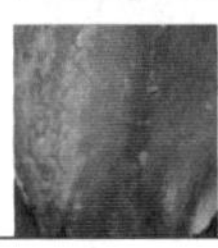	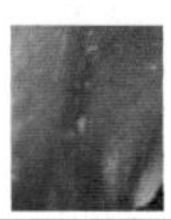	8,3675	0,75102	0,45932	Not suitable for consumption

Table 2. HSV Value for tilapia meat testing image

#	IMAGE	Cropping Image	HSV Value			Result
			H	S	V	
1	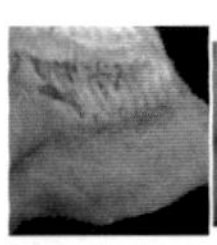	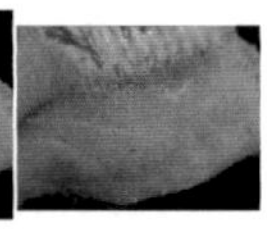	26,4842	0,71565	0,39163	Very suitable for consumption
2	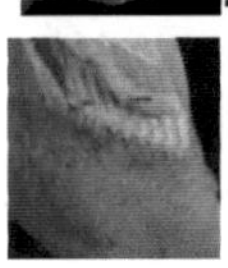		118,9415	0,73051	0,45276	Suitable for consumption

#	IMAGE	Cropping Image	HSV Value			Result
			H	S	V	
3	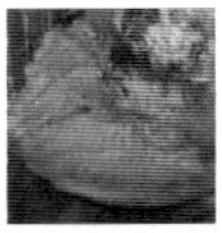	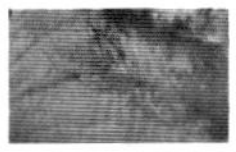	120,5819	0,74068	0,46318	Less suitable for consumption
4			15,917	0,77547	0,50307	Not suitable for consumption

3.2. Fish Scale Measurement

Figure 5 shows an example of fish identification using fish scale images as an input. The steps to perform classification based on the fish scale image is similar to that of classification based on fish meat image.

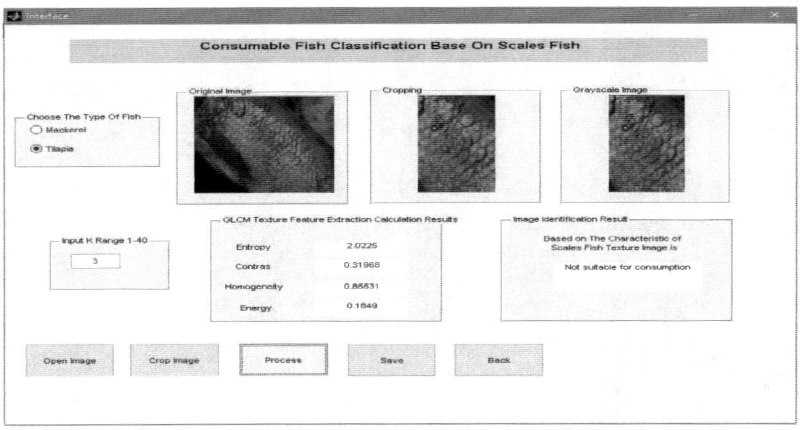

Figure 5. Matlab program for measuring fish scales

Based on the flowchart in Figure 3, the image of the fish scale is converted into grayscale and analyzed using GCLM method to produce four values: entropy, contrast, homogeneity, and energy. **Error! Not a valid bookmark self-reference.** shows the four features extraction values from mackerel and

Table shows four features from tilapia fish scale. These values will be used to calculate the distance in the next step. The result for classification is also displayed directly in the program GUI.

Table 3. Feature extraction for mackerel scales testing image

#	IMAGE	Cropping Image	Feature Extraction				Result
			Entropy	Contrast	Homogenity	Energy	
1	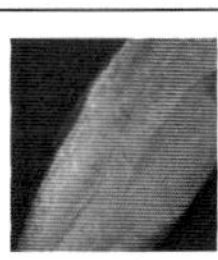		0,64109	0,21189	0,89569	0,32906	Very suitable for consumption

2	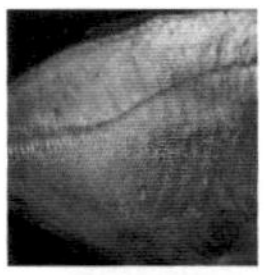	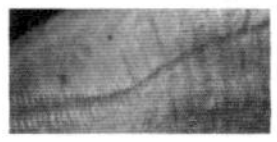	1,0092	0,43618	0,81827	0,13267	Suitable for consumption
3	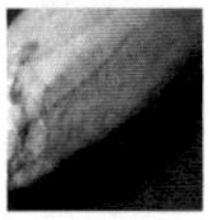	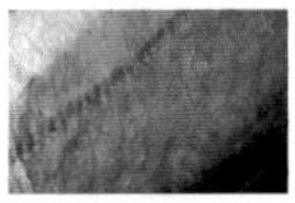	0,99192	0,24722	0,88247	0,16219	Less suitable for consumption
4	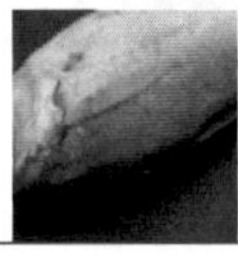	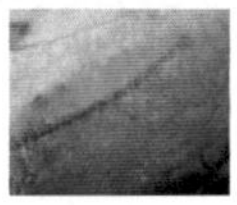	0,95466	0,1631	0,9198	0,15385	Not suitable for consumption

Table 4. Feature extraction for tilapia scales testing image

#	IMAGE	Cropping Image	Feature Extraction				Result
			Entropy	Contrast	Homogenity	Energy	
1	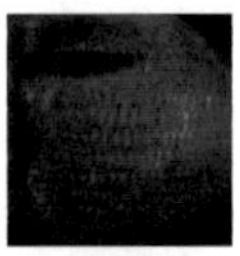		0,45411	0,069207	0,9654	0,47975	Very suitable for consumption
2	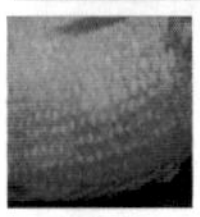		0,85319	0,18405	0,90918	0,18212	Suitable for consumption
3	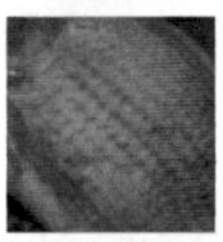	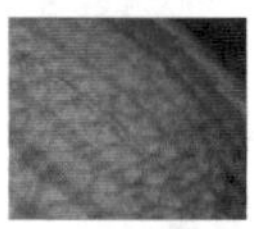	0,68564	0,10422	0,94789	0,26642	Less suitable for consumption
4	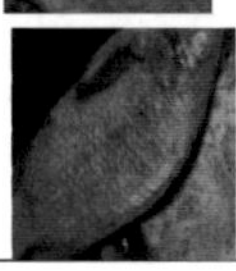	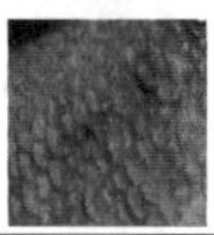	0,78624	0,30847	0,86304	0,23388	Not suitable for consumption

To ensure the results of the classification obtained from the Matlab program, an assessment of experts in the fisheries field is carried out to conduct testing. Table shows testing accuracy between our Matlab program and expert judgment using four scales features. Table 5(a), 5(b), and 5 (c) show testing accuracy for mackerel scales, tilapia scales, and overall accuracy result for all fish, respectively. From Table 5(c) we can see that distinguishing very suitable for consumption and not suitable for consumption category using scales feature results in the most higher accuracy. The lowest accuracy is for suitable for consumption category. The overall accuracy using scales features is 91.25%.

Table 5. Testing accuracy for scales

Clustering Result	Validation Expert Judgment	
	Match	Unmatch
Very suitable for consumption	10	0
Suitable for consumption	10	0
Less suitable for consumption	8	2
Not suitable for consumption	10	0
Accuracy (%)	**(38/40)x100% = 95%**	

(a) Mackerel scales

Clustering Result	Validation Expert Judgment	
	Match	Unmatch
Very suitable for consumption	10	0
Suitable for consumption	5	5
Less suitable for consumption	10	0
Not suitable for consumption	10	0
Accuracy (%)	**(35/40)x100% = 87.5%**	

(b) Tilapia scales

Clustering Result	Validation Expert Judgment		
	Match	Unmatch	Accuracy
Very suitable for consumption	20	0	100%
Suitable for consumption	15	5	75%
Less suitable for consumption	18	2	90%
Not suitable for consumption	20	0	100%
Overall Accuracy (%)	**(73/80)x100% = 91.25%**		

(c) Overall result for all scales

Error! Reference source not found. shows the results of the accuracy testing for meat features between our Matlab program and expert judgment. Table 6(a), 6(b), and 6(c) display the testing accuracy for mackerel meat, tilapia meat, and the overall result for all fish, respectively. Classification using meat features has overall accuracy of 93.5% for all fish. The lowest accuracy result is for not suitable for consumption category. Based on the overall accuracy assessment, an average value of 92.5% accuracy of the Matlab program has been developed.

Table 6. Testing accuracy for meat

Clustering Result	Validation Expert Judgment	
	Match	Unmatch
Very suitable for consumption	10	0
Suitable for consumption	10	0
Less suitable for consumption	10	0

Clustering Result	Validation Expert Judgment	
	Match	Unmatch
Very suitable for consumption	10	0
Suitable for consumption	10	0
Less suitable for consumption	10	0

Not suitable for consumption	9	1
Accuracy (%)	**(39/40)x100% = 97.5%**	

(a) Mackerel meat

Not suitable for consumption	6	4
Accuracy (%)	**(36/40)x100% = 90%**	

(b) Tilapia meat

Clustering Result	Validation Expert Judgment		
	Match	Unmatch	Accuracy
Very suitable for consumption	20	0	100%
Suitable for consumption	20	0	100%
Less suitable for consumption	20	0	100%
Not suitable for consumption	15	5	75%
Overall Accuracy (%)	**(75/80)x100% = 93.5%**		

(c) Overall result for all meat

4. Conclusion

By applying k-Nearest Neighbor algorithm, the user can find information about the classification of fish freshness from fish meat and scale image. The test results of our system show that the identification accuracy for tilapia meat is 90% and 97.5% for mackerel meat. Meanwhile, for the scales, the accuracy is up to 87,5% for tilapia scales and 95% for mackerel scales. In the future, this approach can be implemented in a mobile application so that users will find it easier to identify fish suitable for consumption

References

[1] M. Tomić, D. Matulić, and M. Jelić, "What determines fresh fish consumption in Croatia?," *Appetite*, vol. 106, pp. 13–22, 2016.

[2] L. Gil *et al.*, "Analysis of Fish Freshness by Using Metallic Potentiometric Electrodes," in *2007 IEEE International Symposium on Industrial Electronics*, 2007, pp. 1485–1490.

[3] A. A. Azeriee, H. Hashim, R. Jarmin, and A. Ahmad, "A Study On Freshness Of Fish By Using Fish Freshness Meter," in *2009 5th International Colloquium on Signal Processing & Its Applications (CSPA)*, 2009, pp. 215–219.

[4] R. Jarmin, L. Y. Khuan, H. Hashim, N. Hidayah, and A. Rahman, "A Comparison on Fish Freshness Determination Method," in *2012 International Conference on System Engineering and Technology*, 2012.

[5] S. Guney and A. Atasoy, "Fish Freshness Assessment by Using Electronic Nose," in *2013 36th International Conference on Telecommunications and Signal Processing (TSP)*, 2013, pp. 742–746.

[6] M. K. Dutta, A. Issac, N. Minhas, and B. Sarkar, "Image processing based method to assess fish quality and freshness," *J. Food Eng.*, vol. 177, pp. 50–58, 2016.

[7] A. Issac, M. K. Dutta, B. Sarkar, and R. Burget, "An Efficient Image Processing Based Method for Gills Segmentation from a Digital Fish Image," in *2016 3rd International Conference on Signal Processing and Integrated Networks (SPIN)*, 2016, pp. 645–649.

[8] A. Issac, M. K. Dutta, and B. Sarkar, "Computer vision based method for quality and freshness check for fish from segmented gills," *Comput. Electron. Agric.*, vol. 139, pp. 10–21, 2017.

[9] J. Atanbori, W. Duan, J. Murray, K. Appiah, and P. Dickinson, "Automatic classification of flying bird species using computer vision," *Pattern Recognit. Lett.*, vol. 81, pp. 53–62, 2016.

[10] V. Ashok and D. S. Vinod, "A comparative study of feature extraction methods in defect classification of mangoes using neural network," in *2016 Second International Conference on Cognitive Computing and Information Processing (CCIP)*, 2016, pp. 1–6.

[11] B. Debska and B. Guzowska-Swider, "Application of artificial neural network in food classification," *Anal. Chim. Acta*, vol. 705, no. 1–2, pp. 283–291, 2011.

[12] J. Gu, N. He, and X. Wu, "A New Detection Method for Fish Freshness," in *2014 Seventh International Symposium on Computational Intelligence and Design*, 2014, pp. 555–558.

[13] L. K. S. Tolentino *et al.*, "Fish Freshness Determination through Support Vector Machine," *J. Telecommun. Electron. Comput. Eng.*, vol. 9, no. 2–5, pp. 139–143, 2017.

[14] A. Atasoy, U. Ozsandikcioglu, and S. Guney, "Fish Freshness Testing with Artificial Neural Networks," in *Electrical and Electronics Engineering (ELECO), 2015 9th International Conference on IEEE*, 2015, pp. 700–704.

[15] S. Bermejo, "Ensembles of wrappers for automated feature selection in fish age classification," *Comput. Electron. Agric.*, vol. 134, pp. 27–32, 2017.

[16] N. M. S. Iswari, Wella, and Ranny, "Fish Freshness Classification Method Based on Fish Image using k-Nearest Neighbor," in *2017 4th International Conference on New Media Studies*, 2017, pp. 87–91.

[17] J. Sun *et al.*, "Classifying fish freshness according to the relationship between EIS parameters and spoilage stages," *J. Food Eng.*, vol. 219, pp. 101–110, 2018.

The insertion of confidential information into digital images using blowfish cryptography and end of file steganography

E Aribowo* and A Suryadi

Department of Informatics Engineering, Faculty of Industrial Technology, Universitas Ahmad Dahlan, Yogyakarta, Indonesia

*Email: ekoab@tif.uad.ac.id

Abstract: Identity Theft Resource Centre revealed there were 668 cases of cybercrime occurring with a total number of lost data 22,408,258 started from January to July 2018. It indicates that there was a vulnerability in the defences from attacks carried out through the internet. Thus, the world of information systems should also be accompanied by a high level of security. The stages of system development were started from the analysis of system requirements, the design of the UML, implementation, and testing. The testing process into two-phase. Firstly, system output testing in the form of stego image with various criteria, namely imperceptible, fidelity, recovery, robustness, and histogram. Secondly, system feasibility testing using the SUS method. Based on the results, it revealed that the merging of the blowfish cryptographic method and the end of file steganography was not very effective because, based on the output testing with robustness criteria; it was proven that the message inserted into the image was damaged during extraction. The changes in the size of the original cover with the resulted stego image had increased file size with a ratio of 1: 5.5, meaning that each created stego image was five times the size of the original image.

1. Introduction

The rapid development of the Internet provides complete facilities and easiness for its users, especially for a medium of communication and data transmission. If you look at the existed data, along with the development of the internet and applications that require the internet, the number of crimes in the world of information systems is also increasing [1]. Identity Theft Resource Centre (ITRC) revealed there were 668 cases of cybercrime occurring with a total number of lost data 22,408,258 started from January to July 2018 [2]. Based on the fact from ITRC, the security of information is not a trivial topic. There is much information that should be kept its confidentiality, especially personal information, financial data, and nation data. Along with the rapid development of the internet, the security of information on the internet should be more increased. Various techniques are used to protect the information from the third person in terms of ownership, access, and use. Two of the popular technique are cryptography and Steganography [3]. The cryptography technique will encrypt the messages that will be sent into random information that is meaningless or often called as chipper text using specific algorithms and keywords before being transmitted over the internet. Then, it will be decrypted into meaningful information that can be understood (plaintext) by the recipient when the information or message has been received using the key owned.

However, the security of information by using cryptography has some weaknesses. Firstly, the random information (chipper text) can be seen clearly, so it attracts people's curiosity to break the encryption. Secondly, the chipper text can be solved in a short time, along with the rapid growth of hardware and increasingly efficient optimization software makes computing capabilities faster [4]. Therefore, another technique is needed to address the weaknesses of the cryptographic technique. This technique is called Steganography. Steganography is the study and art of hiding messages into a media in such a way so that no one knows or realized that there is a message contained in the media because it is not visible to the human sense of sight [5]

2. Literature review

2.1. *Review of previous relevant studies*

Based on result of research [6], after conducting various tests such as MSE (mean square error), PSNR (peak signal to noise ratio) and coefficient, it was found that the LSB method (least significant bit) proved to be better in terms of quality while the resistance in the form of salt and pepper attacks MSB (most significant bit) is superior.

Moreover, other research by [7] is an application that can insert messages into images using blowfish cryptographic techniques and End Of File steganography. In this application, the information that will be sent is encrypted and then inserted into an image. However, in this study, it was only limited to the process of inserting messages into a cover without being decrypted into plaintext.

2.2. *Review of literature*

2.2.1. *Cryptography*

In general, cryptography is defined as the study and art of encoding, aiming to maintain the security and confidentiality of a certain message. Cryptography is the study of encryption techniques where random data using certain encryption key, so it becomes hard to read by someone who does not have a decryption key. Cryptography is the study and art of maintaining the security and confidentiality of messages by encoding them into forms that cannot be understood [8]. Furthermore, there is also another definition; it is a study of mathematical techniques related to security aspects such as data confidentiality, data validity, data integrity, and data authentication.

Blowfish is a type of symmetric-key cryptography that has encryption and decryption processes using the same key, which an efficient algorithm security [9]. It's an encryption method similar to the DES cryptographic algorithm. However, after a test was conducted to evaluate the DES algorithm with Blowfish, it can be concluded that the encryption security and encryption process speed of the blowfish algorithm is superior, whereas for the power consumption used for the two algorithms has the same magnitude [10].

Blowfish algorithms are divided into two important phases. They are expansion and encryption data.

a. Key Expansion has a function to change keys (32-bit minimum and 448-bit maximum) into sub-key arrays with a total of 4168 bytes.
b. Data encryption Consists of simple function iterations (Feistel Network) of 16 cycles. Each round consists of key-dependent mutation and key-and data-dependent substitution. All operations are addition and XOR to 32-bit variables. Another additional operation is four indexed array table lookups for each round.

2.2.2. Blowfish algorithm

Since Blowfish uses large sub keys, the key must be counted before encrypting or decrypting data. Blowfish is an algorithm that applies a Feistel network consisting of 16 rounds [6]. For the flow of the encryption algorithm, the Blowfish method is explained as stated:

a. The initials of the P-array are 18 (P1, P2, ... P18), each initial consist of 32-bit. The P array consists of eighteen key, 32-bit sub key keys: P1, P2, ..., P18.
b. 4 S-boxes are 32-bits, each with 256 entries. Four 32-bit S-boxes each have 256 entries:
   S1,0, S1,1, ..., S1,255
   S2,0, S2,1, ..., S2,255
   S3,0, S3,1, ..., S3,255
   S4,0, S4,1, ..., S4,255.
c. The Plaintext to be encrypted is assumed to be an input, the Plaintext is taken as much as 64-bit, and if it is less than 64-bit, then we add more bits so that later the operation matches the data.
d. The results of the retrieval were divided into 2, the first 32-bit is called XL, and the second 32-bit is called XR.
e. Next do the XL = XL xor Pi and XR = F (XL) xor XR operations.
f. The results of the operations above then were exchanged XL to XR and XR to XL.
g. Do the process 16 times. In 16th repetition, then repeat do the XL and XR exchange process.
h. In the 17th process, do XR = XR xor P17 and XL = XL xor P18 operations.
i. The last process is to bring back XL and XR so that it becomes 64-bit again.

Blowfish uses a Feistel network consisting of 16 rounds. To be more precise, the Feistel network scheme can be seen in Figure 1.

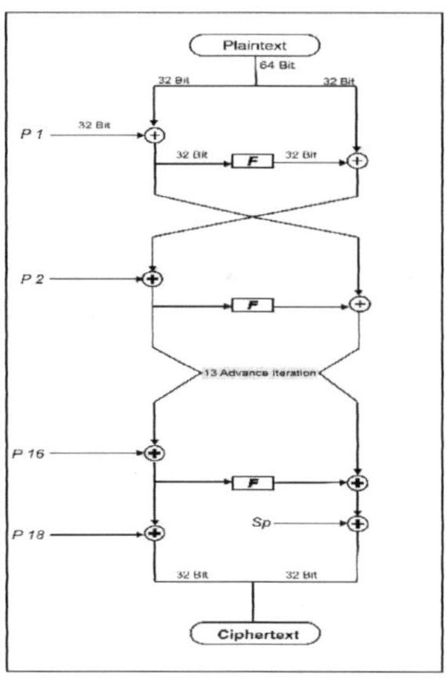

Figure 1. Feistel Network Scheme

The blowfish algorithm is unique in terms of the decryption process. The decryption process is carried out in the same order as the encryption process. However, decryption process P1, P2, ..., P18 is used in reverse order. In the Blowfish algorithm, there is also an F function. F function is:

a. Divide XL into four 8-bit parts: a, b, c, and d.
b. F (XL) = ((S1, a + S2, b mod 232) xor S3, c) + S4, cmod 232

Subkeys are calculated using the blowfish algorithm, with the method as follows:

a. First, initialize the P-array and then four S-boxes in sequence with a fixed string. This string consists of hexadecimal digits of Pi.
b. XOR P1 with the first 32-bit key, XOR P2 with the second 32-bit key, and so on for each bit of the key (up to P18). Repeat the key bits until the entire Parray of XOR is changed with the key bits.
c. Encrypt all zero strings with the Blowfish algorithm by using subkeys as described in steps (1) and (2).
d. Replace P1 and P2 with the output from step (3).
e. Encrypt the output from step (3) with the Blowfish algorithm with a modified subkey.
f. Replace P3 and P4 with the output from step (5).
g. Continue the process, replacing all elements of the P-array, then all four S-boxes in sequence, with the output changing continuously from the Blowfish algorithm. The F function in blowfish can be seen in Figure 2

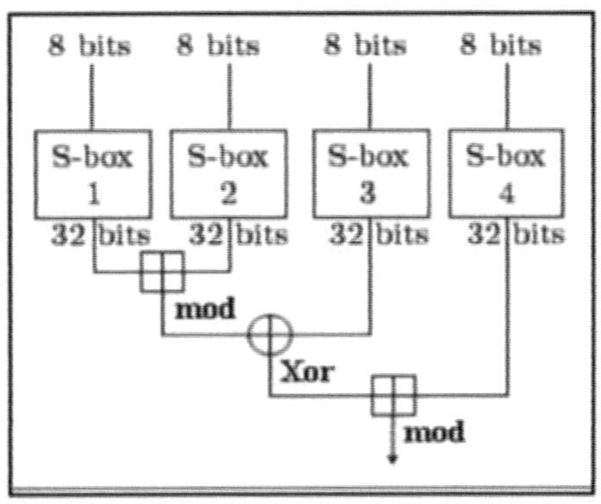

Figure 2. Function in Blowfish Algorithm [6]

2.2.3. Steganography

The word steganography (steganography) derived from the Greek word *steganos*, which means hidden or veiled and *graphics* means writing. It is defined as veiled writing [5]. Steganography is the study and art of hiding confidential messages so that they are not detected by the human senses. Steganography consists of two main processes. They are the process of insertion and extraction. The insertion process is the process of hiding information on the media to be boarded or often with media coverage.

The extraction process is the process of removing secret messages from the stego medium. Data that becomes stego media is usually in the form of general data that is often sent like audio, photo, or video. In an ideal situation, anyone who scans the data will not know that the data contains confidential information so that the retrieval of information can only be done by those who are entitled to receive the information.

In hiding the messages using steganography techniques, the several criteria that must be required [3], as follows:

a. Imperceptible means that every message inserted cannot be perceived by the human senses. If a message is inserted into a picture, then the picture that has been pasted by the message or information must be indistinguishable from the original image by the human sense of sight. For hiding the sound media, the ear must not find the difference between the original sound and the sound that has messages inserted. Similar to them, video as a combination of images and sound, the

IOP Conf. Series: Materials Science and Engineering **821** (2020) 012040 doi:10.1088/1757-899X/821/1/012040

eyes and ears must not be able to detect the difference between the original file and the file that has been pasted.

b. Fidelity means that the quality of the storage media (stego medium) does not change due to insertion.
c. Recovery means that confidential messages can be revealed again. The purpose of steganography is hiding information so that at any time, the confidential information can be retrieved and reused as needed.
d. Robustness means the data or information that hidden must be stand in the media cover manipulation. When the object processing operation is done, the hidden data will not be broken.

2.2.4. End of File algorithm

The end of File (EoF) technique is one of the steganography techniques. This technique is to utilize the attached data in the end of the file. This technique also able to insert a message that sized based on the need. Thus we will able to insert a message based on our need. However, we need to remember that the stego size transformation will increase significantly if the image/cover is bigger than 5 MB. This certainly needs attention to avoid damage in a stego image pixel. In this technique, the message also inserted in the end of the file.

On the digital image processing, histogram has an important role that able to use as a determinate indicator of a digital image quality. These are the roles of histogram in digital image processing:

1. Use to observe the spread of color intensity and can be used for decision making, for example in increasing the brightness or stretching the contrast of an image and color distribution.
2. Use to determine the boundaries in separating objects from their backgrounds.
3. Give a presentation of color composition and texture intensity for the purpose of identification of digital images.

2.2.5. System Usability Scale (SUS)

Usability is a qualitative analysis that determines how easily the user uses the interface application. An application product is categorized or called "usable" if it meets the criteria of useful, efficient, effective, satisfying, learnable, and accessible System Usability Scale (SUS) is an effective and reliable usability testing package for use in a various products and applications [11]. This SUS test has 10 questions using Likert scale of 1 to 5. Each question with an odd number (1,3,5,7,9) is a positive question, while an even-numbered question (2,4,6,8,10) is a negative question.

3. Finding and discussion

3.1. Designing the system

Modelling is the process of designing software before a coding. Creating a model of a complex system is very important because we can understand the system as a whole. Besides modelling can also be used as a visualization and construction in a software manufacturing process. In designing this software, the analysis is done by modelling the problem in the form of UML diagrams.

This application is designed to look as user friendly as possible, making it easier for users to use this application. Moreover, this application is not only targeted for users from the Informatics Engineering department, but for any kind of users. Therefore the appearance of this application is made by selecting the layout of the menu, the choice of color or text is made as attractive as possible.

IOP Conf. Series: Materials Science and Engineering **821** (2020) 012040 doi:10.1088/1757-899X/821/1/012040

3.2. Testing result

3.2.1. System output testing

- Imperceptible & fidelity criteria testing

This test is done in a subjective way, all digital images that will try to insert a message as much as 256 words with a file size of 1.62KB. This test as Table 1 is done by asking for the assessment of 5 respondents who of course have the most important criteria which is to have good vision as.

Table 1. Result test imperceptible & fidelity criteria testing

Image	Original size	Stego image size	Result
	41KB	1024KB	5 respondents do not see the difference
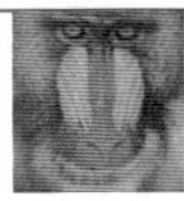	21KB	778KB	5 respondents do not see the difference
	17KB	749KB	5 respondents do not see the difference
	8KB	148KB	5 respondents do not see the difference
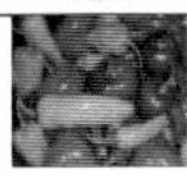	14KB	148KB	5 respondents do not see the difference

- Recovery criteria testing

A steganography can be categorized as good if all messages that have been inserted are able to be expressed again (recovery / extraction) because the purpose of steganography itself is to hide a message. Sometimes at a certain time, a message must be taken for a various purpose in the further use. Therefore, in testing this criterion, the inserted message has a different number of characters, ranging from 16 words, 64 words, 128 words, and 256 words. In this test as Table 2, the indicator used as a reference that the message recovery is successful if all the message characters are saved can be returned in full without any difference.

Table 2. Result test of recovery criteria testing

Image	Words number	Extraction result	Result
	16 words 64 words 128 words 250 words	16 words 64 words 128 words 250 words	Success Success Success Success
	16 words 64 words 128 words 250 words	16 words 64 words 128 words 250 words	Success Success Success Success

IOP Conf. Series: Materials Science and Engineering **821** (2020) 012040 doi:10.1088/1757-899X/821/1/012040

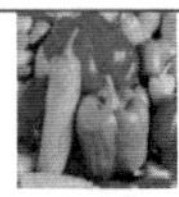	16 words 64 words 128 words 250 words	16 words 64 words 128 words 250 words	Success Success Success Success
	16 words 64 words 128 words 250 words	16 words 64 words 128 words 250 words	Success Success Success Success
	16 words 64 words 128 words 250 words	16 words 64 words 128 words 250 words	Success Success Success Success

- Robustness criteria testing

Robustness criteria are information/ messages that are hidden into digital images that are resistant to a form of manipulation performed on stego images (Table 3). To do this test various manipulations will be carried out on the stegoi mage. Then, the message that has been inserted will be tried to be extracted again.

The testing step is done by making a stego image using the system, where the inserted message consists of 10 words. After successfully embedded/ encoding, the stego image will be manipulated by changing it into a number of patterns, and then it will be extracted again. Of the 6 pieces that have been used, one will be chosen as a robustness test, namely the Lena.png image.

Table 3. Test of robustness criteria

Stego image	Manipulation Types	Extraction Result	Result
	90^{0-} Rotation	The message has been extracted but the chypertext is broken	Cyphertext is failed to decrypted
	50% brightness addition	The message has been extracted but the cyphertext is broken	Cyphertext is failed to decrypted
	Mercury effect addition	The message has been extracted but the cyphertext is broken	Cyphertext is failed to decrypted
	Zeke effect addition	The message has been extracted but the cyphertext is broken	Cyphertext is failed to decrypted

- System usability scale testing

This testing scenario is done by presenting how the system works, how to use the system to respondents, and asking respondents to try the system, the result as Table 4. The number of respondents used in this system test is 10 people from various background/study programs in order to know that this system can be used by all types of users

Table 4. Respondent systems

User	SUS Score	Grade
10 user testing	75.625	Acceptable (B)

3.3. *Respondents calculation result*
By looking at the SUS value obtained at 75.625 and rank with grade B, then this application is categorized acceptable or acceptable.

3.4. *Recapitulate all questions points*
In accordance with the explanation on the theory of the system usability scale (SUS) that questions that are odd numbered are positive-pitched questions, while questions that are even-numbered are negative. This means that the more agreed the respondent is with the odd numbered questions, the higher the positive score obtained. And conversely, the higher score even numbered the respondent disagrees with the question [11]. In accordance with the tests that have been done, the recapitulation of the results of all odd and even question values can be seen in Figure 3

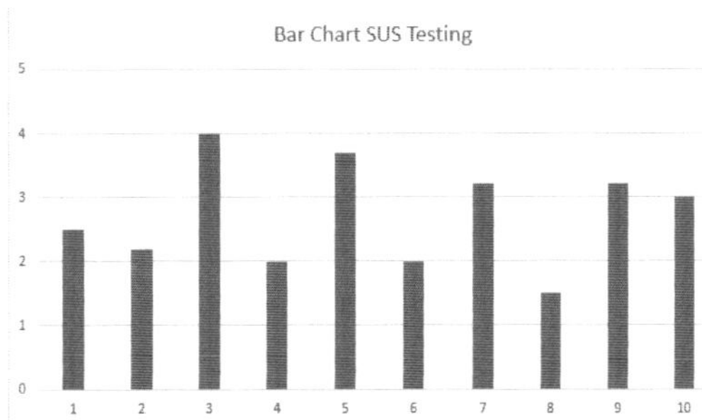

Figure 3. Respondents score diagram

Referring to Figure 3, it can be seen that Question No. 10 has the highest score which means that most respondents gave a negative assessment of the aspects in question number 10. From this, it can be identified aspects that become the lack of this application. So for the development of this application which is the main focus aside from the main function of the application are aspects related to question number 10. If it is further examined questions that are positive in tone also still have less than the maximum value such as question no 1, then this needs to be deepened and needs to be improved in the future.

4. Conclusion
From all the research steps that have been carried out, the following conclusions can be drawn as follows: First, the application is made capable of implementing a combination of cryptographic blowfish and end of file steganography very well. Without any damage to the image which is used as a cover for hiding messages. Second, the performance of the EOF method (Output system testing) was tested using imperceptible, fidelity, recovery, robustness, and histogram criteria. It turns out that this EOF method successfully passed the imperceptible, fidelity and recovery tests but was unable to pass the robustness criteria test. While for the histogram, found a slight difference between the original image histogram with the stego image histogram. Finally, from the usability scale (SUS) system testing conducted on 10 respondents, this application has an acceptable of 72.5%, meaning that this application has a good usability level.

IOP Conf. Series: Materials Science and Engineering **821** (2020) 012040 doi:10.1088/1757-899X/821/1/012040

References

[1] Moreno F M M, Blanco F, Garaizar P and Matute H 2017 Fishing for phishers Improving Internet users' sensitivity to visual deception cues to prevent electronic fraud 2017 *Comput Human Behav* vol **69** pp 421–36.

[2] Hamilton A 2019 Identity theft resource centre (ITRC): Third party data breach becomes focus a hacker [Online] Available: idtheftcenter.org/tag/cybercrime/. [Accessed: 26-Apr-2019].

[3] Chauhan S 2017 Multiple layer text security using variable block size cryptography and image steganography in *2017 3rd International Conference on Computational Intelligence & Communication Technology (CICT)* pp 1–7.

[4] Mishra R and Bhanodiya P 2015 A review on steganography and cryptography in *Conference Proceeding - 2015 International Conference on Advances in Computer Engineering and Applications, ICACEA 2015* pp 119–22.

[5] Hussain M, Wahab A W A, Bin Idris Y I, Ho A T S and Jung K H 2018 Image steganography in spatial domain: A survey *Signal Process Image Commun* vol **65** pp 46–66.

[6] Dwi G, Rumani and Nasrun M 2015 Implementation of cryptography and steganography in image media using the blowfish algorithm and the least significant bit method *e-Proceeding Eng* vol **2** no 2 pp 3762–69.

[7] Abbas T, Beiji Z and Abdullah M Y 2011 Information security technique in frequency domain *Int. J Digit Content Technol its Appl* vol **5** no 12 pp 279–89.

[8] Rwabutaza A, Yang M and Bourbakis N 2012 A comparative survey on cryptology-based methodologies *Int J Inf Secur Priv* vol **6** no 3 pp 1–37.

[9] Abdul N and Mustafa A 2014 The effect of the smoothing filter on an image encrypted by the blowfish algorithm then hiding it in a BMP image *J Next Gener Inf Technol* vol **5** no 4 pp 83–92.

[10] Joshi A, Wazid M and Goudar R H 2015 An efficient cryptographic scheme for text message protection against brute force and cryptanalytic attacks *Procedia Comput Sci* vol **48** no Iccc pp 360–66.

[11] Martins A I, Rosa A F, Queirós A, Silva A and Rocha N P 2015 European portuguese validation of the System Usability Scale (SUS) *Procedia Comput Sci* vol **67** no Dsai pp 293–300.

IOP Conf. Series: Materials Science and Engineering **821** (2020) 012041 doi:10.1088/1757-899X/821/1/012041

Data security improvements on cloud computing using cryptography and steganography

N R D P Astuti*, E Aribowo and E Saputra

Department of Informatics Engineering, Faculty of Industrial Technology, Universitas Ahmad Dahlan, Yogyakarta, Indonesia

*Email: rochmahdyah@tif.uad.ac.id

Abstract. Cloud computing is a service that eases its users to store data on the internet. However, cloud computing has lack, which is related to stored data security specifically to image file that can be manipulated easily. To improve the data securities of JPG file type on cloud computing, the combination of advanced encryption standard (AES) cryptography and least significant bit (LSB) steganography methods is required. This study implemented AES 128 algorithm cryptography process to scramble the messages that will be inserted and LSB steganography to hide the messages that has been scrambled. The data were assessed using Stegspy. The results of this study indicated that cloud computing service of Platform as a Service (PaaS) has resulted ciphertext that was attached to JPG/JPEG images using Least Significant Bit (LSB). From 5 (five) samples, the encryption result testing on JPG/JPEG files was bigger than the original file. The bigger the number of characters of the inserted text, the bigger the size of JPG/JPEG files will be.

1. Introduction

Digital storage media is massively utilized at this time along with the development of technology. One of which is the cloud computing-based storage. Cloud computing provides easy access to store the data on the internet for the users. Various storage capacities are offered by the service providers based on the users' requests [1]. The users do not need to prepare either software or hardware devices to store the data for the data and information will be stored in a physically unknown server. The data on cloud can be accessed by the client using internet network [2, 3]. By utilizing cloud computing as storage media, the users will be able to reduce the operational costs without providing infrastructure and software devices to use it [1, 5].

However, from the security aspect, cloud method comes up with lack that related to data security, which can be stolen [4]. This is linked to the increase of practiced hackers in the whole world that escalates the risks toward cloud computing data security [3, 4]. Image file is one of the most stored types in cloud. Image file format manipulation can convey negative impacts with the change of the original form that can harm the owners by giving them unpleasant impressions [5].

To secure image format file, cryptography technique is encouragingly suggested to be implemented. The data encoded by cryptographic techniques will convert into incomprehensible data [6]. Nevertheless, by only implementing cryptography technique will trigger suspiciousness to the cryptanalysts. To enhance or lessen the suspiciousness, hence steganography process is put into operation. The advantage of steganography technique covers the ability to hide classified data into digital media; therefore, the data are hardly recognized by other people. The Least Significant Bit (LSB) is one of steganography methods most popular [7]. With the change of bit, it will converse the value in the RGB colors equal to 1 out of the 256 available colors, therefore the conversion is scarcely detected by bare eyes [8].

To improve the security of image format file, two processes of combining cryptography technique using AES algorithm with steganography technique using LSB method [9] were implemented in this study. The image format file tested in this study was JPG file type that stored in cloud computing.

2. Research method

2.1. Cloud computing architecture system
There are four layers covered in cloud computing architecture system. The layers consist of hardware device, infrastructure, platform, and application [3] as illustrated in Figure 1.

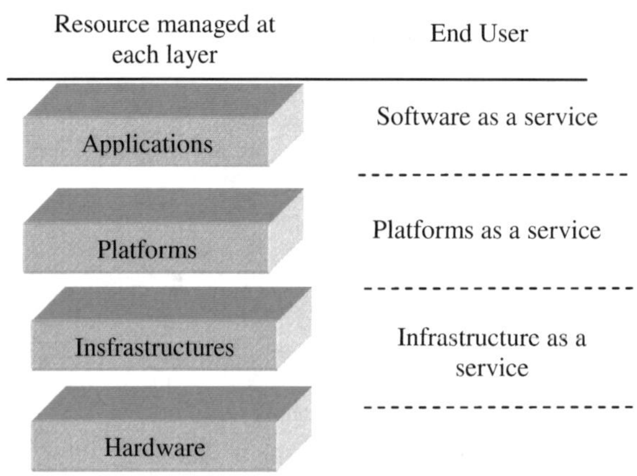

Figure 1. Cloud computing architecture

To assemble the system, a service based on the Platform as a Service (PaaS) of the cloud computing system was utilized. PaaS is a delivery service that consists of two layers. It provides several services such as assembling, testing, and spreading any under development application [3].

2.2. Flowchart system
Flowchart is used to describe the built system flow. There are two processes that will be executed by the system. The first process takes in the cryptography method using AES algorithm, which includes encryption and decryption processes for the file that will be hidden. The second process takes in the steganography method using LSB that used to hide classified file in image file format.

2.3. Construction
Construction refers to the implementation of the system workflows that have been designed. The system will be implemented on cloud computing. The construction process uses PHP and JavaScript programming.

2.4. Testing
To make sure that all methods run well, hence the testing proses applies both functionality and compatibility tests.

3. Experimental result

3.1. Finding and discussion
Cloud computing scheme is a model to give a comfortable request network access from several users to share a configured computing resource (for example a network, server, storage, application, and service) that able to quickly set and release using a minimum management or interaction service provider. The result of developing scheme can be seen in the Figure 2.

The process of cloud computing system that built to support a security process is PaaS with encryption and Decryption process run on cloud computing. With the benefits on PaaS has been provide an operating system, database, and web server. Figure 3 is an illustration of cloud computing system that has been built.

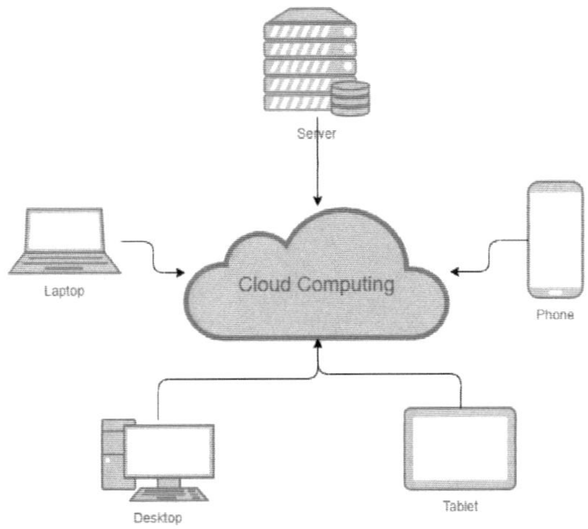

Figure 2. Cloud computing scheme

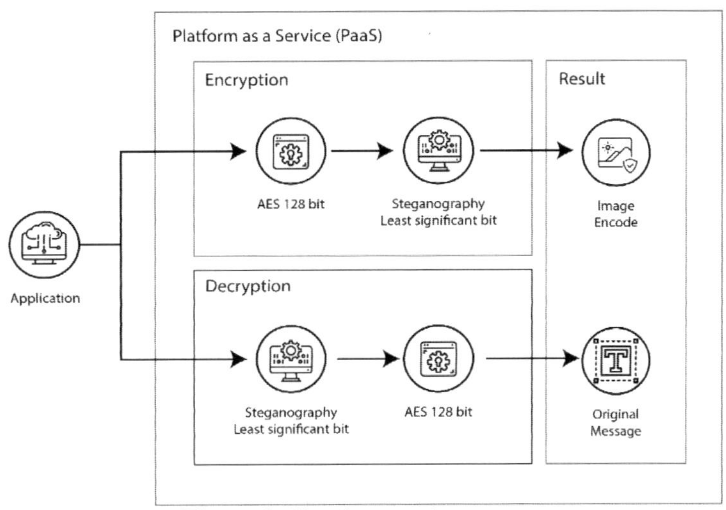

Figure 3. Cloud computing process.

The first process is encrypting using a cryptography hidden picture method. The algorithm that will be used in this process is AES 128 bit. The result of encryption process will be hidden using LSB (Least Significant bit) steganography method. The final result of this process is an image with a hidden message inside.

To open a hidden message, the hidden message is done first using steganography to separate an image file as a place to hide a hidden message. This decryption file will produce two file. After a separation to know the hidden message so second process is decryption using AES 128 Bit Algorithm. It is used to restore a message to its previous form so it can be read by the user.

To know the whole plot system so the flowchart system is built, as well as in Figure 4. The first step is started with input a plaintext and followed by input key that must be 128 bit in length or 16

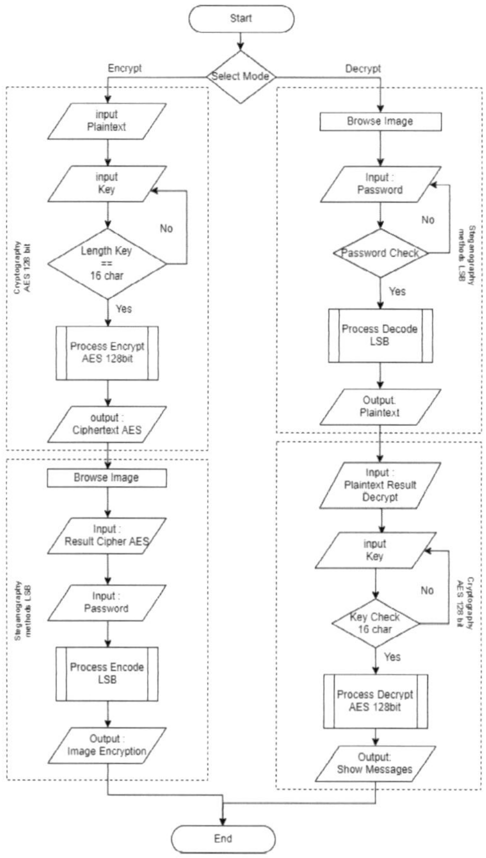

Figure 4. Encryption and decryption algorithm by AES 128 bit and LSB flowchart

character, if the step is different it will go back to input key. After the right key process done, there will be an encryption process that produces a ciphertext. This ciphertext then copied for the next process, which is inserted the result of encryption AES 128 Bit into an image using LSB Steganography method. The following step is uploading the image and then filling the plaintext with the result of ciphertext AES 128 Bit, unlimited character on password input and then the LSB encryption is done and resulting an encrypted image.

The first decryption process is uploading a picture, then putting the password if it is already suit with the decryption process that produce a plaintext, the plaintext will be put on the decryption AES 128 process, and then if the input key is appropriate it will continue with decryption AES 128 that produce an original message that has been encrypted before.

The encryption page using algorithm AES 128 bit by putting a plaintext, a key with 16 character. An encrypted Ciphertext will be inserted using LSB method on the JPG formatted Image. The final result is an image that has been inserted a random message.

3.2. Testing

Tests are carried out using a JPG / JPEG image file for inserting a message. Text or message file that will be hidden. The file to be used is a file that will be stored in cloud computing. Files before being saved to the cloud will be processed encryption and hiding messages. Testing by using five samples of different image files of different sizes and messages to be hidden that have been through the encryption process with AES 128 bits. The test results are shown in Table 1. Based on the testing as Table 1, it shows that the change in the file capacity is very significant, this is because the process of inserting images with different characters.

The next testing process is done five times using the StegSpy 2.1 application. This StegSpy application is an application that allows identifying hidden files that are used to hide messages. The test results are shown in Table 2.

Table 1. The comparison between the size of image file before and after inserting a message

No.	File Name	Dimension size	JPG Capacity	
			Before	After
1	20180709044816.JPG	1200x690px	539kb	1.47 Mb
2	UAD kampus 3.JPG	1024x768px	174kb	1.16 Mb
3	UAD-2.JPG	1303x745px	164kb	1.14 Mb
4	Masjid-UAD.JPG	1080x810px	97.5kb	1.03 Mb
5	Masjid Islamic Center UAD.JPG	1600x1067px	133kb	1.15 Mb

The testing process is done four (4) times with different shots, namely from DSLR cameras, cellphone cameras, from the internet, and from Photoshop, this is intended to determine changes to files before and after encryption. Table 2. Described a comparison of the results of stegspy.

Table 2. A test using image file taken from different tools

No	Before Encryption	After Encryption
1	 Source : Canon EOS 6D Dimension : 5182 x 3456px FileSize : 6.75 MB (7,080,872 bytes) 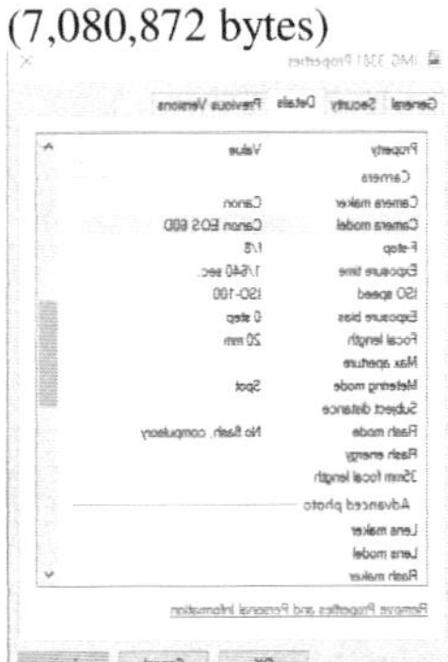	 Dimension : 700 x 466 File Size : 726 KB (743,473 bytes) **Undetected by StegSpy** 
2	 Source : Internet Dimension : 1024 x 768px File Size : 174 KB (178,485 bytes)	 Dimension : 700 x 525px File Size : 804 KB (823,315 bytes) **Detected by StegSpy**

	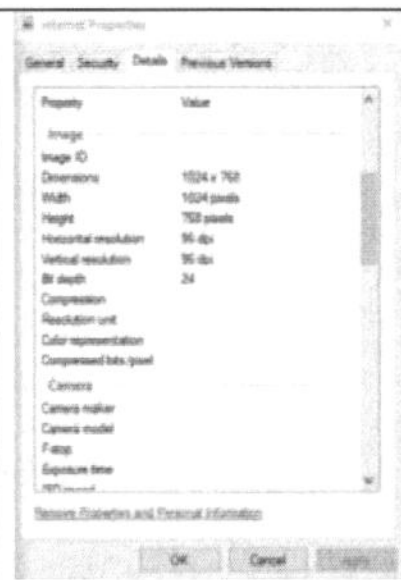	
3	 Source : Smartphone camera Dimension : 3000 x 4000px File Size : 3.29 MB (3,456,020 bytes) 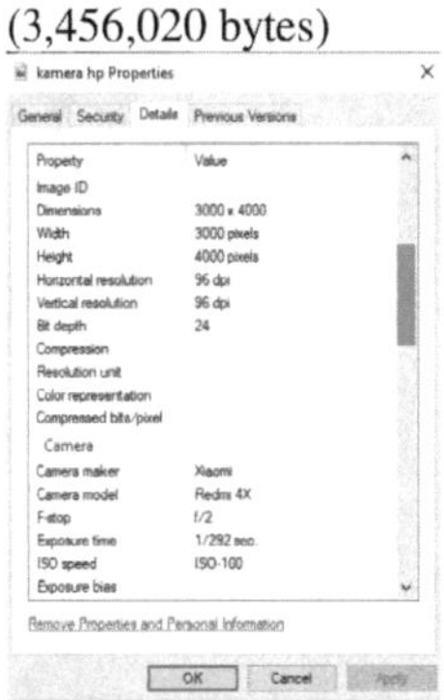	 Dimension : 700 x 525px File Size : 799 KB (818,892 bytes) **Undetected by StegSpy** 
4	 Source : Photoshop Dimension : 1920 x 1200px File size : 189 KB (193,690 bytes) 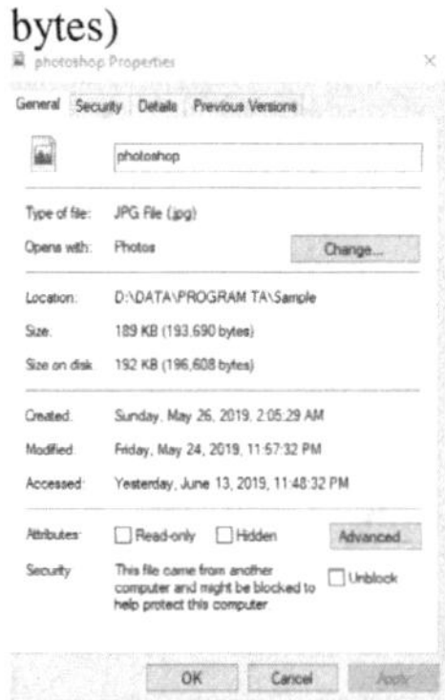	 Source : Kamera HP Dimension : 700 x 437px File Size : 459 KB (470,812 bytes) **Undetected by StegSpy** 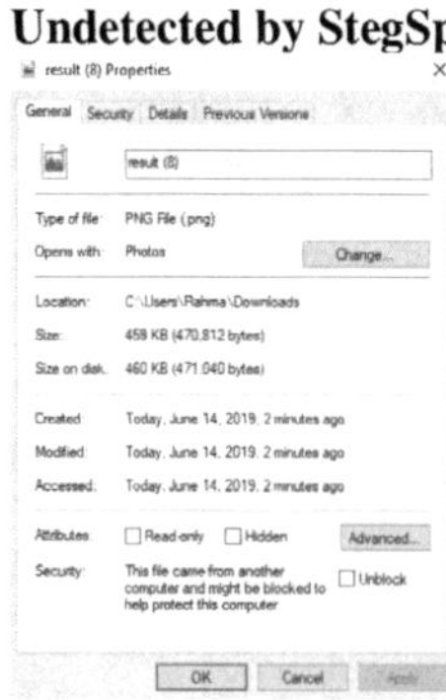

4. Conclusion

The cryptographic application of the AES128 algorithm and LSB method steganography in cloud computing can be used to improve data security in the form of JPG / JPEG image format files. The process of hiding a message successfully done into a JPG / JPEG image using the media canvas as a message insertion process with the LSB method steganography process into the image carried out on the cloud. The Advanced Encryption Standard (AES 128bit) algorithm can be used for message randomization so that messages that will be inserted in the image are in the form of a ciphertext that will be difficult to read.

References

[1] KPMG Advisory N.V 2010 From hype to future: KPMG's 2010 cloud computing survey in *KPMG's 2010 Cloud Computing Survey* p 44.

[2] Webb J, Ahmad A, Maynard S B and Shanks G 2014 A situation awareness model for information security risk management *Comput Secur* vol **44** pp 1–15.

[3] Zhang Q, Cheng L and Boutaba R 2010 Cloud computing: state-of-the-art and research challenges *J Internet Serv Appl* vol 1 no 1 pp 7–18.

[4] Zhou L, Varadharajan V and Hitchens M 2013 Achieving secure role-based access control on encrypted data in cloud storage *Inf Forensics Secure IEEE Trans* vol 8 no 12 pp 1947–60.

[5] Hussain M and Hussain M 2013 A survey of image steganography techniques *Int J Adv Sci Technol* vol **54** pp 113–24.

[6] Mishra R and Bhanodiya P 2015 A review on steganography and cryptography in *Conference Proceeding-2015 International Conference on Advances in Computer Engineering and Applications ICACEA 2015* pp 119–22.

[7] Chia-Chen Lin C C C and Yi-Hui 2009 CLSB-based high-capacity data embedding scheme for digital images *Int J Innov Comput Inf Control* vol **5** no 11 p 4289.

[8] Qazanfari K and Safabakhsh R 2014 A new steganography method which preserves histogram: Generalization of LSB++ *Inf Sci (Ny)* vol 277 pp 90–101.

[9] Juneja M 2014 A covert communication model-based on image steganography *Int J Inf Secur Priv* vol **8** no 1 pp 19–37.

IOP Conf. Series: Materials Science and Engineering **821** (2020) 012042 doi:10.1088/1757-899X/821/1/012042

Correlation of risk factors for coronary heart disease in type-2 diabetes mellitus using the chi-square method

L Syafaah*, D Arisnanda, M Irfan, and R H Hendaryati

Department of Informatics Engineering, Faculty of Engineering, Universitas Muhammadiyah Malang, Malang, Indonesia

*Email: lailis@umm.ac.id

Abstract. The male sex is taken into account a crucial risk issue for coronary heart disease (CHD) within the general population, however, studies on diabetic patients counsel that diabetic girls have a better risk of developing CHD. This study was to determine whether there is correlation between risk factors and the incidence of CHD in patients with type 2 DM. Furthermore, this work also investigates the mathematical model work in calculating the magnitude of the risk of variables against the tendency of CHD complications for patients with type 2 diabetes. This research utilizes current information on the setup of Dr. Soetomo Hospital's medical records in 2015-2016. Samples were all patients with type 2 diabetes with CHD and without CHD reported in the inpatient medical records of Dr. Soetomo Hospital Surabaya. Data processing was carried out using the chi-square test technique. As a result, the system is proof to calculate the worth of the chi-square and confirm the results of conditions on every variable. There is a distinction in price between the calculations of the chi-square on the system with the calculation of the manual chi-square test. This is often caused by taking variety of digits behind completely different commas (,). The variables of gender, age, GDA, GDP, GD2PP, triglycerides, HDL, weight, smoking, and length failed to have a significant relationship on the incidence of CHD in patients with sort two DM. Sterol variables and lipoprotein levels have a highly vital relationship within the incidence of CHD for patients with sort two polygenic disease.

1. Introduction

Diabetes mellitus (DM) is currently one of the medical problems that affect efficiency and reduces the quality of human resources. DM patients worldwide in 2015 increased from 333 million people (5.4%). People with diabetes in Indonesia are continuing to occur. Considering the record of the world health organization in 2004, Indonesia ranked fourth with the highest number of diabetics after India, China, and USA [1]. It was realized that in 1995 there were around 5 million people with DM in Indonesia with a growth of around 230 thousand diabetics consistently so that by 2025 it was estimated that there would be 12 million people. The increase occurred due to expansion in older populations and changes in lifestyle, starting from eating habits, types of food, to decreased physical activity. This happens especially in the category of adults or more for all strata of society. In addition, developments in the number of DM cases occur due to the absence of workers, checking specific equipment and medicines, especially in remote zones, and there is no consistency in supervising DM patients.

The ten countries estimated to have the highest numbers of people with diabetes in 2000 and 2030 are reported in reference [1]. The "top five" countries are the same as those identified for 1995 (India, China, U.S., Indonesia, and Japan). Bangladesh, Brazil, Russia, and Pakistan also appear in the lists for both 2000 and 2030. The similarity of DM arises all over the world, especially in Asian countries. Calculations in 2020 of this disease refer to the death of 7 individuals from each of 10 individuals in developing countries. About 16 million people in America are diagnosed with diabetes, the dominant being 6% to 7% in individuals 45 to 65 years. About 90% of them experience the adverse effects of type 2 diabetes. About 9.7 million women in America have diabetes. Type 2 diabetes occurs at all ages, even at an early age and teenagers.

Type 2 diabetes is the most well-known type of diabetes found in DM patients than type 1 diabetes mellitus, gestational diabetes, and various types of diabetes [2]. Type 2 diabetes is a multifactorial disease with the same genetic and ecological parts as its effect to cause disease. The impact of heredity on this disease can be seen clearly with the high level of diabetics from fathers or mothers who have a profile of diabetes mellitus. Type 2 diabetes mellitus is also often called diabetes as a way of life, considering the causes other than heredity, ecological components including age, weight, insulin resistance, daily consumption, physical activity, and poor individual way of life also influence diabetes. Type 2 diabetes mellitus which is not monitored properly will cause a variety of chronic complications [3], for example, retinopathy and nephropathy and macroangiopathy, for example, heart disease, stroke, and further additional blood vessel infections.

The main driver of mortality and morbidity in patients with type 2 DM is coronary heart disease (CHD) [4, 5]. As demonstrated by the American Heart Association in May 2012, at least 65% of people with diabetes die of heart disease or stroke. In addition, adults who are poor diabetics have a risk of coronary heart disease two to several times higher than individuals who are not diabetics [6]. A slow increase in type 2 diabetes mellitus often causes side effects and unclear signs. The component of type 2 CHD events is complicated and is associated with atherosclerosis which is influenced by various elements, including hypertension, hyperglycemia, dyslipidemia, smoking, CHD family offspring, and body weight.

Type 1 diabetes mellitus is a metabolic problem caused by an immune system response that makes pancreatic beta cells initiated by prolonged hyperglycemia due to the absence of substantial insulin [7]. During the range of DM disease can cause different complications, namely short-term and long-term complications. Short-term complications, namely hypoglycemia and ketoacidosis. Long-term complications occur due to microvascular changes such as retinopathy, nephropathy, and neuropathy. There are 2 types of DM, namely specific type 1 DM which is insulin subordinate diabetes, where the pancreas gives little insulin or does not create insulin at all. While in type 2 DM, the pancreas still gives insulin, but in some cases, the amount is higher than the normal limit, where this event will make the body's structure insensitive to its effect, which causes relative insulin deficiency.

Type 2 DM is a hyperglycemic disease due to a lack of insulin sensitivity of cells. Insulin levels can be reduced slightly or be within reasonable limits. Because insulin is still made by pancreatic beta cells, type 2 diabetes mellitus is considered a non-insulin subordinate diabetes mellitus [8]. Type 2 diabetes is not caused by the absence of insulin emissions, but because insulin target cells become flat or unable to react normally. This condition is called insulin resistance. Variables associated with diabetes risk are those who have polycystic ovary syndrome, patients with metabolic disorders, backgrounds that are characterized by inhibited glucose resistance or fasting blood glucose, and who have cardiovascular backgrounds. According to [9], there was a positive correlation obese patients. LITAF is a TNF-α description of the between obesity and blood glucose levels.

In previous study on the reference [10], employs the Spline Interpolation technique to convert discrete blood glucose data to continuous signal. The validation of interpolated signal is conducted by comparing the pattern of discrete data and continuous signal for both original and clustered data. The experiment showed that both scenarios depicted identical patterns. The Spline interpolation technique indicates that the method is the most appropriate in predicting the interpolated points on blood glucose data. This paper describes a correlation between risk factors of coronary heart disease in type 2 DM

IOP Conf. Series: Materials Science and Engineering **821** (2020) 012042 doi:10.1088/1757-899X/821/1/012042

patients using Chi-Square method. This method was applied to determine whether there is correlation between risk factors and the incidence of CHD in patients with type 2 DM. Furthermore, this work also investigate how does the mathematical model work in calculating the magnitude of the risk of variables against the tendency of CHD complications for patients with type 2 diabetes.

2. Methodology

This study uses existing data on the medical record installation of Dr. Soetomo Hospital in 2015-2016. Samples were all patients with type 2 diabetes with CHD and without CHD recorded in the medical records of the inpatient disease in Dr. Soetomo Hospital Surabaya. Data processing was performed using the chi-square test method. The purpose of processing data using the chi-square test is to find the chi-square test value of each CHD risk factor variable, and then the results will be in the form of a graph that will illustrate the high or low results of the chi-square test value. From the results obtained, it will be seen which risk factors are most influential on the occurrence of CHD in patients with type 2 DM at Dr. Soetomo Hospital in Surabaya in 2015-2016.

Data using patients totaling 50 people along with information from 12 CHD risk factors for type 2 diabetes. The risk factors in question are age, gender, GDA, GDP, GD2PP, triglycerides, cholesterol, HDL, LDL, body weight, smoking, and long DM.

Chi-Square is also called Kai Squared. Chi-square is a type of non-parametric similar test carried out on two factors, where the size of the information from both factors is in the form of values [11]. In parametric insight, the proportion of relationships that are normally used is Pearson's product moment relationship. Among the nonparametric relationships that can be compared with these standard connection coefficients and those commonly used are the Spearman R, Kendal Tau and gamma coefficients. Besides these three methods, chi-square depending on the cross table is also quite well known in estimating the relationship between factors.

Chi-square test technique depends on this cross table is organizing variables in the classification and testing of hypotheses that the observed iteration is the same as the expected iteration [6]. The integrity of the chi-square compatibility test is about the observed repetition and the expected repetition in each class, it must be tested that all groups contain the same value or a test that each group contains a certain value.

The assumption used is an example derived from a random sample. The expected frequency for each group must be more than 1. The expected frequency of values below 5 must not be more than 20% of the group. The chi-square calculation can be expressed in Equation (1).

$$x^2 = \sum_{i-1}^{r} \sum_{j-1}^{c} \frac{(O_U - E_U)^2}{E_U} \qquad (1)$$

where r is number of rows, c is number of column, i is indicates of line to-i and j is column to-j. While O_{ij} is observed frequencies in line to i and column j. And E_{ij} is expected frequencies in line to i and column j.

3. Result and discussions

Based on the results on Table 1, it can be drawn condition decisions based on the results of the chi-square test value of the DF value of each risk factor. If the chi-square test value <DF value, then the H0 logic is accepted, which means there is no relationship between these risk factors with the occurrence of CHD in patients with type 2 DM. If the chi-square test value > DF value, then the H1 logic is accepted, which means there is a significant relationship between these risk factors with the occurrence of CHD in patients with type 2 DM. From the value of the chi-square test in Table 1, A Chi-square test indicated that cholesterol and LDL have a significant relationship in the incidence of CHD in patients with type 2 DM, while gender, age, GDA, GDP, GD2PP, triglycerides, HDL, obesity, smoking, and for a long time DM did not have a significant relationship in the incidence of CHD in patients with type 2 DM in the hospital, Dr. Soetomo Surabaya in 2015-2016.

To find out the level of accuracy of the system that has been made, can calculate with a simple accuracy formula in Equation (2), which is to add up the total value of the chi-square test system, divided by the sum of all calculated chi-square test values. Table 1 shows the comparison result of Chi-Square Test and Measured Chi-Square.

$$\text{Accuracy} = \frac{number\ of\ Chi\ Square\ test}{number\ of\ measured\ Chi\ Square}\ x\ 100\%$$

$$= \frac{30.247}{30.753}\ x\ 100\% = 98.354\% \tag{2}$$

Table 1. Comparison result of Chi Square Test and measured Chi Square

Risk Factor Variables	Chi Square Measured	Chi Square Test	DoF value
Genders	0.673	0.673	3.841
Ages	3.165	3.171	7.814
GDA	1.188	0.855	3.841
GDP	0.608	0.421	3.841
GD2PP	1.235	1.236	3.841
TRIGLISERIDA	0.178	0.17	3.841
Cholesterol	3.846	3.848	3.841
HDL	0.169	0.17	3.841
LDL	11.455	11.459	3.841
Obesities	0.478	0.48	3.841
Smoking	2.802	2.805	3.841
Time of DM	4.956	4.959	5.991

Based on data testing that has been done using 50 patient data, it can be drawn a hypothesis by comparing the value of the chi-square test with the DF value of each risk factor. The conclusions that can be drawn from the mentioned experiments in which risk factors that influence the incidence of CHD in patients with type 2 DM are LDL and cholesterol because the value of the calculated chi-square test and chi-square test system is greater than the DF value each. Therefore the H1 condition applies. LDL risk factors have the highest value among other risk factors.

Risk factors for gender, age, GDA, GDP, GD2PP, triglycerides, HDL, obesity, smoking, and duration of DM indicate that they do not have a significant relationship to the incidence of CHD in patients with type 2 DM, because of the calculated chi-square test or chi test -square¬ system is smaller than each DF value. Therefore the H0 condition applies. There is a difference in value for several risk factors between the calculated chi-square test and the chi-square test system. This is because the retrieval value behind the comma (,) in the system is more than the manual calculation.

4. Conclusions

After designing and testing this web application, it can be concluded that the system is able to calculate the value of the chi-square test and determine the results of conditions on each variable. There is a difference in value between the calculation of the chi-square test system with the calculation of the manual chi-square test, this is caused by taking a number of digits behind different commas (,). The variables of gender, age, GDA, GDP, GD2PP, triglycerides, HDL, body weight, smoking, and duration did not have a significant relationship on the incidence of CHD in patients with type 2 DM. Cholesterol variables and LDL levels have a very significant relationship in the incidence of CHD for patients with type 2 diabetes.

Acknowledgement

The authors wish to express their gratitude to the Faculty of Engineering, Universitas Muhanmmadiyah Malang for supporting this study. To the Dr. Soetomo hospital for providing the research data, the cooperation is highly appreciated.

References

[1] Wild S, Roglic G, Green A, Sicree R, and King H 2004 Global prevalence of diabetes estimates estimates for the year 2000 and projections for 2030 *Diabetes Care* vol **27** pp 1047–53.

[2] Islam R and Rahman O 2012 The risk factors of type 2 diabetic patients attending Rsajshahi Diabetes Association, Rajshahi, Bangladesh and its primary prevention *Food Public Heal.* vol **2** no 2 pp 5–11.

[3] Li Y, Li H, and Yao H 2018 Analysis and study of diabetes follow-up data using a data-mining-based approach in new urban area of Urumqi, Xinjiang, China, 2016-2017 *Comput. Math. Methods Med.* vol **2018**.

[4] Yuliani F, Oenzil F, and Iryani D 2014 Hubungan berbagai faktor risiko terhadap kejadian penyakit jantung koroner pada penderita diabetes mellitus tipe 2 *J. Kesehat. Andalas* vol **3** no 1 pp 37–40.

[5] Chiha M, Njeim M, and Chedrawy E G 2012 Diabetes and coronary heart disease: A risk factor for the global epidemic *Int. J. Hypertens.* vol **2012**.

[6] Langari S H, Bahar A, Asadian L, Abediankenai S, Namazi S S, and Kashi Z 2019 Coronary heart disease and ABO blood group in diabetic women: A case-control study *Sci. Rep.* vol **9** no 1 pp 5–10.

[7] Halilovic E A, Ljaljevic S, Alimanovic I, Mavija M, Oros A, and Nisic F 2015 Analysis of the influence of type of diabetes mellitus on the development and type of glaucoma *Med. Arch. (Sarajevo, Bosnia Herzegovina)* vol **69** no 1 pp 34–7.

[8] Betteng R, Pangemanan D, and Mayulu N 2014 Analisis faktor resiko penyebab terjadinya diabetes melitus tipe 2 pada wanita usia produktif Dipuskesmas Wawonasa *J. e-Biomedik* vol **2** no 2.

[9] Silitonga H A, Siahaan J M, and Anto E J 2019 Correlation between obesity and lipid profile in type 2 diabetes mellitus patients at the endocrine and metabolic polyclinic in general hospital pirngadi Medan *Open Access Maced. J. Med. Sci.* vol **7** no 8 pp 1309–13.

[10] Syafa'ah L, Purnomo M H, and Basuki S 2018 Discrete mean amplitude of glycemic excursion (MAGE) measurement on diabetics with spline interpolation method *Int. J. Electr. Eng. Informatics* vol **10** no 2 pp 259–70.

[11] Jousheghany F, Phelps J, Crook T, and Hakkak R 2016 Relationship between level of HbA1C and breast cancer *BBA Clin.*vol **6** pp 45–8.

IOP Conf. Series: Materials Science and Engineering **821** (2020) 012043 doi:10.1088/1757-899X/821/1/012043

Application of genetic algorithm concept on course scheduling

R Ansari[1, *] and N Saubari[2]

[1]Departement of Informatics Engineering, Faculty of Engineering, Universitas Muhammadiyah Banjarmasin, Banjarmasin, Indonesian

[*]Email: rudy@umbjm.ac.id

Abstract. Genetic Algorithms have parameters such as crossover probabilities and mutation probabilities used based on entering population numbers and number of generations. Of the two entries, nine rules were obtained which would produce crossover probabilities and mutation probabilities. One problem that can be solved using a genetic algorithm is the scheduling of courses. In the preparation of course scheduling, it takes quite a long time and needs a very high accuracy. Therefore, the purpose of this study is to implement genetic algorithms on lecture scheduling problems. So that the accuracy and speed in determining the class schedule can be fulfilled. The test results show that applying a genetic algorithm can obtain the course schedule without any collision in one iteration process.

1. Introduction

In the process of lecturing activities, scheduling lecture activities in a university is a complicated matter and often has difficulty in allocating courses, lecturers and rooms to be used for lectures, to make sure that there is no conflict with the schedule of lectures, lecturers and other rooms in one lecture schedule period. Besides, changes in the schedule of lecturers often make it difficult in allocating time and searching for empty rooms that will be used to schedule temporary changes. In this case, the planned lecture activities should be carried out according to a predetermined schedule if it can be managed properly [1, 2].

The problems that are felt when preparing the lecture schedule are the difficulty in allocating courses with the lecturer and his room so that there is no clash and the number of college participants who adjust the capacity of the classroom. Besides, there is also a change in the temporary lecture schedule because the lecturer is unable to attend, often having difficulty in finding empty rooms for surrogate courses [3], [4]. Based on these problems, it is necessary to apply the algorithm in the preparation of the lecture schedule to facilitate the schedule creation process.

2. Literature review

2.1. Scheduling

Scheduling is the activity of allocating resources or machines that exist to run a set of tasks within a certain period of time. According to Ambuhi, scheduling has 2 meanings, namely:

a. Scheduling as a function of decision making. Scheduling is a process for determining a schedule
b. Scheduling is a theory. Scheduling is a collection of principles, models, techniques and logical conclusions in decision making. Scheduling is part of the shop floor control

According to Ambuhi, if scheduling is constant, then the proper work order will reduce the flow time and average work in the process. Decisions made in scheduling include sequencing, starting time and working time, a sequence of operations for routing (routing). Scheduling problems often arise if there is a set of tasks simultaneously, while the equipment is limited [3].

A more general definition of scheduling is to assign a set of events (lectures) with a limited set of sources from time to time to meet the defined constraints (constraints). These constraints can be categorized as hard constraints and soft constraints, where hard constraints have a higher priority than soft constraints. There are two limitations in the preparation of lecture schedules, namely: hard constraints (must be fulfilled) and soft constraints (attempted to be fulfilled). Hard constraints are boundaries that must be applied to scheduling courses and must be fulfilled. A solution can only be said to be valid if the solution has no violated hard constraint [5, 6].

Common hard constraints in scheduling courses are as follows:

a. A lecturer can only teach courses for one location at a certain time.
b. A student can only attend classes for one location at a time.

In contrast to hard constraints, soft constraints are constraints that cannot always be fulfilled in the process of forming schedules, but even if they do not have to be fulfilled, the resulting schedule must try as much as possible to meet the soft constraint provisions. Examples of soft constraints in scheduling courses are:

a. Lecturers can ask for the specific teaching time they want.
b. Placement of the schedule for the time requested by the lecturer is adjusted to the priorities of the lecturer[7-9].

2.2. Genetic algorithm

The genetic algorithm is a heuristic search algorithm based on the mechanism of biological evolution. In this algorithm, search techniques are carried out at the same time on some possible solutions, in other words, this algorithm can help provide the best and fastest solution to a problem that has many possible solutions known as the population. Individuals in one population are called chromosomes. This chromosome is a solution that is still in the form of a symbol. The initial population is built randomly, while the next population is the result of the evolution of chromosomes through an iteration called generation [10-13].

Components of Genetic Algorithms. There are 6 main components contained in the genetic algorithm, namely:

1. Coding Techniques
2. Initialization Procedure (generate initial population)
3. Evaluation Function
4. Selection
5. Genetic Operators

3. Research method

The steps in the process of making a scheduling using genetic algorithms can be seen in Figure 1. The chromosome used in producing the initial individual in this system is as follows:

Schedule chromosomes contain information:

1) list of courses (m),
2) room (r),
3) time (w),

The schedule chromosome array structure on the new chromosome is: [m, r, w]. The list of course genes includes data on course codes, names of courses, credits, lecturers and study program names. The room gene includes room name and room capacity. While for the time gene includes hours and days.

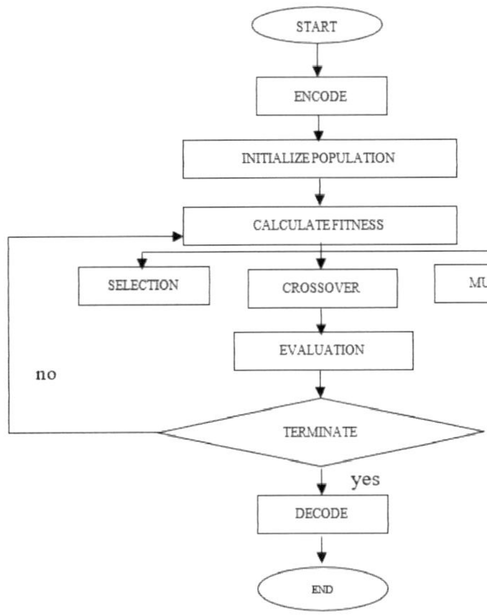

Figure 1. Process of making a scheduling using genetic algorithms

4. Result and discussion

The application of Genetic Algorithms begins with forming the initial population, scheduling limits, soft constraints, GA components. The GA component is divided into the following steps:

4.1. Coding technique

Genes presented here are in the form of arrays of real numbers. The parameters used in each chromosome are a list of courses (m), room (r) and time (w). So that in each one chromosome is filled with data [(m r, w)].

4.2. Initialization of initial population

In the initialization of initial population process, it generates a population randomly, for example chromosomes that are generated as a schedule that will be generated are 4 chromosomes in 1 generation by using 10 experimental data of list of courses (m) and time (w), from the informatics study program.

4.3. Chromosomes evaluation

Chromosome evaluation is a step to calculate the fitness value of a chromosome that was initialized in the previous process. To determine the fitness value, the scheduling value generated by the fitness function will represent how many requirements are violated, so that in the case of scheduling lecture, the smaller the number of violations produced, the better the solution will be. For violations that occur if there is a conflict on lecturers, a value of +5 will be given and conflict on rooms will be given a +1 value. To avoid an infinite fitness value, the total number of all violations will be added by 1. The best standard fitness value for this scheduling is 1 where fitness on the chromosome = 1, means that the optimal solution results obtained were no conflict found on a schedule.

Calculate fitness value with the chromosome formula, (n) = 1 + (BBD + BBR)

Fk1 : 1 + 1 (+5) + 2 (+1) = 8

Fk2 : 1 + 1 (+5) + 1 (+1) = 7

Fk3 : 1 + 2 (+5) + 1 (+1) = 12

Fk4 : 1 + 1 (+5) + 1 (+1) = 7

Total Fitness = Fk1 + Fk2 + Fk3+ Fk4 = 8 + 7 + 12 + 7 = 34

Description:
BBD = Total conflict on Lecturers
BBR = Total conflict on room
n = Chromosome Sequence
Fk = Chromosome Fitness

4.4. Selection
This selection stage is selecting the best chromosome where the most suitable chromosomes that have a higher probability are more likely to be selected for the next generation. To calculate the probability value of each chromosome can be seen as follows:
Pk1 = Fk1/(Total fitness) = 8/34 = 0,23
Pk2 = Fk2/(Total fitness) = 7/34 = 0,21
Pk3 = Fk3/(Total fitness) = 12/34 = 0,35
Pk4 = Fk4/(Total fitness) = 7/34 = 0,21
Total Probability = Pk1 + Pk2 + Pk3 + Pk4 = 0,23 + 0,21 + 0,35 + 0,21 = 1
Description:
Pk = Chromosome Probability
Fk = Chromosome Fitness
Chromosome 1 = [(1,1,7),(2,3,2),(3,4,1),(4,5,9),(5,2,19),(6,6,1),(7,3,7),(8,3,8), (9,3,7),(10,2,13)].
Chromosome 2 = [(1,1,7),(2,3,2),(3,4,1),(4,5,9),(5,2,19),(6,6,1),(7,3,7),(8,3,8), (9,3,7),(10,2,13)].
Chromosome 3 = [(1,3,1),(2,5,15),(3,3,2),(4,2,16),(5,1,2),(6,4,15),(7,6,1),(8,2,3), (9,5,19),(10,4,13)].
Chromosome 4 = [(1,2,19),(2,1,2),(3,3,1),(4,5,20),(5,6,3),(6,3,1),(7,1,20),(8,4,2), (9,5,15),(10,5,15)].

4.5. Crossover
At this stage, a crossover is done by crossing two chromosomes. This technique uses the one cut point method, which is to randomly choose a position of the parent chromosome then exchange sub-chromosomes. In this process, the probability of a crossover (Pc) will control the crossover operator in each generation in the crossover population. the probability of crossover (pc) used is 75% or 0.75. The greater the value (Pc) the faster the new chromosome will appear in the population, but if (Pc) is too large then the chromosomes that are candidates for the best solution may disappear faster in the next generation. If the random number generated is smaller than the crossover (Pc) probability value, a crossover will be done.
Generates random numbers in the range of (0 - 1).
Chromosome 1 = 0,65
Chromosome 2 = 0,81
Chromosome 3 = 0,50
Chromosome 4 = 0,69

Thus, the selected parent will be crossed, namely chromosome 1, chromosome 3 and chromosome 4. the rules of crossing are done randomly (random) following the parents who have been previously selected. The process of this crossing can be seen as follows
Chromosome 1 >< Chromosome 3
Chromosome 3 >< Chromosome 4
Chromosome 4 >< Chromosome 1

4.6. Mutation
From the calculation result, the best fitness value for the new population in one generation is obtained, chromosome 4 as the best schedule because it has a fitness value that is close to 1. The following is the result of schedule output from the best chromosome, chromosome 4, can be seen in the following table (Table 1):

Table 1. Schedule output chromosome

No	Day	Time	Course Codes	Courses	Credits	Lecturer	Class Room	Capacity
1	Senin	08.50-10.30	UMB0002	Pend. Kewarganegaraan	2	Harpani	FT7	30
2	Senin	08.00-10.30	INF6242	Pemrog. Perangkat Mobile	3	Wahyu Ridhoni, M.Kom	Lab Dasar	30
3	Senin	10.30-13.50	INF4223	Sistem Operasi	3	Kamarudin, M.Kom	Lab Dasar	30
4	Senin	08.00-10.30	INF4221	Metode Numerik	3	Mukhaimy Gazali, M.Si	Lab Jaringan	40
5	Senin	08.00-10.30	INF4222	Sistem Pendukung Keputusan	3	Rudy Ansari, M.Kom	FT1	60
6	Selasa	10.30-13.50	INF4220	Interaksi Manusia & Komputer	3	Nahdi Saubari, M.Kom	FT1	60
7	Rabu	03.00-15.30	INF4219	Ilmu Kealaman Dasar	3	Windarsyah, M.Kom	FT2	30
8	Kamis	08.50-11.20	INF2209	Algoritma & Struktur Data	3	Munsyi. S.Kom, M.T	FT1	60

5. Conclusion and recommendation

5.1. Conclusion

Scheduling at Informatics Study Program, Muhammadiyah Banjarmasin University, has implemented the GA method, from the results obtained there is still a conflict of rooms from chromosome 4, this is because the number of initial chromosomes and generation generated is small. If the initial population and generation are increased, then there will be more variations on chromosome and will lead to the most optimal solution, namely the chromosome with a fitness value = 1, which indicates that there is no conflict of schedules.

5.2. Recommendation

There should be experiments using other techniques in selection operators, crossover and mutation. This is to compare the use of techniques for each operator so the most optimal technique is found.

References

[1] Nugraha T A, Putra K T and Hayati N 2017 university course timetabling with genetic algorithm: a case study *J. Electr. Technol. UMY* vol **1** no 2 pp 100–5.
[2] Mauren A O, Gudo C O and Oanda I O 2011 University expansion in Kenya and issues of quality education: Challenges and opportunities *Int. J. Bus. Soc. Sci.* vol **2** no 20 pp 203–15.
[3] Ambühl C, Mastrolilli M, Mutsanas N and Svensson O 2011 On the approximability of single-machine scheduling with precedence constraints *Math. Oper. Res.* vol **36** no 4 pp 653–69.
[4] Ouelhadj D and Petrovic S 2009 A survey of dynamic scheduling in manufacturing systems *J. Sched.* vol **12** no 4 pp 417–31.
[5] Xu Y, Zeng, Liu Q and Wang X 2014 A genetic algorithm based multilevel association rules mining for big datasets *Math. Probl. Eng.*vol **2014** pp 1–9.
[6] Jouya M and Khayati S 2018 *Review local search algorithms in artificial intelligence* vol **4** no 2 pp 81–6.
[7] Science C and Engineering S 2017 *Automatic Time Table Generator 1* vol **7** no 5 pp 204–11.
[8] Duan K, Fong S, Siu S W I, Song W and Id S S G 2018 *SS symmetry Adaptive Incremental*

Genetic Algorithm for Task Scheduling in Cloud Environments pp 1–13.

[9] Sani H M 2016 *Solving Timetabling problems using Genetic Algorithm Technique* vol **134** no 15 pp 33–8.

[10] Shinde S and Gurav S 2018 *Automatic Timetable Generation using Genetic Algorithm* vol **9** no 4 pp 19–21.

[11] Yadav S L and Sohal A 2017 Comparative study of different selection techniques in genetic algorithm *Int. J. Sci. Math. Educ.*vol **6** no 3 pp 251–54.

[12] Salvi A, Khanvilkar O and Balkhande B W 2016 *Automatic Time-Table Generation System using Genetic Algorithm* vol **5** no 3 pp 824–26.

[13] Sari Y, Alkaff M and Pramunendar R A 2018 Iris recognition based on distance similarity and PCA *AIP Conf. Proc.* vol **1977**.

IOP Conf. Series: Materials Science and Engineering **821** (2020) 012044 doi:10.1088/1757-899X/821/1/012044

First time user experience of academic information system: An evaluation of usability

A Setiawan* **and A Primadewi**

Department of Informatics Engineering, Faculty of Engineering, Universitas Muhammadiyah Magelang, Magelang, Indonesia.

*Email: setiawan@ummgl.ac.id

Abstract. The use of information technology continues to grow, one of which is software development. The software does not only focus on functionality, but the ease of use becomes a heuristic development. The interface design cannot be underestimated because the wrong interface design can also result in repeated software development from an initial stage. Software development from the start is certainly not easy and time-consuming to revise it. The specific purpose of this study is to adapt various models of usability testing instruments to measure how easy it is to use the system being tested. This research aims to help the designer evaluate what he has designed. The first stage is the validation of the first user experience instrument of the Academic Information System Universitas Muhammadiyah Magelang. The results of this study found that the KRS system was able to be well received by the first user.

1. Introduction

The software has become a big phenomenon with millions of developers. The pioneer of major software manufacturers is Microsoft. The software development competition is very fast and dynamic to attract users with easy services [1]. Good services in software today must pay attention to the flexible and easily controlled interface [2-5]. So that software development in Indonesia has now begun to focus on Usability. This is indicated by the existence of a job position called UI / UX Designer.

The study of Usability, or what is often called Usability, is an important factor for determining a quality user experience [6, 7]. Usability is a broad part of the branch of Human-Computer Interaction, which involves various branches of science. This knowledge is related to observing human behaviour. Usability is still something important in the development of technology that focuses on user aspects, not on technological aspects, so that the development is in a user-centered design, which is now more commonly referred to as User Experience.

IBM has developed and specifically studied Usability through James R Lewis. IBM developed the usability measurement method by referring to psychometric methods that will measure the quality of a user's psychological size [2]. Lewis is conducting structured research that will take respondents from stage one to the next stage. Starting from the visit to the Laboratory stage to be briefed and a quick look at the equipment and systems to be tested, getting a demonstration of system usage, basic training in using the system, until finally, if they have finished carrying out all the stages, then fill out the questionnaire provided. This filled questionnaire uses a 7 Likert scale.

The research conducted by Brooke developed a Usability measurement tool called SUS (System Usability Scale), which was used to evaluate Usability quickly [4]. The development of SUS has been started since 1986, proving that Usability has a very important role in determining the success of a system in the relationship between humans and computers. Failure at the time of system design, namely at the interface design stage, will certainly make regrets at the end because the development continues to progress towards the implementation of design even to the testing stage. Usability evaluations avoid failure system implementation. It carried out at the stages between interface design and coding implementation. If we read the retrospective of usability Brooke researchers, we will see that the preparation of measuring instruments and their implementation requires observations from all fields. When Brooke developed SUS, it was costly to install cameras on all sides to observe user behaviour. Towards the system that was made from view to monitors, keyboards, and their body movements [4].

Usability Rating is measured by Kortum to find out from the 1000 respondents involved in this study how helpful and easy it is for 14 computer applications that are used daily [8]. Using Usability measuring instruments with 14 questions that have been adapted questionnaires, managed to get an average value of 80.2 out of 100. The highest value still belongs to Google Searching, with a value of 93.4. Moreover, the lowest value is the Excel application that has a value of 56.5. This difference that is too far away is a common phenomenon because Excel has more complex features than Google Search, which also has minimal interaction. This research hopes to become a reference for uniform application development.

In Lewis's latest study, Usability evaluation used three types of usability System Usability Scale (SUS), Usability Metric for User Experience (UMUX), and Computer System Usability Questionnaire (CSUQ) measurement tools to measure four commonly used applications, Excel, Word, Amazon, and Gmail (Lewis, 2018). Research on usability evaluation of information or profiles from an educational institution at the Muhammadiyah University of Magelang was carried out in a study conducted by [9] with the URL address http://ummgl.ac.id. Setiawan has evaluated http://ummgl.ac.id by 95 respondents with a score of 2.77 on a scale of 4, which has a conclusion in the Good category.

Based on several previous studies of Usability, the development of measuring instruments with experiments on several objects to be measured has different characteristics. Usability evaluation was also conducted on games software conducted by Barnett, examining the first experience of the respondents playing the games "Super Mario Run" and "Linia" [10]. Barnett argues that the first experience affects Usability in aspects of information quality, but does not significantly influence Usability in its entirety in the game. Barnett's explanation is that games will be satisfied with the game rather than its functionality [11]. Based on Barnett's statement, the researchers tried to conduct usability evaluations for the first time users to measure their experience in using the academic information system of Muhammadiyah Magelang University on the URL http://krs.ummgl.ac.id.

Previous research has been applied to measure the user's first experience playing the game. So the focus of his research is on resolving challenges that have levels. If on a system other than games, of course, the challenge is to complete an administration as an obligation, so that any interaction that is carried out must be done without knowing the level or game over.

The purpose of this study was to measure the perceptions of first-time users using the KRS system. One's first experience using the system is certainly an important input for system developers to design systems that are easy to learn, use, and handle errors that are easily recovered.

2. Method

The research method used by researchers is the method of observation and interviews (Figure 1). Respondents, in this case, are UMMgl Students as users who have used the Academic Information System (krs.ummgl.ac.id) for the first time. A questionnaire compiled with a list of questions based on usability aspects refers to several instruments that have been developed by experts [3, 4, 6].

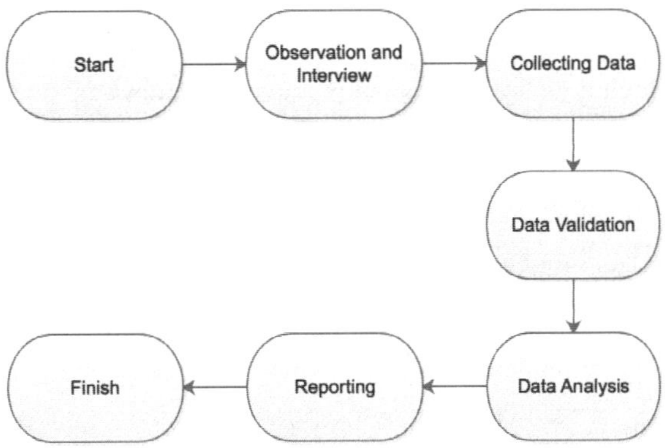

Figure 1. Research sequence

The stages of this study begin with observing the system used by respondents, namely KRS. Respondents are UM Magelang's students as the first users of the KRS system. Students use the system for the first time to enter a choice of second semester courses. At the beginning of the first semester, students are inputted by the academic department for the default course. The second stage is data collection. Data collection is done by interview. The interview method was chosen because of that more flexible to explain questions to the respondents. The third stage is data validation. Data validation is done to ensure that the data is correct and in accordance with the facts. The fourth stage is data analysis. Data analysis is done using a simple method of calculating basic statistics. The data is analyzed based on the average results of the respondents' answers. The analysis examines the anomalies of the system measurement results. The last stage is reporting. Reporting of measurement results using ranking 1 to 7, with a smaller ranking, means better value. This ranking of 1 to 7 has a neutral point, which is at value 4, which is in the middle rank. 1 means strongly agree, and 7 strongly disagrees.

3. Result and discussion

The results of this study are measurements of 76 respondents who can take interviews. But before that, the validation test of the questionnaire used was all declared valid and reliable using SPSS, as shown in the following Figure 2. Respondents freely give grades without any pressure. Here is an answer to the eleven questions that have been asked.

Reliability Statistics

Cronbach's Alpha	Cronbach's Alpha Based on Standardized Items	N of Items
.779	.786	11

Figure 2. Reliability testing

Figure 3 for Q1 Rank shows that the majority of respondents gave a response to agree by giving a value of two. Respondents agreed that respondents were satisfied with the ease of this system to learn how to use it. Figure 3 for Q2 Rank shows that the majority of respondents gave a response to agree by giving a value of two. Respondents agreed that this system was simple in its use.

Figure 4 for Q3 Rank shows that the majority of respondents gave an agreed response by giving a value of two. Respondents agreed that this system was simple in its use. Figure 4 for Q4 Rank shows that the majority of respondents gave an agreed response by giving a value of two. Respondents agree that users can quickly complete tasks and challenges

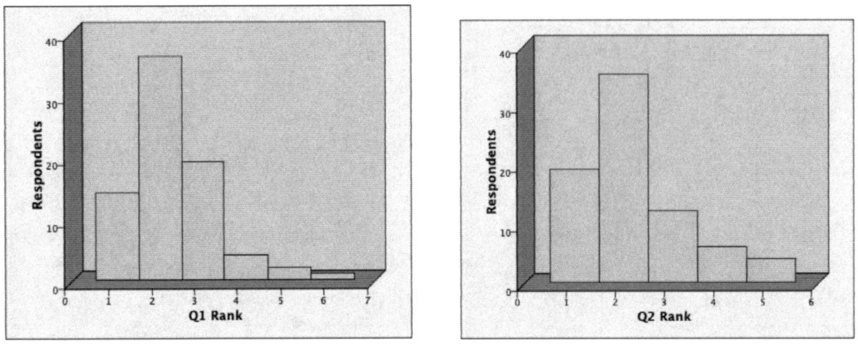

Figure 3. First question graph

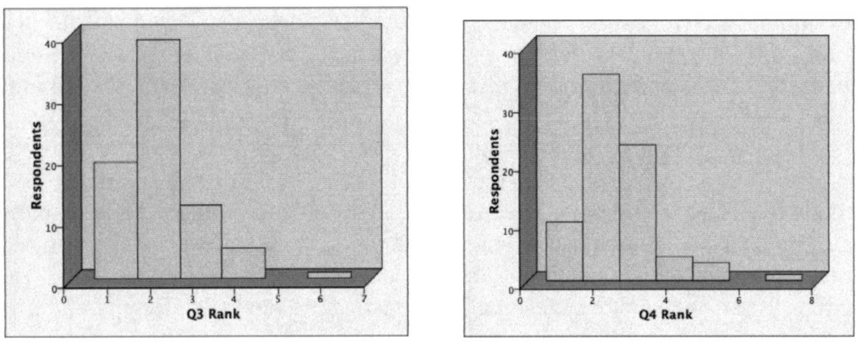

Figure 4. Third question graph

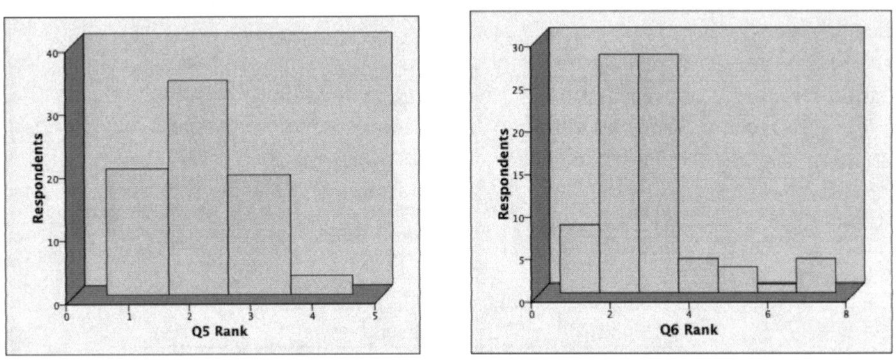

Figure 5. Fifth question graph

Figure 5 for Q5 Rank shows that the majority of respondents gave a response to agree by giving a value of two. Respondents agree that users can complete tasks and challenges efficiently. Figure 5 for Q6 Rank shows that the majority of respondents gave an agreeing response by giving two and three balanced scores. Respondents agreed that users felt comfortable using the KRS system.

Figure 6 for Q7 Rank shows that the majority of respondents responded strongly to agree with giving a value of one. Respondents strongly agree that the system is easy to learn about its use. Figure 6 for Q8 Rank shows that the majority of respondents gave an agreed response by giving a value of two. Respondents agreed that when the respondent made a mistake when using the system, it was easy to fix it.

Figure 7 for Q9 Rank shows that the majority of respondents gave a response to agree by giving a balanced value of two and three. Respondents agreed that the layout of the information displayed was clear. Figure 7 for Q10 Rank shows that the majority of respondents gave a response agreeing by giving a value of three. Respondents agreed that the system interface was fun.

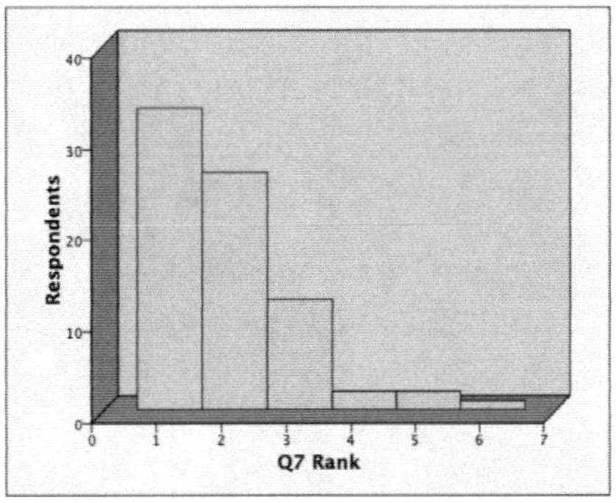

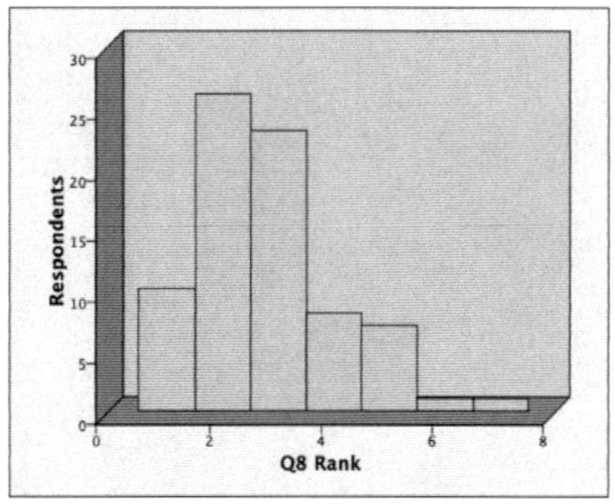

Figure 6. Seventh question graph

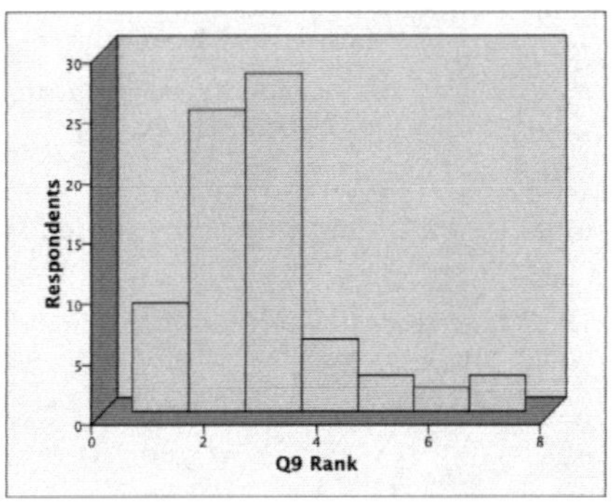

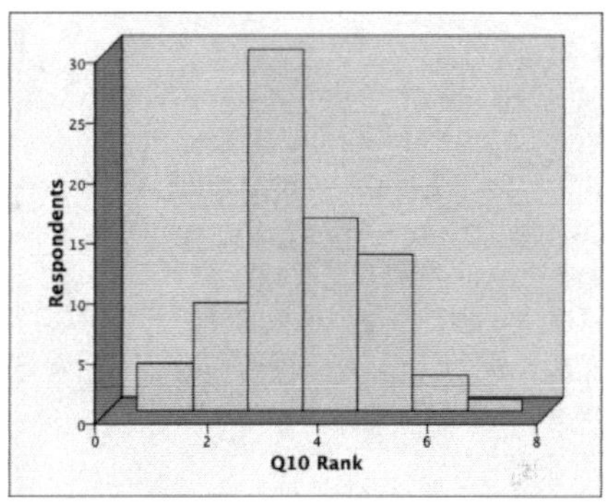

Figure 7. Ninth question graph

Figure 8 for Q11 Rank shows that the majority of respondents gave a response to agree by giving a balanced value of two and three. Respondents agreed that respondents liked the system interface. Figure 9 is the average of eleven questions that have been submitted in the positive acceptance category for KRS system users. KRS users referred to here are students who are using KRS for the first time.

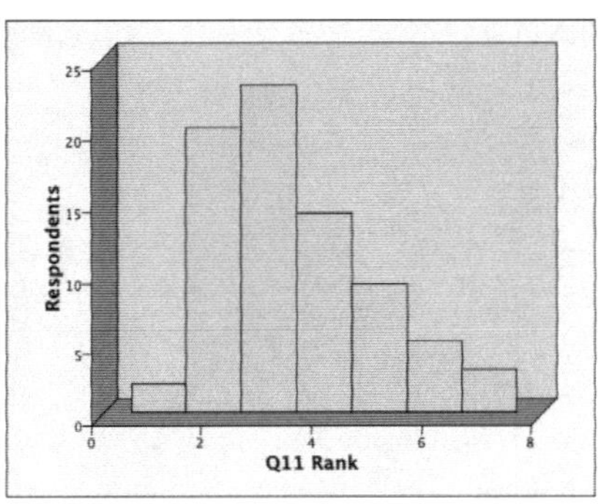

Figure 8. Eleventh question graph

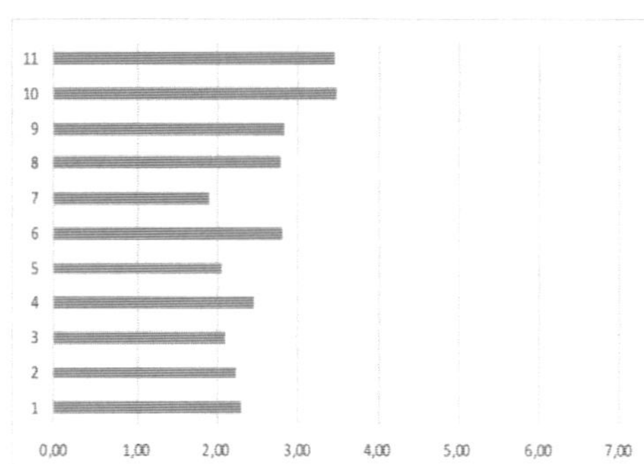

Figure 9. Average of 11 questions

4. Conclusion

First-time user experience in using a new system certainly hopes to get a system that is easy to learn. Users will be easily frustrated in using a system that is difficult to learn, especially to carry out obligations such as filling in the KRS, which is the obligation of students at the beginning of each semester. Measurement of user experience is also more effective using interview measurement by using references from standard questions. Users will have more discussion before answering the questions. The measurement results show that the KRS system feels easy for the first-time users with a mean value of 2.59 on a scale of 7.

Acknowledgment

This paper is part of the project of Smart Campus, which is conducted at the Informatics Laboratory of Universitas Muhammadiyah Magelang. The researchers are grateful to the technicians who have helped data collecting and also the reviewers.

References

[1] Setiawan A, Nugroho E, and Adhipta D 2014 *Perancangan Sistem Informasi Pelacakan Berbasis C Ustomer Relationship Management Untuk Studi Kasus* : *T-COMP JOGJA* pp 7–12.

[2] Lewis J R 1995 IBM computer usability satisfaction questionnaires : Psychometric evaluation and instructions for use *Int. J. Hum. Comput. Interact.* vol **7** no 1 pp 57–78.

[3] Lewis J R. 2013 Measuring perceived usability: SUS, UMUX, and CSUQ ratings for four everyday products *Int. J. Human–Computer Interact.* vol **00** no 00 pp 1–16.

[4] Brooke J 2013 SUS: A retrospective *J. Usability Stud.* vol 8 no 2 pp 29–40.

[5] Setiawan A and Widyanto R A 2019 Evaluasi website perguruan tinggi menggunakan metode usability testing *J. Inform. J. Pengemb. IT* vol **3** no 3 pp 295–99.

[6] Hartson R and Pyla P S 2012 *The UX book process:process and guidelines for ensuring a quality user experience* Waltham MA 02451 USA: Elsevier Inc.

[7] Setiawan A, Artha E U, Arumi E R, Sunarni A. P, and Nugroho S 2019 Task analysis of facebook users on frequently used menus *J. Phys. Conf. Ser.* vol **1179** p 012019.

[8] Kortum P T and Bangor A 2013 Usability ratings for everyday products measured with the system usability scale *Int. J. Hum. Comput. Interact.* vol **29** no October 2014 pp 67–76.

[9] Setiawan A and Widaryanto R 2018 Mobile CRM student-parent information system *IOP Conf. Ser. Mater. Sci. Eng.*vol **403** p 012076.

[10] Barnett L, Harvey C, and Gatzidis C 2018 First time user experiences in mobile games : An evaluation of usability *Entertain. Comput.*vol **27** no April pp 82–8.

[11] Barnett L, Gatzidis C, and Harvey C 2017 An investigation into usability and first time user experiences within a mobile gaming context no. April 2018.

IOP Conf. Series: Materials Science and Engineering **821** (2020) 012045 doi:10.1088/1757-899X/821/1/012045

Redesign and modification of torque measurement instrument for centrifugal pump

E Widodo*, F Nurwenda, A Akbar, Mulyadi, N Fediyanto and A Fahruddin

Department of Mechanical Engineering, Faculty of Engineering, Universitas Muhammadiyah Sidoarjo, Sidoarjo, Indonesia, 601271

*Email: ediwidodo@umsida.ac.id

Abstract. The main function of a pump is moving fluid through a pipe from one place to another. The dominant characteristic of a centrifugal pump is its output torque and power. Torque is obtained from the measurement of force using a dynamometer, then the result is multiplied by the moment measuring arm (L). Dynamometers are used to measure weight, which is influenced by gravity, which is mass (m) multiplied by the acceleration of gravity (g). The design concept of this equipment consists of a free moving motor body, dynamometer layout, safety pump, and fluid flow pipe. The pump is designed to be equipped with laboratory scale test kits and sketches of images made using ZWcad. The product design test is carried out using a comparison of the torque value to the centrifugal pump both in filled and unfilled condition. The difference in the value of the output torque is not too far from the theoretical calculation, so this torque measuring instrument is worth using.

1. Introduction

The addition of a torque measuring device is necessary. The sliding seat engineering research on the observation pump for laboratory effectiveness requires [1]. The use of several forms of standard mechanisms for the application of torque to various implant components is inevitable, especially in vital tools both in industrial activities and household needs [2]. Today, there is no torque measurement system used by industry that is capable of fulfilling the expectations [3]. The design of proper torque measurement systems is needed to solve problems in the industries. The analysis focuses at pump characteristics involing torque and rpm rotation that has been produced. To measure torque and rotation, the addition of a measuring device mounted on the pump is necessary [4]. The pump is driven by an induction motor to produce torque. The obtained torque formulation is as follow:

$$\mathrm{T} = \frac{5250\,.P}{n} \tag{1}$$

Where,
T = torque (Nm)
N = rotation perminute (rpm)
P = power (watt)

The pump as a tool transfer fluid, converts the mechanical energy of the shaft that moves the pump blades into kinetic energy and fluid pressure [5]. Pump specifications are categorized by the amount of fluid that can be flowed per unit time (capacity) and lift energy (head) of the pump [6, 7]. Torque is obtained by measuring the force using a dynamometer, then the result is multiplied by the moment measuring arm (L). The unit of torque is Nm [8]. The addition of a torque measuring device is applied by making a measurement mechanism that utilizes the torque force that is forwarded and read the force gauge balance apparatus. This addition is needed considering the need for pump torque measurements for data collection purposes. Pump torque data is needed in calculating pump performance [9, 10].

2. Method

The pump is designed with the modification of an electric motor torque successor system. With the addition of this tool, the drive motor can move freely according to the torque. The mechanism of pump movement is obtained from the addition of bearings on each side of the pump, made as a rail / path / for motor movement /rotation. With the addition of bearings on both sides of the electric motor and ring covering the motor, it allows the motor to rotate along the ring path. The motor rotation is held by a torque measuring balance. The magnitude of the rotating force / motor torque reads torque meter as the torque produced by the electric motor, as shown in Figure 1.

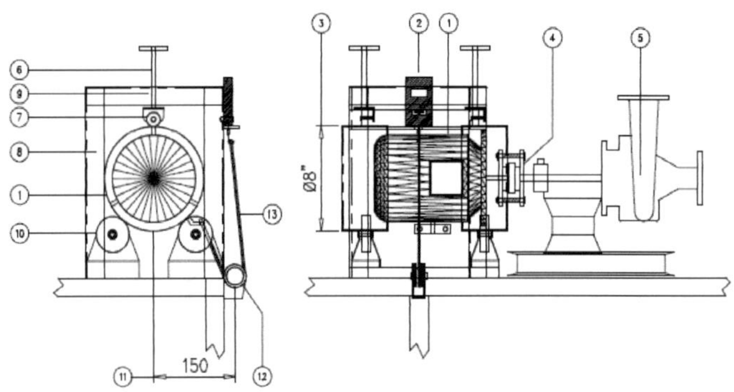

Caption

1. 3 phase induction motor
2. Digital dynamometer
3. 8 inch pipe for path
4. Clamp plate
5. Centrifugal pump
6. Thread for stabilizator pump movement
7. Bearing 6201 for smoothing motor movement /rotation
8. Vertical elbow as Pump holder frame
9. Horizontal Elbow 50 x 50 SA.36 as the frame of the pump holder
10. Bearing 6306 as a support for the bottom so the motor can rotate smoothly
11. Catrol
12. Cable for dynamometer device

Figure 1. Design of pumps and torque gauges

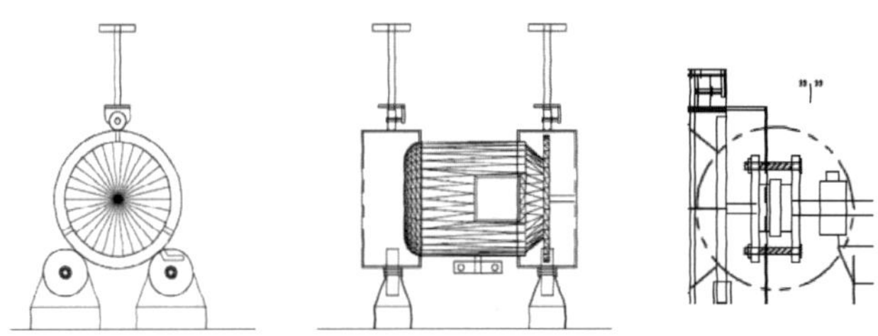

Figure 2. Design of motor movement mechanism

Figure 2 shows the mechanism of a free moving motor, which plays the role of providing additional tools that support the motor which allows it to rotate in all directions.

IOP Conf. Series: Materials Science and Engineering **821** (2020) 012045 doi:10.1088/1757-899X/821/1/012045

Product concept design :

1. Mechanism design
   The mechanism of motor movement is made by adding bearings, plates for the path mechanism of the movement of the bearing and torque to resist motor movement / rotation. The equipment is made of a metal plate that is bent to form a cylinder encircling the motor. On the outside of the motor are installed two bearings that connect the motor body and the plate surface. Thus the motor can rotate with the outer plate rotating plane. To hold the electric motor rotation, one side is connected with a torsimeter. Thus the energy of the pump rotation is transferred to the torque meter into a tensile force. This tensile force is read as the magnitude of the tensile force of the motor according to the reading result of meter torque.
2. Design of dynamometer
   The dynamometer is mounted on one side of the holder and connected to the motor body with a cable. Dinamometer facing forward to facilitate reading the scale. The connecting cable of the dynamometer functions as a connector so that the dynamometer can be read when the motor rotates clockwise or counterclockwise. Dynamometer holder uses steel elbows, because of the tough character.
3. Pump safety
   Figure 3 shows the design of pump safety when it is running with the addition of spesific device. The tool in the form of clamp coupling that functions as a binder between the motor and the pump, so that when the motor is operated, it will not be separated from the holder. Coupling clamps are made of two metal plates and four sets of binding bolts.

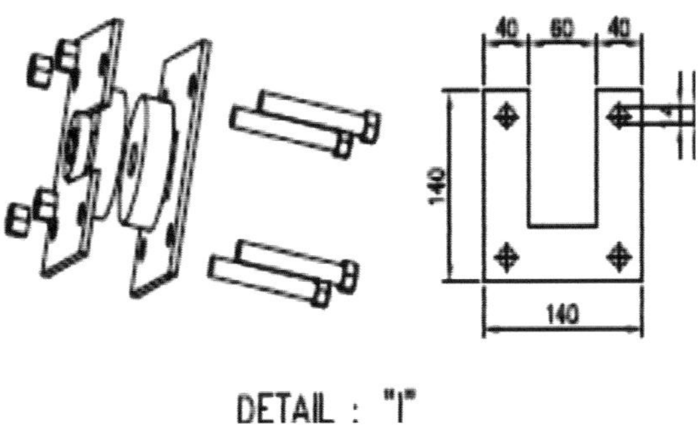

Figure 3. Design of pump safety

4. Pipe instalation
   Pipe installations are arranged using PVC pipes, with input pipes : 1-1 / 4 "and the output is : 1". The output pipe uses a smaller diameter to increase the output pressure (Figure 4). Figure 5 shows a tool for motor movement that can rotate in any direction with a bearing 6306 at the bottom of the cylinder with a bolt as a connecting pipe to the motor body.

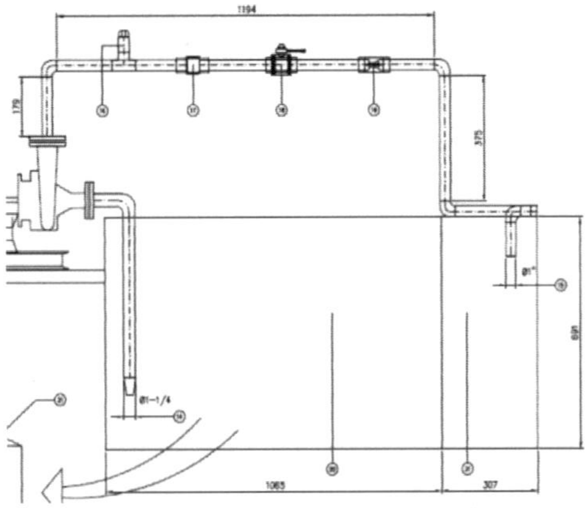

Figure 4. Pipe instalation design

Figure 5. Pump and movement mechanism design

3. Result and discussions

Figure 6 shows the design of the dynamometer support using steel elbows and the installation of coupling clamps using 2 plates that tie the coupling between the motor and the pump with 4 sets of bolts. The principle and method of work of designing centrifugal pump torque measuring devices as follows: When the electric motor is turned on, the motor body rotates clockwise spontaneously. It happens because the motor rotates following its torque and direction. The movement of the motor body automatically rotates the rope as a connection to the dynamometer that is pulled down through the pulley. The body of the motor becomes the support of the rope so that the dynamometer installed at the top of the motor will issue an output value in units of kg when the motor is moving and pulling the rope dynamometer. Then the rotating motor body is held by the rope and the rotation is forwarded to the motor to be forwarded to the pump shaft with the opposite rotation to the motor stator. The motor stator is connected with a coupling with a plate as security so that the motor is inseparable from the pump while rotating. When the pump shaft rotates automatically, the fluid will be sucked and enter the pump house in the flange section and pushed out into the flange out section.

Figure 6. Design of torquemeter and coupling clamp

- The Test Analysis of Centrifugal pumps torque

$$\tau = \frac{5250\ x\ P}{n} \tag{2}$$

Where P is 2 HP, n is 2900 rpm. So

$$\tau = \frac{10500}{2900} = 3.62\ \text{lb.ft} = 4.91\ \text{Nm}$$

IOP Conf. Series: Materials Science and Engineering **821** (2020) 012045 doi:10.1088/1757-899X/821/1/012045

- Fluid velocity obtained

$$v = \frac{Q}{A} \tag{3}$$

$$v = \frac{3{,}06\ x\ 10^{-3}}{5{,}064\ x\ 10^{-4}} = 6.04 \text{ m/s}$$

P : 1,5 kW = 1500 W

$$F = \frac{P}{V} \tag{4}$$

$$F = \frac{1500}{6{,}04} = 248.34 \text{ N}$$

- Torque

$$\tau = F\,.\,L \tag{5}$$

$$\tau = 284.34 \text{ x } 0.150 = 42.65 \text{ Nm}$$

Pump torque without load is obtained from calculating torque pump equipped by stator that is designed (Table 1).

Table 1. Pump performance without load

	Pump performance	
Experiment	**Mass *m* (kg)**	**Gravity Acceleration *g* (m/s²)**
1	2.430	9.806
2	2.495	9.806
Rate	2.463	9.806

- Forces (F)

$$F = m.g \tag{6}$$

$$F = 2.463 \text{ x } 9.806 = 24.15 \text{ N}$$

- Torque (τ)

$$\tau = F\,.\,L \tag{7}$$

$$\tau = 24.15 \text{ x } 0.150 = 3.62 \text{ Nm}$$

Torque pump with load is obtained by experiment conducted to measure mass and pressure (Table 2).

Table 2. Pump performance with load

	Pump performance	
Experiment	**Mass *m* (kg)**	**Pressure of fluid (mBar)**
1	7.815	1200
2	6.400	1200
Rate	7.108	1200

Water debit from the pump is calculated using the V-notch calculation equation :

$$Q = \frac{8}{15}\sqrt{2.g}\tan\frac{\theta}{2}H^{\frac{5}{2}} \tag{8}$$

$$Q = (0{,}53)\sqrt{2.9{,}81}(\tan\frac{60}{2})(0.06^{\frac{5}{2}})$$

$$Q = 1.2\ x\ 10^{-3}\ m^3/_s$$

Velocity of water debit is calculated using formula:

$$v = \frac{Q}{A} \tag{9}$$

$$V = \frac{1.2\ x\ 10^{-3}}{5.064\ \text{x}\ 10^{-4}} = 2.37\ m^2$$

Force of pump derivated from velocity of fluid flow with the equation of:

$$\text{F} = \text{m} \cdot \frac{V^2}{l} \tag{10}$$

$$\text{F} = 7.10 \cdot \frac{2.37^2}{0.150} = 266.19\ \text{N}$$

Torque of pump obtained from pump shaft measurement to the arm of torque meter. We obtained from Equation (5):

$$\tau = 266.19\ \text{x}\ 0.150 = 39.93\ \text{Nm}$$

Table 3 and Figure 7 show that torque without load between theoretical result and the experiment conducted has a difference of 4.91 Nm – 3.62 Nm = 1.29 Nm.

$$\frac{The\ difference}{Theoretical}\ \text{x}\ 100\% = \frac{1.29}{4.91}\ \text{x}\ 100\ \% = 26.27\ \% \tag{11}$$

Figure 8 shows that torque with load between theoretical result and the experiment conducted has a difference of 42.65 Nm – 39.92 Nm) = 3.03 Nm.

$$\frac{The\ difference}{Theoretical}\ \text{x}\ 100\ \% = \frac{3.03}{42.65}\ \text{x}\ 100\ \% = 7.1\ \%$$

Table 3. Torque comparison

The torque without load (Nm)		The torque with load (Nm)	
Theoretical	**Real**	**Theoretical**	**Real**
4.91	3.62	42.65	39.92

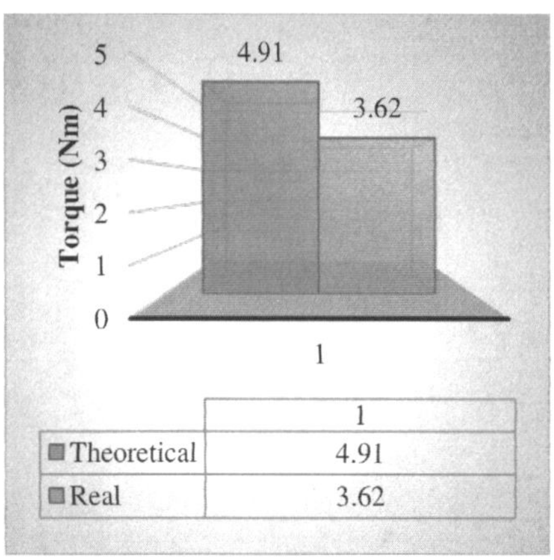

Figure 7. Torque without load

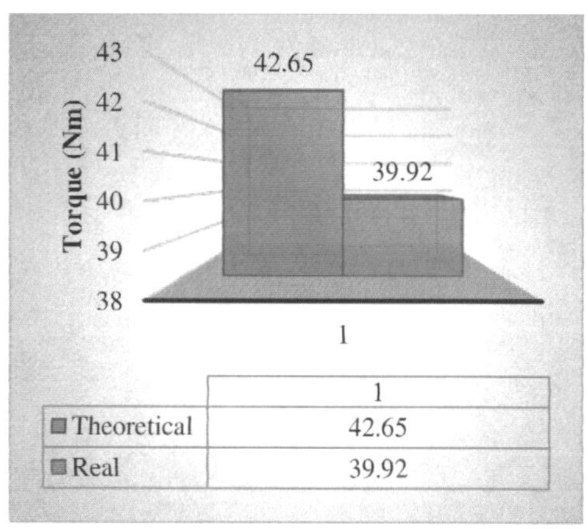

Figure 8. Torque with load

The value of torque with a ratio between 0 to 5 Nm. It can be concluded that the design of a torque meter on a centrifugal pump can work well and it can be concluded that the design of a torque meter on a centrifugal pump can work well, thus helping the performance of the centrifugal pump torque measurement tool be better.

IOP Conf. Series: Materials Science and Engineering **821** (2020) 012045 doi:10.1088/1757-899X/821/1/012045

Some suggestions can be given after conducting this research so that in the future there will be better research. The rope should use a rope with a low stress so that it can reduce the force lost, and Dynamometers should be connected to a computer so that the resulting graph is more accurate.

4. Conclusion

The addition of support frame to hold the pump provides a strong force to maintain pump stability and reduce vibration. The addition of holder aims to increase the measurement accuracy of the torque meter. The torque meter does not get disturbed by engine vibration, so the scale reading can work well. The addition of plate clamps provides an operator safety factor and acts as vibration damper. This torque measurement can work well if it is ensured that its tools does not get interference from external factors. The design of this gauge is able to eliminate the vibration factor and provide more easy readings to the operator. Finally the result of pump torque measurement is match to the pump specifications.

Acknowledgements

The writers would like to thank to Mechanical Engineering Department Universitas Muhammadiyah Sidoarjo, Mechanical engineering Laboratory Chief, Mulyadi, ST, MT, for his wise counsel and constructive comment. The writers also owe gratitudes to all the faculty members of Mechanical Engineering Department for their kind co-operation. Finally, the writers would like to acknowledge the assistance of all parties that have suppprted process of this work.

References

[1] Widodo E G D P 2018 Rekayasa sliding seat pada pompa observasi untuk efektifitas laboratorium *Turbo* vol **7** no 1 pp 65–73.

[2] Goheen K L, Vermilyea S G, Vossoughi J and Agar J R 1997 Torque generated by Handheld Screwdrivers and mechanical torquing devices for osseointegrated implants *Int. J. Oral Maxillofac Implants* vol **9** no 2 pp 149–55.

[3] Karayel D and Yegin V 2016 Design and prototype manufacturing of a torque measurement system in *2nd International Conference on Computational and Experimental Science and Engineering (ICCESEN 2015)* vol **130** no 1 pp 272–75.

[4] Janevska G 2013 Mathematical Modeling of Pump System," in *Electronic International Interdisciplinary Conference* no. **September 2013** pp 455–58.

[5] Sharma S R A K S 2016 Design optimization of centrifugal pump impeller by varying blade angle - A review *Int. J. Appl. Eng. Technol.* vol **3** no 2 pp 321–26.

[6] Walter P T, PE, Scott L, David M 2018 Unappreciated challenges in applying four quadrant pump data to waterhammer simulation part 1: Fundamentals in *13th International Conference on Pressure Surges 2018* pp 741–54.

[7] Pradana E W R Y 2018 Analysis of pipe diameter variation in axial pumps for reducing head loss in *2017 1 st International Conference on Engineering and Applied Technology (ICEAT)* pp 1–8.

[8] Volkov A V, Parygin A G, Naumov A V, Vikhlyantsev A A and Druzhinin A A 2017 Application of methods of the optimum control theory for development of high efficiency centrifugal pumps *Int. J. Appl. Eng. Res.* vol **12** no 19 pp 8768–78.

[9] Pradhana R Y and Widodo E 2017 Analisa pengaruh variasi diameter pipa tekan pvc pada pompa aksial untuk kecepatan gaya dorong air *R E.M. (Rekayasa Energi Manufaktur) J.* vol **2** no 1 pp 37–43.

[10] Ansori F and Widodo E 2018 Analysis on centrifugal pump performance in single, serial, and parallel *JEMMME (Journal Energy, Mech. Mater. Manuf. Eng.* vol **3** no 2 pp 79–86.

IOP Conf. Series: Materials Science and Engineering **821** (2020) 012046 doi:10.1088/1757-899X/821/1/012046

Energy analysis desalination of single slope solar still

M A Siregar* and W S Damanik

Department of Mechanical Engineering, Faculty of Engineering, Universitas Muhammadiyah Sumatera Utara, Medan, Indonesia

*Email: munawaralfansury@umsu.ac.id

Abstract. The surface of the evaporator glass in the seawater desalination process by utilizing solar energy as the main energy source is a phenomenon that has been widely studied and developed in order to get increased performance efficiency to produce fresh water. The heat energy from the sun will enter through the surface of the evaporator glass and absorbed by water inside the evaporator, while the evaporator wall is insulated using aluminium foil with a thickness of 20 mm in the hope that the heat energy received will not be wasted around the environment. The process of evaporation of seawater begins when the water has reached the point of evaporation without pressure and occurs naturally. When seawater starts to evaporate, it can be calculated the amount of heat energy absorbed by the water until it reaches the point of evaporation, in order to determine the increase in performance and efficiency of the desalination process. Efforts to improve continue to be done if it is still possible to increase the efficiency and performance of desalination devices with various energy utilization methods. After testing, the highest solar intensity obtained on the test day reached 420.85 W/m^2, and the lowest solar intensity on the second day was 96.89 W/m^2. The highest evaporation heat transfer coefficient reaches 10.54 $W/m^2\,^{\circ}C$ and the lowest is 4.42 $W/m^2\,^{\circ}C$. The highest convection heat transfer coefficient reaches 51.15 $W/m^2\,^{\circ}C$ and the lowest heat conversion coefficient is 15.28 $W/m^2\,^{\circ}C$. The total absorbed energy reaches 6.26 kW and the lowest absorbed energy reaches 1.37 kW. Energy efficiency reaching 85% and the lowest energy efficiency is 35%.

1. Introduction

The development of seawater processing technology into clean water is based on the importance of clean water as a human need in everyday life and agricultural and industrial needs. About 97% to 97.5% of the water on the earth's surface is saltwater and about 2% is stored as ice in the polar region and 1% is freshwater available for the needs of plants, animals and human life [1, 2]. Every day the availability of drinking water from natural sources is shrinking due to the rapid growth of the human population and the poor processing of dirty water produced from industrial wastes using chemicals that pollute clean water sources [3]. Some desalination technologies have been developed including, Reverse Osmosis (RO), Thermal Vapor Compression (TVC) Mechanical vapor compression (MVC), multi-stage flash distillation (MSF), multi-effect distillation (MED) and Vaccum Desalination (VD) [3, 4]. But some researchers say the desalination process using solar energy is the most efficient, economical and environmentally friendly technology [3, 4, 5, 6]. In the last 40 years, desalination has been developed and produced ideas in developing the model and methods including single slope models and double slopes by the workings of passive and active systems [6, 7]. However, the amount of clean water produced from a single slope model desalination device is greater than the amount of water produced in the double slope model [3]. Tiwari et al. (2016) conducted a single slope and double slope desalination model experiments and obtained a number of clean water production on a single slope of 499.1 L/year, and the amount of clean water production on a double slope model 464.68 L/year or around 1.3 liters per day [3].

Many variations in the slope of the glass have also been carried out starting from 15°, 30° and 45° as well as the depth of seawater inside the evaporator ranging from 0.01 m, 0.02 m, 0.03 m, 0.04 m, 0.05 m, 0.08 m 0.12 m, 0.16 m and 0.18 m [3, 8, 9, 10]. D.B. Singh (2016) conducted an experiment on the desalination of a single slope model of the total power system received by the evaporator 1159.43 kW and the amount of clean water production of 1739.15 kg/year or about 1.73 Kg per day [11]. In the following year, D.B.Singh (2017) added a copper and heter electric collector whose photovoltaic electricity flow to a single slope desalination model pasiv system and total energy obtained by 3389.00 kW and the amount of clean water production obtained 4519.54 kg/year or about 12.55 kg/day [12]. In addition to the use of solar energy, wind and geothermal energy are also considered to aim at efficiency and also the energy used is environmentally friendly [13, 14]. Of the three types of energy that are often used in desalination devices, namely the sun, wind and geothermal energy, solar energy is the most widely used up to 73% compared to other energy [15]. Before the development of the use of solar energy, the use of electrical energy was also carried out, around 10.000 tons of fuel was spent on the desalination process [16]. The energy received is very influential on the amount of the rate of production of clean water produced in desalination devices, it requires the use of good energy to reduce the amount of energy wasted during the process carried out. This study aims to review the overall heat energy absorbed by the single slope seawater desalination process by adding thickness and insulation material

2. Method and equipment

Tests carried out for 8 days with the location of the evaporator and glass evaporator to the south to be able to receive heat energy from the sun perfectly. Six thermocouples were placed in areas that wanted to know the temperature, namely the temperature of the outer glass, the inner glass temperature, the basin temperature, the water temperature, and ambient temperature. The data was recorded every hour and was used to find the heat transfer coefficient and the total energy absorbed in the desalination device (Figure 1). The following are dimensions of the solar distillation device and the physical properties of the materials used. In Table 1 we can see the specifications and physical properties of the materials used in the desalination device.

Table 1. Physical properties of distillation equipment [3]

Symbol	Value	Symbol	Value
Ag	1 m^2	Li	0.02 m
hbw/f	100 $W/m^2 K$	Ki	0.039 W/m °C
σ	$5.67x\ 10^{-8}\ W/m^2 K^4$	α_b	0.8
Ab	1 x 1 m	α_w	0.6
L_g	0.003 m^2	α_g	0.05
K_g	0.78 W/m^2 °C	ϵ_{eff}	0.9048
M_w	20 Kg	ϵ_g & ϵ_w	0.95

Figure 1. Experimental scheme single slope distillation

The dimensions of the desalination device are adjusted according to previous research D.B.Singh (2016). The workings of the desalination tool used are to open the inlet valve from the reservoir tank into the basin until the water level reaches 0.02 m from the bottom of the evaporator. After the water in the evaporator is filled enough, then operate the logger temperature data to start recording the temperature of the outer glass (Tgo), the temperature of the glass (Tgi), the water temperature (Tw), basin temperature (Tb), and ambient temperature (Ta), while the intensity data The sun is recorded using a solar intensity sensor. Test data obtained is stored and calculated using Excel 2010 M.S.Office.

The data obtained during the test are analyzed so that the calculation temperature of the outer glass and the inside of the glass are also known and also to calculate the amount of heat transfer coefficient that occurs during the test. There are two parts that will be calculated the heat transfer coefficient value on the evaporator, namely the heat transfer coefficient outside and the heat transfer coefficient inside. The outer heat transfer coefficient includes the convection heat transfer coefficient from the glass to the environment and the heat transfer coefficient from the glass environment. Whereas the heat transfer coefficient is in the convection, radiation and evaporation heat transfer coefficients. Calculating the temperature of the outer glass T_{go} evaporator can follow the following equation [11, 17, 18].

$$T_{go} = \frac{\frac{K_g}{L_g} T_{gi} + h_{1g} T_a}{\frac{K_g}{L_g} + h_{1g}} \tag{1}$$

Calculating the temperature of the inner glass T_{gi} can use the following equation [11,17,18].

$$T_{gi} = \frac{\alpha'_g I_S(t) A_g + h_{1w} T_w A_b + U_{c,ga} T_a A_g}{U_c g_a A_g + h_{1w} A_b} \tag{2}$$

To get the total energy absorbed by the evaporator, first, calculate the heat transfer coefficient outside or inside. Calculating the convection transfer coefficient outside h_{cg} the evaporator can use the following equation [3]

$$h_{cg} = \begin{bmatrix} [5.7 + 3.8 \times V] : \leq 5 m/s \\ [6.15 \times V^{0.8}] > 5 \text{ m/ s} \end{bmatrix} \tag{3}$$

Wind speed is very influential on the large heat transfer coefficient of the evaporator because it relates to the environment. To calculate the heat transfer that occurs in the evaporator h_{rg} can use the following equation [3]

$$h_{rg} = \frac{\in_g \times \sigma \times (T_g^4 - T_{sky}^4)}{(T_g - T_a)} \tag{4}$$

After getting the total external heat transfer coefficient, it calculates the inner heat transfer coefficient. To calculate the amount of heat transfer coefficient outside the water evaporator h_{rw} into the air in the evaporator, the following conditions can be used [3,11,12,17,18,19]

$$h_{rw} = \in_{eff} \times \sigma \times \left[(T_w + 273)^2 + (T_{gi} + 273)^2 \times (T_w + T_{gi} + 546) \right] \tag{5}$$

Effective emissivity $\in$ and Stefan-Boltzman numbers σ can be obtained from table 1. To find the convection heat transfer coefficient in water h_{cw} can use the following equation [3, 11, 12, 17, 18]

$$h_{cw} = 0.884 \left[(T_w + T_{gi}) + \frac{(P_w - P_{gi}) \times (T_w + 273)}{268.9 \times 10^3 - P_w} \right]^{\frac{1}{3}} \tag{6}$$

To calculate the amount of evaporation heat transfer coefficient that occurs in the evaporator h_{ew} can use the following equation [3, 11, 12, 17, 18].

$$h_{ew} = 0.034 \times 5.67 \times 10^{-8} \left[(T_{giE} + 273)^2 + (T_{giW} + 273^2) \right] \times \left[T_{giE} + T_{giW} + 5546 \right] \tag{7}$$

After getting the heat transfer coefficient outside and inside, we can find out the total energy absorbed by the evaporator every day. The energy received and absorbed by the evaporator every hour can be calculated using the following equation [3, 11, 12, 17, 18].

$$En_{hours} = \left[h_{ef} (T_{bf} - T_{gi}) \right] \tag{8}$$

After getting a large amount of energy absorbed every hour, energy efficiency can be sought from the desalination device using the following equation [3, 11, 12, 17, 18].

$$\eta_{En} = \frac{\sum (m_{ew}) \times L}{(A_g - I_S(t)) \times 3600} \times 100 \tag{9}$$

The above equation is used to find the efficiency of the evaporator.

3. Results and discussion

The test was carried out for 7 days by observing the temperature of the glass evaporator outside and also the temperature of the water and air in the evaporator by recording the clock. There are 6 thermocouples installed on each side that you want to observe the temperature.

After analyzing the test data, it was found that the heat transfer coefficient in the average evaporator can be seen in Figure 2 below. From Figure 2 can be seen on the fifth day the test of solar intensity is very high, which is 420.85 W/m^2 so that the evaporation heat transfer coefficient that occurs on that day is very high compared to the previous day which reached 10.54 W/m^2 °C, this is

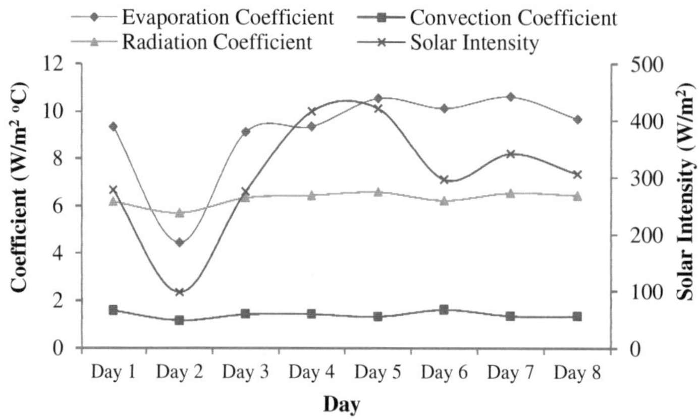

Figure 2. Graph of the solar intensity of the heat transfer coefficient in the evaporator.

due to heat energy enter high and absorbed by water so that the water evaporates faster. While on the second day the intensity of the sun looks low only reaching 96.89 W/m^2 and the evaporation heat transfer coefficient also looks low only reaching 4.42 W/m^2 °C this is because the intensity of the sun is very low so that the energy entering and absorbing also causes the evaporation rate the water in the evaporator is not as fast as on the fifth day. With the same solar intensity, the convection coefficient obtained on the fifth day is 1.32 W/ m^2 °C and the lowest convection coefficient on the second day is 1.16 W/m^2 °C. While the radiation coefficient obtained the highest data on the fifth day which is 6.58 W/m^2 °C and the lowest on the second day is 5.7 W/m^2 °C, while the glass temperature is low, the rate of evaporation in water is very fast.

The solar intensity and the wind speed during the test greatly affect the rate of heat transfer from the glass of the environment, the heat transfer coefficient that occurs on the outside of the evaporator can be seen in Figure 3.

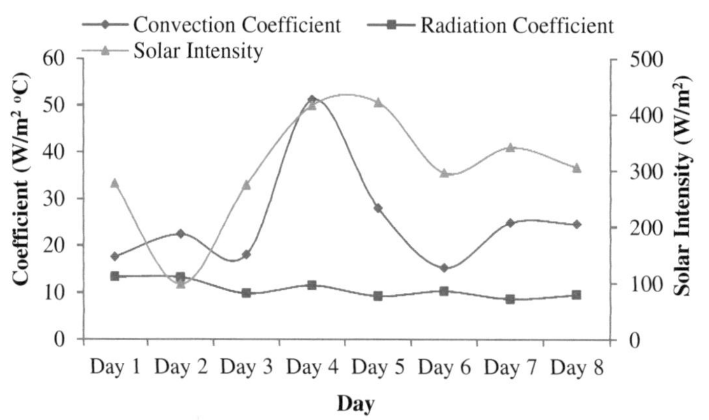

Figure 3. Graph of the solar intensity of the heat transfer coefficient outside the evaporator

In the picture above, it can be seen that the highest convection coefficient occurs on the fourth day which is 51.15 W/m^2 °C and the lowest coefficient is seen on the sixth day 15.28 W/m^2 °C. The high solar intensity affects the magnitude of the convection coefficient from the outer glass surface to the environment, this is also because the wind speed on the fourth day is much higher, reaching up to 14.11 m/s compared to the wind speed on the sixth day only reaching 2.76 m/s. While the highest radiation heat transfer coefficient was seen on the second day at 13.25 W/m^2 °C and the lowest on the seventh day 8.62 W/m^2 °C.

Even though the amount of energy absorbed by the water inside the evaporator is large, then the rate of consumption of water will be greater and this is expected to increase the amount of clean water production. The amount of energy absorbed must be in accordance with the amount of water produced to minimize losses due to wasted energy. The total energy absorbed by water and energy efficiency in the evaporator can be seen in Figure 4 below.

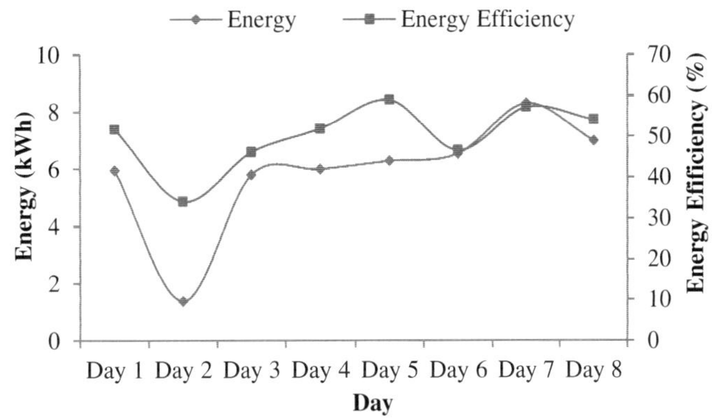

Figure 4. Graph of energy towards energy efficiency absorbed by the evaporator

The highest total absorbed energy was seen on the fifth day of 6.28 kWh and the lowest total absorbed energy on the second day was 1.37 kWh according to the high intensity of the sun shown in Figure 2 and Figure 3. The highest energy efficiency was also obtained on the fifth day to 58.89% and the lowest efficiency obtained on the second day is 34.05%.

4. Conclusion

With the highest solar intensity 420.85 W/m^2 obtained by total heat transfer coefficient in the evaporator by summing the evaporation heat transfer coefficient, convection heat transfer coefficient and radiation heat transfer coefficient which is 18.44 W/m^2 °C, and total heat transfer coefficient in the lowest evaporator with sun intensity 96.89 W/m^2 which is 11.31 W/m^2 °C. While the total heat transfer coefficient is 62.63 W/m^2 °C and the lowest total heat transfer coefficient is 25.57 W/m^2 °C. The highest energy absorbed by the evaporator was obtained on the fifth day of 6.28 kWh and the lowest energy was absorbed by the evaporator on the second day, namely 1.37 kWh. The highest efficiency on the fifth day is reaching 58.89% and the lowest energy efficiency on the second day is 34.05%.

References

[1] Bindra S P and Abosh W 2001 Recent developments in water desalination *Desalination* vol **136** no 1–3 pp 49–56.
[2] Vinoth Kumar K and Kasturi Bai R 2008 Performance study on solar still with enhanced condensation *Desalination* vol **230** no 1–3 pp 51–61.
[3] Tiwari G N and Sahota L 2017 Review on the energy and economic efficiencies of passive and active solar distillation systems *Desalination* vol **401** pp 151–79.
[4] Sahota L and Tiwari G N Effect of nanofluids on the performance of passive double slope solar still: A comparative study using characteristic curve *Desalination* vol **388** pp 9–21.
[5] Singh D B, Tiwari G N, Al-Helal I M, Dwivedi V K and Yadav J K 2016 Effect of energy matrices on life cycle cost analysis of passive solar stills *Sol. Energy* vol **134** pp 9–22.
[6] Sahota L, Shyam and Tiwari G N 2017 Energy matrices, enviroeconomic and exergoeconomic analysis of passive double slope solar still with water based nanofluids *Desalination* vol **409** pp 66–79.
[7] Sahota L and Tiwari G N 2016 Effect of Al_2O_3 nanoparticles on the performance of passive

double slope solar still *Sol Energy* vol **130** pp 260–72.

[8] Tiwari A K and Tiwari G N 2006 Effect of water depths on heat and mass transfer in passive solar still: in summer climatic condition *Desalination* vol **195** no 1–3 pp 78–94.

[9] Kumar S and Tiwari G N 2009 Life cycle cost analysis of single slope hybrid (PV/T) active solar still *Appl Energy* vol **86** no 10 pp 1995–2004, 2009.

[10] Dev R and Tiwari G N 2009 Characteristic equation of a passive solar still *Desalination* vol **245** no 1–3 pp 246–65.

[11] Singh D B, Yadav J K, Dwivedi V K, Kumar S, Tiwari G N and Al-Helal I M 2016 Experimental studies of active solar still integrated with two hybrid PVT collectors *Sol Energy* vol **130** pp 207–23.

[12] Singh D B and Al-Helal I M 2018 Energy metrics analysis of N identical evacuated tubular collectors integrated double slope solar still *Desalination* vol **432** pp 10–22.

[13] Dwivedi V K and Tiwari G N 2010 Experimental validation of thermal model of a double slope active solar still under natural circulation mode *Desalination* vol **250** no 1 pp 49–55.

[14] Siregar C A 2019 Pengaruh jarak kaca terhadap efisiensi alat destilasi air laut yang memanfaatkan energi matahari di kota Medan *J Mech Eng Manuf Mater Energy* vol **2** no 2 p 51.

[15] Eltawil M A, Zhengming Z and Yuan L 2005 A review of renewable energy technologies integrated with desalination systems *Renew Sustain Energy Rev* vol **13** no 9 pp 2245–62.

[16] Kalogirou S A 2005 Seawater desalination using renewable energy sources *Prog Energy Combust Sci* vol **31** no 3 pp 242–81.

[17] Singh D B 2018 Improving the performance of single slope solar still by including N identical PVT collectors *Appl Therm Eng* vol **131** pp 167–79.

[18] Singh D B and Tiwari G N 2016 Effect of energy matrices on life cycle cost analysis of partially covered photovoltaic compound parabolic concentrator collector active solar distillation system *Desalination* vol **397** pp 75–91.

[19] Raju V R and Lalitha Narayana R 2018 Effect of flat plate collectors in series on performance of active solar still for Indian coastal climatic condition *J King Saud Univ – Eng Sci* vol **30** no. 1 pp 78–85.

IOP Conf. Series: Materials Science and Engineering **821** (2020) 012047 doi:10.1088/1757-899X/821/1/012047

Simulation of air flow loss in triangle pipe construction

S Lubis*, **C A Siregar**, **F Abdilah**

Department of Mechanical Engineering, Faculty of Engineering, Universitas Muhammadiyah Sumatera Utara, Medan, Indonesia

*Email: sudirmanlubis@umsu.ac.id

Abstract. The purposes of this research are to find out how the simulation of flow velocity, mass flow rate and pressure that occur in each variation of Reynold numbers on the triangle pipe arrangement of solid work software. By using test and simulation analysis methods, using a number of tools and materials so that they can be simulated and simultaneously analyze the resulting design. The tools and materials used in the simulation and analysis process in this design are (1) computers with specifications of the Intel (R) Xeon (R) CPU E3-1246 v3 @ 3.50Ghz 3.50 GHz (2) a solid work keel device for analyzing pressure. This fluid mass flow rate is influenced by the velocity of fluid flow and the temperature at this fluid can influence fluid density and specific gravity. This fluid speed can affect the pressure that hits the pipe. In the simulation results, there is a significant pressure change due to the speed that changes over the sidelines of the same pipeline. This change raises the loss of friction in the arrangement of the pipe pipeline due to the pressure factor that hit the pipe. The pressure that occurs is influenced by the velocity of fluid flow, fluid mass flow rate and cross-sectional area on the triangle pipe arrangement.

1. Introduction

People are always faced with fluid every day without being struck by it. This fluid flow causes many natural symptoms such as hills, canyons that are in the aftermath of the forces that are caused by months of fluid flow. But in the utilization of many losses that arise one of them is the disadvantage in the installation of heat exchanger equipment. The instrument of the heat exchanger has been very often met in daily life. One of them is on the power plant. Tubes compiled in triangular form as a zigzag arrangement will result in gaps in the installation of the Calor Exchanger. The longitudinal direction of stacking arranged with adjacent distances aims to make the pipe arrangement have a higher flow of turbulence so that the heat transfer occurs more and more. Doing modelling is a way to study the system and the model itself and also the various differences in behaviour [1].

But the consequence of the density of the tube file is the occurrence of a larger pressure drop, resulting in large friction as well. This will result in a loss of flow to the triangular pipe arrangement [2]. The more complex an installation arrangement of the heat exchanger is used, the more loss of flow can be found. The fluid that hits the pipe arrangement on the installation of the heat exchanger can weaken the current pipeline. Unless the thickness on the pipe wall is operating enough to maintain the pressure working on the pipe installation.

Knowing the loss of flow on a system that utilizes fluid flows as a medium, determines the level of efficiency. If you want to know the losses that occur on the pipeline then it should be done laboratory research. The amount of flow loss on the pipe arrangement is influenced by several factors, such as pressure hitting the pipe, broad cross-section, fluid temperature, viscosity, and pipe roughness, fluid type weight, and pipe length. Then based on the background, this research is done using SolidWorks software to facilitate the design and calculation process. To understand the fluid flow, you must know some basic properties of the fluid. The basic properties of the fluid are density, specific gravity, pressure, a viscosity [3]. Adjustment of the volume flow rate in the fluid can cause a decrease in the quality of the broadcast that is in the pipe, for that some models are needed when simulating the fluid flow in the program [4]. The deformation of a pipe construction can cause an imbalance between pressure and gravity and will cause a small disturbance in the visual waves of the pipe flow [5]. Several studies have been carried out using a double gamma densitometer to measure fluid retention in a three-phase slug flow and pressure reduction measurements used using a transducer mounted at the end of the pipe, obtained superficial fluid flow velocity of 0.24 m / s for air, 0.05 for oil and water [6]. Much earlier in the year, a slug flow experiment with low viscosity oils of 10 MPa to 5 MPa was drained on a transparent acrylic pipe in a horizontal position with a pipe length of 29.7 m and an internal diameter of 38.1 mm at 2 bar pressure. The lowest speed in the experiment was 1.5-4.3 m / s for air 0.26-1.36 m / s for oil and 0.19-2.08 for water [7]. From some experiments that have been done by researchers, there was actually interest that was carried out related to the flow model of each fluid that moved in a pipe to be able to engineer with simulation methods. The approach using simulation is not new in analyzing, using solidwork the writer will start testing the flow rate of fluid in the pipe and compare the results of experimental testing with the results of the simulation conducted.

2. Research purposes

This research was conducted to find out how the simulation form of flow speed, mass flow rate, and pressure occurred on each variation of the number Reynold on the triangular pipe arrangement of solid work Software.

3. Theory

Simulation is a technique of imitating operations or processes that occur in a system with the help of computer devices and based on certain assumptions so that the system can be learned. Simulation is also a great tool to use if required to experiment in order to find the best comments of the components used in the system.

The simulation approach begins with building the system model. The model should be able to demonstrate how the various components in the system interact with each other so that it really describes the behavior of the system. Once the model is created then the model will be informed to the computer making it possible to be simulated

4. Piping system

The pipe is a medium where the fluid from one unit to the other. In general, the characteristics are determined based on the constituent material. The size of the pipe diameter is based on the "Nominal" diameter of the outer diameter (OD) or inner diameter (ID). Tubing is a pipe with a diameter smaller than the pipe, its usefulness (in general) is for the liaison between the measuring instrument with the process of pipe an instrument to the control system. The standard size for tubing is always the outer diameter (OD) [8].

5. Fluid

Fluid flow or liquid substances (including water and gas vapors) are distinguished from solid objects due to their ability to flow. The fluid is easier to flow because the bonding of molecules in the fluid is much smaller than the molecular bonds in solids, as the fluid has a relatively small barrier to change of shape due to friction.

6. Research and method

This research was conducted on the arrangement of triangular pipes at the technical faculty building of the Universitas Muhammadiyah Sumatera Utara. The implementation time is about 4 months. The research process starts from the process of simulating, designing and analyzing objects using several tools and materials so that they can be simulated and also analyze the resulting design.

The tools and materials used in the simulation process and analysis of this design are (1) computer with specification Intel (R) Xeon (R) CPU E3-1246 v3 @ 3.50 GHz (2) Solid work software to analyze the pressure.

7. Research procedure

This research procedure is made into several steps:

1. Preparing the simulation design of the flow made on the triangular pipe arrangement.
2. Determine the size to be made in the triangular pipe arrangement, pipe length, pipe diameter, fluid type, speed flow and the mass flow rate used in the simulation
3. Designing a simulated flow using a DESIGN solid work that has been made
4. Conducting simulations
5. Observation of the simulation while recording all observations of flow velocity on the triangular pipe arrangement and performing numerical simulation
6. Perform documentation on testing/simulation

8. Result and discussion

The process of evaluation and simulation of the pipe arrangement and using solid work, there are known numbers Reynold that varied up to 10 variations are analyzed and produce the speed value in each experiment and flow rate of the mass. External simulation results are varied based on the Reynold numbers 1000 – 10000 obtained comparisons between the speed and the mass flow rate can be seen in Figure 1.

It is known that the higher speed of incoming fluid hit the pipe then the mass flow rate is getting bigger so that each flow level causes the airflow loss on the triangular pipe arrangement. Further, there is a comparison between speed and pressure can be seen in Figure 2.

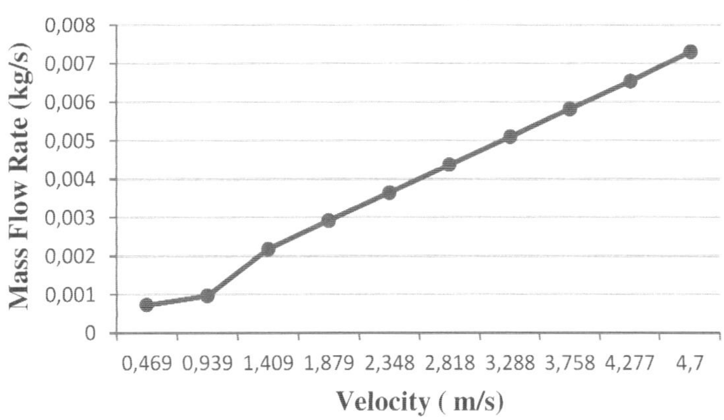

Figure 1. Comparison of graphs of acceleration against mass flow rates

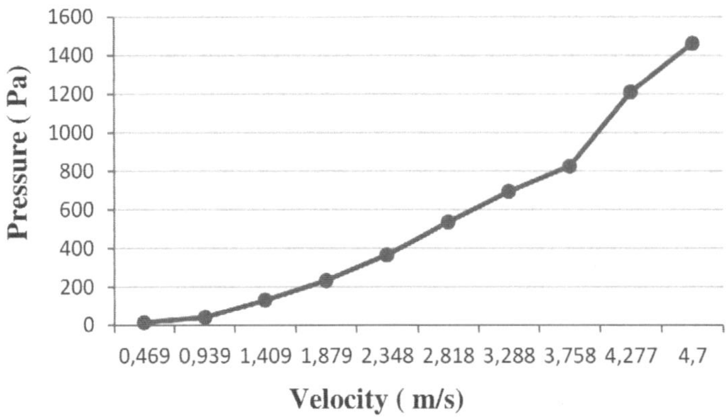

Figure 2. Comparison of speed on pressure

The Data on the graph provides information that the higher the speed of incoming fluid hit the pipe, and then the pressure received pipe from the air will be greater. Further to the comparison between speed (V) and fluid friction loss (f) on the above test result data can be seen in the following chart.

Known the higher the speed of incoming fluid hit the pipe, then the friction losses that occur are not always greater. It is influenced by the fluid speed factor and the fluid type of weight. This fluid mass flow rate is influenced by fluid flow velocity and the fluid temperature is capable of affecting the density and weight of fluid types. This fluid velocity is able to affect the pressure hitting the pipes. In simulated results, there is a significant pressure change due to the speed that changes through the sidelines of the triangular pipe arrangement. This change raises the friction loss of the triangular pipe arrangement due to the pressure factor hitting the pipe. The pressure that occurs affected by fluid flow velocity, fluid mass flow rate and cross-section of triangular pipe arrangement. the ratio between the friction losses of the pipe to the fluid will be shown in figure 3.

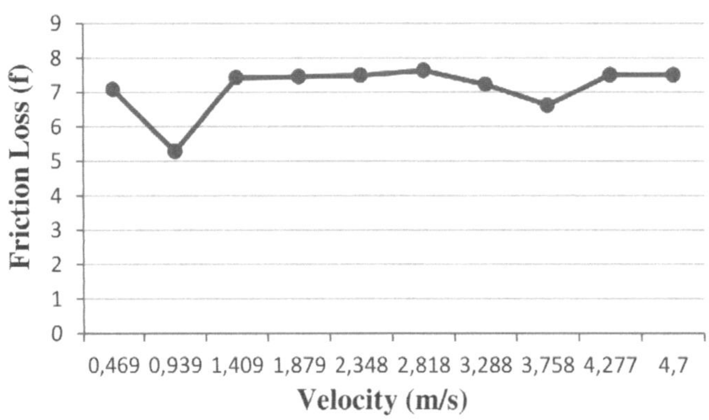

Figure 3. Comparison of velocity on friction loss

From the graph above it can be seen the influence of pipe construction, the intersection will affect the velocity of fluid flow. Pipe roughness will reduce the speed of fluid flow, and affect the travel time of the fluid to be delivered. The energy used will also be greater if the losses in the pipe are greater. The highest roughness when testing was obtained up to 8 at a speed of 3,758 m/s and at a speed of 4,277 m/s. while the lowest roughness is obtained at speeds of 0.939 m/s and 3,288 m/s.

9. Summary

This research provides several conclusions. First, the fluid mass flow rate is influenced by fluid flow velocity, the fluid used is air with temperatures 300 ° K (26.85 °c). The temperature in the fluid affects the fluid density and weight of the fluid type. Second, fluid velocity affects the pressure that will hit the pipe. In simulated solid work, the color-changing part is a representation of the significant pressure change due to the changing speed when passing a gap-between the triangular pipe arrangement. Third, friction losses occur due to pressure factors that hit the pipe. The magnitude of pressure is influenced by fluid flow velocity, fluid mass flow rate and a broad cross-section of the triangular pipe arrangement.

10. Suggestion

Based on the results of this study, the researcher gives some suggestions, i.e: The process of arranging should pay more attention to the air temperature that hit the pipe correctly. Subsequent research should pay attention to the details of the selection of materials and pipe sizes used to avoid the loss of friction that is not too large.

Acknowledgments

I thank a lot to the university and my students who provide support morally and help me in completing this research so that the purpose of this research can be achieved. I hope I can develop this research with more active student involvement so that it is able to enrich student's and universities ' knowledge.

References

[1]. Holman RY Baha'ullah, C Imron, Simulasi Numerik Aliran Fluida pada Saluran T-Junction 90': PLTA Tulungagung, Jurnal Sains dan Seni ITS, 2016.
[2]. Holman J P 1997 *Heat TransferTenth Edition (*USA: John Willey and sons, Inc).
[3]. Zainudin, I Made Adi Sayoga, I Made Nuarsa, Analisa Pengaruh Variasi Sudut Sambungan Belokan Terhadap *Head Losses* Aliran Pipa, Dinamika Teknik Mesin, Volume 2 No.2 Juli 2012.
[4]. Surbhi sharma, Manisha arora, Kuldeepak kaushik, Analysis Of Liquid Level Control Of Coupled Tank System By Pi, Pd, Pid Controller, International journal of scientific & technology research volume 4, issue 11, November 2015.
[5]. R.I. Issa, M.H.W. Kempf, Simulation of slug flow in horizontal and nearly horizontal pipes with the two-fluid model, International Journal of Multiphase Flow 29, 2003.
[6]. M. Bonizzi, R.I. Issa, On the simulation of three-phase slug flow in nearly horizontal pipes using the multi-fluid model, International Journal of Multiphase Flow 29 (2003) 1719–1747.
[7]. Malinowsky, M.S., An experimental study of oil-water and air–oil–water flowing mixtures in horizontal pipes. MS Thesis, The University of Tulsa, US, 1975.
[8]. Munson R B, Okiishi H T, Huebsch W, and Rothmayer A P 2012 *Fundamentals of Fluid Mechanics Seventh Edition* USA: John Willey and sons Inc.

IOP Conf. Series: Materials Science and Engineering **821** (2020) 012048 doi:10.1088/1757-899X/821/1/012048

Engineering of motorcycle exhaust gases to reduce air pollution

A M Siregar*, C A Siregar and M Yani

Department of Mechanical Engineering, Faculty of Engineering, Universitas Muhammadiyah Sumatera Utara, Medan, Indonesia

*Email: ahmadmarabdi@umsu.ac.id

Abstract. Air pollution from motor vehicle exhaust emissions is increasing, so it is necessary to attempt to control its exhaust emissions. Given the dangers of exhaust emissions, efforts need to be made to control and reduce air pollution so that negative impacts on people and the environment can be reduced. One engineering technology as a manifestation of vehicle emission control is engineering and modification of the exhaust gas channel. Engineering and modification are expected to be able to make a tool that will be able and function to reduce the danger of exhaust gases. The researcher will conduct experiments and manufacture of tools and initial testing on one motorized vehicle to see and observe the composition of the exhaust gas produced from the exhaust. The elements to be observed are CO values, HC values, and CO_2 values as comparative data. The tool that will be used to observe and see these elements is the Gas Analyzer tool. This tool is one instrument that is useful for measuring the portion and composition of a combined gas. From the results of testing and analysis obtained data on the exhaust emission test with an average engine speed of 500 rpm, and with the temperature of the exhaust tube 40 °C to 45 °C. After testing the standard exhaust model, then testing the engineering exhaust model which added 50 gr, 70 gr, and 90 gr scrap of stainless steel obtained the best conclusion to reduce and reduce the danger of exhaust emissions is engineered exhaust which added 70 g scrap of stainless steel and when compared to a standard exhaust, CO elements dropped to 71.09% and HC elements dropped to 48.26% and CO_2 elements dropped to 66.35%.

1. Introduction

Transportation today still uses fossil fuels as the main fuel, and the increasing use of oil-fueled vehicles, such as cars and motorbikes, results in increased levels of air pollution caused by exhaust emissions. Some types of emissions include Carbon Monoxide (CO), Hydrocarbon (HC), and Carbon Dioxide (CO_2) which has a bad impact on the health of the human body and erodes the ozone layer in the atmosphere [1]. High air pollution makes people and the world concerned, this makes people want alternative friendly and environmentally friendly transportation such as electric-powered motorized vehicles, but electric-powered vehicles in Indonesia are still not mass-produced, so the problem of air pollution caused by this exhaust gas is still not finished.

Given the dangers of exhaust emissions, efforts need to be made to control and reduce air pollution so that negative impacts on human beings can be reduced and minimized. In accordance with the Environment Sustainable Transportation (EST) program or better known as environmentally friendly transportation, there are 12 programs or approaches that can be carried out to reduce air pollution

IOP Conf. Series: Materials Science and Engineering **821** (2020) 012048 doi:10.1088/1757-899X/821/1/012048

problems originating from the transportation sector, one of which is Vehicle Emissions Control which will be the focus of research studies. One engineering technology as a manifestation of Vehicle Emission Control is the modification of the exhaust gas channel [2].

This research will discuss the problems related to the manufacture of alternative tools for reducing exhaust gas in 100cc motorcycle vehicles. Tests were conducted to find the results of 100cc motor vehicle exhaust emissions with standard exhaust as comparable data. Testing is done to find the results of exhaust emissions of 100 cc motorized bicycle vehicles with an exhaust that has been added in the tool which is the result of engineering and modification.

Based on the orientation of the rotor rotation, the wind turbine is divided into two types namely horizontal and vertical axis. The horizontal axis of rotation is parallel to the direction of the wind, while the vertical axis is opposite to the direction of the wind. Horizontal axis-based axis lift, slim blade, and high rotational speed. The rotor-based vertical axis drag force, wide blade and low rotational speed. There are many technologies developed to reduce the number of air pollution sourced from vehicle exhaust gases. So far the use of Catalytic Converter is the best way to control vehicle exhaust emissions [3].

2. Theory

Exhaust gas emissions resulting from the combustion process in a vehicle engine is one source of air pollution. The resulting exhaust gas emissions are in the form of carbon monoxide (CO), carbon dioxide (CO_2), hydrocarbons (HC), and nitrogen oxides (NOx). Fuels generally contain elements of carbon, hydrogen, oxygen, nitrogen, and sulfur. In perfect combustion, exhaust gas from combustion results in the form of carbon dioxide (CO_2) and water (H2O) and air that is not involved in combustion. But perfect combustion is difficult to achieve so that there are exhaust gases from other combustion products such as CO, HC, and also NOx because 79% of air for combustion consists of nitrogen. This allows the reduction of CO, HC, NOx and SO_2 levels. Catalytic converters are believed to be able to meet future emissions requirements and emissions regulations [4, 5].

2.1. Motor vehicle pollution sources

There are four sources of pollution originating from motorized vehicles, namely:

1. Exhaust pipe (exhaust) is the most important source (65-85%) and emits burning or unburned hydrocarbon (HC), various nitrogen oxide (NOx), carbon monoxide (CO) and alcoholic mixtures, aldehydes, ketones, phenols, acids, esters, ethers, epoxides, peroxides and other oxygen.
2. The oil bath is the second source (20%) and releases hydrocarbon (HC) that burns or not.
3. The fuel tank is a factor caused by hot weather with crude hydrocarbon evaporation losses (5%).
4. Carburetors are another factor, especially when driving in a position of congestion with hot weather, with evaporation losses and raw fuel (5-10%).

2.2. Calculating exhaust emissions

The formula below will be used for;

a. Average value when testing exhaust emissions;

$$\text{Average value} = \frac{\text{Amount of Value}}{\text{Number of Data}} \quad (1)$$

b. Emission percentage;

$$\text{Percentage of exhaust emissions} = \frac{\text{average emissions with scrap}}{\text{average emissions without a scrap}} \text{x } 100\ \% \quad (2)$$

c. Emission reduction percentage;

$$\text{Emission reduction percentage} = 100\% - \text{emission percentage} \quad (3)$$

2.3. Impact of air pollution

Pollutants that are mainly found in the exhaust gases of motorized vehicles are carbon monoxide (CO), various hydrocarbon compounds, various oxides of nitrogen (NO_x) and sulfur (SO_x), and dust particulates including lead (PB). Air pollution can be explained by 3 processes, namely (attrition, vaporization, and combustion) [6].

2.4. *Exhaust line*

The exhaust gas channel is not merely a function of delivering residual combustion. The exhaust gas channel is still one unit of the exhaust process. In this exhaust gas channel, the effects of turbulence are constantly maintained. With the exhaust line, the exhaust gas turbulence flow is changed to the piston driving force. Another function is the exhaust gas channel as a vibration damper. Vibration due to rising and falling of the piston from the cylinder head is forwarded to the exhaust body, frame, and chassis, so that engine vibration is not too much.

2.5. *Exhaust parts*

The following are parts of the exhaust on motorized vehicles:

1. Header: The header is the exhaust tip installed on the engine. The number of headers on the exhaust depends on how many cylinders the engine has. The main function of the header is to connect the entire exhaust system with the exhaust system owned by a motor vehicle.
2. Resonator: The second part of the exhaust is a resonator or commonly known as the exhaust filter. Many resonators are owned by motorized vehicles that function to process the noise produced by the results of engine firing.
3. Silencer: Silencer also has a function similar to a resonator, to help minimize the noise produced by the combustion of a motor vehicle.

2.6. *Pertalite fuel oil*

The fuel that will be used in this research is PERTALITE. The specifications owned by Pertalite, the official PT. Pertamina is based on the decision of the Director General of Oil and Gas No.313.K/ 10/ DJM.T/ 2013 concerning the Standards and Quality of Gasoline Fuel 90 which is marketed domestically (Table 1).

Table 1. Specifications of fuel oil Pertalite [7]

No.	Content	Information
1.	Octane levels	90 - 91
2.	Maximum sulfur content of	0,05% m/m ((equivalent to 500 ppm)
3.	Lead content	There is no
6.	Metal content	There is no
5.	Maximum residue	2.0%
6.	Specific gravity	Maximuml 770 kg/m^3 Maximum 715 kg/m^3 (at 15 ^{o}C)
7.	Visual appearance	Clear and bright

3. Methods

The method begins with designing and drawing engineering techniques (Figure 1), then make modifications to the standard pot according to the design. Continued to test the exhaust emissions on the standard pot and on the engine pot, and make analysis and conclusions from experiments and research.

IOP Conf. Series: Materials Science and Engineering **821** (2020) 012048 doi:10.1088/1757-899X/821/1/012048

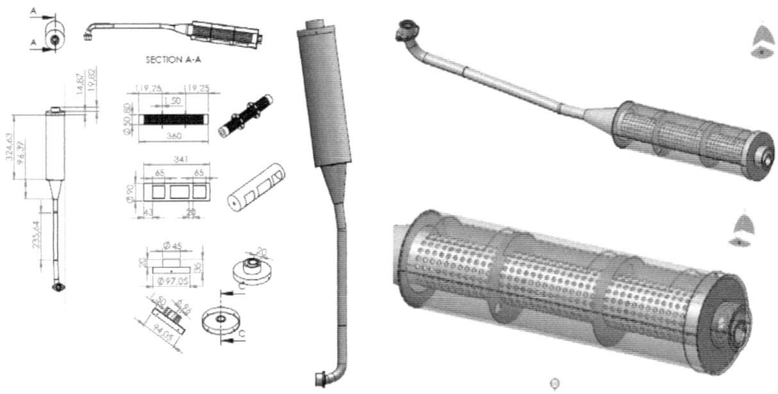

(a) (b)

Figure 1. Image of exhaust engineering (a), engineered exhaust and 3D modification (b).

4. Research procedure

The study was conducted using experimental research methods with different treatment of gram or scrap weight variables (utilization of residual turning called gram or scrap to reduce exhaust emissions).

Tests and analysis of data on the exhaust emission test will be regulated by the dependent variable with an engine speed of ± 500 rpm, and with the temperature of the outer tube with 40 °C to 45 °C. After testing the standard model, then the engineering model that is added 50 gr, 70 gr, and 90 gr scrap of stainless steel (Figure 2 and 3) [8].

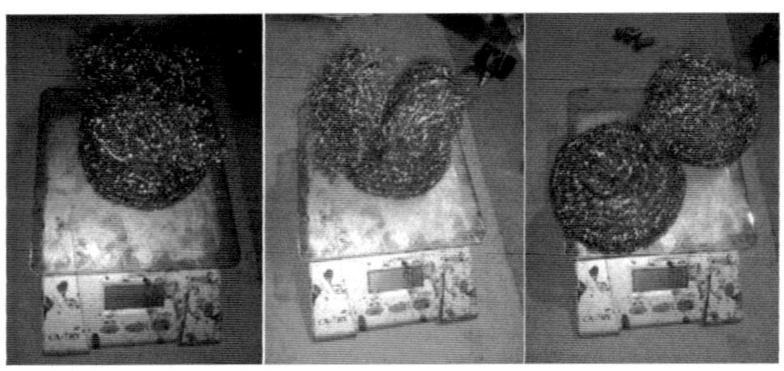

Figure 2. Stainless steel scrap added to the engineering exhaust is weighed 50 gr, 70 gr, and 90 gr.

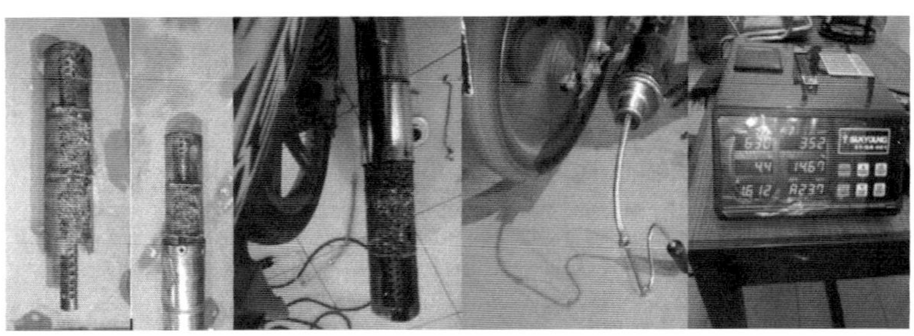

Figure 3. Testing with a gas analyzer on the engineered exhaust is added to scrap stainless steel weighing 50 gr, 70 gr, and 90 gr.

5. Results and discussion

The average value of the exhaust gas data obtained from standard exhaust testing, exhaust testing with 50 gr of added scraps, 70 gr of added scraps, and 90 gr of added scraps tabulated and put together, as can be seen in Table 2;

Table 2. Comparison data of waste gas emissions testing on the standard exhaust and on modified exhaust with additional stainless steel scrap.

No.	Exhaust	Engine rotation	Wind Speed Exhaust gas	Exhaust tube temperature	CO	HC	CO_2
		rpm	m/s	°C	%	ppm	%
1	Standard	500	28	42.6	6.33	513.3	4.16
2	With scrap 50 gr	500	23.8	40.7	3.66	282	2.4
3	With scrap 70 gr	500	23.3	44.3	1.83	265.6	1.4
4	With scrap 90 gr	500	21.5	44	3.57	335	2.7

By using Equation 1, Equations 2 and 3, the percentage of emissions is calculated as well as the percentage of emissions reductions that occur;

a. The decrease in the element of Carbon monoxide (CO) in exhaust emissions with a 50 gr and 70 gr scrap added exhaust decreased 42.18% to 71.09% but for the added scrap 90 gr of CO elements in exhaust emissions only dropped 43.60 %, as shown in Figure 4;

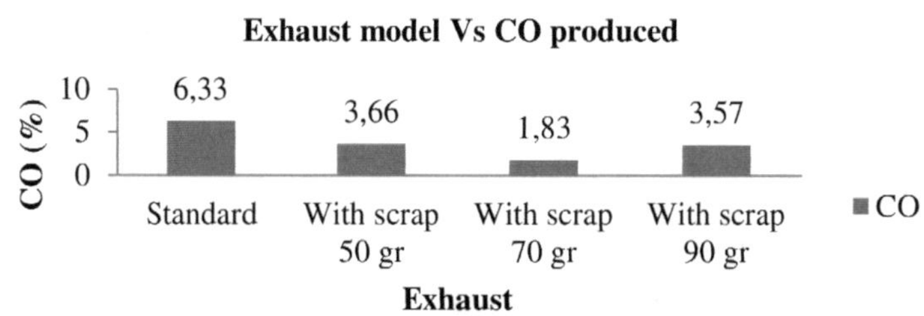

Figure 4. Exhaust model graph with CO produced

b. Decreasing the element of Hydro Carbon (HC) on exhaust gas emissions with 50 gr and 70 gr scrap added exhaust has decreased by 45.06% to 48.26% but for the added scrap 90 gr of HC element in exhaust emissions only decreased by 34.74 %, as shown in Figure 5;

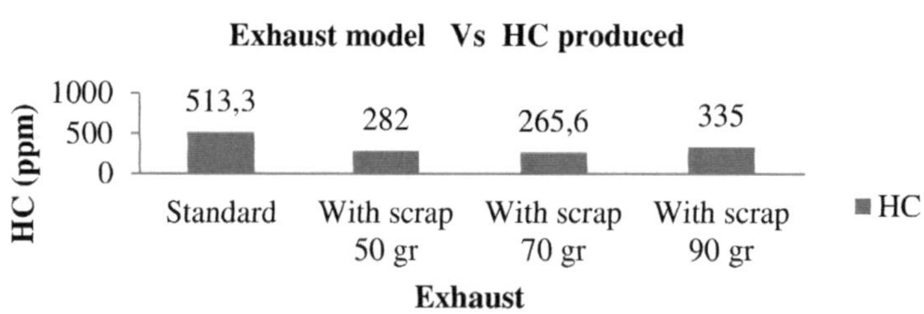

Figure 5. Exhaust model graph with HC produced

c. The decrease in the element of carbon dioxide (CO_2) in exhaust gas emissions with scrap added 50 gr and 70 gr occurred a decrease of 42.31% to 66.35%, but for those who added scrap 90 grams of CO_2 elements in exhaust emissions only dropped 35, 10%, as shown in Figure 6;

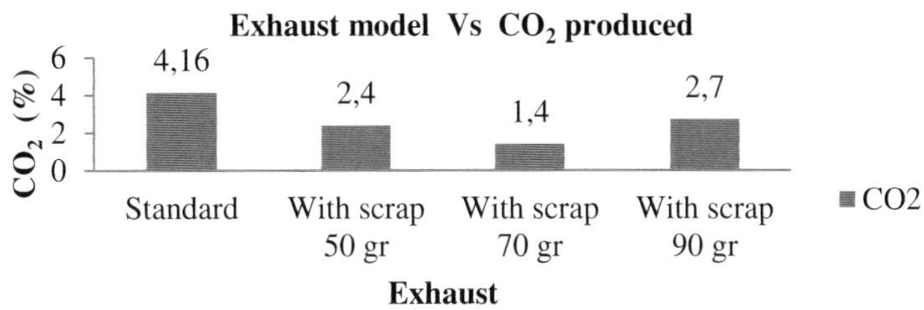

Figure 6. Exhaust model graph with CO_2 produced

6. Conclusion

From the results of testing and analysis of data on the exhaust emission test with an average engine speed of 500 rpm, and with the temperature of the outer tube, the 40 °C to 45 °C. After testing the standard exhaust model, then the engineering exhaust model which was added 50 gr, 70 gr, and 90 gr scrap of stainless steel was concluded; the best way to reduce and reduce the danger of exhaust emissions is engineered exhaust which added 70 gr of stainless steel scrap. And when compared to a standard exhaust, CO elements dropped to 71.09% and HC elements dropped to 48.26% and CO_2 elements dropped to 66.35%.

References

[1]. Ali M 2014 *Pembakaran sisa Hidrokarbon (HC), Karbon Monoksida (CO), dan nitrogen oksid (NO_X) yang semula berbahaya akan berubah menjadi senyawan yang stabil sewaktu melewati Catalytic Converter berupa CO_2, H_2O, N2, dan O_2*. Universitas Muhammadiyah Malang.

[2]. Irawan Bagus Rm 2012 *Rancang bangun Catalytic Converter dengan bahan katalis Tembaga-Mangan untuk unjuk kemampuan dalam mengurangi emisi gas buang*. (Portalgaruda.org article=4740), diakses 27 oktober 2013.

[3]. Pardiwala Julie M, Femina P, Sanjay P 2011 *Review paper on Catalytic Converter for Automotive Exhaust Emission* (Ahmedabad: Institute Of Technology, Nirma Unversity)

[4]. R. Zhang A, Villanueva H, Alamdari and Kaliaguine S 2006 Reduction of NO by CO over nanoscale LaCo1−xCuxO3 and LaMn1-xCuxO3 perovskites *Journal of Molecular Catalysis A: Chemical* vol **258** pp 22–34.

[5]. Iwakuni H, Shinmyou Y, Yano H, Matsumoto H and Ishihara T 2007 Direct decompostion of NO into N2 and O2 on BaMnO3 – based perovskites oxides *Applied Catalysis B: Environmental* vol **74**.

[6] Mukono 1997 *Pencemaran Udara dan Pengaruhnya Terhadap Gangguan Saluran Pernapasan.* Surabaya: Airlangga University Press.

[7]. https://www.otosia.com/berita/ini-kandungan-detail-pertalite-menurut-standar-mutu-bahan bakar-bensin.html.

[8]. Muharnif M, Septiawan R 2018 Analisa Pengujian Lelah Material Stainless Steel 304 Dengan Menggunakan Rotary Bending Fatigue Machine *Jurnal Rekayasa Material, Manufaktur dan Energi* vol **1** No 1 e-issn 2622-7398.

IOP Conf. Series: Materials Science and Engineering **821** (2020) 012049 doi:10.1088/1757-899X/821/1/012049

Implementation and design gas leakage detection system using ATMega8 microcontroller

N Evalina* and A Azis H

Department of Mechanical Engineering, Faculty of Engineering, Universitas Muhammadiyah Sumatera Utara, Medan, Indonesia

*Email: noorlyevalina@umsu.ac.id

Abstract. In Indonesia and many countries for economic reasons and green energy LPG is widely used for cooking. This paper show implementation and design gas leakage detection system. The main objective of the project to build a gas LPG leakage detector using an LPG gas sensor and micro-controller. It developed a security system by providing an early Warning System to give a sign if there is a smell of gas around Home. If this system has been the existence of leakage and smell of LPG gas, then the system will give a sign of alarm or buzzer where if there is leakage of gas that smelling the LPG system will work and the early warning alarm of the system. That the presence of gas leakage signals on the sensors to work, the MQ-6 is the gas Sensor used to detect leakage gas LPG. The tool is designed to detect gas leaks that are then legible on the LCD screen and alarm, and in certain circumstances, the buzzer will sound. This MQ-6 sensor test is performed by measuring the sensor output with the Atmega8 microcontroller when the LPG gas is detected and then writes it on the LCD screen. This test was conducted to determine the contribution of the MQ-6 sensor. It becomes essential to protect gas leakage from damage and accident.

1. Introduction

It is very important to bring technology into home, by making home smart and safety, it becomes important to protect it from damage and accident [1], many countries for economic reason used LPG (Liquid Petroleum Gas) is a mostly used fuel in the world, people cannot know that the gas is leaking, it is a scary thing for most of the gas users. To avoid gas leakage is needed the fire detection and security system is needed as well the gas leakage detection system, of automatic security system, can make safe people from dangerous blast accidents and sometimes prevent the accident [1, 2].

In essence, the explosion can be avoided when the prevention is done early, when the gas comes out or when the gas leakage occurs through the tube, regulator, hose or the stove itself. Along with the development of more modern science and technology, then a security system is developed by providing an early warning system to give a sign if there is a smell of gas [2-4]. Especially from LPG gas. This design has a goal as a deterrent or early warning system where the goal is to prevent casualties or loss of material or non-material caused by a gas leak where the gas leakage can cause Fire when the LPG gas is not used carefully so that the level of accidents from the leakage of LPG gas can be reduced.

Companies or industries that are engaged in the oleochemical indeed booming, because community needs will be incessant oil. PT. Musim Mas is one company that has been the focus of the industry moving in the oleochemical or Palm oil processing industry into chemicals such as fatty acid (fatty acid) and Glycerin. The industry, in the process of doing the processing of palm oil, has equipment that supports the company's goals in order to produce a good product can be achieved. As for the equipment are a number of industrial machinery.

Along with the development of Science and Technology in particular automation system, PT. Season Mas has largely adopted the Mechatronics-based industrial machinery and automatic. However, the many automatic machines that are implemented on palm oil processing jobs in this industry are on the other side of which, according to researchers less effective in the workmanship of a process that is located on the process control of the heating system of the water jacket.

The water jacket is a system that functions as a water heater using the heat energy of the steam generated from combustion system where water is heated above that will be pumped to the outside pipes throughout the channel streamed palm oil that has processed previously in which case this is done to prevent freezing in the pipe oil its channels. On the process of this work still do manual work processes so less effective done because to do this job should still stand by in-room control and is monotone remains mindful of how the State monitors on display temperature of the water jacket that would do the process of opening or closing of the valve with the degree suggested by the needs of the system. In addition, the process of opening and closing a valve made by workers also spent a long time, while the needs of the heating systems should be immediately done. In addition, when arriving the distribution workers' room must be wary of the challenge of a steam explosion overpressure sound reason can lead to a heart attack for the workers concerned.

2. Literature review

The mouth of the cylinder and the pipe are placed in front of the sensor. The solenoid is used to close the gas flow, while Arduino acts as a controller, Wi-Fi module uses ESP module, so Arduino Uno is connected, IoT devices are connected with Arduino signal. This process can prevent gas leakage because it is connected to Twitter, so the neighbors can find out if there is a gas leak [5].

Sahil Adsul, et al [6] describes that development of leakage detection system, gas leakage occurs because the equipment used is old, this paper presents leak detection with several sensor systems namely sound detection, pressure, temperature, gas leakage, zigbee wireless module used to transmit data to feel the gas leak. The MQ6 sensor module is used, 3 sensors are used to detect cigarette and cooking smoke whose concentration is 200-10,000 ppm LPG namely 20 kHz microphone, 40 kHz ultrasonic receiver and 40 kHz ultrasonic module.

Katole [7] study is based on a system used to detect many dangerous gases with Arduino coordination. We can change the systems currently available that are regulated by industrial estates and we can use systems at home and factories. The main objective of the activity is to design a gas leak-based detection system with Arduino, using gas sensors, the toxic gases like butane (also known as LPG), methane and carbon monoxide are sensed and displayed on the LCD display.

Study present the MQ6-used gas sensor is controlled with Arduino Uno R3 as the case of gas sensor gas leakage MQ6 will detect the gas leakage occurring so that there will be a voltage change in the gas sensor and will be read by the Arduino and will Be Ordered the LCD and GSM signal there is a gas leakage [8].

Anindya Nag, et al discusses novel sensing approach for LPG leakage detection: Part I-Operating Mechanism and Preliminary results, the paper present a new approach in the detection of LPG at ambient conditions, temperature and humidity conditions were maintained fixed during experimentation, the fabricated integrated sensor was used to test different concentrations (in ppm) of LPG undercoated and uncoated conditions to analyze the difference in response in two situations [4].

Onengiye M, et al discuss detection and monitoring gas leakage with implementation SMS Based, this system is designed and implemented using MQ-9 gas sensor and embedded microcontroller test was conducted to verify if the sensor can trigger, from the test gas detector sensor the presence of gas leakage and sms alert sent to the service station or to the owner [2].

3. Methodology

Design system takes two major part by hardware module and software module, the hardware module is designed by schematic diagram, the software module is developed using C-language, in case of gas leakage [9], the gas sensor will detect the gas leakage and will then make the sensor output have a certain voltage value (analog voltage) (Figure 1). When the output voltage of the curtain has exceeded the specified limit value (settings) then this condition will microcontroller automatically activated the Buzzer to sounding to mark the people closest to the place is accompanied by an alert on the LCD screen of the device. This paper will be discussed how the MQ-6 gas sensor using the ATmega8 microcontroller is used as the detection of LPG gas leakage.

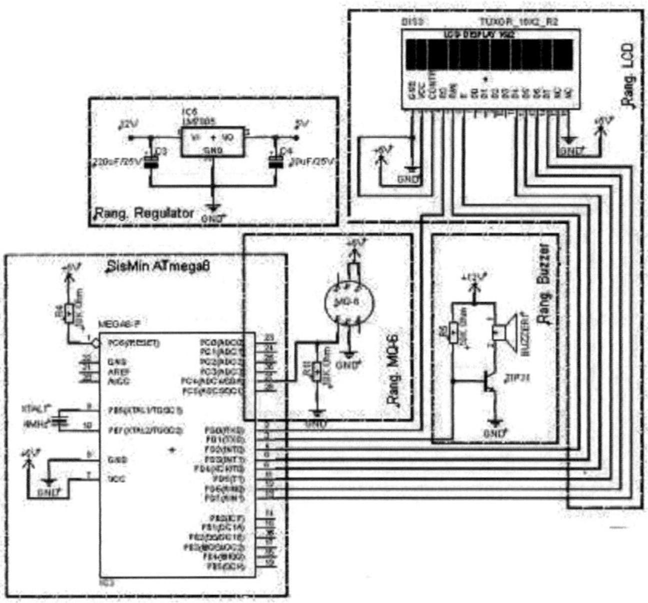

Figure 1. Overview of the system

***Power supply unit*:** power supply is a device used to supply power to all chip and component of the system, the supply is regulated for constant +5 volt DC [10, 11], the Arduino Uno microcontroller based system requires a 9 ~ 12 V power supply with a maximum current of 1A. On the Arduino board itself 9 ~ 12 V voltage will be changed to + 5V voltage.

***MQ-6 sensor unit*:** The MQ-6 gas sensor is a sensor that has a fast response to leakage gas LPG (liquid petroleum gas) and can be used in a simple set of drives, MQ-6 sensors commonly used in equipment detecting gas leaks in household and industrial activities [1].

***Microcontroller ATmega8*:** A microcontroller is often called a single microcomputer chip, it means that in a microcontroller IC there is used for microcontrollers to work, namely include microprocessor, ROM, RAM, I/O and clock as it is owned by a computer PC, given the packaging is only a chip with a relatively small size. Atmega8 is low power with 8 bit CMOS, a microcontroller that enables system designers to optimize power consumption versus processing speed. ATmega8 incorporates many instruction sets with 32 registers, all register amounting to 32 are directly connected to the logic unit arithmetic, which allows two independent registers to be accessed in one single instruction and executed in one cycle [2].

***Buzzer*:** is a device that can emit a loud noise when active. Generally, buzzers are used to give signals to indicate certain conditions. In this experiment, Bel is used to indicating the condition of the detector has detected an LPG leak that [2], on this circuit, the buzzer will sound if there is a gas leak.

IOP Conf. Series: Materials Science and Engineering **821** (2020) 012049 doi:10.1088/1757-899X/821/1/012049

Resistor: Resistors are electronic components that function to inhibit or limit the flow of electricity flowing in an electronic circuit. As the function of the resistor which is as the name implies is resistive and includes one electronic component in the category of passive components. The function of the resistor here is to adjust the lighting on the LCD.

LCD**:** is a display the text of leakage gas, The LCD used is a 2 x 16 character (2 lines 16 columns), with 16 connector pins.

4. Result

The implementation a design gas leakage detection system takes two major part which is hardware and software as early explained, in hardware section power supply unit was designed, built and tested with multimeter to ensure supply constant of +5 volt DC to system, The MQ-6 sensor was tested by connecting the output pin to analog pin input port of ATmega8 microcontroller, the cigarette lighter gas was used to test the performance of the gas sensor when the gas is detected LCD show text leakage and security alarm triggered by microcontroller, can be seen in Figure 2.

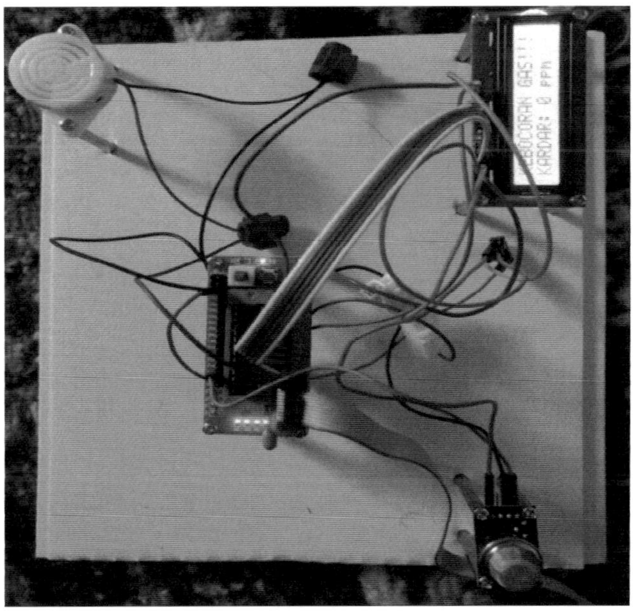

Figure 2. Testing of gas leakage detection

From the results of the experiment, if the MQ-6 detects the gas level exceeds 2000 ppm, the buzzer will sound along with it being detected as a leakage gas. But when the gas levels are reduced to less than 2000ppm, the buzzer will stop sounding, it is seen in Table 1.

Table 1. Gas leakage detection

Gas level (ppm)	Buzzer	LCD
0	off	Off
1598	off	Off
1765	off	Off
1870	off	Off
2540	on	Leakage
3450	on	Leakage
5087	on	Leakage
9077	on	Leakage
1970	off	Off

5. Conclusion

After testing on a tool designed to detect the leakage of LPG gas by using an MQ-6-based ATMEGA8 sensor, the conclusions are, Tools designed to detect gas leaks on the LCD screen and the buzzer will sound. This MQ-6 sensor test is done by measuring the sensor output with the ATmega8 microcontroller when the LPG gas leakage is detected and the gas level is 2000 ppm, it will be automatically written on the LCD screen and the buzzer will sound. However, when the gas levels are reduced to less than 2000 ppm, it will be automatically will disappear on the LCD and the buzzer will stop sounding. This test was conducted to determine the contribution of the MQ-6 sensor. It became important to protect gas leakage from damage and accident.

Acknowledgments

Thank internal grant aid that has been given by APB UMSU by contract number: 125/II. 3AU/UMSU-LP2M/C/2019, so the research can be executed.

References

[1] Shingan G G, Sambhare S V, Bhokare V S, Nikam A L and Shinde H D 2017 Smart Gas Cylinder: Leakage Alert and Automatic Booking in *2017 International Conference on Energy, Communication, Data Analytics and Soft Computing (ICECDS)* pp 1127–30.

[2] Georgewill O M and Ezeofor C J 2016 Design and implementation of SMSs-based industrial/ homes gas leakage monitoring & detection alarm system *Int J Eng Trends Technol* vol **35** no 9 pp 410–16.

[3] Ann E A B, Torstein O S, Roy E H, Rolf B P and Andreas A 2016 Automatic detection of marine gas seeps using an interferometric sidescan sonar pp 1–13.

[4] Nag A, Zia A I, Li X and Mukhopadhyay S C 2016 Novel sensing approach for LPG leakage detection: Part I — Operating Mechanism and Preliminary Results vol **16** no 4 pp 996–1003.

[5] Vasantakumaar N 2018 Intelligent gas leakage detection system with IoT using ESP 8266 Module no December.

[6] Adsul Sahil M. R. G., Kumar S A 2016 Development of leakage detection system in *International Conference on Automatic Control and Dynamic Optimization Techniques (ICACDOT)* pp 673–77.

[7] Katole P K R 2016 Hazardous gas detection using Arduino vol **2** no 10 pp 534–38.

[8] Tamizharasan R S K S V, Ravichandran T, Sowndariya M 2019 Gas level detection and automatic booking using IOT in *5th International Conference on Advanced Computing & Communication Systems (ICACCS)* pp 922–25.

[9] Nag A, Zia A I, Li X and Mukhopadhyay S C 2016 Novel sensing approach for LPG leakage detection-Part II: Effects of particle size composition, and coating layer thickness *IEEE Sens. J* vol **16** no 4 pp 1088–94.

[10] Somov A, Baranov A, Spirjakin D, and Passerone R 2014 Circuit design and power consumption analysis of wireless gas sensor nodes: One-Sensor Versus Two-Sensor Approach vol **14** no 6, pp 2056–63.

[11] Tharo Z A, Siahaan P U and Evalina N 2016 Improvisation analysis of reactive power energy saving lamps based on inverter *Int. J Eng Tech* vol **2** no 5 pp 141–45

IOP Conf. Series: Materials Science and Engineering **821** (2020) 012050 doi:10.1088/1757-899X/821/1/012050

Design of control system expand valve on water heating process air jacket

F I Pasaribu[*] and I Roza

Department of Mechanical Engineering, Faculty of Engineering, Universitas Muhammadiyah Sumatera Utara, Medan, Indonesia

*Email: faisalirsan @umsu.ac.id

Abstract. Many automatic machines implemented in the processing of palm oil in the industry are less effective in working on a heating control process in the jacket water system. With the ATMega, 328 Arduino Uno microcontroller control system as an Arduino program input and output controller and "C" language, a servo motor as output (valve) is controlled, and LCD as a display device for LM35 temperature detection can make it easier for workers to control the process of controlling water temperature on the jacket and can avoid failure in the process of melting the palm oil from the previous processing so that the oil in the pipeline continues to flow. The method used in this research is the design method that consists of several stages, namely, (1) Requirement analysis, (2) Design, (3) Implementation of the circuit, (4) Testing Procedures and (5) Testing tools. The level of accuracy of the rotation degree of the servo motor (as a valve) qualifies 100% (>35 °C and ≤ 49 °C, the motor rotates 40°), the relationship between temperature and the rotation speed of the servo motor qualifies with 100% accuracy (>50 °C and ≤ 53 °C then the motor rotates 80 °C) The overall performance of this tool can work well and calibration of the detection temperature value between the LM35 sensor with a thermometer has a difference of 1 mV so that it can be implemented by adjusting devices in the industry.

1. Introduction

Companies or industries that are engaged in the oleochemical indeed booming, because community needs will be incessant oil. PT. Musim Mas is one company that has been the focus of the industry moving in the oleochemical or Palm oil processing industry into chemicals such as fatty acid (fatty acid) and Glycerin. The industry, in the process of doing the processing of palm oil, has equipment that supports the company's goals in order to produce a good product can be achieved. As for the equipment are a number of industrial machinery.

Along with the development of Science and Technology in particular automation system, PT. Season Mas has largely adopted the Mechatronics-based industrial machinery and automatic. However, the many automatic machines that are implemented on palm oil processing jobs in this industry are on the other side of which, according to researchers less effective in the workmanship of a process that is located on the process control of the heating system of the water jacket.

The water jacket is a system that functions as a water heater using the heat energy of the steam generated from combustion system where water is heated above that will be pumped to the outside pipes throughout the channel streamed palm oil that has processed previously in which case this is done to prevent freezing in the pipe oil its channels. On the process of this work still do manual work

IOP Conf. Series: Materials Science and Engineering **821** (2020) 012050 doi:10.1088/1757-899X/821/1/012050

processes so less effective done because to do this job should still stand by in-room control and is monotone remains mindful of how the State monitors on display temperature of the water jacket that would do the process of opening or closing of the valve with the degree suggested by the needs of the system. In addition, the process of opening and closing a valve made by workers also spent a long time, while the needs of the heating systems should be immediately done. In addition, when arriving the distribution in the worker's room must be wary of the challenge of a steam explosion overpressure sound reason can lead to a heart attack for the workers concerned.

2. Review of literature

2.1. Control system

The system controls the degree of valve opening and closing the steam distribution system is a design tool that is composed of electronic components and software to control a system on the process of heating the water jacket in an industrial kitchen, where is the system that controlled the process of heating the water jacket is the capacity of the hot steam from the boiler flowing on the water jacket which is controlled manually by using human hands in the space distribution of steam temperature based on air jacket seems from your computer screen. With this tool, it will automatically do the work alone without human help more fully to set the degree of opening of the capacity of the hot steam that uses a servo motor based on information from a sensor water jacket temperature detector [1].

2.2. Arduino Uno

Arduino Uno is the Arduino ATmega328 microcontroller using the board. Arduino Uno has 14 digital pins (6 pins can be used as PWM outputs), 6 analog inputs, a 16 MHz crystal oscillator, a USB connection, a connector voltage source, an ICSP header, and a reset button. Arduino Uno contains everything needed to support a microcontroller. By only connecting it to a computer via USB or DC voltage from the battery or the AC adapter to the DC was able to make it work. Arduino Uno using ATmega16U2 programmed as a USB-to-serial converter for serial communication to the computer via the USB port. Looked up from the Arduino Uno can be seen in Figure 1.

Figure 1: Arduino Uno

2.3. Sensors LM35

LM35 is a special IC used for temperature/temperature sensors that the results are quite linear. LM35 does not require external calibration or special timing. LM35 is the most widely used temperature sensor for practice because in addition to its fairly cheap price the linearity is also pretty good. This sensor can operate in the range of -55°C to + 150°C and has a linear characteristic at 10 mV/ °C.

This sensor is very simple but has fairly high precision. This sensor has only three legs. The first leg of the IC LM35 is connected to the power source, the second leg as the output and the third leg is connected to the ground. Figure 2 here is a picture of the physical shape and leg arrangement of the LM35 sensor.

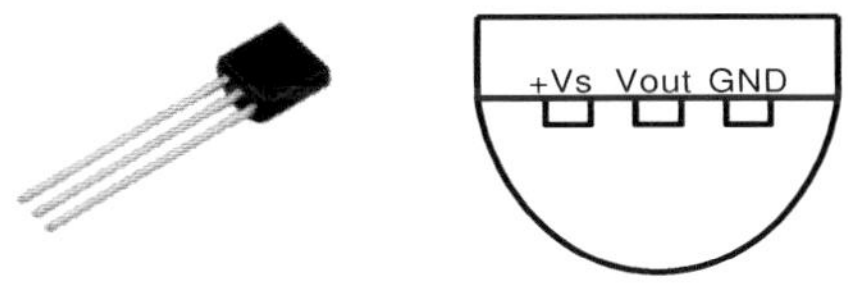

Figure 2. (a) Physical form LM35; (b) Look at the bottom LM35

Besides its very simple physical shape, the LM35 sensor also has several advantages. The advantages of LM35 are as follows:

- It can be calibrated directly inside the Celsius scale.
- Linear scale factor + 10 mV / °C.
- The accuracy level of 0.5 °C when room temperature (25 °C).

The working principle of this sensor circuit is based on changes in temperature or heat that occurs in the vicinity. Where if the temperature around the sensor is higher, then the value of resistance in the sensor circuit that serves as a regulator of the current to input non-inverting will be greater and vice versa. So if the resistance value of the current regulator to the non-inverting is small then the current from the previously priories transistor will be grounded and non-inverted, so as the value of the current regulator's resistance causes the incoming voltage to the amplifier (A2) is getting bigger. Thus the transistor serves only as a large regulator of the auxiliary current (A1), and the symbol (i) is only a custody controller.

Although the voltage of this sensor can reach 30 volts, but given to the sensor is 5 volts, so it can be used with a single supply provided that the LM35 sensor requires only a current of 60 μA. This means that LM35 has the ability to generate heat (self-heating) of the sensor that can cause a low error reading that is less than 0.5 °C at 25 °C

The LM35 output voltage (VOUT) has a working range of 0 volts up to 1.5 Volts with an LM35 sensor operating voltage that can be used between 4 Volts to 30 Volts. LM35 sensor output will rise by 10 mV every degree Celcius so that Equation 2.4 as follows.

$$V_{LM35} = Temperature \ x \ 10 \ mV \quad (1)$$

where Temperature is temperature changes that occur around the sensor.

In principle, the LM35 sensor will do the sensing at the temperature change every temperature 1°C will show a voltage of 10 mV. In the placement LM35 can be affixed with adhesive or can also be cemented on the surface but the temperature will slightly decrease about 0.01°C because it will be absorbed at the surface temperature [2].

2.4. Standard servo motor

Servo motor is a DC motor equipped with a control system. This control system will provide motor rotation position feedback from 0 to 180 degrees. Besides, this motor also has relatively strong enough torque. Figure 3 shows the cross-section and wiring of the servo motor. Servo motor cabling system consists of 3 parts, namely Vcc, Gnd, and Control (PWM = Pulse Width Modulation). Giving PWM to the servo motor will make the servo move to a certain position and then stop (position control).

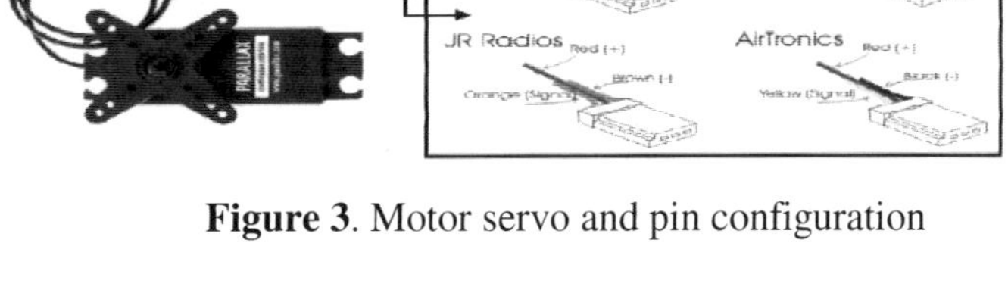

Figure 3. Motor servo and pin configuration

IOP Conf. Series: Materials Science and Engineering **821** (2020) 012050 doi:10.1088/1757-899X/821/1/012050

The main principle of servo motor control is the assignment of PWM value to its control. The PWM frequency used on the servo motor controller is always 50 Hz so that the pulses are generated every 20 ms. The pulse width will determine the desired servo position. Giving 1.5 ms pulse width will make the servo motor rotate to a neutral position (90 degrees), 1.75 ms pulse width will make servo motor rotate approach 180-degree position, and with pulse width 1.25 ms servo motor will move to the 0-degree position. Figure 4 shows the relationship between pulse width PWM with rotation direction servo motor.

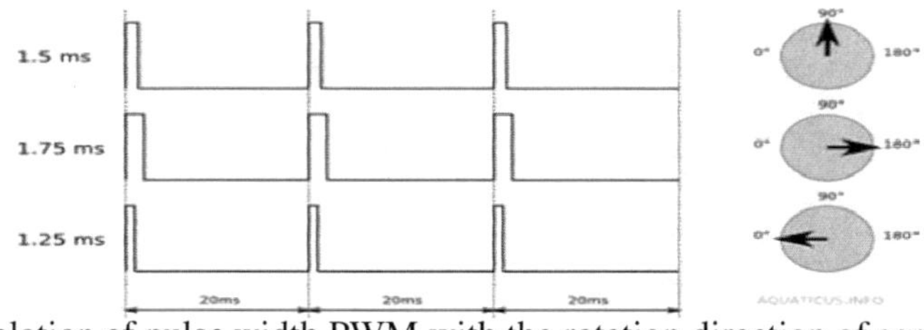

Figure 4. Relation of pulse width PWM with the rotation direction of servo motor

An industry primarily engaged in the processing of liquid, of course, has a piping system that serves as a place to flow liquid. Each pipe circuit must have a device used to adjust the amount of flow for the processing process that can run in accordance with the specified. The tool is called a valve or often also called a valve. Valve or valve is a device installed in the piping system, which functions to regulate, control and direct the flow rate of fluid by opening or closing part of the fluid flow (Figure 5).

The valve can be operated manually, either by using the handle, pedal lever and so on, besides manually operated valve can also be operated automatically by using the principle of change of flow, pressure and temperature. The changes will affect the diaphragm, spring or piston so that it will automatically move the valve with an open system close [3, 4].

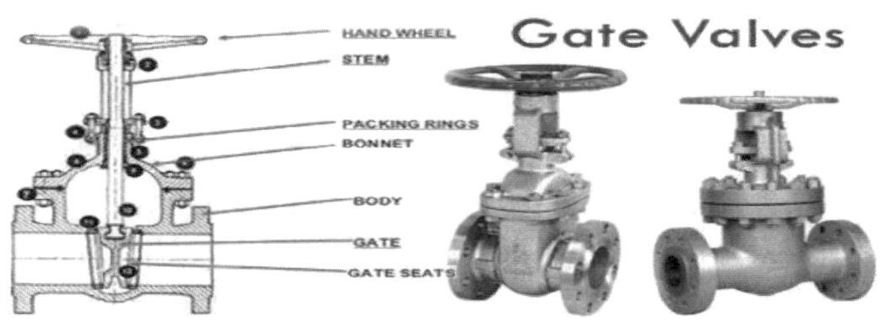

Figure 5. Gate valve

The water jacket or water jacket is a process of heating the water in the normal temperature of the pam water, where the water is heated in a water bath, then the water in the bath is heated by steam [5]. The function of the jacket water is to heat the oil through a coated pipeline of cashing, aims to prevent the oil from freezing. Here is Figure 6 which shows the physical shape of the water jacket.

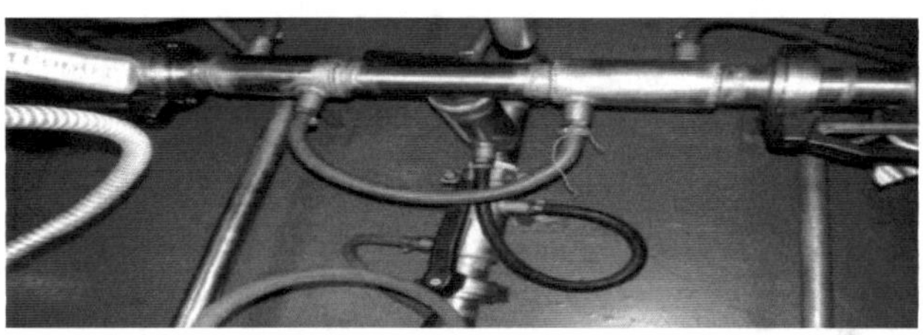

Figure 6. Water jacket

How to warm-up jacket water at the company where researchers get the idea of this final title is to open and close valve steam manually, where high and low-temperature water jacket bias seen on the computer monitor screen. Below is Figure 7 which shows the window screen of the application:

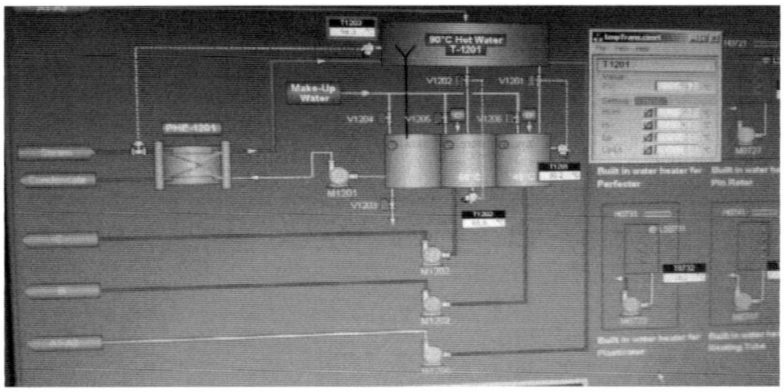

Figure 7. Window display application monitor temperature water jacket

3. System planning

The scheme of merging the entire installation of a tool-making system is a scheme that shows the layout of the connection of each tool-forming component in this study. The scheme can be seen in Figure 8 below.

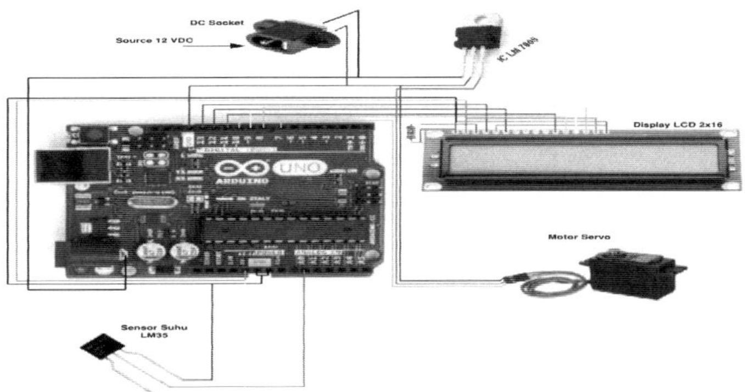

Figure 8. Scheme of the whole system

4. Methods

The test was carried out for 7 days by observing the temperature of the glass evaporator outside and also the temperature of the water and air in the evaporator by recording the clock. There are 6 thermocouples installed on each side that you want to observe the temperature.

4.1. Testing method

After all, components forming the system control degree opening and closing valve steam distribution system on the heating water jacket process is completed as in Figure 3 above, then the next to do the testing process. But before doing the test as a whole, in this research first tested is the function of the LM35 sensor with the aim of whether the sensor can work to detect the temperature well before it is connected with the minimum Arduino Uno system.

As a sensor that detects temperature, IC LM35 is one of the main components that must be tested condition and level of reliability. Because the work of LM35 will greatly affect the work of the system as a whole. This test is performed to test the response time of the LM35 sensor to the ambient temperature changes and compared with the ordinary (standard) thermometer.

IOP Conf. Series: Materials Science and Engineering **821** (2020) 012050 doi:10.1088/1757-899X/821/1/012050

This test is done through two stages, namely testing the LM35 sensor response to the surrounding temperature rise and testing the LM35 sensor response to the decrease in ambient temperature. Further, for each stage, three experiments were conducted to obtain cumulative data to obtain more accurate average results. The results of this experiment will be compared with the same experimental results performed on a regular thermometer.

In accordance with the characteristics of the LM35 sensor that every 1 °C temperature will show 10 mV voltage on its output, then the process of measuring the temperature change of the LM35 sensor is done by measuring any voltage changes at the sensor output. The measurement of the output voltage is performed using a measuring device by connecting the probe (+) Voltmeter with VOUT foot and probe (-) Voltmeter with Ground feet on the sensor. Figure 9 shows the scheme and form of LM 35 sensor testing:

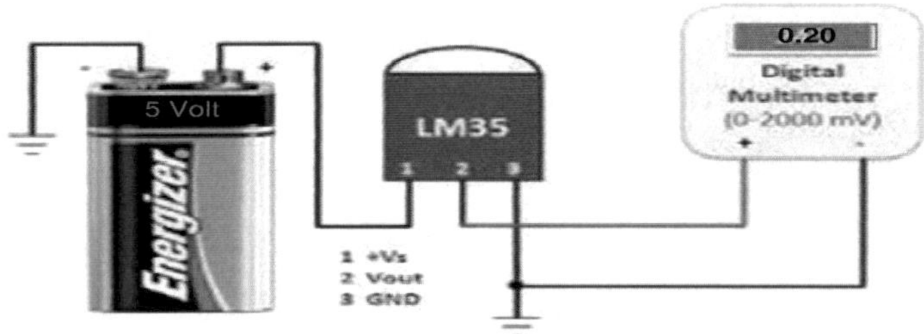

Figure 9. LM35 sensor testing scheme

As for how the overall test which is the goal in making this tool is:

- Distributing power source to all systems by means of power supply (AC to DC power supply) to source PLN 220 VAC.
- Then, heat the LM35 sensor by bringing the electric soldering temperature closer or closer to the fire temperature from various sources of fire (for example: from a fountain, match, or the like).
- Taking into account the temperature displayed by the display or LCD, then pay attention to the following parameters:
  a). If the temperature displayed on the LCD is ≤ 35 °C then the motor rotates 0°
  b). If the temperature displayed on the LCD is > 35 °C and ≤ 49 °C then the motor rotates 40°
  c). If the temperature displayed on the LCD is > 50 °C and ≤ 53 °C then the motor rotates 80°

5. Results and discussion

5.1. Tool design results

After the whole process of designing all components forming the control device the degree of opening and closing valve steam distribution system on the jacket water heating process using microcontroller controller Atmega 328 is done then the following in Figure 10 which shows the results of the design tool:

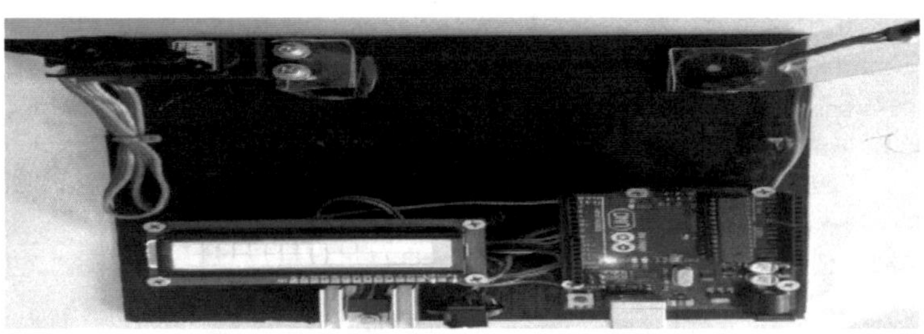

Figure 10. The design result of the tool

5.2. Overall test results

In the case raised in this study is how to control the degree of opening and closing of the steam distribution valve in the case of jacket water heating, wherein this jacket water in the industrial process is as a medium to keep oil production ingredients melt in the jacket pipe along the channel the process of printing oil until the oil is finished in print in solid form.

Since the oil in the pipeline of the production process must be kept at a temperature, the jacket water must have a hot temperature so that the oil in the pipes does not freeze and can still flow along the pipeline of the production process for the final printing. It is important that the temperature of the oil be kept hot along the pipe is the main focus in the process of setting the degree of valve opening and closing, since the heated water medium uses steam from the boiler and the manual arrangement is from the vapor distribution chamber manually. After understanding the workflow of the control process about the explanation of the problem in Chapter I is the background of the research problem, it departs from the problem that is done an automatic valve regulation process. Answering the question above has been designed the automatic tool and has been tested with results as in Table 1 below:

Table 1. The results of testing the system as a whole

No.	Display temperature (°C) on the LCD	Degree of servo motor rotation	Temperature (°C) on thermometer	Explanation
1	33°C	0°	32°C	True
2	34°C	0°	33°C	True
3	35°C	0°	34.5°C	True
4	36°C	40°	35°C	True
5	37°C	40°	36°C	True
6	38°C	40°	37.5°C	True
7	39°C	40°	38°C	True
8	40°C	40°	39°C	True
9	41°C	40°	40°C	True
10	42°C	40°	41°C	True
11	43°C	40°	42.5°C	True
12	44°C	40°	43°C	True
13	45 °C	40°	44°C	True
14	46 °C	40°	45.5°C	True
15	47 °C	40°	46.5°C	True
16	48 °C	40°	47.5°C	True
17	49 °C	40°	48°C	True
18	50 °C	80°	49.5°C	True
19	51 °C	80°	50.5°C	True
20	52 °C	80°	51.5°C	True

6. Conclusions and recommendations

6.1. Conclusions

From the results of the research conducted it can be concluded as follows. First, the system of rotation control motor servo degree (as a valve) able to fulfil the requirement of research purpose with an indicator of circle arc as a calibration value of rotation degree of the motor. Second, the degree of accuracy of motor rotation servo degrees (as a valve) meets 100% requirement. Third, the relationship between the temperature with the rotation degree of the servo motor is eligible with 100% accuracy. The last, calibration of detection temperature value between the LM35 sensors with an ordinary thermometer only has a difference value of 1 mV.

6.2. Recommendation

Some advice on this research is, first, for the perfection of this tool in the future it would be a nice tool design system that can be implemented directly in the industry. Finally, If you want to change the dimensions of servo motors used both in terms of physical size and input voltage values, then adjust it with the power used on this Arduino with a record no more than 5 Volt DC.

References

[1] Emerson Automation Solutions Edition F (N D) Control Valve 2019 *Fisher Controls International Llc* D101881x012/ Jun19.
[2] Pasaribu FI, Roza I and Efendi Y 2019 Memanfaatkan panas exhaust sepeda motor sebagai sumber energi listrik memakai thermoelectric. www.ojs.uma.ac.id/index.php/jesce/article/view/2787
[3]. Stations G 2015 Draft Guide for motor-operated valve (mov) motor application protection.
[4]. Power N 2014 Draft guide for motor-operated valve (mov) motor application, protection control, and testing in nuclear power generating stations *Committee E & Society E.*
[5]. Chemical Processing. Jacketed Heating. (N.D.).www.Pickheaters.Com

IOP Conf. Series: Materials Science and Engineering **821** (2020) 012051 doi:10.1088/1757-899X/821/1/012051

Workability of coconut fiber concrete with rice husk ash as a sand replacement

F V Riza, D Karnelia, and J Hadipramana*

Department of Mechanical Engineering, Faculty of Engineering, Universitas Muhammadiyah Sumatera Utara, Medan, Indonesia

*Email: josef@umsu.ac.id

Abstract. Fiber concrete has been widely known for its ability to prevent cracks especially micro crack. Natural fiber concrete was one way to accommodate the use of waste natural fibers and at the same time producing sustainable materials. This research tried to examine the workability of Coconut Fiber Concrete (CFC) that incorporated 2% of Coconut Fiber (CF) with Rice Husk Ash (RHA) as a sand replacement with ratio 10, 20 and 30% respectively. Three types of water-cement ratios were used 0.38, 0.41 and 0.44. The result showed that slump value ranging from 3.5 cm and 4 cm. The addition of 20% of RHA resulted in the steady slump value for all water-cement ratio. The CFC is on the same par compare to the steel fiber concrete which incorporated 1.28% fiber.

1. Introduction

Natural material has increasingly been investigated and used in the construction area as a result of depleted of unrenewable construction material [1]. Hence the need to incorporated green, sustainable and environmental-friendly components has become essential in the millennials era. By substituted or additions concrete materials with natural waste, the aim of green concrete are achieved. As exhibit naturally that concrete is brittle and has low tensile strength [2]. These weaknesses of cementitious materials can be improved by incorporated any kind of fibers whether it steel fibers, polymer fibers or natural fibers. Incorporated more than one fibers usually called hybrid Fiber Reinforced Concrete (FRC) [3].

This paper incorporated Coconut Fibers (CF) to reinforce and enhance tensile strength and its ductility, where Rice Husk Ash (RHA) was integrated as sand replacement. CF are local waste that abundantly readily available in Indonesia which is estimated produced 2.2 million ton per year [4]. RHA has found not quite an ecological safe since it is not decomposed easily [5] which made a strong point to recycle it. RHA burnt in uncontrolled temperature can contain silica up to 89.90% and the amorphousness can vary depending on its burning temperature [6].

From microstructural observation, fibrillation is the process of forming fibers which are taken place when fibers in mortar would bound the aggregate in a concrete matrix [2]. Fibers in concrete would experience fibrillation which the fibers would split and start to bind materials around it. This binding would reduce the micro-crack in concrete [7]. Micro-crack could lead to a bigger crack and eventually caused structure failure [4].

2. Material and mix proportion

The mix proportions were designed according to SNI 03-2834-1993 with design compressive strength 23 MPa. Three types of water-cement ratios are incorporated as presented in Table 1. Water used is ordinary tap water with a balanced pH and Ordinary Portland Cement (OPC) was used as the main binder. RHA was incorporated into every water-cement ratio mix design with a ratio 10%, 20% and 30% respectively.

Control specimen with zero addition of CF or RHA was manufactured as a milestone to compare to other proportions. Two percent of CF was incorporated for every mix proportion. It was cut at 2.5 cm and obtained from Binjai, Sumatera Utara. Figure 1 shows the cut CF used in the concrete mixture.

Table 1. Coconut Fiber Concrete (CFC) mix design

Water Cement Ratio	Water	Cement	Coarse Aggregate	Fine Aggregate
0.38	0.372	1	3.247	1.605
0.41	0.409	1	3.094	1.428
0.44	0.432	1	3.309	1.599

Figure 1. Coconut fibers cut at 2.5 cm

Fine and coarse aggregate characteristics used in fresh CFC such as density, mud content, bulk density, water content, water absorption and fineness modulus (FM), are shown in Table 2.

Table 2. Fine and coarse aggregate characteristic

Test	Fine Aggregate	Coarse Aggregate
Density	2.571 gr/cm^3	2.716 gr/cm^3
Mud content	3.3%	0.767%
Bulk density	1.165 gr/cm^3	1.511 gr/cm^3
FM	2.775	7.086
Water content	2.145%	0.604%
Water absorption	1.730%	0.752%

The free flow test utilized slump test was conducted according to ASTM C.143 Standard Test Method for Slump of Hydraulic Cement Concrete to investigate the workability of fresh CFC.

3. Result and discussion

The slump test result showed that all mix proportions of fresh concrete did not collapse and exhibited true slump shape. Every proportion showed a good slump as shown in Figures 2a, 2b, and 2c

2a. w/c ratio 0.3 2b. w/c ratio 0.41 2c. w/c ratio 0.44

Figure 2. The slump of fresh CFC

respectively. The mixtures are quite dense as the result of adding CF and RHA. For the control specimen, the higher the water-cement ratio leads to a higher slump value. The addition of a water-cement ratio from 0.38 to 0.41 caused the concrete mixture flowier.

From Table 3, concrete with 10% RHA showed slump value slightly decrease at water-cement ratio 0.44 from 4 cm to 3.5 cm. whereas concrete with 20% RHA exhibited the same level of slump value for every water-cement ratio. Concrete with 30% RHA demonstrated the smallest slump value at a 0.41 water-cement ratio.

Table 3. Slump value of CFC mix proportions

No	RHA Percentage	Slump Height (cm)		
		w/c 0.38	w/c 0.41	w/c 0.44
1	0%	3 cm	4 cm	4 cm
2	10%	4 cm	4 cm	3.5 cm
3	20%	3.5 cm	3.5 cm	3.5 cm
4	30%	4 cm	3.5 cm	4 cm

The highest slump value for water-cement ratio 0.38 achieved by concreted with an addition of 10% and 30% RHA. Concrete with 0.41 water-cement ratio showed the more RHA addition then the slump value will decrease. But for the concrete with water cement ratio 0.44 exhibit different result that the addition of RHA from 20% to 30% will cause the slump value increase.

Compared to the slump value of steel fiber concrete of 3.7 cm that incorporated fiber 1.28% [8], these CFC slump value is quite in the same range slump value.

4. Conclusion and recommendation

Some conclusions can be drawn from the results of the experiment. First, addition of 20% RHA as sand replacement demonstrating steady slump value for all water-cement ratio. Second, the slump test of fresh CFC demonstrated the workability of concrete that incorporated 2% of coconut fibers have value comparable to the 1.28% addition of steel fiber concrete. The last is the splitting tensile test and the flexural test should be done to evaluate how the Coconut Fiber would contribute to the strength of concrete.

Acknowledgments

This research was supported by Universitas Muhammadiyah Sumatera Utara through the Internal Basic Research Scheme. We thank our students from the Department of Civil Engineering, Faculty of Engineering who greatly assisted the research.

IOP Conf. Series: Materials Science and Engineering **821** (2020) 012051 doi:10.1088/1757-899X/821/1/012051

References

[1]. Morel J C, Mesbah A, Oggero M and Walker P 2001 Building houses with local material: means to drastically reduce the environmental impact of construction. Building and Environment. vol **36** 1119-26.

[2]. Bentur A and Mindess S 2007 Fibre reinforced cementitious composites. *Second ed Bentur A, Mindess S,* editors New York Taylor & Francis.

[3]. Pakravan HR, Latifi M and Jamshidi M 2017 Hybrid short fiber reinforcement system in concrete: A review. *Construction and Building Materials* **142** pp 280-94.

[4]. Hadipramana J, Samad AAA, Zaidi AMA, Muhammad N and Ali N 2013 Contribution of polypropylene fibre in improving strength of foamed concrete *Advanced Materials Research* vol **626**(2013) pp762-8.

[5]. Beagle EC 1978 Rice-husk conversion to energy. FAO Agricultural Services Bulletin 31.

[6]. Riza FV 2011 Application of RHA pozzolanic properties in CEB production *International Journal of Sustainable Construction Engineering and Technology* (IJSCET) vol **2**(2).

[7]. Kakooei S, Akil HM, Jamshidi M and Rouhi J 2012 The effects of polypropylene fibers on the properties of reinforced concrete structures *Construction and Building Materials* vol **27**(1) pp 73-7.

[8]. Figueiredoa ADd and Ceccatob MR 2015 Workability analysis of steel fiber reinforced concrete using slump and Ve-Be test *Materials Research* vol **18**(6).

AUTHOR INDEX